THEOLOGICAL ETHICS

THEOLOGICAL ETHICS

Volume 1: Foundations

THEOLOGICAL ETHICS

Volume 1: Foundations

by

Helmut Thielicke

edited by

William H. Lazareth

WILLIAM B. EERDMANS PUBLISHING COMPANY
GRAND RAPIDS, MICHIGAN

This is an abridgment and translation of *Theologische Ethik,
I Prinzipienlehre* (2. Ausgabe; Tübingen: Mohr, 1958), and
Theologische Ethik, II¹ Entfaltung: Mensch und Welt (2.
Ausgabe; Tübingen: Mohr, 1959).

Reprinted, February 1984

© 1966 by Fortress Press

This edition published 1979 by Wm. B. Eerdmans Publishing
Company through special arrangement with Fortress Press
and J. C. B. Mohr.

Printed in U.S.A.

Library of Congress Cataloging in Publication Data

Thielicke, Helmut, 1908-
    Theological ethics.

    "An abridgement and translation of Theologische Ethik."
    Vol. 1-2 are reprints of the 1966-1969 ed. published
by Fortress Press, Philadelphia; v. 3, of the 1975 ed.
published by Baker Book House Co., Grand Rapids, Mich.,
under title: The ethics of sex.
    Includes bibliographical references and indexes.
    CONTENTS: v. 1. Foundations.—v. 2. Politics.—
v. 3. Sex.
    1. Christian ethics. 2. Sociology, Christian
(Lutheran) 3. Christianity and politics. 4. Social
ethics. 5. Sex (Theology) I. Lazareth, William
Henry, 1928- II. Title.
BJ1253.T5213 1979   241   78-31858
ISBN 0-8028-1791-2 (v. 1)

To the memory of my friend and translator

John W. Doberstein

EDITOR'S FOREWORD

Helmut Thielicke emerged from the air raid shelters of Germany as one of the most compelling voices of the resistance movement within the Christian church. He was able to address the judgment and mercy of God relevantly to both the perpetrators and the victims of Nazism and World War II. In an endless stream of sermons, speeches, essays, newspaper articles, and radio talks, this modern prophet deflated the proud, comforted the broken, inspired the hopeless, and challenged the bewildered and the skeptical in his own distinctive style.

For the past twenty years Thielicke has fused these wartime and postwar experiences together with his deep insights into the fields of theology, philosophy, modern literature, and drama. His semesters of lectures to packed halls of enthusiastic university students in Tübingen and Hamburg have resulted in the most extensive study of theological and social ethics ever written on Reformation soil. In the multivolume German edition he offers us a fresh, evangelical approach to the biblical and theological bases of Christian ethics, and then proceeds to analyze many currently crucial issues in the areas of sex and marriage, politics, economics, culture, and recreation.

It is a tribue to Thielicke's "ethical realism" that one of his grateful foreign students so gladly assumed the "unethical" responsibility of condensing the first 1,000 pages from volumes I and II[1] of the original German edition. Our aim was to produce a manageable, one-volume introduction to Christian ethics for English readers. To achieve it we endeavored to reduce the materials of the original by roughly a third, and to restructure some of its more extended chapters and subsections. At the same time every effort was made to preserve the theological integrity and organic unity of the total work. While the central thrust of the argument thus remains intact, many of the nuances and illustrations have been thinned out considerably. This is especially true of Thielicke's ongoing dialogue with classical German philosophy and the arts, along with more recent developments in modern European intellectual history. Scholars who want these added features are strongly urged to consult the original edition.

Thus the present condensation has become possible only by "pruning" the text to its essentials and reducing the (almost exclusively foreign) critical apparatus to a bare minimum. In short, this book aims to present only "the heart" of Thielicke's theological ethics. But for the English reader 700 pages is still a healthy-sized heart at that!

It should also be recognized that such strict editorial demands can often play havoc with an author's style and flow of argumentation. The fact that the German edition itself, however, is organized in such a way as to allow the general reader to skip the technical paragraphs set in small type, happily helps to make our numerous deletions far less obvious and offensive to the English reader than would otherwise be the case.

What should emerge from these pages is a clear call for Christian discipleship and decision making. Thielicke's ethics in this introductory volume is an authentic, contemporary restatement of Paul's proclamation of responsible Christian freedom. Later volumes consist largely of situational analyses of ethical "borderline cases" drawn from the various areas of concrete experience. These illustrative models repeatedly show how a free, responsible Christian may live out his faith in love and in justice as a citizen of both the old age in Adam and the new age in Christ.

May Thielicke's motto inspire among us a whole new generation of Christian ethical thought and action: "Theological ethics is eschatological or it is nothing!"

Lutheran Theological Seminary, Philadelphia   William H. Lazareth

PREFACE TO THE ENGLISH EDITION

By the time this first part of the *Theological Ethics* is published in English, the final volume of the work will have appeared in Germany. Extending to over three thousand pages, it is probably the most extensive work on ethics that has ever been written. I began the preliminary work on it during the war, when I was removed from the university by the Nazi government and interned in a remote part of southern Germany. Hence the work has gone on throughout the lifetime of my eldest son, namely, twenty-one years. It is not surprising then that on the completion of so extensive a work one stops to inquire about the course that has been traversed and about how the work fits into the over-all theological enterprise of our time.

It is fitting that I should share my convictions on this question with my English-speaking readers. Except for the specialists, they know me chiefly through collections of my sermons and essays. I have always been grateful for the many letters I have received from readers of these collections. They have enabled me to see that here and there I have indeed been able to give some help, and in many cases contribute to the deepening of Christian life. It is now a great joy that my major scholarly work is also to be presented to the English-speaking world. By way of introduction I should like to say two things, first, concerning my purpose in seeking to enter the contemporary theological debate by way of this work, and second, concerning the method and materials I have chosen to use.

I can answer the question concerning purpose only in the form of a confession. When I first began the work, I had certain definite ideas as to what I intended and what I did not intend. But in fact, my true intentions have only become clear to me in the course of the work itself.

To deal first with the negative side, I did not plan to present a doctrine of what the Christian has to *do*, i.e., a book of morals. Such a book would be possible only on the basis of a legalistic orientation which would conflict with the message of the Gospel, and hence with Reformation theology. The Christian stands, not under the dictatorship of a legalistic "You ought," but in the magnetic field of Christian freedom, under the empowering of the "You may." "Love, and do what you want," says Augustine.

But this freedom is not caprice. It does not mean that *what* we do is a matter of indifference. We are "empowered" for freedom. We stand under a Lord who does this empowering. Hence we live under his eyes and inquire concerning his will. The implied dialectic of freedom and bondage, as discussed by Paul in Romans 6, is really the central problem of this theological ethics.

The question of how we can live under the eyes and will of God becomes a problem for us only because we must live in a reality which limits the scope of our action and does not give us the breadth of choice we should like. This restricting reality makes itself felt particularly at the point where we encounter the orders [*Ordnungen*] of life which appear to be forcing us into fixed channels of action. Take the business man, for example. In his private life, in marriage or personal friendships, he knows fairly well how to fulfill the command to love his neighbor. But what about his work? Here he is involved, for example, in the competition which goes with being in business. Can love of neighbor mean here that he must consider only the interests of his competitors and not his own? Should he—out of love!—cease to compete with them in quality and price? Is not the very idea of a "competitive neighbor" a contradiction in terms? Quite obviously, the autonomy of business life has to be taken into account at this point. The business man cannot just "love blindly" without regard for the structure of the economic world. In his work he has to take into account other laws than those which rule in the more immediate man-to-man relationship, i.e., in the personal sphere.

I have intentionally overdrawn the picture to make the issue clear. A more carefully defined discussion will be necessary in the course of the work. My present concern, however, is to make clear to the reader what is in my view the true problem of ethics, namely, that man with his existence is integrated into a particular structure of reality and that he has to do with this structure in his daily work. Many books on ethics treat man as a private being who lives his life in the quiet chamber of a devotional relationship to God and of contact with his wife, children, and immediate neighbors. I have attempted rather to cover the *whole* of the reality in which modern man finds himself, including that of his work, and of political, social, and economic life, where most people—indeed all but the very young and the very old—experience their real problems of conscience, their conflicts and personal difficulties.

I can now turn to the positive intention behind the work, which is to give a Christian interpretation of human and historical reality in general, and to do this in a comprehensive and systematic way. This aim impels me to inquire concerning the shape of the social, political, and economic structures in which we must live and act. But such a question cannot be construed in terms of a purely phenomenological interest, as in sociology or philosophy. It is concerned rather with the relation of these structures to God, and therewith to man as the creature of God. Are there "orders of creation" in which the will of God is declared? Does our obedience to God—our conformity with him—depend on our aquiescence in the claim of these orders? Or are the orders at the same time objectifications of man and of his ambiguity? Is the model of our world's structure the Garden of Eden or the Tower of Babel? Or both? And if both, what is the interrelation of these conflicting models?

These questions suggest the kinds of problems which are at the heart of this systematic interpretation of reality. So far as I can see, they constitute a new subject matter which Reformation theology, to its hurt, has thus far avoided—though Roman Catholic theology has always found a place for it. The need now is to make up for several centuries of neglect. I know of course that this first attempt to build something new will produce only modest structures and that future generations will have to take up the task afresh with better resources. Nevertheless, I could not help but make a first and provisional attempt to fulfill a task which to my mind is unavoidable.

I can understand the reasons why Reformation theology has not previously tackled the question to any serious extent. The doctrine of justification, which Luther rediscovered, is in fact the heart of theology. The heart, however, must pump blood into *all* the members of the body. The task is not just that of a theological cardiology. The whole bloodstream must be made the theme of theology. For understandable reasons, Luther could not address himself as fully to the wider task. This larger theme involves such questions as these: What are the implications of justification by grace alone, freedom from the dominion of the Law, and the polarity of sin and grace for the existence of man in his whole earthly life, and especially in the orders, in marriage, politics, social intercourse, economic competition, labor-management relations, etc.?

These questions all reduce themselves to the one task of declining the doctrine of justification through all the case forms in which it appears within the grammar of our existence. For the man justified by grace is not just a private individual. He is also the man who has a job and who is woven into the web of the orders. If he knows himself to be justified and endowed with Christian freedom, what must this mean for him as a family man or a bachelor, as a business man or artist, as a senator or union executive?

If the liberating significance of justification for all these dimensions of life is not indicated, the Christian is in danger of succumbing to schizophrenia. For he will live a life that is divided into different compartments. In his private life he will be a believer living, as it were, supernaturally in a kind of superworld. But as a man of the world he will follow the laws of the world. We know very well the highly unsatisfactory forms that such a divided, half-Christian humanity can take. We often refer to them—and not without cause—in terms of "hypocrisy." The very same man who comes away from the solemnity of divine worship on Sunday forgets it, or regards it as irrelevant, when he sits at his desk on Monday. His motto now is, "Business is business." The heart may beat for God, it may honestly feel it is a redeemed heart, but it does not pump blood into the extremities. There are limbs which are cold and clammy, not yet connected to the heart.

My aim in this interpretation of reality, at least the goal of my modest endeavor, is to liberate Christian consciousness from this cleavage and to establish its unity. I should like to make the Christian credible, and to do so on the basis of faith. For this very reason I would seek to champion the thesis of a "worldly Christianity," rescue Christian dogmas from the sphere of the otherworldly, and bring the church out of the ghetto and back to earth, to the place where man actually lives in his secularity and where he "may" live with his faith.

I have accordingly written this work, not just for Christians, much less theologians, but for educated and thoughtful men generally. Such men can be shown by way of what they share with Christians, namely, a secular existence within the orders, just how the Christian understands and interprets this world which is common to us all. As long as the Christian speaks about heaven and hell, the agnostic, atheist, and secular idealist may not care to listen. They may just withdraw and say, "What do these 'spheres' mean to me?" But the moment the Christian speaks of that which is common both to himself and to non-Christians, things such as marriage and the state, society and economics, they will listen. Christian preachers must surely regard it as a matter of no small consequence when even secular men are forced to concede, "Here is someone who is speaking about my problems, indeed about me; I must listen to what he has to say."

Now this brings me to what I can only describe as a confession. I have roughly described the goal of the work as I first saw it. But gradually it became clear to me that in this scholarly undertaking what I was basically trying to do was to lay a new foundation for Christian preaching. My sermon collections, as I now realize, all draw their life from this attempt at a comprehensive interpretation of reality. Let me indicate briefly what I mean.

With many colleagues in the theological enterprise of our generation I share a concern for both the substance and the form of preaching. With them I also share by and large the diagnosis of the present sickness of preaching, namely, that it moves for the most part in an otherworldly and sacral sphere which has no relation to man's secular existence and hence leaves him helpless and alone, and that it has very largely ignored the problems posed by the change in our modern picture of the world. This diagnosis, and the corresponding attempts at a new approach, have brought me into a certain proximity to such pioneering spirits as Dietrich Bonhoeffer, Rudolf Bultmann, and Paul Tillich. The popularization of their ideas in J. A. T. Robinson's *Honest to God* has now made these conceptions familiar also in the English-speaking world. It would take us too far afield if I were to try to show here how in spite of our similar intentions my way, as pursued in this theological ethics, has had to take a somewhat different course from theirs. This much at least I may say.

We owe to Bonhoeffer the fascinating slogans, "a nonreligious interpretation of the Christian message" and "secular proclamation."

Because of his early martyrdom, however, he was never able personally to develop what he meant by these phrases. He has simply left the slogans behind as thorns in our soul to keep us salutarily disturbed. Consequently I do not know whether he would have regarded my interpretation of reality as consonant with what he had in mind. As they stand, the fragmentary slogans are so ambiguous and imprecise that they can be cited in support of a variety of views.

Bultmann has given to his attempt to renew Christian preaching the attractive and sensational title "demythologization." [1] His remedy is really a new hermeneutical principle, the new method called "existential interpretation." I believe that such an approach to the problem in terms of method is most important, even though I have several fundamental objections to the particular method chosen.[2] Discussions of method are very popular today, and the pointed formulation of the question in terms of "demythologization" has evoked a world-wide echo. Yet I do not believe that the attempt to make modern man realize the significance, relevance, and existential affinity of the Christian message for himself can succeed merely by way of methodological and hermeneutical discussion. With the hypertrophy to which it is often subject, such discussion can carry with it the danger of obscuration—a danger which Cardinal Newman had in view in another connection when he said that there are people who can no longer see the goal because of the multiplicity of ways that lead to it. I personally am alarmed when I see the younger generation of theologians practically drowning in hermeneutical reflection. Long before they have seen and handled the bedrock of the biblical teaching concerning creation, the fall, and the return of Christ, they are busily occupied with mineralogical analysis and with optical researches in religious microscopy. Alongside them are those who shy away from the heresies that could result, and who often go on preaching with an artificially preserved naïveté as if we were still living in the sixteenth century.

What we are to preach as our message and how we are to put it is indeed the crucial question. But I am persuaded that we cannot

---

[1] See my essay "The Restatement of New Testament Mythology" in *Kerygma and Myth*, ed. H. W. Bartsch (London: Macmillan, 1954), pp. 138 ff.

[2] See my essay "Erwägungen zu Bultmanns Hermeneutik" in *Theologische Literaturzeiting*, XII (1955), cols. 705 ff.

even begin to answer it unless we first see man in his secularity, unless we interpret the world of man, and therewith lay bare the theme which is of concern both to Christians and to secular men. Only thus can our message acquire a new worldliness. Only thus can there be a new incarnation of the Word which seeks out man in his earthly relationships.

We are constantly asking what is the missionary point of contact with our contemporaries outside the church. Where can we find a bridgehead for our message in their souls? My firm conviction is that it is to be found here, precisely here, and in my own preaching I have repeatedly tried to put this conviction to the test. For me, preaching was never intended to proceed independently of theological endeavor. On the contrary, in intention at least theology and preaching are one in substance. As it seems to me, however, the fundamental error in modern preaching is that we hurry directly from biblical exegesis to preaching and skip right over the whole field of ethical inquiry. Surely it goes without saying that this field is not to be regarded merely as that sphere of morals in which practical applications of faith are to be made. As I have shown already, it is rather the place where under the eyes of God we learn to understand our worldliness. It is the place where we interpret reality.

I hope that these brief introductory remarks have made one thing clear to the reader, namely, that the Christian interpretation of reality is not an extra task over and above that which is the theologian's true calling, one to which he can devote himself—perhaps in demonstration of his openness to the world!—after he has completed his proper work of teaching and proclaiming the faith. Nothing could be further from the truth than to understand ethics as a kind of intellectual dessert, an added luxury for theologians on vacation. For the faith of which theology treats is always the faith of living men, men who stand in the reality of this world of ours and are subject to its constant pressures. The believer cannot inquire concerning God's will without drawing into the focus of that inquiry the whole complicated web of his own being. He cannot believe "in" God without believing "against" the reality in which he finds himself, that reality which seems to be opposed to God and in face of which he must struggle through to the "nevertheless" of faith (Ps. 73:23). He can never fulfill the obedience of faith without a readiness to do justice to the human, social, and material tasks which are set before him. But these he never encounters merely as earthly reflections of the eternal. They do not confront him merely as the temporal form of the divine commandments. On the contrary, the mandates which "goods and kindred" seem to put before him often appear in such a form that he falls precisely in fulfilling them. His disobedience sets in at the very point of his giving himself to them, because they come to stand *between* him and God. (Indeed this is exactly what Luther had in mind when he spoke of goods and kindred in his hymn, "A Mighty Fortress is Our God.") It may also be that the several mandates individually contradict one another, so that the believer is involved in a conflict of values. For, after all, he lives out his faith precisely in this aeon between the fall and the judgment, in this world which is no longer whole, no longer transparent for God.

Hence the form which the obedience of faith is to take, and therewith faith itself, is always an open question for pilgrim man. The notion that this question is answerable, that the ways of obedience are clear-cut and unequivocal, is only a variation of that righteousness by the Law which feeds on the illusion that man is capable of satisfying the claim of God. At the very point where the possibility of obedience reaches its limit and is engulfed in crisis, namely, where the question of forgiveness arises, reflection concerning the mode of obedience in this crisis comes to an end. For the crisis is an impasse which shows us that the reality of this aeon, like our own Babylonian heart, can of itself produce no real righteousness. Hence there arise at this point too both the question of forgiveness and the awareness that all our action stands in need of such forgiveness. Thus dogmatics and ethics are essentially saying the very same thing about one and the same theme. They have a common root in the doctrine of justification.

Now it is the correlation between revelation, faith, and reality thereby indicated which demands of us a particular method. I would therefore say yet a word concerning this second point to which I made reference at the outset. Methodologically we pursue a dialectical movement of thought which involves both the erection according to abstract principles of a systematic scaffolding which will embrace and interlace the indicated correlation in all directions, and the detailed discussion within this framework of specific situations in terms of "models" or cases which are representative of the problem in question. For example, the problem of conflict is represented by the opposition between truth and love, which in turn finds illustration in, for example, the particular question of truthfulness at the sickbed.

As regards method, it should be noted that these model cases which occur throughout the ethics are something more than mere illustrations. Many reviewers have mistakenly lauded the graphic character of the work on this account, and their praise has not brought any particular pleasure to the author. The function of these models is the substantive one of displaying in concrete detail the whole complicated web of reality, and of thus averting the danger that ethics will simply propound normative principles under which the individual cases are then presumably to be subsumed. To attempt that would be a mistake, for an ethics so conceived would inevitably produce nothing but "hypothetical cases" which as ideal types are so grossly oversimplified that in reality they hardly ever occur in just this way. (In medicine too it is usually regarded as highly extraordinary for a doctor actually to encounter the kind of case that the textbooks call "normal.")

As a matter of fact, the problems of ethical decision arise precisely because actual cases do not conform to the theoretical norm. The hardest thing about ethical decision is usually not to muster up a readiness for obedience, for action commensurate with the norms, but to decide what is in fact demanded, or in Christian terms, what the will of God *is* in this specific case. For the norms are not usually so clear-cut and unambiguous—this points us to the theological background—that we can subsume the concrete case under them. On the contrary, they usually confront us as part of a web of conflicting norms among which we have to choose.

The real chore in the matter of ethical decision is not that of performing the act itself, but that of searching and struggling to discover what *is* right, what *ought* to be. Thus the norm which requires that I do my job honestly and well never arises in isolation, but always in conflict with other norms which set forth an equal claim, such as my duty to my family which also has a claim on me, or various other conflicting demands which the business world itself imposes. So it is not without reason that we have devoted large sections of the present work to an analysis of the "conflict situation," thus daring to put at the center of everything a problem which is usually treated, if at all, only on the margin of ethics, as a kind of borderline case. This central ethical question can be dealt with, however, only if a concrete but representative part of reality is analyzed in such detail as to make clear the complicated web of conflict. These detailed analyses are for their part incorporated in turn into an abstract systematic scheme. This conjoining of deductive and inductive methods is intended to prevent the ethics from falling apart aphoristically into a discontinuous series of individual cases, and also to assure that it will not become a mere system of hypothetical cases far removed from reality.

This method of using models has also a second substantive function, namely, to guard the ethics against any legalistic tendency. Such a tendency would arise if the ethical task were taken to be that of giving clear ethical directives, even to the point of casuistry. It would be a legalistic tendency because it would take decision out of the hands of the reader by prejudicing his decision and urging him to accept the decision of another, in this case the author. It would debar rather than engage the self and its adult autonomy.

The method here adopted blocks off this legalistic bypath by making it clear, in the detailed analyses, that every case is different, that there can be no exact repetition, and that the case which cannot even be fixed with precision certainly cannot be subsumed under ready-made norms. The path of ethical reflection must follow a quite different course. The analysis of concrete cases can only lead to a clarification of the various claims which enter into a particular ethical decision, an evaluation of their weight and merit in the given instance, and an indication of the theological background of the conflict of values thereby disclosed.

Thus the intention is quite different from that which underlies a casuistical ethics. In the first place, the task is to elicit individual decision, not to anticipate it but to provide, as it were, the materials for making it. Secondly, the aim is to shatter the illusion that there is an unequivocally "correct" form of action which can be clearly delineated, as if there were such a thing as "*right*eousness." Attention is drawn instead to the form of the world in this aeon between the fall and the judgment, which of itself cannot effect a fulfillment of the will of God in the sense of legal righteousness, and to the fact that the Sermon on the Mount is right when it eschatologically calls in question this world of ours. This points to the cosmological horizon of ethics, though not in the sense that the world becomes a constricting destiny which involves me in undeserved frictions and "makes me guilty," after the manner of the old harpist in Goethe's *Wilhelm Meister*. On the contrary, that world which cannot of itself produce righteousness is "my" world; it is the objectification of my own Babylonian heart. This statement has momentous consequences for theology's doctrine of the orders, which accordingly can no longer be conceived as a doctrine of the orders of creation, much less in terms of natural law. We are thus forced to engage in a comprehensive combining of anthropology and cosmology.

With this reference to the structural arrangement of the entire work and to the method used in this interpretation of reality, I have, as it were, been anticipating. What follows in this first English volume is primarily a consideration of such fundamental matters as the distinction between theological and philosophical ethics, the debate between Reformation and Roman Catholic theology on the interpretation of reality, and the relation between dogmatics and ethics. Only after the discussion of these basic matters does the work go on to treat human reality in the broadest sense. The remainder of the work, as it is now published in German, carries forward the presentation here begun in a preliminary way. It deals specifically with the ethics of politics, society, economics, law, sex, and art.

At the conclusion of this introduction it is my pleasant duty to thank Dr. William H. Lazareth of Lutheran Theological Seminary, Philadelphia, for accepting responsibility for the critical editorial tasks. These tasks imposed more than ordinary demands since the work itself had to be greatly shortened, one and one-half volumes of the German edition being compressed into this single English volume. Through correspondence I have come to see what delicate considerations had to be taken into account in making the present edition relevant to the situation of the English-speaking world, and I am particularly pleased that Dr. Lazareth undertook the task, not least of all on account of our friendship of long standing. It was when I was living in miserable conditions in Tübingen shortly after the war, and gradually getting back to a regular schedule, that Dr.

Lazareth came and studied with me. It was quite a sensation that an American should come to us in this situation to share our meager existence and to renew our feeling of kinship with his great country and with American Christians. We have never forgotten his solidarity with us in this dark period of our history. There was little I could do in return—except on one occasion to hold his hand over at the clinic while he had a painful wisdom tooth extracted. And his American affirmation of life displayed itself even then as he was able to laugh despite a sore and swollen cheek. He even ate patiently the pap which for some four hundred years (!) has been the traditional Tübingen diet for students with a toothache. A foreign land generally means to a person only as much as the friends he has there. If I love America, it is not least of all because Bill Lazareth is there!

I should like also to express a word of gratitude to my revered colleague James Luther Adams of Harvard Divinity School, who was the first to suggest the translation of the *Theological Ethics,* and who has for many years strengthened and encouraged me in my work.

1966   Helmut Thielicke

PREFACE TO THE FIRST GERMAN EDITION

If appearances do not deceive, the center of theological enquiry is now shifting to the problem of ethics. This shift would be wrongly understood if we were to see it only as a sign of man's helpless frustration with respect to his public tasks, i.e., his economic, political, and social action, or with respect to his personal life in such matters as sex, work, and leisure. The impulses towards new ethical reflection actually lie at a much deeper level. What is really called in question is man himself, and his place in the world. This is not surprising in view of the fact that man has so obviously lost himself, and the Christian—at least the Protestant Christian— has to so great an extent lost the world. Hence the task of theological ethics today, so far from involving questions of casuistry, must be that of constructing a doctrine of man and the world. What must be made plain is that in man's every relation to the world, in every step he takes in it and in every sphere of life in which he is involved, the theological theme is primary.

Our primary concern, then, is not to give a Christian commentary on the matter of marriage or on the foundations of law or of art, or to offer the church's counsel either to its members or to "secular man." Ours is rather the more fundamental concern to explicate the theological theme of "being-in-the-world" and of what it means in all its forms.

Ethics must thus begin by posing basic *theological* questions. It must itself be a part of the first principles of which dogmatics treats. The fact that this has usually not been the case is what has brought ethics into disrepute and caused it to be adjudged an illegitimate child of theology and at best a kind of guidebook on the journey of life, a guidebook which may be totally misused—as travel guides are wont to be—by people who would legalistically adopt the recommendations of others, do their sightseeing through the eyes of others, and take over lock, stock, and barrel in an impersonal way the decisions somebody else has reached.

This is not the place, however, to speak further about the theological character of ethics. We shall deal with that matter in the book itself. But attention may be drawn here to an important result which follows when the task is seen in this way.

Treating of man and of his "being-in-the-world," theological ethics takes up a theme which is not handled by it alone, but is expressly dealt with also by philosophy, various of the sciences, and especially Roman Catholic moral theology. Hence treatment of the theme involves constant reference to other partners in the discussion. If theology rightly understands its business, it is thus compelled to steer clear of esoteric monologue and to recapture its true character as dialogue. The theme of ethics can be a big help in liberating theology from the sterility of a purely internal and ecclesiastical monologue and freeing it for fruitful forms of utterance; and it can do so purely in terms of its subject, quite apart from any missionary or apologetic pretensions.

If we are right about the theological nature of the theme of ethics, and if this theme is related to the very core and center of the Christian message, then during the course of the discussion the debate between Roman Catholicism and the Reformation will naturally come to the fore quite of itself. Indeed whether it does so is itself a good test of whether theological reflection is truly to the point, determined by that which is its own proper subject matter.

In this sense the present book is at many points explicitly and at all points implicitly a contribution to the debate between Evangelical and Roman Catholic theology. The author cherishes the conviction that in the next decades the two confessions will find precisely in the field of ethics new contacts, points of dispute, and areas of discussion, and that these will lead to new skirmishes. One cannot say that the Reformation churches—including Lutheranism in which the present ethics is rooted—are particularly well equipped for this task. They are indeed alarmingly unconcerned. The powerful and impressive efforts Roman Catholic theology is making—whether successfully or not is another question—to understand the presuppositions of the Reformation, to correct its previous picture of Luther, and to redefine its theological relation to Protestantism put us Protestants to shame. We simply have nothing comparable to show. The Una Sancta movement has the merit of at least keeping open and keeping in remembrance the spiritual wound of separation. Beyond that, however, it simply degenerates into a mere concern for cordiality and tolerance. But the use of smothering cushions and concealing veils has never been able even to clarify serious dissent. In any case, we Protestants go largely unprepared into the debates which may already be clearly descried on the horizon. The devastating fact is that the theological and ecclesiastical problems with which we occupy ourselves are in many instances of quite secondary importance.

One consequence of our emphatic orientation here toward the intraconfessional debate is the fact that we shall have to define our terms very precisely and engage in a good deal of detailed analysis. For our partner in the discussion is of a heavily freighted and well thought through tradition. Even such Reformation principles as that of the *sola gratia* have long since—at least since the Tridentine confession—been taken up into, and given their place within, its system. It is thus no simple matter even to find the crucial points of doctrinal agreement and disagreement. Still, it is the fate of Reformation theology to have to attain to a position of its own under the constant scrutiny and control of this particular partner. If the reader does not keep constantly in mind the presence of this partner, the scope and difficulty of many of the analyses which follow, e.g., of the doctrines of the *imago Dei*, the indicative and imperative, Law and Gospel, will often seem to be superfluous. A purely theoretical development of the contents would be clearer and simpler. But it would not do justice to the tasks which are posed on the longer view.

In this respect the author has a secret wish that Roman Catholic theologians who study this work will be stimulated by their encounter with Reformation theology and will be moved to test themselves, as we have tried to do, in the fire of the partner's questioning. The author is not without concern lest these representatives of the other confession should feel they have been misunderstood. This concern is justified in view of the fact that we must question, observe, and understand Roman Catholic theology from a standpoint different from its own, and cannot go along with its particular approach to the problems involved. We often feel in turn that we ourselves are misunderstood, e.g., in Roman Catholic interpretations of Luther, even those that are concerned to be truly objective. So we know how hard it is for someone to feel that he has been properly understood "from without." This difficulty of understanding is of course connected with the ultimate problem of hermeneutics and is indicative of the basic crisis in which the concern for objectivity is engulfed. The author is also aware that in this work he has taken only a first step and made only a first attempt at such an intraconfessional debate on ethics, and he would be glad if this effort to delineate certain themes were to stimulate many others to further and more specialized efforts.

That the author has personally made an honest and serious effort truly to understand his partner, and that his attitude is not one of mere opposition to Roman Catholic theology, should become plain in the course of the argument to any sympathetic reader of the other confession. It is hoped that this effort derives from the ongoing self-criticism of Reformation Christianity, e.g., in the treatment of the doctrine of Law and Gospel, or in the discussion of Luther's doctrine of the two kingdoms.

Because there are so many partners in the discussion, recruited from the whole world of philosophy and from general intellectual as well as ecclesiastical history, and because the individual themes are so complicated, it is inevitable that the arguments at some points will intersect. Through considering the same thing from a variety of aspects, however, it is hoped that added distinctness will accrue. At any rate, the more carefully the way is prepared now, the more carefully the course is plotted here in terms of theory, the more one may hope that what follows later may concentrate specifically on matters of practical implementation, and thus appeal to a wider circle of readers.

I should like to thank my colleague Hans Schönweiss for his unwearying help in making corrections, and also my young friend Heinz Schladebach for his careful work on the indexes.

1951   Helmut Thielicke

CONTENTS

xxviii

ABBREVIATIONS

BC   — *The Book of Concord.* Translated and edited by
Theodore G. Tappert
(Philadelphia: Fortress Press, 1959).

LCC   — *Library of Christian Classics.* John T. McNeill
and Henry P. van Dusen, General Editors
(Philadelphia, 1953-   ).

LW   — American Edition of *Luther's Works*
(Philadelphia and St. Louis, 1955-   ).

PE   — *Works of Martin Luther.* Philadelphia Edition
(Philadelphia, 1915-1943).

WA   — *D. Martin Luthers Werke.* Kritische Gesamtausgabe
(Weimar, 1883-   ).

WA, Br   — *D. Martin Luthers Werke.* Tischreden
(Weimar, 1930-   ).

WA, TR   — *D. Martin Luthers Werke.* Briefwechsel
(Weimar, 1912-1921).

WA, DB   — *D. Martin Luthers Werke.* Deutsche Bibel
(Weimar, 1906-   ).

Part One

CHRISTIAN ETHICS IN THE AGE OF SECULARISM

1.

The Challenge of Secularism

Anyone writing on ethics today must begin with the seemingly embarrassing business of having to consider whether and how far his enterprise is even possible. The very ground he moves on shakes beneath him. He is not like the man who lives in a secure theological and academic tradition and who, by simply adding his stone to the building, contributes to a work that has been growing for centuries. It may be that a Roman Catholic colleague can work on the basis of this kind of ethos. But in the Reformation tradition a theologian today must first look to the very foundations on which he would build. He has seen spires fall, and he sees others standing only as odd and isolated monuments, alien to the world and—which is worse— to the kingdom of God.

The state of upheaval in which the ethicist must now work may —like any other such state of upheaval—augur the onset of complete collapse; it may be a "sickness unto death." If that be the case, it would really be advisable to begin an ethics of this sort with the confession of an individual soul in upheaval. I myself, however, would not for a moment claim to have the courage to put such a confession in first place—even if I felt compelled to interpret the crisis of ethics in this fashion.

To say this, however, is to suggest already that I do not regard the fact that evangelical ethics has become such a problem today as indicative of—shall we say—doddering old age and decay. On the contrary, I see in this fact the outcropping of a positive and productive idea. What that idea is may be stated in at least a preliminary way as follows: Secularization has radically called in question the concept of reality long held in the Christian tradition.

Demonization of the World

For the moment it will suffice if I refer only to the following developments: With the progress of secularization the world has

3

been able to cease its chafing under the yoke of commandments and the strait jacket of so-called "Christian" states and customs. It has attempted to organize and constitute itself exclusively on the basis of factors already inherent within itself. As a result, we have been able for the first time to see clearly what the world really is. To be sure, its most sinister possibilities have not yet been fully unleashed. However, we are beginning to get some idea of the monstrous nature of its demonic potential, and we have the sneaking suspicion that the visions in the Johannine apocalypse are not too far removed from reality. Hence it is not by accident that there has developed among us a new readiness to take demonic powers seriously, as is attested in the repeated attention now given to this matter by such serious writers as for example Tillich, Schmid-Noerr, and Würtenberg. Of course, knowledge of the demonic power never did completely disappear in Christianity, but full knowledge was always limited to the pages of the Bible. There it remained concealed, as it were, against that suitable time at the end of the age for which it had been appointed. Furthermore, knowledge of the demonic power heretofore has always been only "in part," and in respect of this part of reality we ourselves do not yet walk "by sight."

If with increasing secularization we enter more and more the stage of excessive emancipation—one might call it the stage of increasing disclosure of the world's most sinister possibilities—one thing at least is clear: this does not imply the emergence of any new factor in our basic understanding of reality. It is not a question of the old categories having become so inadequate that we are now forced to shift our co-ordinates quickly in order to accommodate the old system of Christian values to this new factor. On the contrary, it is altogether possible to set forth our "new" knowledge of the reality of the demonic precisely in the name of the biblical view of reality. The times which war against one another on the battlefield of our spirit, and which "cause us to see demons" (C. F. Meyer), do not yield anything new. They simply provide us with certain aids by which we may see biblical reality afresh. And that is all they do. What we have discovered, as a consequence of secularization, about demonic reality and about the emancipated world is wholly along the lines laid down in the Bible. Our discovery simply adds color, as it were, to the biblical contours; it brings into the open a picture previously concealed. In the re-emergence of the picture of the demonic we thus have a true "reformation," a return to the "Word."

History does not, through this emancipation, now begin to be a second source of revelation. It simply takes on illustrative force in relation to the text. Indeed, without the text the illustration itself would be unintelligible. The stratum of historical concretions becomes transparent only "when and where God pleases" [*ubi et quando visum est Deo*].[1] It does so, however, not to point to itself (as some kind of a repository of divine mysteries) but to point right through itself to the very text, which the "wisdom of the world" can of itself no more control than it can control the historical illustration.

The chief point of all we have been saying is this, that in this world where Christian values have been discarded Christian action as such has itself been called into question. In view of these defections, is it still possible to define at all in a normative way the Christian's "being-in-the-world"? Better yet, can it be defined in any other way than to say it is practically a total contradiction of anything and everything—of the orders of this emancipated world, of the good and evil, the true and false, the lofty and base within it? To carry that a step further, can it be defined other than as the suspension of action altogether, and that with the eschatological finality and radical negation of this whole aeon which led Kamlah to say that, because of this primitive Christian radicalism, self-assertion of any kind—and therefore historical existence as such—becomes completely impossible?[2] Is *this* the "crisis of Christian ethics" about which everybody is speaking with such alarm?

Be that as it may, this situation of complete alienation in which Christian ethics finds itself must in any case be weighed and evaluated theologically. Very simply put: In what relationship does the demanding will of God stand to our whole human-historical reality, e.g., to the created orders? This apparently is the question which has to be faced before we come to that other problem, the one men have repeatedly called the cardinal question of Christian ethics: What shall we *do*? That is, how are we to act within the framework of this human-historical reality?

[1] The Augsburg Confession, Art. V; *BC* 31.
[2] See Wilhelm Kamlah, *Christentum und Selbstbehauptung. Historisch-philosophische Untersuchungen zur Entstehung des Christentums und zu Augustins "Bürgschaft Gottes"* (Frankfurt, 1940).

If it be true that we find the relation between God's commands and our human-historical reality so very problematical today, and if it is secularization that has made us thus aware of the problematical nature of that relationship, then it would appear to be altogether possible that in our experiencing of this awareness quite different depths of knowledge should be disclosed to us, and quite different impulses of action mediated to us, than in past times. Insights and motives today may vary greatly from what they were before, when men tried to conceive of Christian ethics as merely the transcendent extension of Kant's "categorical imperative" or as a reservoir of religious powers enabling us to do what is already known to be "the good," [3] when the sphere of God's commands on the one hand and the sphere of our human-historical reality on the other were accordingly thought to be attuned to one another in perfect harmony.[4]

In any case, we seem to be justified in regarding the crisis in the concept of reality, and thereby also the crisis in Christian ethics, as a wholly productive challenge. For it prevents our identifying the question of ethics with the specific question, "What shall we do?" and compels us to see it instead embedded in the far more comprehensive question, "How are we to understand that reality within which our doing must take place?" In other words, what does our "being-in-the-world"—in the emancipated world!—really mean? What the New Testament calls being "in the world" but not "of the world," living "in the flesh" but not "according to the flesh," thus becomes for us the most pressing problem of ethics.

Autonomous Spheres of Life

There is a second form of the crisis of evangelical ethics which, strictly speaking, is only a specific instance of the first. It results from the fact that life now seems to be lived out in autonomous spheres and that the longer the process of secularization continues, the more firmly it is proclaimed—under the slogan of "autonomy" [Eigen-

[3] See, for example, Wilhelm Herrmann, *Ethik* (2nd ed.; Tübingen-Leipzig, 1901).

[4] See Richard Rothe, *Theologische Ethik* (2nd ed.; Wittenberg: Koelling, 1869-1871).

6

gesetzlichkeit]—that the world is self-contained and, by virtue of its own immanent laws, self-regulated.

The concept of autonomy implies that each of the various spheres of life (politics, economics, art, etc.) has its own particular set of laws inherent in the very nature of the matter which is its peculiar concern, and that these inherent laws forbid us to bring these spheres under the judgment of ethics. For any attempt to do so would mean imposing on the sphere in question the external norm of an alien sphere, such as that of morality or of religion. This would lead, however, not to the establishment but to the chaotic disruption of a normative ethical order. For what would happen—so the proponents ask—if politics were conducted no longer on the basis of its own inherent laws regarding the friend-foe relationship, and hence no longer with the help of power and cunning? What would happen if men in politics took their cue instead from certain ethical ideals requiring that all they do be "open and above board," that they earnestly seek to rule out egoism and finally even deny the law that the strong march over the graves of the weak?

Politics seems to be so much a law unto itself that it tries to shake off any attempt at ethical heteronomy. On the other hand, the values of good and evil cannot be thrown out altogether. Consequently political action always leads in practice to the questionable compromise of "political morality," a highly specialized and very much reduced form of morality. How extreme and undeviating are the attempts of these autonomies to assert themselves we have learned from the experience of war—which insists on being a law unto itself. We have witnessed numerous efforts to humanize war, e.g., to protect civilian populations from being visited by its furies. At the same time we have seen it develop by its own autonomous weight to the point where it becomes all-out war, excluding everything outside its own inherent law. War rides roughshod, for example, over the law of humanity, even though one of the warring parties may seek for its part to abide by humanitarian tenets.

Does not this mean that ethics is once again shown the door, lest it should address a world which is blatantly ready to listen only to its own voice, a world which wills its own fate? For what is involved in these autonomous spheres is actually the beginning of a new belief in fate, the consciousness of an utterly suprapersonal force which no man can direct.

We cannot ask this rhetorical question about ethics being rudely booted out the world's back door without having to add at once that theology itself has contributed not a little to what has happened. There have been and are certain theologians who willingly acquiesce in the expulsion, in order thereby to purchase continuing toleration on the inside for themselves. Particularly in some circles in recent German Lutheran theology, there have been thinkers who have affirmed these autonomies in the name of the "two kingdoms." [5] They have not infrequently begun their discussion of ethics by stating expressly that we should first listen to the voice of the political economists and politicians. This in itself would indicate that in their view there are indeed spheres of life which are autonomous, not subject to ethics, and hence not subject to theology either. It may be that among the different interpretations of these autonomies, e.g., the socialistic or the capitalistic understanding of the economic process, there are some views which these theologians find more acceptable than others, and which they would simply bring to the stage for a curtain call by means of their applause. But just as it is the performer and not the applauding listener who makes the music in the concert hall, and just as the musician can make his music quite independently of the friendly gesture of hand clapping, so a theology which limits itself to applause and makes no contribution of its own cannot actively exercise any constructive function. Such a theology can never be in a position to give to the church the intellectual and spiritual equipment whereby it may proclaim God's claim to lordship over the entire world and all the various spheres of life. On the contrary, it will represent a laissez faire position in relation to the world. It will favor the restriction of God's claim to the private sector of life, to "inwardness." It will in effect seek to withdraw the lordship of God from the "worldly" sphere, trying, as it were, to make this lordship private rather than public.

We cannot consider the crisis into which Christian ethics is betrayed by these autonomies without discussing at least briefly how

---

[5] For a discussion of this topic see Chapter 18 below. And for a critical review of the positions of Ernst Troeltsch, Friedrich Gogarten, and Wilhelm Stapel see Harald Diem's excellent book, *Luthers Lehre von den zwei Reichen* (Munich, 1938).

it could come about that the way toward their sanctioning was prepared within the church and its theology—and still pursued even after the terrible consequences of such a course had been manifested! The fact that this could happen is due to a decisive misunderstanding, rather a complete and callous re-evaluation, of Luther's concept of the "orders" [*Ordnungen*].

The False Theological Sanctioning of Autonomy

During the course of the Reformation Christian ethics came under the fateful influence of Philip Melanchthon, and therewith into philosophical dependence on Aristotle. This is apparent primarily in the fact that the motivation of our "works" is thought to be determined not so much by faith as by a virtuous ideal, namely, the "idea" of moral perfection. It is significant that at the heart of Melanchthon's ethical thinking there stands the idealistic concept of *societas*, the autonomously defined form of fellowship which derives from the law of nature [*lex naturae*] and provides the norm for my relation to my neighbor, and therewith for my ethical standard of goodness [*virtus*].

The fateful character of this development is to be found not merely in the fact that a non-theological concept, namely, a value deducible by logic, is here made the supreme commandment. It lies also and more pre-eminently in the fact that, in this context at least, faith is now assigned the task of overcoming the obstacles to be found in sin, and thus of providing the possibility and the power of attaining to "the good" which is itself already known quite apart from faith. This is the development which in the nineteenth century reached its climax in the ethics of Wilhelm Herrmann, except that he took his ethical *virtus* from Kantian rather than from Aristotelian philosophy. In the twentieth century, it issued in what was ultimately the only possible slogan for such a development [the slogan of the Nazified "German Christians"]: "Germany our goal, Christ our power."

One may well suppose that this philosophical takeover would necessarily have a profound influence not only on the doctrine of justification but also on ethics, and would constitute an attack on the very substance of the Reformation. That this is in fact the case may be seen in relation to the concept of "order," a concept that is bound to be particularly imperiled when the view of some ideal society is imposed upon faith.

9

It may be said that Luther's doctrine of the created "orders" arose in consequence of his doctrine of justification, and within the context of his theological work on the First Article of the Creed. It thus stands in most intimate connection with his concept of faith. To put the matter sharply, one might say that Luther's doctrine of the orders is a by-product of his doctrine of justification. In no case did it arise out of an independent, secular interest in the nexus of generations, the nation, politics, or economics, or out of merely a general interest in the orders which sustain and organize "creation." The doctrine of the orders, especially of estate [*Stand*] and calling [*Beruf*], arose in closest connection with the struggle between Luther's doctrine of justification and the theory and practice of monasticism.[6]

Luther attacked the very basis of monasticism on two principal points. First, the monastic life rests on the division of the divine commands into "precepts" and "counsels," a distinction which fosters the idea that monks are exceptional men. Second, in order to realize this exceptional life in practice, monasticism resorts to a divine worship and service of its own choosing; it summons men to abandon the orders, to practice renunciation, and in this covert way to seek a righteousness of "works." As Luther understands the Law, there can be no escaping its judgment. The Law can lead us only to despair, to defiance, or to false confidence. To serve God by ways of our own choosing, in a vain attempt to satisfy the Law, can only bind us the more closely to ourselves. Indeed, such worship and service is particularly dangerous since it seeks to establish a claim on God and in the strict sense is thus diametrically opposed to the true worship and service of God. It is quite impossible for men to escape the solidarity of human sin; there is no such thing as a place apart, outside the sphere of sin.

Luther's doctrine of the orders is developed, then, from this starting point. First, it rests on the negative position that there is no peculiarly qualified place outside ourselves or above our world to which we can resort in order to please God. As a result we have to stay precisely where God has placed us. Just as it is impossible for men to co-operate in their own salvation by means of doing good works, it is logically also impossible to co-operate by means of striving

6 See Karl Holl, *Gesammelte Aufsätze zur Kirchengeschichte*, I (Tübingen, 1923), 198-203.

for or living at certain levels of existence which are supposedly more open to God.

Second, the doctrine of the orders rests on the positive consideration that through the orders (the generations, marriage, politics, etc.) we are brought into relationship to the concrete Thou of our neighbor. It is here, at the point of this relationship, that we have to manifest our complete and undeviating obedience.

Third, the doctrine rests on one final reassuring thought. Service to the order in which God has placed us, the order in which he has ordained that we shall be of service to our neighbor, while it does indeed "sanctify" us, cannot "save" us. It is incapable of ordering afresh our fellowship with God.[7] It is precisely with reference to the works we do in our particular estate and calling that Luther says, "All works have only as much worth as God reckons them to have."[8]

Hence the orders have no independent dignity which they can impart to us if we but fulfill them. The "alien righteousness" [aliena iustitia] which alone can help us lies exclusively in the reckoning of God which we lay hold of in faith. Just as there is no justification through what is "extraordinary," e.g., the monastic enterprise, so there is no justification through the "ordinary," i.e., through fulfillment of the orders. The orders tie us down perpetually to the sphere of works, which leads no further. The maid's sweeping, to which Luther occasionally refers, is service of God not because it achieves cleanliness but because it is done in faith. Conduct within the orders has theologically the character of "works" and must be treated accordingly.[9]

This shows plainly that Luther's doctrine of the orders is part and parcel of his doctrine of justification. Conduct within the orders is justified, not by its conformity to the orders, but by the forgiveness which comes to me in an "alien Word." To sunder the orders from this "alien Word" is to misunderstand Luther to the point of cari-

---

[7] "However, none of these orders is a means of salvation. There remains only one way above them all, viz., faith in Jesus Christ. For to be holy and to be saved are two entirely different things. We are saved through Christ alone; but we become holy [in the sense of the sanctification of good works subsequent to faith] both through this faith and through these divine foundations and orders." LW 37, 365.

[8] *Luther: Lectures on Romans (1515-1516)*, translated and edited by Wilhelm Pauck (Philadelphia: Westminster, 1961); LCC 15, 303.

[9] LW 37, 363-365.

cature. But to base an ethics on the Aristotelian concept of *societas* adopted by Melanchthon is an even greater distortion. What we have here is a perfect illustration of a principle that has repeatedly been at work in the history of theology. Whenever a biblical truth— Luther's doctrine of justification is exactly that—is combined with a secular ideology, such as the concept of *societas*, the latter always prevails; it breaks loose and in so doing distorts the original proposition to the point where it is no longer recognizable.

This is exactly what has happened in the present instance. With increasing secularization, Luther has been increasingly hailed as the discoverer of the orders of this world. He is the discoverer and champion of nationalism, nature, and marriage, and the one who has enthroned freedom of conscience and opened the way for political autonomy. This view sees the by-products but misses the process itself. In other words, men no longer inquire in whose name Luther was speaking when he discovered and championed these rights and orders. Men are no longer aware of the background out .of which everything was said and done—the doctrine of justification. Men have forgotten the Word, which is the real theme of all these processes.

The most grotesque distortion of Luther in the secularized movement to which we refer is perhaps that of Arno Deutelmoser.[10] Here the legend of Luther's supposed emancipation from the transcendent authorities of the Middle Ages is decked out in such glowing colors that Luther's Christianity itself is called in question. Its ambivalence may be seen, it is alleged, in Luther's concept of the state and in his purely immanental approach. The development which ends with this distorted view of Luther is undoubtedly typical. As secularization proceeds, the orders, loosed from the Word, increasingly acquire a life of their own. Their autonomy is discovered within the sphere of immanent values. And the fundamental struggle is no longer about the binding authority to which these values are ultimately obligated, i.e., about the way in which they stand under the judgment and grace of God; it is rather about their relative rank in the hierarchy of values and about the co-ordination of the individual autonomies.

Autonomy in the Morally Independent Subject

The same process of secularization which in terms of the orders

---

[10] Arno Deutelmoser, *Luther, Staat und Glaube* (Jena, 1937).

unfolds in the world of objects is also going on in the sphere of the personal subject. My attitude to this world of objects and to the Thou whom I encounter in it is regulated by the moral autonomy which I bring forth out of my own ego, an autonomy which does not need to be confirmed by any other authority outside myself. Indeed, any power which lays claim to moral authority, e.g., the Decalogue, must vindicate itself before the bar of my autonomy. Thus the final authority which is binding beyond everything else resides in me, or, more precisely, in my intelligible ego. Man no longer needs to receive the Word which binds and looses him, because man now speaks that word himself.

There have been and are theologians[11] who have built their doctrine of orders on the basis of their having observed the autonomies of the orders. By so doing they have been enabled to sit down in the company of sociologists and ethicists of every description, unhappily for too long a time and without ever giving offense. In the same way there have been and are theologians who base their doctrine of the Law of God on the axiom of the autonomous conscience, that intelligible ego which believes it no longer needs to ask the question of Law and Gospel inasmuch as the children of this world are quite as rich and wise as the prophets themselves in having and knowing the directives which are of ultimate validity. This theology of "what men have in common" is to be found with occasionally different accents in Schleiermacher, in the German moralists who followed Kant (Albrecht Ritschl and Wilhelm Herrmann), and during the 1930's in those champions of the identity of the Law of God and the law [nomos] of the German people (Wilhelm Stapel, Emanuel Hirsch, and Friedrich Gogarten).

This brief glance at the ethical problem of the autonomies, a problem which will occupy us again and again later, may suffice for purposes of an initial orientation. At this point it is enough to assert that the autonomies of the various spheres of public life evade the grasp of the Christian ethos and limit Christian action to the ghetto of private life.

This crisis too can be productive, however, if it makes clear that the emancipation of the various spheres of life from God's claim to lord-

11 I am thinking especially of Georg Wünsch, *Evangelische Ethik des Politischen* (Tübingen, 1936).

ship causes everything to disintegrate. Negatively at least it will thus make clear that life which is left to itself, i.e., to its own autonomies, can only move towards self-destruction. And unless appearances are wholly deceptive, we in our time have seen man in the economic, political, military, and aesthetic spheres of life foundering on the rock of divine providence in almost unprecedented fashion.

This does not mean that we have been given a negative proof of the existence of God. It does mean, however, that where the process of hardening has not yet set in—it usually goes hand in hand with this process of foundering and destruction[12]—there might be a new readiness to hear a divine claim which comes with full authority and with its own Law to confront the autonomies of the world and give them tit for tat. Certainly the problem of the autonomies must loom large in any treatment of evangelical ethics.

[12] On this problem of the theology of history see Edmund Schlink, *Die Gnade in Gottes Gericht* (Gütersloh, 1946) and Helmut Thielicke, *Gericht und Heimsuchung. Untersuchungen zur Frage der konkreten Gerichtspredigt sowie darüber, ob eine theologische Geschichtsdeutung erlaubt sei* (Tübingen, 1946), esp. pp. 18 ff.

2.

The Challenge of Humanism

In addition to the challenges posed by the demonization of the world and by the asserted autonomy of the various spheres of life and of the individual himself, we must list yet another challenge which also has its source in secularization. The process of secularization has taught us that Christian existence itself is by no means so self-evident and unequivocal a matter as was sometimes apparently supposed in ages of greater naïveté. What we mean by that will become clear in the ensuing discussions.

The Interchangeability of Faith

Faith too is susceptible of interpretation in terms of immanent categories, such as those of psychology for example. With the help of secular history men have been able to reduce the figure of Jesus of Nazareth to the level of general religious history. In the same secular way, with the help of psychology it is possible to reduce faith to the level of the spiritual processes inside a man. It is characteristic of those who thus view Christian faith as only a subjective, psychological matter that they employ many expressions which receive their stamp from secularization. For example, the typically employ the term "credulity" in order to suggest that faith does not primarily refer to and receive its character from its object, but is simply one of the many products of man's creative subjectivity. Credulity as a property or disposition [*habitus*] of subjectivity is primary. Instead of being determined by its object, credulity itself determines on its own the object to which it wishes to refer, e.g., a certain world view of a particular religious confession.

It was Schleiermacher who in his book *On Religion: Speeches to its Cultured Despisers* developed in a classic way this notion of a creative credulity which fashions its own object. He held that those things which religion cherishes in the way of objective elements, and in which religion has set down a record of itself, e.g., as "Word" (in

15

such things as holy scriptures, dogmas, or doctrines), are not things which encounter human subjectivity from without or from above, addressing and claiming it. On the contrary, it is this subjectivity itself which is the source of all these things. Religious feeling, because it is so powerful, produces a need to communicate: "Out of the abundance of the heart the mouth speaks" (Luke 6:45). This means, however, that there is a need for verbal expression, since this is the medium of communication. It is thus that the "music" of the heart is transposed into verbal form; it is—to use the famous phrase of Rudolf Otto—"schematized" by the Word. Dogma and doctrine are thus nothing but ossified religious feeling. It is of course possible that the record of this feeling may have a stimulating effect on the subjectivity of others, especially when what is recorded is such classical music as that of prophetic charismatics. But even here what is involved is solely and exclusively the outward-streaming subjectivity, the *habitus* of credulity.[1]

An important consequence follows. Within the sphere of credulity all forms of faith become in principle co-ordinate and of equal worth. Only the objects produced are different. Since these objects are of secondary rank, however, the differences among them are obviously of quite secondary importance as compared with the far more decisive fact that faith is nothing more than a psychical factor. It may thus be seen that Christian existence, to the degree that it is expressed in faith, loses its distinctiveness. Christian faith is "interchangeable" with the credulities of general subjectivity, just as its object, Jesus Christ, is interchangeable with the leading personalities in the general history of religion and of philosophy.

Here too one may again see the heuristic fruitfulness of secularism —for of course secularism did not itself create interchangeability, e.g., that of faith. This interchangeability has always been present. It was obviously known already by Luther, who said, "Do not rest on yourself or on your faith; creep into Christ . . ."[2] Clearly Luther knew how questionable is the attempt to base faith itself on faith, and therefore on subjectivity. Here as elsewhere Luther binds faith to the

---

[1] See particularly the second of Schleiermacher's "Speeches." The relation between objectivity and subjectivity becomes incomparably more complicated in his later *Christian Faith*, so the point we would make here has reference only to the younger Schleiermacher:

[2] WA 10I,1, 126.

object of faith, to Jesus Christ. I repeat, secularism has not created interchangeability, but by means of secular psychology and history it has carried the thing to such an extreme that it can no longer be overlooked or denied.[3]

The Interchangeability of Ethics

The same phenomenon of interchangeability observed in connection with the concept of faith has, with the progress of secularism, also come to exist in the field of ethics. It has called in question the possibility of a specifically "Christian" ethics. For any particular deed of "Christian action" is not unequivocal; it is quite interchangeable with acts having a completely different motivation. After all, once it has left the inner sphere of motives and become an objectified "work," even the Christian deed enters into a framework of values which is comprehensible quite apart from considerations of any particularly Christian motive.

One has only to recall the slogan "Christianity of action" which was mouthed so often in the conflict of world views in Nazi Germany. The point of the slogan was that what matters is not the dogmatic or doctrinal basis of our action, but the two far more important facts that there should *be* action, and that generally recognized values should be created by it. Among these values, which were acknowledged to be in accord with Christian motives, there was, for example, this, that everything be done in keeping with the principle that "the common good comes before my own."

When regard is given only to these objective values which are to be ethically actualized, the motive of the action recedes into the background. It could thus happen that under the increasing anti-Christian pressure of the period the slogan "Christianity of action" could secretly and surreptitiously change over into a very different slogan, namely, the more popular "Socialism of action." Historically this is exactly what happened. The very fact that this change in slogan could take place makes it evident that behind the actualization of any given value there may lie a whole world of divergent motives. As Herod and Pilate, with their wholly different political motivations,

---

[3] See my article *"Das Problem des autoritätsgebundenen Denkens,"* in *Theologische Literaturzeitung* (August, 1947), reprinted in my *Theologie der Anfechtung* (Tübingen, 1949).

could become one in their common hatred of Christ, so Christian and anti-Christian motives can unite when it comes to common ethical goals.

These facts should suffice to clarify our statement that "Christian" action is interchangeable with action of a totally different kind. The interchangeability reaches its climax when non-Christian ethics is regarded simply as a secularized form—and therefore a continuation—of Christian ethics, as, for example, in Kant or in Idealistic humanism.

Implied in all this is an explicit challenge to Christian ethics. This challenge might be expressed in the following formula, which is really a rhetorical question: Cannot the goals of Christian action (service to one's neighbor, establishment of community, unselfishness, obedience, etc.) be as well achieved—or at least sought—"without Christ," simply within the framework of secular ideologies? Is not Christian conduct very largely identical with conduct based on ethical humanity? Or, in Kantian terms, is not Christian behavior distinct from purely secular behavior only to the degree that it construes the commands which arise within the autonomy of conscience to be commands of God, thus subordinating them to a divine subject and thereby giving them a theonomous denominator? This would, of course, be an expression of the fact that the commands themselves are in both cases identical, and that the Decalogue as natural law cannot therefore be distinguished in principle from the directives of the "categorical imperative." It would mean too that in reality there is no longer any basic distinction whatsoever between Christian and secular morality.

If I see the matter correctly, however, this challenge of secularism to evangelical ethics again represents a productive crisis; it can give us vital help in arriving at fundamental definitions. For by the very fact that ethical action is interchangeable, and that the level of overt deeds no longer tells us as a matter of course anything about the underlying level of motives, we are most urgently compelled to consider what point there is in attempting a specifically "Christian" ethics at all. It must be obvious by now that it is not a matter of particular programs or "ideals of virtue," for what is involved in both these areas is the level of overt acts.

This is why it always happens that Christian ethics, whenever it begins with these two concepts of world betterment and virtuous ideals, promptly loses its distinctive content. Indeed one might even go so far as to say that this distinctive content is lost from the very outset because such moralists—quite logically in terms of their own view of the matter—find before they even begin that they are compelled to take over their concepts of the ethical goal from a-Christian systems of thought. And they can do that with comparative unconcern because, when the field of vision is restricted to the level of acts, the result is that natural ethics and Christian ethics are brought into harmonious relation with one another and the profound differences between them are either concealed or manifested only in distorted form.

Melanchthon's concept of virtue is a classical example of this. In his ethics he took over, by way of the Scholastic tradition, the Aristotelian—earlier Stoic—concepts of *societas* [society] and ἀρετή [virtue]. How he could do so can only be explained by the fact that in very large measure he abstracted the sphere of acts, e.g., the system of orders, from that of the acting person and established it on the basis of its own inherent worth. We can also see why for this reason an ethics built on such a base must follow to the bitter end the law by which it began. The pagan-humanistic foundation will necessarily come to the surface again and again, and with ever increasing dominance. It will inevitably come to overlay and suppress the Christian content which fills the non-Christian frame. In fact in some respects Melanchthon became the first secularizer of the theology of Luther, though it goes without saying that one cannot summarily subsume Melanchthon's entire work under such a simple formula. Nevertheless it was he who gave the decisive impulses to the continuing process of secularization which reached its first climax in the broadly conceived doctrines of Protestant Orthodoxy—on general revelation, the orders, and natural law.

The specifically "Christian" element in ethics does not emerge at the level of acts and in ethical programs, though this is not to say that for Christian ethics deeds and programs are a matter of complete indifference. It is true of course that there are no acts and programs on which Christian ethics can be said to have a monopoly or a patent, as if they were beyond the competence of natural ethics to require. But it is equally true, as we shall show, that certain acts and programs are wholly excluded from Christian ethics; they are forbidden in no uncertain terms. The degree to which this negative accent has theological significance will concern us later when we examine the negative form of the Decalogue.[4]

The Distinctiveness of Christian Ethics at the Level of Motives

The specifically "Christian" element in ethics is rather to be sought explicitly and exclusively in the motivation of the action. This may be seen clearly in the polemics of Jesus against the Pharisees (Matt. 23:2 ff.). For the hypocrisy of the scribes and Pharisees consists precisely in the fact that they are blameless at the level of acts, i.e., "outwardly" (v. 28). All that they do is designed to be seen by men (v. 5). Since they act on the basis of ossified legalism and feverish casuistry, they are walking on slippery ground, for example, in the use of oaths (vv. 16 ff.). They are testifying that their actions spring from a heart which is apostate from God.

This, of course, is obvious only to him who evaluates the whole man in his relation to God, who looks not only at the outside of the plate but also at the inside, and who is not deceived by the whitewash on the tombs but knows full well of the dead men's bones and uncleanness within (vv. 26 ff.). It is no accident that it is Jesus—who knows what is in man—who must first unmask this hypocrisy. It certainly does not unmask itself. To be sure, the corrupt tree brings forth corrupt fruit (Matt. 7:17 ff.; 12:33), and the life of action is infected by the sickness of the root from which it springs. Nevertheless, that the fruit is in fact corrupt is not apparent as a matter of course. That fact can be seen only by him who knows why the tree is there and what it is supposed to do. It can be seen only by him who has the necessary criteria, for whom a comparison of what is with what ought to be is possible, and who is capable of taking in the tree as a whole.

Except in very extreme cases of hypocrisy, the natural man simply cannot penetrate the secret of the Pharisee. For the natural man the world of fruits, the level of acts, remains veiled, because he does not know men's inner disposition toward God. In other words, observation of the life of action will not lead him to unmask the secret of the Pharisee, or to know the nature of the divine requirement as it really

4 See below pp. 440 ff.

20

is. The natural man can do neither, even if confronted by a more perfect obedience than that of the Pharisees.

The level of acts is in principle unsuited for disclosing that which is distinctive about obedience to God. To be sure, it is unconditionally true that words cannot be believed without deeds. But it is equally true that deeds without words are ambiguous. The emancipated ethics of secularism has given us some impressive examples of this too.

Hence the specifically "Christian" element in ethics is found only at the level of motives. This is pointedly expressed in the saying in Colossians 3:23: "Whatsoever ye do, do it from the heart (ἐκ ψυχῆς), as to the Lord, and not unto men." For in this fundamental imperative of Christian conduct the action is characterized, not in terms of its material content, but according to its basis and its goal.

The basis of Christian action consists first in the fact that the action is genuine, i.e., that there is agreement between the inner side and the outer (cf. the expressions "inside" and "outside" in Matthew 23:25). To the extent that I act from the heart, the motive of mere outward show is indeed excluded, and expression is also given to the fact that the ultimate binding authority of the "inner man"[5] must play its role also in a man's conduct, for only so can a man be fully present in his action.

But since that which is ultimately binding for a man is his fellowship with the Lord, the goal of a man's action is thereby posited at the same time, namely, that what I do I should do to the glory of the Lord, the kurios. Just as I myself must be wholly present in my action, if I am to act genuinely, so this kurios is also truly present in it. For after all it is he who comes to meet me in the form of the hungry, imprisoned, thirsty, or homeless human brother (Matt. 25:35 ff.). It is the Savior who goes about the earth masked as my neighbor. This indeed is the "alien dignity" which distinguishes my neighbor. Jesus Christ has died for him (Rom. 14:15; I Cor. 8:11); that which stands back of every man, giving him infinite worth, I see to be Jesus. As a result, He is present not merely in sign or symbol but in reality, when for His sake I enter into contact with my neighbor or take my

[5] Luther often uses *interior homo* interchangeably with *anima* and *persona* as synonyms. See *LW* 31, 344-345, and *WA* 9, 83 (lines 30-33) and 85 (lines 10-13). On this same matter cf. also *LW* 40, 90-91.

place concretely in the affairs of men—including, for example, the various events of history which are so intimately related to the personal sphere.

The relationship to the *kurios* is not restricted to the basic reflection which precedes my action, and which is accordingly outside it. It is not restricted to the sphere of the "prolegomena" to conduct. The relationship rather becomes part and parcel of the action itself. Here is to be seen, then, the extent to which the inner structure of the action is "Christian." The action may be called "Christian" if in both its basis and its goal it is determined by the *kurios*. (The question whether the "may be" in this formulation implies mere potentiality or unreality will be discussed later.)

This is particularly well expressed in Luther's explanations of the Ten Commandments in his Small Catechism. As is well known, they all begin, "We should fear and love God, so that . . ." In other words, obedience to the commandments is rendered on the level of motives. All right action springs from the fact that I fear and love God. And any action which does not spring from this fact is to be disparaged either as an attempt to earn salvation by merit or as hypocrisy. When the explanations continue, ". . . so that we do not kill . . . commit adultery . . . steal, etc.," these are all consecutive clauses which make it plain that certain very definite acts necessarily follow from the love of God and the fear of God.[6] True obedience consists in fear and love. But this act of obedience to God, as will be explained later, is identical with a very definite kind of behavior toward men. (Even Jesus' command to love in Matthew 22:36-39 links these two—love of God and of man—together.)

Now if it is true that we should fear and love God "so that" we do not steal, kill, commit adultery, etc., if it is true that the conduct resulting from Christian motivation is that prescribed in the Decalogue, then it is also true that the logical sequence of these consecutive clauses cannot be reversed. That is to say, one cannot possibly deduce from a correct observance of the Ten Commandments that God is truly feared and loved. Seen from the standpoint of the motivation, the action is unequivocal; but seen from the standpoint of the action, the motivation is highly equivocal.

[6] On this point, see pp. 69 ff. below.

Differentiation from the Ethics of Disposition

At this point, however, we must be careful to avoid a possible misunderstanding. When we say that the Christian character of an action arises only on the level of motive, it might be thought at first glance that we are thereby making Christian ethics simply an ethics of disposition, comparable to that found in Kant. (To be sure, the widespread nature of this misunderstanding often suggests that it is by no means restricted to a "first glance" but has something of a chronic character.) For the level of motives is surely the same thing as the sphere of disposition.

Does not the ethics of disposition appear to agree with Christian ethics in saying that the area of acts as such is equivocal? Indeed this is why, at least for dispositional ethics, there can never be an ethics of material values, one that would take as its starting point "goods" which presumably are to be sought because they are of objective value. For these goods might conceivably be sought for purely eudaemonistic and therefore immoral reasons. If someone endows an orphanage, this tells us very little about the real nature of his act. For he could be doing so merely out of a desire for fame or status, especially if he insists that the orphanage be named after himself—in immortal memory of so noble and unselfish an action. This is why Kant was so concerned to exclude all non-moral motives and to discover in the categorical imperative itself a formula which would represent moral autonomy, as it were, in a pure and crystallized form.

Does this mean that if Christian ethics is thus referred to the sphere of motive and disposition, it thereby forfeits its distinctive character? That depends entirely on whether the nature of the disposition proclaimed by the Christian ethos remains wholly distinct from that asserted in other ethics, such as the Kantian. Ethicists like Wilhelm Herrmann who construct their ethics on the basis of the categorical imperative—and hence in the dispositional dimension—agree with this, even if only in the sense that Christianity offers to the disposition which is reduced to despair by the impossibility of moral fulfillment the possibility of liberation, the possibility of becoming a disposition redeemed for action and in action. All the same, to subsume Christian ethics under the general term "ethics of disposition" is to reduce it to a common denominator, pervert it into an immanent this-worldly phenomenon, and deny its transcendent origin. It is to effect in the area of ethics something analogous to what has happened elsewhere, where Christian revelation has been subsumed under the general term "religion" and so robbed of its distinctive character.

In actual fact, our proposition that the specifically "Christian" element in ethics appears only at the level of motivation does not mean that we are making Christian ethics into a mere ethics of disposition. For disposition is something which is first imposed upon me as a task, over which I therefore have control. This is how Kant views the moral subject, as something that is not simply a given fact. For man is also a member of the sensible world, and consequently in all his endeavors the object of various drives and impulses. "By nature" man is a eudaemonist and an egoist. He must first find his way to ethical disposition if he is not to be dominated by these drives and impulses. For Kant there is only one impulse that is legitimate, namely, the readiness so to act that "the maxims of this action" can be made "the principle of universal legislation." In consequence, the ethical disposition as a motive of action is not an actuality already given; it is a task yet to be accomplished.

Furthermore, because he is a moral subject (an intelligible ego), man is not perfect, complete, and whole. He must first lay hold of himself and set himself free. One cannot say, therefore, that for Kant man's ethical disposition is the necessary presupposition of all moral action; it must rather be understood as something which has first to be attained. It becomes the presupposition of moral action only insofar as it is actually made to be that, only insofar as it is so constituted. The moral ego must first posit itself as such. Only then will man be conceived as a being that is in control of itself, that rules itself. And in no way does this control mean simply that man prescribes for himself the law of his action and is therefore autonomous— at least to the extent that he can also fulfill that law ("You can, because you ought"). This sovereignty over himself is exercised at a much deeper level, namely, at the level where man first determines himself as subject at all, that subject which prescribes its own law and then resolves to fulfill it. This sovereign autonomy in the breadth of its scope amounts to a kind of self-creation of the ego. In its later development Idealism, especially as represented by Fichte, took up this matter of self-creation with a proper though one-sided understanding of it, and carried it to the extreme.

Disposition and Christian Existence

It is supremely important to see correctly the decisive point that the ethical disposition is not something given but something yet to be attained, to be attained—paradoxically—by itself. This immediately poses the possibility of actually attaining the ethical disposition. In other words, man is fundamentally in a position where he can attain the right starting point for moral conduct in keeping with the command.

Here it becomes clear why Christian ethics can never be subsumed under the general term "ethics of disposition." If we assume for a moment that the decisive imperative of Christian ethics, namely, "we should fear and love God," expresses a disposition—which for many reasons is already a mistaken assumption—it would be a "disposition" of a qualitatively different kind from what was described above with the help of philosophical terminology as "disposition." For the "disposition" whereby I fear and love God is not something which I can lay hold of by myself. If I could, that would mean that I am capable of mastering and controlling God. In fact I can no more love God on my own than I can know him on my own.

When it comes to the matter of my attitude to God, the only options open to me which I can of myself attain are very different from those of fear and love. To put it in Luther's terminology, one possibility is that I may think I really am capable of keeping the commandments of God, especially if I believe them to be present in the voice of conscience or in such representative values as my nation or people, civilization, etc., in which case my attitude has the character of false security (*securitas*). Again, I may rebel against the condemning Law of God and proclaim my own autocratic autonomy, in which case my attitude has the equally wrong character of haughty defiance (*superbia*). Finally, under the overwhelming pressure of God's demands I may sink into despair and resignation (*desperatio*), or even into a consuming hatred of God (*odium Dei*). How can I have a "disposition" of love towards someone I do not even know, or towards someone who encounters me as an overpowering enemy approaching me in wrath (the *ira Dei*)? Obviously I can love only where I know that I am loved, where I am dealing with the heart of God. But how can I know God's heart apart from Jesus Christ, the figure who mirrors the heart of God? Apart from this mirror what else is there to see but "fate"? Outside the sphere of Christ, what other attitude can I adopt but that of anxiety, or reserve, or illusion (e.g., "Up above the starry sky, A loving Father dwells on high")?

In other words, the disposition of fearing and loving God is not under my control as if it were an attitude which I could on my own lay hold of and adopt. It is not under my control because in the strict sense what is involved is not a "disposition" at all, but a totally new existence. This "new existence" the καινή κτίσις of II Cor. 5:17 and Gal. 6:15) before the author of the Law and the Gospel can take place only as God calls me, i.e., discloses himself to me as the one who judges and who in his judgment blesses. Thus it takes place only in the miracle of the divine history of salvation. I am forced to understand my fearing and loving of God as the act of sanctification, and this act is necessarily preceded by his act of justifying the sinner. (We shall discuss later the entire question of the significance of justification and sanctification for Christian ethics).

It is thus evident that the distinctiveness of Christian ethics does not consist in the fact that it is an ethics of disposition. For every ethics of disposition presupposes a basic inner attitude of man which he can autonomously grasp on his own. We noted earlier that the unique element in Christian ethics is to be found not at the level of acts, with its interchangeability and ambiguity, but at the level of motives. The level of motives, however, is not simply to be equated with disposition. What is really involved is rather a wholly new existence. Even the word "motive," as it has come to be used in our secularized age, is not a clear and untarnished reproduction of the factual situation we have been considering. For it always carries with it an undertone which suggests one's own initiative behind the motion or movement [motus]. We have insisted, however, that man must first be put into a motus or history by God, if he is to be able to do anything "Christian." Man must first be the object of justification, if he is to be the subject of sanctification. Only in this context can we speak of the level of motives as that in which the unique element in Christian action and in Christian ethics is to be seen.

3.

Dogmatics and Ethics

To integrate ethics, as we have done, into the prevenient history of God with us, is to confront it with a most radical challenge. It is to raise the crucial question whether there can even be such a thing as Christian ethics at all, in the sense of a discipline distinct from dogmatics.

If we so much as attempt to speak of "the" good in the sense of ethical goals, or to give a concrete answer to the question "What shall we do?" does not this imply that we are relegating the question of conduct to an area which is simply the level of acts, where there can never be a true reflection of the divine will and certainly not of the divine command? For who among us could dare to construe his own conduct as real obedience to God's command? Indeed, who among us would presume to interpret and affirm even some theoretical conduct, some ideal form of behavior, as real obedience to God's command and hence distinctively "Christian"? Is not an ultimate inconsistency involved in even entering upon this area? After all, this is the arena where ethics of all types are already engaged. And do we not discredit and cheapen ourselves through such contact with all these others, however much we may seek therein to demonstrate the uniqueness of Christian ethics by drawing fine lines of distinction?

For the good with which Christian ethics has to do does in fact lie on a very different plane! It is basically not an "ideal" to which one may more or less approximate. It is rather, in terms which for the moment are intentionally tentative, something which is related to a prior history which God pursues with us in Law and Gospel. This history, being on the one hand the action of the author of the Law towards us, confronts us in the first place with the impossibility of every good, and with the absolute incompatibility of our existence with the command of God. But since it is just as much the action of the author of the Gospel towards us, this history in addition imparts to us the power for and the possibility of living before God in new obedience.

But what has all this to do with an ethics which is oriented to ideals, and which believes that the good is representatively expressed in these ideals? Will not such an ethics be informed by the conviction that these ideals are at least approximately attainable? And how can Christian ethics even be seen in the company of such an ethics without compromising itself? Does not Christian ethics thereby become a quite impossible enterprise? Since "good" is not really possible and every good action is ambiguous, can there even be such a thing as a Christian directive concerning conduct? What form could it take other than that of the paradoxical refusal to set up any concrete norm or to sanction any kind of theoretical conduct? Could it ever go beyond the vacillating answer, "Perhaps, but perhaps not"?[1]

The Traditional Incorporation of Ethics into Dogmatics

To begin with, it seems on the face of it as though there can be no special discipline of ethics existing alongside that of dogmatics, and that we would be well advised to incorporate ethics into dogmatics. This advice seems to be strengthened by the fact that a review of the history of ethics would appear to teach us clearly that the movement of ethics towards autonomy is explicitly a phenomenon of secularization.

The early church at any rate had no independent ethics. It restricted itself to such paraeneses as were essential to catechetical instruction, and to the monographic handling of disputed questions

---

[1] In all of this we are deliberately carrying this challenge to Christian ethics to the point fixed by Karl Barth in a now famous passage where he is commenting on Romans 8:12-13, namely, that point from which Christian freedom appears to be identical with the kind of ethical relativism that simply resolves itself away: "It is this very questioning of both the nature and the fact of our existence [our "life according to the flesh"], of the whole business, our familiar paths and our escape routes, our serious approach to things and our half-hearted indifference, our right deeds and our sins, our beliefs and disbeliefs and our uncertainties—it is this ultimate 'perhaps, but perhaps not!' that is laid like an ax to the root of the trees—this is the freedom which we have in God, beyond the Law, the freedom from which we cannot escape because it is the truth, because it is the freedom of God himself." Karl Barth, Der Römerbrief (fifth impression of the revised edition; Munich: Kaiser, 1929), p. 275; cf. the English translation by Edwyn C. Hoskyns, The Epistle to the Romans (London: Oxford University Press, 1933), p. 292. See also the criticism by Friedrich Gogarten in Gericht oder Skepsis. Eine Streitschrift gegen Karl Barth (Jena, 1937), p. 40. This challenge to Christian ethics, deriving as it does from the very nature of the matter itself, is most serious and will receive our immediate attention.

which urgently needed treatment. Thus Tertullian, for example, writes on the question whether a Christian can attend the public spectacles of the heathen world (*De spectaculis*), or whether a workman can take part in the construction of idolatrous images (*De idololatria*), or whether, in view of the dubious nature of a second marriage, his wife may marry again after his death (*Ad uxorem*).

Similarly, Augustine deals with the question whether there may be remarriage after divorce (*De adulterinis conjugiis*), and whether there is such a thing as a white lie (*De mendacio*), etc. Reference may also be made to Augustine's moral tractates against outbreaks of libertinism, e.g., against signs of moral collapse in monasticism, especially laziness and vagabondage (*De opere monachorum*).

Finally, there is often occasion for ethical discussions in the controversial writings, where the quarrel with heathenism or with specific heresies may often be traced back to ethical causes or issues. In this category may be placed, for example, Augustine's writings against Pelagianism, which contain some important foundational principles of Christian ethics.

For the most part, however, these problems which go to the very root of the matter are dogmatic in character. More precisely, they belong to that part of dogmatics which embraces in germ the ethical problem. Thus the Pelagian question of the freedom of the will, while it belongs to the central dogmatic nexus of the doctrine of grace and predestination, has implicit within it also a wide range of ethical problems, especially the questions of responsibility, of conscience, and of autonomy—just as a question such as that of "justification and sanctification," while it is a purely dogmatic problem, naturally figures decisively in the foundation of every ethics. Hence, if we agree with Emil Brunner in describing the anti-Pelagian writings of Augustine as the point where "fundamental ethical reflection begins," [2] then we must add at once that this kind of reflection does not establish an independent ethical enterprise, but still remains within the discussion of dogmatic principles.

In other words, the theology of the early church when it deals with ethical problems either restricts itself to actual cases or, where the question has to do with ethical principles, remains within the sphere

---

[2] See Emil Brunner, *The Divine Imperative,* trans. Olive Wyon (Philadelphia: Westminster, 1937), p. 94.

of dogmatics. In neither case is there the slightest interest in the independence of ethics; the notion does not even occur to the early theologians.[3] The same is true of Scholastic ethics, which is integrated into the dogmatic *Sentences* (e.g., Peter Lombard) or the theological *Summae* (e.g., Albertus Magnus, Thomas Aquinas). The exception is where it pursues a relatively independent existence in the form of commentaries on the Nicomachean ethics of Aristotle.

Reformation theology, however, is full of new ethical thinking. One has only to recall Luther's doctrine of the "two kingdoms," [4] or the ethical questions connected with his view of the estates and callings of life (e.g., *Whether Soldiers, Too, Can Be Saved*).[5] One has only to think of the way in which the monastic ideal of holiness was destroyed and man was freed to live in the temporal estates and to work in the world, or of the discussion of the problems involved in the ruler-subject relationship, or finally of the whole complex of questions in social ethics and political ethics which were forced on Luther by the Peasants' Revolt and the war against the Turks.[6] Economic ethics too comes within the scope of this new thinking during the Reformation.[7]

But even this rich and varied complex of ethical themes does not lead to any attempt to establish theological ethics as an independent discipline. For the ethical discussions are in the strict sense polemical in nature, being directed against the Scholastic tradition or against the Sectarians. They thus have the significance only of "cases" which demonstrate the new Reformation understanding of grace, or which demonstrate negatively the uselessness of meritorious works and therefore of places, e.g., monasteries, or estates or orders in this world which are preferred because they are thought to be sacral. From this standpoint, the ethical utterances of Luther are simply examples of how Luther makes the Gospel the new starting point of his

[3] On the history of ethics, see Christopher E. Luthardt, *Geschichte der christlichen Ethik*, I-III, (Leipzig, 1923-1926).
[4] A brief bibliography on this topic is given below at p. 359, n. 1.
[5] PE 5, 32-74.
[6] See Helmut Lamparter, *Luthers Stellung zum Türkenkrieg* (Munich, 1940).
[7] See Ernst Troeltsch, *The Social Teaching of the Christian Churches*, trans. Olive Wyon (New York: Macmillan, 1931), Vol. II; Werner Elert, *Morphologie des Luthertums* (Munich: Beck, 1932), Vol II; and Georg Wünsch, *Evangelische Wirtschaftsethik* (Tübingen, 1927).

theology. This is a great deal, to be sure, but it is not yet the establishment of an "ethics."

We have thus to ask in all seriousness whether the nature of evangelical ethics is not most precisely and appropriately expressed in this very self-restriction, this "implicit" form. For how else can the ethicist answer the question "What shall we do?"—and still avoid the casuistry of monasticism—except by referring every matter of conduct to the evangelical point of departure, and hence seeing it in terms of the dogmatic center constituted by Law and Gospel? Here again, do not Luther's explanations of the Ten Commandments indicate the only possible and utterly legitimate way, in that they constantly refer the many cases (the problems of murder, lying, adultery, stealing, etc.) to the one command that "we should fear and love God"? And is not Luther thereby following in an authentic way the master himself, who described love for God and one's neighbor as the sum of all the commandments (Matt. 22:37 ff.)?

We can perhaps formulate the question even more sharply. In the strict sense, does not all ethical reflection consist in a movement of concentration whereby all the various cases are subsumed under this one rule of love? Is it not therefore essentially reflection on "the extent" to which I can demonstrate love for God and neighbor in this or that concrete action, in marriage, in politics, in my calling?

In any event, this movement of concentration in ethical reflection does find precise expression in the story of the rich young ruler (Mark 10:17 ff.). Here Jesus points beyond the young man's many acts— he had observed all the commandments from his youth—to the "one thing" which is lacking and which for him is therefore "needful" (Mark 10:21; Luke 10:42). He must sell all that he has and thus loose the bonds of egocentricity which have held him fast even in the multiplicity of his ethical exertions. Thus Jesus seeks to help him free himself from bondage to self and from its central bastion, his wealth, and allow himself to be bound "self-lessly" to God. In other words, Jesus aims to lift from him the mortgage of self-love which burdens and depreciates everything he does, and free him for the love of God.

Jesus does not give the man some kind of new and supplementary commands with a view to his becoming perfect. He does not add "counsels" to the "precepts" of the Decalogue, as in Roman Catholic perfectionism. What Jesus wants is that the young man should keep the commandments otherwise than in the past. Because the man was previously enmeshed in feverish self-enslavement, every ethical fulfillment of the commandments simply served to strengthen that bondage to self. This ethical path leads only to a Pharisaic consciousness of one's own ethical security, which falls victim to "boasting" and sublimates personal egoism by the most refined and secret means. In the future, obedience to the commandments must be rendered in such a way that the man serves God rather than himself, so that in whatever he does—or leaves undone—he is actually fearing and loving God.

Jesus thus helps the young man to execute that movement of concentration which must underlie all conscious ethics. This is the reduction of all specific imperatives and individual forms of obedience to the single formula that one should "fear, love, and trust in God above all things." What is at issue is simply the clear subordination of the second table of the Law to the first.

We must therefore put the question quite pointedly. Does not all ethical reflection always involve an act whereby ethics really does away with itself by reducing the ethical question to a problem that is essentially dogmatic? Or is the matter of fearing and loving God something other than an express concern of dogmatics, a grounding of our love in the preceding love of God (I John 4:19)? Is not therefore the incorporation of ethics into dogmatics, as has been typical in the church's tradition, the only appropriate solution? In short, does not the solution of the ethical problem lie in the dissolution of ethics? As we have already said, this supposition appears to find support in the observation that the movement of ethics towards autonomy is obviously linked with, and has been advanced by, the history of secularization.

The Way to Ethical Autonomy

To answer the question of how ethics achieved its autonomy, we should do well to turn again to the ethics of Kant, where the independence of ethics is the dominant theme. We do this not only— better, not at all—for the sake of maintaining the prestige of ethics as an academic discipline, or on the innocuous basis of the conflict between the faculties. Kant lays such great stress on the autonomy of the moral sphere because he wishes to differentiate it sharply from all spheres of motivation outside its own.

For example, I may do something because a certain utility or usefulness attaches to it. That may be quite justifiable. But the fact that an action is useful tells me nothing about whether it is moral. Again, I may do something simply because I want to; I am so moved. Perhaps I have founded an orphanage, and the deed has objective worth, and in my eyes moral value as well. But the fact that I have acted in accordance with my inclination, was driven to it, as it were, by my own impulses, is a doubtful basis on which to establish the morality of the action—it might even serve more appropriately to call that morality in question. For the unconditional moral imperative mediated to me by practical reason sets itself in opposition to the natural "sensible" tendency of my inclinations and impulses. The moral law invades the natural tendency whereby I am moved by my impulses and brings to light the new and intelligible reality of the moral.

What is involved in the sphere of the moral is thus a conflict between two worlds, the sphere of natural impulses and the invading sphere of the imperative, the sensible world and the intelligible world, I myself being the battlefield. "Virtue is moral disposition militant." [8]

It is indeed a proof of the moral worth of an action when that action involves struggle and is wrung out of me despite my natural inclination to do the opposite. On the other hand, a by no means negligible element of suspicion attaches to it if the action is performed too easily, e.g., out of the natural impulse of a generous heart, for then the possibility of eudaemonism lurks dangerously near.[9] The "beautiful soul" [10] in which Schiller sees the harmony between duty and inclination to be perfect is for Kant a thought which, like his own doctrine of immortality, can at best only be an eschatological postulate; in face of concrete existence Kant views it very skeptically.

The crucial thing for Kant is thus to present the categorical imperative in all its purity and to cleanse it from all eudaemonistic and

[8] Immanuel Kant, *Critique of Practical Reason* ("Great Books of the Western World," ed. Robert Maynard Hutchins [Chicago: Encyclopedia Brittanica Inc., 1952]), Vol. 42, p. 327a.

[9] *Ibid.*, pp. 321-325.

[10] On the questions involved in this topic see my book *Das Verhältnis zwischen dem Ethischen und dem Aesthetischen* (1932).

utilitarian stains. In consequence it is all-important that the imperative should be marked off from all "heteronomous" spheres.[11]

Linked with the autonomy of the moral is the fact that I cannot simply trace the moral back to God as the author of the Law. For if I then did what is moral, I would merely be obeying a divine command. Obedience of this sort, however, says nothing about my inner attitude to such a command, e.g., whether I affirm it out of conviction or fulfill it only reluctantly and out of compulsion. In the moral we are confronted, not by a compulsion (which on the contrary is a mark of the laws of natural causality), but by an imperative which "in freedom" resists natural compulsion and impulse. Moreover, such slavish obedience to the divine command would necessarily arouse the suspicion that I desired to stand well with the Lord of the Law only for eudaemonistic reasons, and that the morally heteronomous forces of fear and hope were the true motives behind my action.[12]

The Ten Commandments are thus valid for Kant, not because the compelling authority of the divine lawgiver stands behind them, not because God has made himself "the condition of the moral Law," [13] but because independently and on its own authority practical reason could have formulated similar commandments.[14] Indeed, the authority of the Decalogue must first vindicate itself before the forum of moral autonomy. Only insofar as its underlying maxims of action could be made "the principle of universal legislation" [15] in the sense of the categorical imperative, can we say that it is in agreement with practical reason and hence ethically obligatory. The supposed theonomy of the Decalogue is thus only a subsequent reflection of the autonomy. It simply represents a particular "conception" of this autonomy, a conception which Kant defines as the mere acknowledgment of all "true duties as divine commands." [16]

As an avowed disciple of Kant, the theological ethicist Wilhelm Herrmann advances a corresponding formulation of this immanent

[11] See Kant's definition of autonomy and heteronomy in *Critique of Practical Reason, op. cit.,* pp. 304, 308.
[12] See Kant, *Critique of Practical Reason, op. cit.,* pp. 321-325.
[13] *Ibid.,* p. 291.
[14] *Ibid.,* p. 326.
[15] *Ibid.,* p. 302.
[16] Immanuel Kant, *Religion within the Limits of Reason Alone,* trans. Theodore M. Greene and Hoyt H. Hudson (Chicago: Open Court, 1934), p. 180.

starting point of the ethical *nomos*. He too argues that its theological character as Law of God is, as it were, a subsequent interpretation in which the given moral *nomos* is related to another given factor, "God."

It is true, of course, that we should not oversimplify or exaggerate this pre-eminence of the autonomous ego over the claim of transcendence. Kant's primary concern is certainly not to assert this pre-eminence. Otherwise he would have had to distinguish his view very differently from that of traditional Christianity; his presentation would have had to assume more Voltairean features. Undoubtedly for Kant it was rather also a question of how the authority of the divine commandments can be established in the true sense, how to keep them from being misunderstood in terms of a transcendent dictatorship. For an authority, i.e., an entity to which I bow freely and which I do not confront in mere servile timidity and fear, is possible only if it constantly shows itself to *be* authorized. Only when I am able to verify the legitimacy of an authority can I maintain my existence over against it. Otherwise I should simply be the functionary of an alien will, the mere product or effect—like a performing animal in relation to its trainer. For instance, blindly to parrot Luther would be to treat him not as an authority but as a dictator, and thus to misunderstand him. Only as I am ready and able to verify his authority in the court of Holy Scripture, which we both accept as binding, can I maintain my own existence over against him and in freedom admit his authority.

In like manner Kant tries to establish the authority of God's commandments. Like Lessing, he is ready to bow to them if he can do so with integrity, i.e., without surrendering his own existence and freedom. He is willing to be "convinced." But in his view, as in that of Lessing, this is possible only as the court competent in matters of conviction, namely, the court of reason and conscience, is first satisfied. Any obedience which bypasses this court violates the authority of God and the existence of the ego, and is thus unworthy of credence. For it makes God a dictator and the ego a mere functionary.

We completely misunderstand this radical integrity, and its rejection of any compromise which would obscure the issue, if we see in the concern for human autonomy nothing but rebellion and hubris. Perhaps it will become that later, for reasons which are yet to be investigated. In the first instance, however, one may say that there is in these particular thinkers a true humility, a readiness to bow before an entity which is greater than they—if it is clear to them that it really is greater.

The Relation of Dogmatics and Ethics

Thus far we have been speaking of the challenge posed by secularism, with its asserted autonomy of the various spheres of life, the morally independent subject, and even ethics itself. These are problems which cannot simply be subsumed within the framework of Christian dogmatics as it has traditionally been understood. For what is involved here is not merely a consideration of certain points of doctrine but also a statement about reality itself, about that human-historical reality which secularization has warped almost beyond recognition but nonetheless brought squarely to the center of attention. What is required in the analysis of reality, at least in its early stages, is reflection on the part of more than just the theologians. Once the problem of reality has been raised in this elemental way, what is needed is a co-operative endeavor involving all the other faculties as well.

It is true that the field of theological ethics has been expanding at many points and in many directions. Methodological reasons alone would suggest that it accordingly be constituted an independent discipline, lest mastery of the field become wholly out of the question. This does not mean, however, that ethics should exercise exclusive sovereign rights over its entire domain. The domain falls rather under the over-all suzerainty of dogmatics. To anticipate for a moment, we shall have to show, for example, that in an ultimate sense—*coram Deo* [before God]—there is no such thing as autonomy of the various spheres of life. Neither are the created orders dispensed from the sphere of responsible existence. On the contrary, the orders and the spheres of life are all included within that realm where man must act responsibly. Hence all that must be said here belongs within the orbit of Law and Gospel, under which man himself stands. If man cannot exculpate himself *coram Deo*, if he finds here forgiveness and the new ordering of his existence, then this can only mean that he does so as the bearer of that "autonomous" structure of his existence from which he himself cannot be separated and within which he fulfills his existence. Man must say concerning the reality dominated by autonomies: "I am this reality." [17]

The analysis of reality, which secularism made a theme and which ethics is now required to undertake, must therefore ultimately cease to be a purely immanental analysis and become a statement about man *coram Deo*. Accordingly, the battle group of ethics, which had to make a sortie out of the dogmatic citadel, must also return to its base once its mission is accomplished, laden it is to be hoped with booty. It cannot be content merely with looking out through the narrow peepholes in the fortress, much less with pure theorizing on the inside. It must rather go out into the open. For the way and the manner in which the whole sphere of man's historical life and existence is, with man himself, included in the ego, and therewith in the zone for which man is responsible, is something that has first to be shown. Ethics in its quality as dogmatics (!) has to understand itself as a task yet to be accomplished rather than as something already given. The followers of Barth think this quality is easy to come by. To proceed simply in terms of the traditional framework of dogmatics, however, is to run the risk of shirking our task with respect to reality. And to do that would be to truncate the sovereign claim of evangelical dogmatics to be a message concerning all reality spoken in the name of the Lord to whom all authority is given both in heaven and on earth.

For faith is equally the content of both dogmatics and ethics. Faith is the content of dogmatics inasmuch as dogmatics is faith reflecting on its object, and treating that object with methodical rigor, i.e., in the manner appropriate to it. Faith is the content of ethics inasmuch as ethics is faith inquiring as to the conduct faith posits for man towards himself, his neighbor, the world, and its orders.

It would thus be a mistake to construe ethics as a discipline in which the "practical consequences" of faith are drawn, and in which a practical epilogue is thus added to the theoretical enterprise of dogmatics. This is impossible for the very reason that faith is of itself the fulfillment (the practical fulfillment!) of the First Commandment, and therewith of all the commandments. For faith, and faith alone,

[17] This theme that I am identical with reality and that the autonomous structure of reality is really an objectification of myself ("I am my history") is further developed in my book *Geschichte und Existenz* (1935), pp. 37 ff.

lets God be the Lord, thereby showing him the honor which the commandment requires. Hence faith does not call for a second, supplementary, "addable" act of "practical-ethical" obedience in which it must itself be "actualized." Either faith is "real", i.e., it exists in self-actualization, or it does not exist at all and is nothing but a "lie" (I John 2:4).

Similarly ethics is not an additional or supplementary discipline. It is rather, if you will, an interpretation of the believer's existence in faith with particular reference to the question of what it means in this connection that this existence is also an "existence in the world." For man as the subject of faith always stands in the orders of the world, in its autonomies, etc. In any case he does not stand in a vacuum. Hence there is need for strict methodological concentration on the question of how the dimension of worldly reality is related to faith, since from the very outset the subject of faith is in fact "related" to this reality. Nor does this question refer merely to specific tasks through which what is given in faith is supposed to be developed and completed. On the contrary, it refers quite expressly and exclusively to the interpretation of that given factor, but from the specific standpoint of its implications for the believer's "existence in the world."

That there is here no challenge to the primacy of dogmatics may be seen from the fact that we propose to eliminate from ethics any apologetic character; we have no intention of setting up against the secular understanding of reality a Christian rival or opponent.[18] In this context, the task of Christian ethics must consist exclusively in putting questions to the secular understanding of reality, in demanding responsibility from it, and in showing it to be a system by means of which man hopes to protect himself against the divine attack. In discharging this function, ethics fulfills the office of a watchman in the church. This, then, is the positive and negative task of evangelical ethics, whereby it both advances its own cause and also recalls itself to its proper limits.

[18] On this point see my book *Fragen des Christentums an die moderne Welt* (Tübingen, 1948).

4.

The Place of Evangelical Ethics
between the Two Aeons

The biblical motto which is traditionally given to theological ethics as it sets out on its journey is to be found in Acts 2:37, which contains the question: "Brethren, what shall we do?" This question expresses the reaction of Peter's hearers to his sermon at Pentecost. What the question actually means is suggested by Peter's answer to it in v. 38: "Repent, and be baptized every one of you in the name of Jesus Christ for the forgiveness of your sins; and you shall receive the gift of the Holy Spirit."

Apparently the significance of the question was not primarily ethical. Peter's hearers at Pentecost were not inquiring about norms and programs of action. Their concern was not for a key to the actualization of their Christianity. On the contrary, what they asked was simply another way of asking the other more radical question of how it might come to pass—and what contribution my conduct could make —that the resurrection of Christ attested by Peter should become an event involving me [*quoad me*]. The question is: How do I become a member of the new aeon, that aeon which came into existence through the events of Christmas, Easter, and Pentecost? This is the basic question which Peter is answering when he refers to repentance and baptism.

The Christian as a Citizen of Two Aeons

It would thus be a real mistake to understand the people's question at Pentecost "ethically." The question refers rather to the initiatory act by means of which I first come into Christian existence as such, and so arrive on the plane where, in a *second* act, there again becomes urgent the question concerning what I have to "do" in order to relate my conduct to my faith. (In the strict sense, one ought not to speak here of two "acts" which succeed one another in time. We do so only provisionally and in order to give the sharpest possible emphasis to the question of the "initiatory" act.)

Thus the question "What shall we do?" really arises twice, in two forms, each of which has reference to the other. In the first form the question runs, "What must I do to come to faith?" Peter's answer does not imply a co-operative action on my part, but that I must rather let something happen to me: I must expose myself to the Christ-event by being baptized.

In the second form the question runs, "How do I move from faith to action?" That is, how do I make my Christianity concrete? What is life in the new aeon to be like? For to be baptized is, after all, to let oneself be called into God's salvation history, and hence out of the old aeon. But to be called out in this way can mean only that we are delivered from the ruling powers of this aeon and set under the dominion of a new and different Lord. It means, for example, to acquire a new relation to the god Mammon, and to the powers of property and possession (Matt. 6:24; Luke 16:13; 12:16-20; Mark 10:21, 24 f.). It means also that I have to revise my relationship to my body (I Cor. 6:19) and its passions (Phil. 3:19; I Cor. 6:16), to the things of this world (I Cor. 7:29 ff.) and anxiety concerning them (Matt. 6:25 ff.), to the Thou of my neighbor and to the groups to which I belong. It implies, in fact, the total revision of my existence in all its dimensions, since Christ is ruler of the entire cosmos and not just Lord of my inwardness. The orientation of my existence—and this means concretely my life in the plenitude of its relationships—is completely transformed because I am now the member of another history and of another aeon.

On the other hand I am simultaneously—by virtue of a mysterious *simul*—a member of the old aeon. For Christ did not pray the Father that he should take his own out of the world, but that he should keep them from alliance with wickedness (John 17:15). After all, they are no more "of the world" (in terms of origin and destiny) than is Christ who, even though he walks in it, still is not "of" the world (John 17:16).

Hence believers in Christ stand to the old aeon in a relationship of both continuity and discontinuity. The relationship is one of continuity insofar as they eat and drink, marry and are given in marriage, laugh and cry, stand under authorities and within orders, etc. It is one of discontinuity because they no longer receive their orientation from all of this. Their relation to that which is relative can no longer be something absolute (to put it in Kierkegaardian terms). They live "in the flesh" to be sure, but no longer "according to the flesh." We shall see later to what degree Luther's well-known phrase "at once righteous and sinful" [*simul justus et peccator*] reflects this relationship to the two aeons, especially when it is seen to involve an interrelating of *res* and *spes*, of present and future, of this aeon and the coming aeon: "sinful in fact, righteous in hope" [*peccator in re, justus in spe*].

The problem of ethics is all wrapped up in this *simul*. For the *simul* cannot be taken to mean that the *res* form of sin is, as it were, quite unaffected in the state of justification, as if justification made no difference whatever *de facto* in the existence of a Christian or in a nation comprised largely of Christian people.

The false reasoning behind such a thesis runs as follows. As long as we are never anything but sinners who have received mercy, nothing really changes in our existence as sinners. Inasmuch as we have to pray each day "forgive us our trespasses," we must continue to be perfectly intact trespassers. The only thing that changes, according to this view, is our relationship to our trespasses and sin: they can no longer separate us from God. We may therefore accept—in a sense in which Luther definitely did not intend it—the tranquilizing imperative: "Sin boldly!" [*Pecca fortiter!*].[1]

The miracle of the Holy Spirit would then relate only to the sphere of man's "inwardness"; outwardly everything would be unaltered. Christianity would have no norms, much less revolutionary impulses, for the economic, political, cultural, and social sectors of life. I might be a Hitler, a Machiavelli, a Ghandi, or the petty head of a utopian-pacifist community without its having any effect whatsoever on my "Christianity." Which of these I would actually be would depend on my general outlook, my world view, or my particular assessment of the concrete situation, none of which need have any kind of affinity with the commandments of God. Capitalism and socialism, democracy and dictatorship, marriage and free love—all these alter-

---

[1] Cf. the total context of Luther's letter to Melanchthon (August 1, 1521) including the clarifying admonition immediately following this passage: ". . . but believe even more boldly and rejoice in Christ." See below p. 504, n. 19.

natives would then be matters of indifference so far as the Law of God is concerned. The normative criterion for choosing between them could then be found only in considerations of conformity to political and economic ends or, in respect of marriage and free love, in biological and hygienic considerations.

We would thus find ourselves on a plane which the church and its theology, particularly since the days of the Enlightenment, have repeatedly occupied. It would involve farming out "the kingdom of this world" to alien ideologies, and beating a retreat into the sanctuary of "inwardness." This inwardness has repeatedly been extolled, especially by numerous nineteenth-century theologians, as the zone which is immune to the assaults of scientific and historical criticism, and which cannot be affected by the power struggles among the ideologies and the political and economic pressure groups.

When we realize, however, that on this view salvation history ceases to be "history" at all and is instead spiritualized into inwardness—in keeping with Luther's unfortunate mistranslation of Luke 17:21: "The kingdom of God is within you [*inwendig in euch*]"—then there is even greater urgency in the ethical problem. The question is: How is our naturalization in the new aeon by baptism to be worked out concretely in the *res* form of our life? We shall give a provisional and admittedly generalized answer to this question in the form of two series of statements.

First, the fact that Jesus does not pray the Father to take his own out of the world, the fact that he allows us to continue in the world, implies as regards its form that our life persists in a certain continuity with the laws of the old aeon. We must continue to exist, for example, in political and economic systems, and to a certain extent (about which we shall speak later) we are therefore still subject to the laws of these systems. To state the matter negatively, this continued existence means that a totally new form of world, one that would be absolutely different, discontinuous, and revolutionary, is not demanded of Christians or of Christianity. Efforts to move in that direction have been made nevertheless, and they are all marked by a quasi-fanatical anticipation of eschatology in which the continuance of this aeon is no longer taken seriously but is despised or ignored.

Ethics is thus concerned with what is essentially only a temporary ensconcement in this world. Its task is decisively determined by the fact that the Parousia is postponed and has yet to come. It is thus in the peculiar position that it cannot draw directly, in the manner of biblicism, on the directives found in the life of the early church (I Cor. 7 for example). For the early Christian community lived in immediate expectation of the end of the world, which was supposed to come with the imminent Parousia. For them the problem of ensconcement in this world consequently played little or no part. The early church's attitude to the world was one of emphatic and one-sided discontinuity. Such one-sidedness has for us been made quite impossible by the delay in, or postponement of, the Parousia. Ethics is normatively determined by this postponement. It is determined by the fact that God is delaying the end of the world, that his patience is still operative, and that he wills that the orders of this passing world should still be respected. The problem of ethics thus arises out of the tension between continuity and discontinuity.

In a relation involving nothing but continuity, the ethical problem would disappear inasmuch as the way we go about settling down and living in this world would ultimately call simply for pragmatic directives. In this sense Nietzsche is the most consistent exponent of continuity, the most consistent opponent of ethics. For in the place of the will to good he sets the will to power, to self-assertion, to whatever enhances life and is useful. On the other hand, in a relation involving nothing but discontinuity, the ethical problem would disappear inasmuch as we would be living in an anticipated eschaton and there would therefore no longer be any problem of integration into the ongoing course of the world, e.g., no question as to conduct within the orders.

Ethics has its place therefore precisely in the field of tension between the old and the new aeons, not in the old alone, nor in the new alone. If this field of tension is to be recognized at all it is necessary for us to assert at the outset that we stand in a relation of relative continuity with the old aeon because of the postponement of the Parousia. This is the first thing that must be said.

But there is also a second. The fact that the old aeon is approaching its end, the fact that the new aeon has already dawned in the Christ-event and has broken concretely into our world in the form of real history, and the fact that in a very real sense all authority in heaven and on earth has been given to Christ already here and now

43

(Matt. 28:18)—this makes it quite impossible for our incorporation into the new aeon to remain concealed and private, limited simply to our own inwardness. It means rather that a real encroachment into the current aeon has been made which must work itself out in the concrete *res* form of Christian action.

We thus maintain that the problem of ethics lies in the fact that the two aeons run concurrently during the "last time," i.e., the time between the Ascension and the Last Day. In other words, it lies in the resultant relation between continuity and discontinuity in respect of "our world." This means that in the strict sense the problem of ethics is a *theological* problem. For it is posed by the interrelation of the two aeons—which intersect in "me."

Theological ethics, therefore, has to do with a problem which is not essentially "moral." Its purpose cannot be to set up a hierarchy of ethical values with a corresponding casuistry of moral action, as in the strongly non-eschatological moral theology of Roman Catholicism. Neither is it a mere ethics of disposition such as that conceived by Kant; its quest is not for timelessly valid maxims for such a disposition. Theological ethics is concerned rather with the pressures posed by time: it is concerned with the time of this aeon which is passing away, for though we can no longer relate ourselves to this aeon in any ultimate way we do nonetheless have to take up our station within it. And ethics is also concerned with the dawning "time" of the new aeon, for this unsettles our world, though it does not "clothe" (II Cor. 5:2, 4) or absorb it but rather lets it continue on its way to dissolution and death (II Cor. 5:1 ff.).

The Eschatological Character of Christian Ethics

The result of all this can be expressed only in the paradoxical statement that Christian ethics is an impossible possibility. What this statement means becomes clear the moment we view the matter in terms of the eschatological perspective sketched by the Sermon on the Mount. The Sermon on the Mount radicalizes the Mosaic *nomos*. It imposes on us a total demand by forbidding not merely the results of anger but anger itself, not merely the act of adultery but adulterous thoughts (Matt. 5:22, 28). In so doing it calls in question not merely our conduct but our very being. It insists not simply that we *do* differently but that we *be* different.[2] It lays claim to us, so to speak, as if the fall had never taken place, as if our whole existence were not determined by this aeon. It imposes its demand upon us as if we were still in the primal state, as when we first came from the hands of God. It imposes its demand upon us as if the new aeon had already come and replaced the old. As a result, everything that we do, and hence also every ethics which would determine our action, is challenged eschatologically.

Christian ethics is an impossible enterprise inasmuch as it lies under the disruptive fire of the coming world. Yet it is also a necessary enterprise inasmuch as we live in that field of tension between the two aeons and must find a *modus vivendi*. In order to characterize this tension by which the whole ethical enterprise is sustained—and at the same time repeatedly shattered—we would list three of its distinctive features.

First, the mystery of Christian ethics is an eschatological mystery. It rests on an irresolvable tension between time and eternity, between this aeon and the coming aeon. This tension cannot be removed by means of a parallelogram of forces which in effect is a compromise. To seek a time-less solution of that kind, in terms of "resultants," would require that time be no longer viewed as hastening toward its end. It would mean looking for that which "lasts," searching for those things that can be accepted as of "permanent" validity, even if only by way of compromise.

Second, the mystery of Christian ethics is a christological mystery. It rests on an irresolvable tension between deity and humanity in Christ. Just as I cannot represent logically the togetherness of the divine and the human natures in Christ, co-ordinating them with one another in a static and time-less way in terms of logic, so I cannot find a formula for the unity of the Christian's existence, which on the one hand is lived out in this aeon and yet at the same time participates in the heavenly commonwealth. Ethics cannot and must not try to resolve this tension by providing rules which on the surface appear to be wise and prudent, satisfying both the law of continuity and that of discontinuity. In such an attempt it would come up against the same impossibility as does the doctrine of the two natures when it tries to show that Christ does justice to the claims of both

2 See our discussion of the Sermon on the Mount in Chapters 17 and 18 below.

natures, when it goes so far as to try to present Christ as the true middle-point between the two natures, and hence as a kind of demigod.

The fact that the tension cannot be resolved is a sign that what takes place when we are called into the kingdom of God, and so exist in the zone of tension between the two worlds, is an inscrutable miracle whose mode is concealed from us. We can no more explain how it happens than we can explain the how of the incarnation and the miracle of Christ's person. In like manner we cannot explain how the new ego is related to the old. "I live, yet not I, but Christ lives in me" (Gal. 2:20). Here it is said of both the ego "according to the flesh" and the ego "according to the Spirit" that it is "I." Here an identity is asserted similar to that declared by the *est* in Luther's understanding of the identity of the eucharistic elements with the body and blood of Christ. But this identity can be maintained only in faith. To state the matter negatively, it is impossible to make the identification in a statistically objective sense, either by speaking of a congruence of the two forms of the ego or by referring to them as intersecting circles in which the zones that overlap and the zones that do not can be determined. One can speak only of a paradoxical identity which may nonetheless be differentiated in terms of perspective: the ego of the first part (the "I live") is seen *coram se ipso*, and the ego of the second part (the "Christ lives in me") is seen from the divine standpoint.

There is first the mystery of the new being. This mystery of being imparts also to the action which proceeds from it the character of mystery. This is true not merely in the sense that the action takes place in the power of the Holy Spirit—and hence in the name of yet another mystery!—but also in the sense that the norms of this action participate in the mystery of the situation between the two aeons.

Third, the mystery of Christian ethics is a sacramental mystery. It rests on an irresolvable tension between the sign and the thing signified, like that we know in the sacraments. It is impossible for me in my concrete existence, by loving, by acting or allowing myself to be acted upon, to express the new existence in adequate *res* form. I can express it only by means of signs, through demonstrative actions which point beyond themselves. As the body of the Lord is hidden under the signs of bread and wine, so the obedience involved in true love of neighbor is hidden in the highly complex act of my loving, an act made up of such diverse motives as sympathy, generosity, and a self-interest which is on the one hand egoistic no matter how sublimated, and on the other a true seeking of the neighbor. Indeed, my act is complex because it includes "opposition to" as well as "support of" (think for instance of the problems involved in expressing love within the orders, e.g., in war or litigation, as touched upon in Luther's doctrine of the "two kingdoms").

Consequently the aim of ethics cannot be to overcome the tension by suggesting compromises which would supposedly do justice to both elements in the tension. Ethics must rather follow the way which leads into and through the tension. With respect to concrete tasks requiring action, ethics can show wherein that tension consists. Beyond that it can show, in a context where every action stands in need of and under the promise of forgiveness, what those actions should be like which are to demonstrate the fact that a Christian is the citizen of a new aeon and that at the same time he also honors the old aeon as the *kairos* of God, as the "acceptable time" which by the patience of God is still permitted to continue. The theme of ethics is this "walking between two worlds." It is in the strict sense the theme of a "wayfarers' theology," a *theologia viatorum*. It lives under the law of the "not yet" but within the peace of the "I am coming soon" (Rev. 22:20). Theological ethics is eschatological or it is nothing.

Part Two

THE FOUNDATIONAL PRINCIPLES OF ETHICS

A.

JUSTIFICATION AND SANCTIFICATION

5.

The New Obedience

Evangelical ethics is completely different from all natural or philosophical ethics. Indeed the two lie on wholly different planes and must be sharply differentiated from one another, however much the theme of "obedience" may be common to both. The difference between them may be stated provisionally in the form of three propositions.

First, philosophical ethics takes as its starting point the goal of the ethical act, whether that be some objective value outside the ego (as in an ethics of material values) or some kind of self-realization within man (as in Kant's categorical imperative, which aims at the realization of "the intelligible ego," of "man-ness," or of "the idea of personality"). For philosophical ethics proceeds from the discrepancy between what *is* and what *ought* to be; it distinguishes between the *de facto* situation and the ideal situation. Evangelical ethics, on the contrary, takes as its starting point not the goal but the presupposition of the ethical act. It proceeds from the fact of justification as accomplished and given (though of course the Christian is constantly thrown back upon it and cannot simply take justification for granted as the given ground of an existence on which he may henceforth proceed to build for himself). The ethical act, then, is simply an expression of the prior fact of justification; it is, as it were, a "subsequent" [1] demonstration of the given justification.

Second, in philosophical ethics the goal, e.g., the intelligible ego, has merely the significance of a heuristic symbol; the ethical act alone has reality. In evangelical ethics, on the contrary, only the given justification has reality; the ensuing ethical acts in which it is expressed have symbolical significance. They are the symbols in which

[1] We have placed the term "subsequent" in quotation marks to suggest that it does not refer to temporal priority. The ethical act is to be interpreted not in terms of its sequential relationship to the fact of justification in point of time, but in terms of its subordination to justification with respect to rank as regards the very substance of the matter involved.

51

the reality expresses itself. The so-called "works" done in the state of justification serve only by way of demonstration.

Third, in philosophical ethics the ethical acts are determined by the "task" to be performed [*Aufgabe*]. In evangelical ethics they are determined by the "gift" already given [*Gabe*].

Justification as The Presupposition of Evangelical Ethics

In accordance with this outline of the distinction between the two planes, we may now proceed to elaborate the basis of evangelical ethics. There can no longer be any doubt as to our initial task. It is the task of determining the relationship between the "gift" and the ethical acts implied in that gift. In other words, we have to consider the relation between justification and sanctification, gift and task, indicative and imperative.

In order to show clearly at the outset the importance which attaches to this question in the New Testament, we may recall that it is found there in the most varied forms, especially in the Pauline epistles. Here expositions of the doctrine of justification are usually followed by a concluding paraenesis in which Paul exhorts his hearers to take on the task of expressing in their actions what the fact of Christ implies for them. That is, they should live out *de facto* the "new creation" which has been given them. They are not to allow the old existence to continue any longer, like an unremoved alien body in the magnetic field of this new creation.[2]

How explicitly and unconditionally the promise of the gospel precedes the ethical claim in the teaching of Jesus may be seen most clearly in the Sermon on the Mount. Here the radicalization of the Mosaic Law, and therewith of the ethical claim, is introduced by the various Beatitudes (Matt. 5:3-11) and by the repeated statement that—in virtue of the fact that *I* am here, here for *you*—you *are* (!) the light of the world and the salt of the earth (Matt. 5:13, 14).

The point of everything that follows in the nature of a claim is that this crucial, prior, given fact should now also be put into action and brought to concrete realization: Even when men revile and persecute you, you *are* blessed; therefore "rejoice and be glad" also *de facto* (Matt. 5:11-12). You *are* the light of the world; therefore

[2] Cf. Rom. 6:1, 15. See our discussion of indicative and imperative later in this chapter (pp. 70 ff.) and the next.

shine out indeed and beware of the bushel which might obscure
and isolate you (Matt. 5:15-16). You *are* sons of your Father who
is in heaven, who makes his sun rise on the evil and on the good;
therefore love your enemies (Matt. 5:44-45). The Father in heaven,
who has become *your* Father, is perfect; therefore you be perfect
(Matt. 5:48).

It is important for everything we have yet to do that we should
define as precisely as possible this relation between gift and task,
between justification and the acts which follow it. For to say that
works have to actualize and express the fact of justification is not to
solve the ethical problem of Christian action, but in effect really to
pose it for the first time. At best it only suggests in a most general
way the outlines of the solution.

The Dual Motivation of Good Works in the Justified

We would do well to start with a detailed exegesis of Article VI
(The New Obedience) of the Augsburg Confession, which brings out
with particular pregnancy the problems which now confront us. The
first and for us most relevant statements in the article are as follows:
"Our churches also teach that this faith is bound [*debeat*] to bring
forth good fruits and that it is necessary [*oporteat*] to do the good
works commanded by God. We must do so because it is God's
will . . ." [3]

That what is involved here is really a fundamental transformation
of the entire ethical situation is expressed in the very title of the
article which speaks of the "new" obedience. That new thing which
so revolutionizes the structure of obedience had already been described in the preceding articles, which speak of the Christ-event
which overturns the old ethos (The Son of God, Art. III), and of
the justification which appropriates this event ("freely . . . for Christ s
sake through faith," Art. IV), and of the means of grace which effect
justification ("the ministry of teaching the gospel and administering
the sacraments," by which "as through instruments the Holy Spirit
is given," Art. V). It may be seen from the very sequential arrangement of this first group of articles that the "new" obedience is not
determined by the "old" goal of fulfilling the Law. It does not look

[3] *BC* 31.

to any goals at all, but refers back to a given event, namely, the fact of justification.

There are two decisive standpoints from which the relation of justification and works is here described. At first glance they seem to stand in mutual contradiction. On the one hand, faith is *"bound* to bring forth good fruits"; and on the other hand, these works must be done "because it is God's will," because they are "commanded" by God.

The feature common to both these motivations for good works is that in both the works are equally referred back to the fact of justification. On the first view, good works are characterized as the "fruits" of this fact. On the second view ("because it is God's will"), the backward reference is expressed in the following sentences where the clearest possible precautions are taken against the misinterpretation that it is these works which first merit justification (". . . not because we rely on such works to merit justification before God"). Rejected is the misinterpretation which, following the analogy of philosophical ethics, would make justification the goal rather than the presupposition of our ethical acts.

The distinction between the two motivations is clear. In the first case, good works arise, as it were, "automatically" out of the event of justification, without human co-operation. The Augsburg Confession uses the metaphor of the tree and its fruits to indicate that the process which here takes place actually takes place in man, though there is no intervening imperative connected with it. There is certainly no need to command a tree to bear fruit. By virtue of the laws of biology the tree cannot do otherwise. It might even be said—though without sure etymological foundation—that the *debere* [bound], on the basis of which there must necessarily be "good fruits," expresses a natural causality rather than a moral compulsion. In the second case there is, on the contrary, a clear command to do good works ("works commanded by God"— "because it is God's will"). On the surface it might appear that what we have here is a logical contradiction, one which involves a negation of the first motivation, that of natural causality. It appears to involve nothing less than a restoration of the Law, whereby the ethical act is again viewed from the standpoint of goal rather than from the standpoint of a previously given reality, justification.

How does this apparent contradiction arise? Do we have here two heterogeneous motives conjoined simply by the exigencies of history rather than by anything inherent in the matter itself? Or do we rather have here a true paradox, one that in the nature of the case cannot be resolved?

The question is all the more urgent in view of the fact that the so-called "third use of the Law" [*tertius usus legis*] in the life of the Christian believer plays only a very secondary role in Luther's thought. For the most part—though not so exclusively as is often maintained [4]—Luther speaks only of a "twofold use of the Law" [*duplex usus legis*], that is, a "civil" use and a "spiritual or theological" use [*civiliter und spiritualiter bzw. theologice*].[5]

In any case, this question of an imperative in the sense of a "third use" of the Law must be examined carefully, since the foundation of ethics depends on a careful answer to it. Our basic concern is to formulate an evangelical ethic for which the fact of justification is decisive. The alternative would be merely the legal and moral code of a so-called "Christian world view," a code which would belong in principle and with only slight ideological variations to the plane of philosophical ethics, the ethics of "task" [*Aufgabe*] rather than "gift" [*Gabe*].

In order not to get off on the wrong track at this crucial point, we shall first examine the motivation of good works as the "fruits" of justification. After that we shall analyze the motivation of good works in terms of an imperative. In each case, we shall consider the degree to which the particular motivation is both based on Holy Scripture and rooted in the Reformation tradition. Once we have considered the two independently, we shall then attempt systematically to understand and to justify their juxtaposition.

[4] We shall return to this question below (p. 134, n. 7). See concerning it especially Werner Elert, "Eine Theologische Fälschung zur Lehre vom Tertius Usus Legis," *Zeitschrift für Religions- und Geistesgeschichte*, 1948[2], pp. 168 ff. [Here Elert corrects his earlier, influential interpretation of Luther's ethics in *Morphologie des Luthertums* (Munich: Beck, 1932), II, 27 f. After exposing some anachronistic forgeries on the part of Luther's later editors, Elert claims in his article that there is no remaining evidence of any teaching by Luther (as is found in Calvin and Melanchthon) on the Law's alleged "guiding function" for the Christian insofar as he is already righteous (*justus*), above and beyond its undisputed "accusing function" insofar as he is still sinful (*peccator*).—Ed.]

[5] See WA 26, 15 f.; 39[I], 441; also 40[I], 429. [Civilly, the Law judges crimes among men; theologically, the Law judges sins before God.—Ed.]

The Motive of "Good Fruits" in The Bible

The New Testament uses biological and chemical metaphors in many and varied ways to show how the new obedience of the justified man arises out of the fact of grace which underlies his life. What is involved in the production of this new obedience is hence a process analogous to processes found in nature.

Thus Jesus says that a good tree can bring forth only good fruit, and a bad tree evil fruit (Matt. 7:16-20). He says the same of the vine (John 15:4-5, 16) and of the grain of wheat (John 12:24). But if the process of growth is natural and its fruits appear symptomatically as a matter of course, what place is there for commands or other kinds of willful tampering with the process? What we are dealing with here is, as it were, the realm of the necessary and automatic, where nothing is arbitrary; in this realm there can be no moralizing. The zone in which man is here addressed lies wholly outside the dominion of the Law. If I must be commanded by the Law, this is a sign that I am not yet "free," that I have not yet died and risen again with Christ, that I do not yet have the spontaneity of the new existence.[6]

Similarly, Galatians 5:19-24 speaks of the "works of the flesh," those which indicate symptomatically what this flesh really is. The point is the same as in the metaphor of the fruit: in these works the flesh comes to outward expression. The works belong "essentially" to that existence which is "according to the flesh." They are part and parcel of it, as surely as an expression and the thing expressed belong together. The same point is made even more clearly in the positive thesis concerning the "fruit of the Spirit" (v. 22). Here love, joy, peace, etc. are understood as necessary and automatic expressions of the gracious event which has broken through into the life of the believers.

For the sake of our subsequent correlation of the motive of fruit with the motive of the imperative, it is important to note already that the necessity by which good works are done refers in the strict sense not to the justified man but to the Spirit who helps him. The key term is not "fruits of a life apprehended by the Holy Spirit," but simply "fruits of the Spirit." Hence these fruits can arise out of a life only insofar as that life itself belongs to the Spirit. This distinc-

---

[6] See on this point the fifth series of theses in Luther's Third Disputation against the Antinomians, September, 1538, esp. the 10th objection, WA 39$^\text{I}$, 519 ff.

tion could perhaps prevent us in our exegesis of the concept of fruit in Matthew 7:17-20 from interpreting that necessary and automatic process psychologically, as an emanation of the religiously stirred psyche, rather than interpreting it theologically, as an event of the Holy Spirit.

Evil works are referred to in the same sense, not as emanations of the fallen ego, but as symptoms of the flesh. They are also symptoms of the ego, but again only insofar as that ego is determined by its flesh-ness, by the fact of its existing "according to the flesh." Flesh and Spirit are thus understood as factors "outside" the ego, exterior entities to which the ego relates itself. This subtle but highly significant nuance is pushed to the limit by Paul in Romans 7:20, where the evil work is traced back not to the ego but to the "sin which dwells within me." Here the ego and sin are separated for a moment; and it may be seen with a kind of ultimate and exaggerated clarity from which tree these fruits come.

Paul is alluding to the same interconnection when in Galatians 5:6 he speaks of "faith working by love." Here again the subject of the "evangelical" process is not "man." It is "faith," and hence a fact to which man is continually summoned. Only to the degree that he is a believer is man the subject of self-expressive love. (Cf. also Hebrews 9:14, where the "blood of Christ" is the subject which effects our purification from "dead works," and where for the believer these dead works go by the board "automatically" only to the degree that he allows himself to be purified by this blood, and so "draws near.")

Consequently we must not misunderstand the automatic and necessary nature of the process by which these fruits arise. The automatism of works must be seen in relation to an act which needs constantly to be repeated, namely, that personal—and hence by no means "automatic"—act of decision in virtue of which I turn either to the flesh or to the Spirit, in order to receive the orientation of my existence from either the one or the other.[7] Depending on how this decision comes out, what becomes operative is the automatism either of the flesh or of the Spirit.

This gives us a first and vital indication of the way in which we shall later have to correlate the motive of fruit with the motive of the

---

[7] The same is naturally true also of orientation by faith (Gal. 5:6).

imperative. For along the lines already mentioned it is evident that they stand in co-ordination; they do not simply "contradict" one another. The production of good works after the manner of a biological process (the motive of fruit) is linked with the perpetual insistence that a decision be made (the motive of the imperative). This is brought out by Paul's reference to the fact that the ultimate subject of all works, good and bad, is either the Spirit or the flesh, and that it is to these two that man must relate himself through an act of decision which has always to be made again and again. But this raises, of course, our greatest difficulty. For that act of decision must itself be more exactly described, and protected against the misunderstanding that what is involved in it is simply a form of human co-operation or even autonomy.

After this brief reference to the deeper problems of the fruit motive, which will occupy us later, we now turn again to deal with the generally pertinent biblical passages. We are reminded, for instance, of James 3:12, where the analogy to natural causality in the areas of biology and chemistry is again elaborated with reference to works: "Can a fig tree yield olives, or a grapevine figs? No more can a fountain yield both salt water and fresh." The problem of human existence is precisely this, that what is contrary to the laws of nature is exactly what occurs repeatedly in the life of man. What takes place is the perverse, the unnatural. Thus, blessing and cursing can issue from the same mouth (v. 10). (For the moment we shall not pursue the question how this may be, and how ethics must thus fulfill the role of watchman against what is contrary to nature. Here we would simply take cognizance of what the New Testament calls "natural"—in this highly specific sense—and what is therefore also expressed in "natural" images.) The parable of the leaven in Matthew 13:33 may also easily be fitted into the context of our present discussion. According to Romans 6:10, the sin to which man has died is "automatically" excluded from our acts, since the death he died, "he died to sin, once for all (ἐφάπαξ)." The word ἐφάπαξ can here be interpreted only in terms of this automatic exclusion.

The Johannine writings point to the fruit motive primarily by showing that the new life and therewith the new obedience are expressions of a new basis of existence. This new basis can be described in terms of being "born anew" (John 3:1 ff.)—again a biological figure—or in terms of the "light" in which we stand and by which our walk and works are accordingly determined. Our existence in the light and the expression of that existence in our works cannot be in contradiction (I John 1:7; 2:8-11; John 3:21; 12:36; cf. also II Cor. 6:14).[8]

The new foundation of existence can also be expressed in terms of Jesus' indication to the rich young ruler that riches are for him a bondage to the flesh, and of the command that he should sell all that he has (Mark 10:21) and thereby give to his existence a new and non-fleshly basis. As we saw earlier,[9] Jesus' purpose was doubtless to renew the foundation. He was sure that, once that bondage to the flesh represented by riches was broken and replaced by bondage to the Spirit in the form of an emancipated love for God, the works of this life would also be transformed. For these works are simply the expression of the basic reality that underlies them. They automatically reflect the existence out of which they proceed, an existence which itself stands in bondage either of one kind or of the other. This is why the externally correct behavior of the rich young ruler, his superficial obedience to the Decalogue, should not blind us to the fact that his works were not good but bad. For whatever does not proceed from faith, i.e., from an existence rooted and grounded in the Spirit, is sin (Rom. 14:23), no matter what its external appearance may be—and it will indeed have an external appearance![10] It is only the new creation (II Cor. 5:17; Gal. 6:15) of the children of God, those who have been set free but have also entered into a new bondage (Rom. 6:18), which will find expression in new works —again no matter what their external appearance may be.

---

[8] Cf. Luther's Theses of 1520 on the question whether works contribute to justification (WA 7, 231). Theses No. 7 and No. 8 emphasize with dialectical precision the contradiction between what man by virtue of automatic natural causality *must* really be and what *de facto* he actually *is*. For on the one hand we are told that he who is born of God does not commit sin, indeed cannot sin (I John 3:9). On the other hand we are told that whoever says he does not commit sin is a liar and the truth is not in him (I John 1:8).

[9] See above, p. 31.

[10] It is in terms of this deceptive appearance that we are to understand the well-known reference of Augustine to the supposed virtues of the heathen as "rather vices than virtues." *The City of God* xix. 25; *A Select Library of the Nicene and Post-Nicene Fathers of the Christian Church* (First Series), ed. Philip Schaff (14 vols.; Buffalo and New York: Christian Literature Co., 1886-1889).

The Motive of "Good Fruits" in Luther and the Confessions

The "automatic" connection between the fact of justification and good works was bound to be taken up again and again in the theology of the Reformation. For one thing, its opposition to Roman "work-righteousness" could be formulated with particular precision by means of this concept. Luther has many well-known passages in which he expresses the antitheses very sharply in formulations which are pregnant with meaning. His statements constantly reflect the basic thought that it is not good works which make a good man, but a good man who does good works.[11] The moment we consider the view Luther was opposing, it becomes quite evident that in such statements the works are regarded as an "automatic" expression of the person. For the Roman ideal of sanctification which Luther had in view implies ultimately that the person is the sum total of the works which (with the help of grace) he has performed. Hence it is not the works which express the person, as Luther maintained, but the person who expresses the works. That is, the person is the expression and representation of his acts. For after all, it is he, the person, who can attribute the works to himself, chalk them up as merits, and so be in effect what his works are. Luther's thesis is the direct opposite: the works are what the person is.

Nevertheless, it will not suffice simply to say of Luther that for him it is the person who does the works and thus finds expression in them. For then the further question has to be put immediately: Who and what is this person herein expressed? We cannot pursue this problem in detail here, but we may at least say this, that the ego is characterized by the fact that it either believes or does not believe. The works which this ego performs, and to which it lends its quality as person, are accordingly determined by the fact that they spring either out of faith or out of unbelief. Luther can say, "As nothing makes righteous but faith alone, so nothing commits sin but unbelief alone." Again, "If adultery could take place in faith, it would be no sin. If you try to worship God in unbelief, you will perform an act of idolatry." [12] Now this sounds dreadful and shocking. But Luther can of necessity see no other alternative since there can be no question

[11] See WA 39I, 283; cf. WA 39I, 48.

[12] Luther's Theses of 1520 on the question whether works contribute to justification, Nos. 1, 10, and 11 (WA 7, 231).

of combining the two extremes by weakening them in some kind of compromise. The moment it is rigorously held that the work is characterized by the person, and that the person for his part is determined either by faith or by unbelief, i.e., either by God or by God's opponent, then we have no choice but to accept Luther's harsh either-or.[13] The alternative could be synthetically blunted in its sharpness, and even smoothed over by compromises, only if there could be ascribed to works some independent value, a dignity divorced from the person. But this would imply a hierarchy of autonomous values which had value in and of themselves, rather along the lines of an ethics of material values. But this is impossible. Hence, if works belong to an ego which stands outside justification, they are blasphemous even though they may be materially correct. They stand under an anti-godly sign, for "whatever does not proceed from faith is sin" (Rom. 14:23). Even the most seriously intended human ethics cannot escape this verdict. For wherever ethics does not understand works with a backward reference as the expression of justification, but instead refers them forward to the ethical goal and understands them as a means to actualize that goal, it is bound to subscribe to the blasphemous idea of a "self-creation of the ego" according to which works will create the good, the pious, the well-pleasing ego. All natural ethics stands ultimately under the sign of this self-creation, of opposition to the First Commandment, and therefore under the sign of hubris and of the deification of the creaturely.

Thus we see that the interrelation of person and work touches on the basic concern of an evangelical ethics grounded in justification.[14] It marks the decisive boundary which separates evangelical ethics from all philosophical ethics. The difference between the two is not merely quantitative, in the sense that the one is a higher and purer ethics, but "infinite and qualitative," in the sense that what is at issue here is the alternative of salvation or perdition, God or

[13] In images and figures to which we shall have to refer later (e.g., some from The Bondage of the Will), Luther can speak of man, along the lines of this alternative, as a horse and again as a battlefield, i.e., as one who is always ridden or possessed by either one or the other of these two antagonistic forces (either God or Satan). The battlefield is never unoccupied or the horse riderless. See The Bondage of the Will, trans. J. I. Packer and O. R. Johnston (Westwood, New Jersey: Revell, 1957), pp. 103-04.

[14] On this whole problem, see Luther's A Treatise on Good Works, 1520 (PE 1, 184 ff.).

Satan, faith or unbelief. Blasphemy is not just a matter of crude tirades in which the name of God is abused; blasphemy can involve all kinds of human and morally exalted sublimations. The Antichrist is not only able to perform miracles; he can also don the philosopher's cloak. He invades not only the temple of God, but also the classrooms of the moralists. This is why we must not be deceived by the aura of "good works"; it could be no more than such a cloak.

In the present context, however, where our primary concern is with the "automatic" nature of good works, we have to stress above all the way in which works are an "expression" of the person, be he the justified man or the blasphemer. This character of works as being expressive of the person may be formulated most sharply in Luther's statement that the person "exists" in his works: "Hence it is that . . . [man] . . . wills, desires, and does, as he himself *is* [*qualis ipse est*]." [15] This idea may even go beyond that of the automatism of works by suggesting that works are not merely related to the person as a "subsequent" product (however much they may arise as a matter of course) but that the very being [*Sein*] of the person *consists* in the works, in what he "does, wills, and desires."

It was inevitable that Luther, like the New Testament, should use the metaphor of the tree and the fruit to describe this relation. For the tree too consists to some extent in its fruits. Similarly, the good will, the will that is determined by justification, consists in that wherein it is actualized, as surely as man always has his being in action. [16] In every case what is involved is one and the same reality, a reality which may be viewed, however, only from different sides, from within and from without. [17] Thus Luther says, "For when he has a good will [*voluntate bona*], the whole man is good, even as a tree that has a good root is a good tree; at least what it bears is good

[15] *The Bondage of the Will, op. cit.,* p. 204.

[16] "The being and nature of man cannot for an instant be without doing or not doing something, enduring or running away from something (for, as we see, life never rests) . . ." (*PE* 1, 198-199).

[17] Article IV of the Apology to the Augsburg Confession expresses this identity by treating the heart and good works as two sides of the same thing: "We mean to include [in the keeping of the Law] both elements, namely, the inward spiritual impulses and the outward good works" (*BC* 126). The early German version of the Apology brings out even more clearly than the original Latin this matter of there being two sides to the one thing: "And when we speak of the keeping of the Law or of good works, we comprehend both [in one], the good heart inwardly and the works outwardly."

fruit, unless the fruit is not according to its kind." [18] "We confess that good works are bound [*debere*] to follow faith—no, not 'are bound' [19] but simply follow naturally [*sponte*], just as a good tree is not bound to bear good fruit but does so of its own accord [*sponte*]." [20]

We lay particular emphasis on the term *sponte* because it describes most felicitously the directness of the relation between justification and works. The term perhaps takes on an even higher degree of pregnancy when Luther combines the concept of spontaneity with that of immediate readiness [*promptitudo*], as in his definition of will. The will is for Luther, in distinction from Scholasticism, not one part of man in contrast to other parts (sensuality, reason, etc.), but an expression for the total movement of the whole person.[21]

It is important to note in this connection that Luther sees no contradiction between *promptitudo, libertas,* and *necessitas* (in terms of natural causality). The new obedience is "automatic" in the sense that it cannot be otherwise (necessity!) than that there should be this obedience. But this necessity is of such a kind that the man who in faith lays hold of justification also *wills* this obedience. To use Fichte's illustration of this total movement of the ego—set forth to be sure in the very different context of his teaching on genius—the needle of the compass "must" and "will" point to the north.[22]

Now the motive of fruit may seem here to be furthest removed and most explicitly different from, indeed, in sharpest antithesis to, that of imperative. For the imperative addresses our will (construed in the narrower sense!) and summons it to fight against inclination, impulse, sloth, etc. The imperative necessarily splits the ego into two parts. There is the one part which in the name of practical reason, to put it in Kantian terms, has to be the ruling authority, and

[18] That is to say, when the good will becomes operative it produces, in conformity with its nature, nothing but good works (WA 3, 25).

[19] In discussing Article VI of the Augsburg Confession we interpreted *debere* in terms of the necessity of natural causality (see above pp. 53-54). However, its juxtaposition with *sponte* in this quotation from Luther suggests that it has here a quite different meaning; it is to be understood rather in terms of *oportere*, the "ought" which we have referred to as an imperative.

[20] WA 39I, 46. Thesis No. 34.

[21] WA 3, 30.

[22] Cf. the similar expression of Luther: ". . . freedom of the kind in which, as it is necessary that the will [*voluntas*] to live well continues, there is also the voluntary and cheerful compulsion [*necessitas*] to live well and never sin" (WA 6, 27).

which is thus invested with the dignity of the intelligible ego. And there is a second part which is subject to this authority. To this extent Kant—and he was not the only one to see this[23] —was surely right when he saw that the imperative involves a deep and radical cleavage of the ego, and when, as was indicated above (see p. 33), he went so far as to regard the sharpness of this cleavage as an indicator of the degree to which our action is the authentic expression of duty as over against inclination or impulse.

Luther, however, does not describe the will as a component *part* of the ego, but as the representative of the *whole* ego which presents this ego in a total, unbroken, and therefore unitary movement.[24] But this is possible only if the Law has no part in the origin of this total movement. For the Law can only divide the ego into two hostile fronts, as in Romans 7:22 f., whereas what is involved here is an automatic and spontaneous movement of the ego which expresses itself quite naturally and as a matter of course in works, so that the works are part and parcel of the movement itself.[25]

This movement devoid of Law and imperative is possible only in love. Hence it is a self-contradiction to command love in the form of a law. For such a law could never lead to this total movement. For this reason Jesus' command to love (Matt. 22:37 ff.; cf. also Rom. 13:9; Gal. 5:14; Jas. 2:8) is not to be regarded as a law just because it "happens" to have the external form of a commandment. Like the command to forgive (Matt. 6:15; 18:21, 32), it involves in its very nature a reference to the fact that in Jesus Christ, the giver of the commandment, our forgiveness is already present fact, and that from this fact certain consequences follow in regard to our own debtors

[23] The cleavage arising under the Law is portrayed as clearly by Paul in Romans 7:14 ff., and by Luther in, for example, his commentary on Romans 2:12 (*LCC* 15, 49).

[24] This total movement is indicated also by Luther's use of the term *anima,* which for him is almost synonymous with *voluntas* (*WA* 3, 296).

[25] Cf. Luther's phraseology in the 1513-16 *Dictata super Psalterium* where he is commenting on Psalm 119:10: ". . . with my whole heart, not half of it" (*non dimidio, WA* 4, 282), and on Psalm 119:69: ". . . with the whole heart, in every disposition" (*omni affectu, WA* 4, 290), and on Psalm 40:8: ". . . within my heart . . . in the very center, the whole of my inmost disposition" (*inmedio, in intimo et toto affectu, WA* 3, 17). Cf. also the great commandment which speaks of loving with "all your heart" (Mark 12:30, 33). See Karl Holl, *Gesammelte Aufsätze zur Kirchengeschichte,* I (Tübingen, 1923), 158 ff.

(Matt. 6:12). (Note that we deliberately speak of results which "follow," not of consequences to be "drawn.")

Thus it is not the Law but the Gospel alone which can release that love, and therewith initiate this total movement of the ego. When the event of grace which is promised to us and allotted to us is also accepted by us, that total movement of love, and hence the total expression of this love in works, is in fact present. For the ego is never *otiosus*, motionless and idle. On the contrary, by its very nature it moves out, unfolds, and hence performs works. It is *actuosus*, in action. The so-called "ego in itself," i.e., the ego insofar as it stands behind its works as the performing subject separable from the works themselves, is an artificial abstraction.[26]

This unleashing of love by the Gospel, by that which is the very opposite of the Law, is expressed in the Lutheran confessional writings most classically in the section of the Apology called "Love and the Keeping of the Law." [27] Here three points are clear.

First, the love which actualizes itself in good works is really posited in and with the love of God which is shown to us. In order to come into being it needs no supplementary intervention on the part of the Law.

Second, we cannot understand the automatic operation of this process in terms of temporal succession. It is not as if there were *first* the act of justification in which the former estate is liquidated, and then love and the works issuing from it followed in a *second* act.

Third, good works are not to be regarded simply as products of the renewed subjectivity, as though one whose heart was full now pressed on to action, as though he were "endowed" with a new dynamic and now sought to apply this dynamic and let it unfold. Possibly this latter process does play a role; but those strains within Pietism and the revival movements which find in it the essential connection between justification and sanctification are quite mistaken. The decisive process is something very different. It takes place in an objective relation to God, who in loving us becomes for us totally different. As the one who loves us he becomes, as far as we are

---

[26] Cf. Luther's statement: "[Where the seed of the Word] . . . takes root in the heart . . . there arises a wholly different man, different thoughts, different words and works. You are thus *completely* transformed" (*WA* 12, 298).

[27] *BC* 124 ff.

concerned, the one who is loved by us. Our love and the acts which flow from it are not just a response to the preceding love of God (even if they take place promptly, spontaneously, and as a matter of course). They are not just a reaction, a second act. On the contrary, they are actually the reverse side of God's love. Or, to speak in terms of "active" and "passive," our love and good works—in short, our active ego—is the theater in which God does his acting. Luther says, "It is true to say concerning ourselves that, inasmuch as God works in us, we work—though 'work' here means actually that the one doing the acting is himself acted upon, moved, and led." [28]

As applied to our interpretation of love, this means that what is involved in our loving is really a being loved and a letting ourselves be loved. We have here simply two sides of the same thing, as in the case of the righteousness of God. Seen from one side the *justitia Dei* means God's righteousness, and seen from the other side it means our righteousness vis-à-vis God. In any case it is quite impossible here to divide righteousness into two acts, with God as the subject of one and man as the subject of the other, with one representing the divine initiative and the other man's reaction to it. On the contrary, God's righteousness consists precisely in the fact that we are acceptable to him. Similarly, the "love of God," as a love accepted by us, consists precisely in the fact that we for our part love him. We are so inclined to the erroneous division of this one love into two separate acts because we have forgotten to regard man in his indissoluble connection with God and have preferred instead to take as our starting point the false conception of "man in himself," man as one who has his being in himself.

All this is expressed in the afore-mentioned section of the Apology which says that in his mercy God himself becomes for us a lovable object: "We cannot love God until we have grasped his mercy by faith. Only then does he become an object that can be loved." [29] In the very fact of his being revealed to us as the loving subject, the one who does the loving, in that very fact God *is* the lovable object; that is, we for our part are now subjects who love. We are so in virtue of this fact. We do not have to become loving subjects, nor demonstrate that we are grateful (at any rate this command

[28] *WA* 5, 144; cf. *WA* 5, 176.
[29] *BC* 125.

belongs to a different context): we simply *are!* Our love appears here as the reverse side of the love of God.

If we might be permitted to put the matter quite pointedly, we can say that what is involved here is not primarily a transformation of the ego whereby it is stimulated to love and hence to do good works. What is involved is rather an event of transformation in God. For us he ceases to be the author of the Law (an object of fear and hate driving us to despair) and becomes instead the author of the Gospel (the epitome of the fatherly heart made palpable and objective in Jesus Christ).

In this transformation in God there is reflected Luther's understanding of the tension between Law and Gospel, of which we shall speak later at greater length. Karl Barth has passionately protested against this understanding of the tension and change in God in his essay on "Gospel and Law." [30] Here he argues that the Law is the "form" of the Gospel. The Law lies in the ark of the covenant, and thus has reference to the Gospel of the covenant. Hence Law and Gospel are indeed to be distinguished, but they are not to be set in tension as Luther sets them. In asserting this tension, however, Luther's theological concern, as we shall see, was primarily to understand God's wrathful reaction in the Law as the "natural" reaction appropriate to his majesty, and to understand the Gospel as the absolutely unheard-of miracle whereby God overcomes God. This is why for Luther there looms up as an enigmatic backdrop behind the God of the Gospel him who is the author of the Law, the *Deus absconditus* or hidden God.

Thus the "automatism" to which we have referred, whereby works are produced automatically, does not involve a process within human subjectivity which runs its course over a period of time and can conceivably be traced psychologically. It is not a case of cause and effect, God's love being the cause and my responsive love and attendant works being the effect. We must really insist that our love simply comes to light as the reverse side of the love of God.

That what is involved is not a subjective process such as can be fixed in point of time, is brought out in the Apology very well. It never

---

[30] See Karl Barth, *Community, State, and Church: Three Essays* (Garden City, New York: Doubleday, 1960) where the essay on "Gospel and Law" has been translated by A. M. Hall.

understands man's love as acting subject, either in the sense that it is capable of effecting justification or in the sense that it is the cause of those good works through which we then fulfill the Law. Over against the first misunderstanding it must be asserted that our love does not justify us by any means; it simply appears as the reverse side of that which happens to us, the event of justification and the fact of the lovable object. "We do not receive the forgiveness of sins through love or on account of love, but on account of Christ by faith alone." [31] Over against the second misunderstanding it must be said that our love is not the first cause which leads to the fulfillment of the Law. It is in itself already the fulfillment of the Law. Good works simply express this fulfillment; they constitute, as it were, its outward form.

There is another equally important indication that in the emergence of love and its fruits what is involved is a metaphysical event that can only be conceived theologically. This is to be found in the fact that the confessional writings, like Paul, do not regard the "fruits" as products of the converted ego but as fruits of the Holy Spirit, and hence of an entity outside ourselves. Clearly the objective conception, namely, that the point of the salvation event consists in God's change towards us rather than in our change towards God, is consistently maintained right down the line in all the theological argumentation.

If we pause here for a moment and look back on the train of thought which we have followed, we see that our interpretation of the new obedience in terms of the motive of fruits has come into increasingly sharper focus. We first understood the "automatic" connection between justification and good works analogously to a process of "natural causality" whose initiation needed no intervention of the Law, indeed as a process in relation to which such an imperative could only have the effect of disturbing and splitting the ego which has become "total" in love. As we pursued the matter further, we were forced to understand this automatic element in such a way that the justifying love of God on the one hand, and our love and the fruits rooted in it on the other, could ultimately be regarded only as two sides of the same thing, and consequently as posited simultaneously.

By thus giving the sharpest possible focus to this train of thought, we have worked out the extreme antithesis to any grounding of the
[31] BC 127.

new obedience in an imperative. If the very process of "natural causality" (by virtue of which a stone lying in the sun becomes warm quite on its own) has no need of the impulse provided by an imperative, then a legal imperative would seem to be completely out of the question. It would make no sense, if the connection between justification and the new obedience exists so assuredly as a matter of course that they are seen to be simply two sides of the same thing.

This is why it must seem all the more strange that Article VI of the Augsburg Confession grounds the new obedience not merely in this "automatic" element but also in a supplementary imperative, so that the Law still exerts an influence on the new obedience and men are still summoned to obedience "because it is God's will." The question obviously arises: Is this imperative a pre-Reformation relic (and is the similar element in Paul a vestige of Jewish legalism)? Is there here a compromise with Roman "work-righteousness," or a corrective device? Or do the two motive complexes simply not lie on the same plane at all? Is the motive of the imperative perhaps a kind of pastoral counsel designed to keep men vigilant and on the alert, so that they do not rely on the "automatic" element or become quietists out of fidelity to the fact of man's utter passivity with respect to salvation?

All these points of view have at any rate cropped up in the history of theology, and been used to try to arrive at some kind of co-ordination between the motive of fruits and that of the imperative, one which would be meaningful in terms of the matter itself or even one which would be purely "psychological" or "historical." Before we make our own attempt to find the co-ordinating link within the substance of the matter itself we shall first assemble the necessary materials relating to the motive of the imperative, even as we have tried to do in the case of the motive of fruits.

The Motive of the Imperative in the Bible

We have already shown that in the preaching of Jesus, especially in the Sermon on the Mount, there are imperatives which appear to be strange. And that strangeness is not dissipated simply by pointing out that the "task" [*Aufgabe*], the radicalized Law of Moses, refers back to the existent "gift" [*Gabe*], the fact that Jesus Christ has come and that in him those imperatives are *de facto* fulfilled (cf. the passages cited: Matt. 5:13, 14, 11 f., 16, 44, 48; also Matt. 10:21). Indeed the strangeness consists in the fact that any additional imperative should be necessary in face of this gift, and that this imperative therefore reminds us of the ethics of "goals," an ethics which Christianity must repudiate.

Why does it not come about "of itself" that I forgive my brother seventy times seven, once I have been forgiven (Matt. 18:22)? Is not this "of itself" a cardinal quality of the Gospel over against the Law? How is it that the disciple has to be commanded to love his enemies (Luke 6:27, 35), when he knows that he is himself an enemy yet beloved of God, when he stands in the light of the same sun which shines on the just and the unjust, on the good and the evil (Matt. 5:45)? How is it that the disciple must be assigned the task of loving his neighbor, when he has already been given the neighbor himself in a totally new way, being obliged to recognize Christ concealed in his neighbor (Matt. 25:35 ff) and to know that Jesus Christ died for this neighbor (Rom. 14:15; I Cor. 8:11)?

Does not the mere fact of the command involve a vote of "no confidence" in the perfect tense of the gift? Is the gift not sufficiently powerful and effective to assert itself in the new existence? Does it need the help of subsequent co-operation? Does not this mean that "work-righteousness," which was thought to have been overcome in Reformation theology, simply rises up again with new vitality and glee at another point? Does it not perhaps signify (with biblical sanction!) that the *sola fide* indeed involves after all the one-sided absolutizing of a principle, a principle which stands self-condemned by the fact that it is forced to introduce an imperative of this kind as a corrective?

The problem becomes even more acute when we find the same phenomenon in Paul.[32] It is not merely that Paul's explicitly theological sections are as a rule followed by practical paraeneses, but above all that interspersed in his discussions of the new existence given us in Christ there are repeated demands that we should lay

[32] See Rudolf Bultmann, "Das Problem der Ethik bei Paulus" (pp. 123 ff.) and Hans Windisch, "Das Problem des paulinischen Imperativs" (pp. 265 ff.), both in *Zeitschrift für die neutestamentliche Wissenschaft und die Kunde der älteren Kirche*, 24/25 (1925-26). See also Paul Althaus' volume entitled *Der Brief an die Römer* in the commentary series *Das Neue Testament Deutsch* (5th ed.; 1946), VI, 53 ff.

hold of this existence and from it draw the appropriate consequences. In this way Paul makes the linking of the indicative and the imperative a problem which obviously stands at the very center of his theology and which we cannot possibly overlook. For the immediacy of the juxtaposition shows us that the tension is undoubtedly intentional. We cannot assume that Paul is involved in an inconsistency or is shifting his point of view in a psychologically explainable way. The passage we think of primarily is Romans 6-8 (especially 6:1-7 and 8:1-17, but also Gal. 5:13-25 and I Cor. 6:9-11), in which indicatives and imperatives stand in the closest possible proximity to one another. On the one hand it is said, with reference to what has already been given and taken place in baptism, that we are really "dead to sin" and members of the new aeon. On the other hand there follows at once the summons to "die," which postulates that we must nevertheless still regard the old aeon as present and active. On the one hand the "putting on" of Christ appears to be something which has already taken place in baptism, with all that this implies for the form in which the new creation manifests itself (Gal. 3:27). On the other hand this same "putting on" is the object of unceasing moral exertion on the part of Christians (Rom. 13:14). The juxtaposition is pointedly expressed in Galatians 5:25: "If we live by the Spirit, let us also walk by the Spirit." Here the eschatologizing and the de-eschatologizing of existence seem to stand cheek by jowl in the same statement, formally dismembering it.

Since these "contradictions" are so striking, the question might arise whether the theoretical enthusiasm which brought the apostle to believe in Christian existence as perfected—"perfect" in the same sense in which the history of the death and resurrection of Christ and also the effect of this history on believers is "perfect" [33]—was not perhaps followed by practical resignation. This brings us back, however, to the prior question: Is there a real contradiction between the two? Is it a contradiction which corresponds to the incongruity which seems to obtain between the sacramentally conditioned "genotype" of Christian existence on the one hand and its "phenotype" on the other? Is the contradiction one between existence understood in consistent eschatological terms on the one hand and existence as concretely realized in this world, this continuing aeon, on the other?

[33] See The Augsburg Confession, XX, 23; BC 44.

71

On any such view, the continuity between the old man and the new would be broken in every respect: as regards the transition from one to the other, as regards the identity of the subject in both instances, and as regards the perspective within which one might be able to view this subject now one way and now the other. Since we should thus be led into interminable contradictions, we are compelled to look for some other way of stating the relation between the indicative and the imperative.

The Motive of the Imperative in Luther and the Confessions

Before we pursue more closely the difficult problem of co-ordinating the indicative and the imperative we may point out that, in keeping with Article VI of the Augsburg Confession, this co-ordination is stated in innumerable ways in Luther and the confessional writings. In order to give even a few examples, we are unfortunately compelled to make an arbitrary selection.

As far as Luther is concerned, it is enough to recall the explanation of the Ten Commandments in his Small Catechism. The imperatives, on his view, are characteristically not oriented to the casuistry of concrete action, but possess rather the thematic significance that we are bound to keep the First Commandment in fear and love, "so that" we do not kill, commit adultery, etc. This means that the imperative, "We should . . . ," definitely does not interrupt the automatic process of new obedience as we have described it but is rather the pre-condition of it. Only when we keep the First Commandment—to be sure in the sense of our being called to do so rather than our resolving autonomously to do so—is there posited the automatic operation of the new obedience. This is why on Luther's view, as we have pointed out already, the nature of the matter is such that the corresponding clauses are to be understood as consecutive clauses. If we were to regard them consistently as final clauses ("in order that we . . ."), this would mean that the fear and love of God given in faith, and therefore faith itself, would simply be religion's means to ethical ends. Wilhelm Herrmann might say this—though even he only with important reservations—but never Luther.

With regard to the confessional writings, it may simply be noted that the grounding of good works in the imperative, as set forth in Article VI of the Augsburg Confession ("because it is God's will," "because commanded by God"), is variously repeated, and even in the same terms. Finally, the same requirement is laid upon believers, that they should fulfill the new obedience in good works. However varied the formulations in which it is expressed, the requirement itself is in substance always the same.[34]

[34] See, for example, Art. XX on "Faith and Good Works" in the Augsburg Confession and Art. IV on "Justification" in the Apology of the Augsburg Confession.

6.

Co-ordination of the Indicative and the Imperative

One-sided Emphasis on the Imperative

What happens when particular emphasis is laid on the imperative? The Apology draws attention to this problem in a polemical section of Article IV on "Love and the Keeping of the Law." According to the Thomistic doctrine of justification the imperative, although not exactly isolated and absolutized, is nonetheless accorded an autonomous significance. For justification is linked with the "keeping of the Law," and the imperative, i.e., the requirement of good and meritorious works, has the significance of co-operation in the attainment of justification. The Apology finds the reason for this primacy of the imperative, or at least for the high degree of emphasis laid upon it, in the Thomistic concept of *prima gratia*.

In opposition to this accentuating of "initial grace," the Apology maintains that Christ does not cease to be the mediator after we are renewed. "All those err who maintain that he [Christ] has merited for us only the 'initial grace' and that *we* then subsequently attain acceptance and earn for ourselves eternal life by our keeping of the Law. Christ remains the mediator, and we must always maintain that on his account we have a reconciled God, even though we ourselves be unworthy." [1]

By way of interpretation, it should be noted that the expression "Christ remains the mediator" is an exaggerated formulation which is to be taken with a grain of salt. For it goes without saying, as the Apology realizes well enough, that Thomism does not present the doctrine of justification in such crude and deistic fashion that Christ is, as it were, only the initiator of justification, and that then, having started the movement, he withdraws, after the manner of Deism, and leaves everything to the human action thus "cranked up" and released. Thomism cannot mean this, since it regards all the "merits" attained

[1] Apology of the Augsburg Confession, IV, 162 (German version). An English reading of the Latin version is to be found in BC 129.

by man as merits only through grace, and hence only for the sake of Jesus Christ. Hence we must not allow this polemical formulation to give us too simple a view of Thomism.

Nevertheless, the Apology does use this polemical formulation; and if we cannot think that it is simply caricaturing its opponents in order to ease the task of refuting them, we must interpret it as follows. The concept of *prima gratia* involves a decisive infringement upon and restriction of the mediatorial significance of Christ. For when justification is linked with the *prima gratia*, this initial grace is regarded as the basis which makes possible our doing of the meritorious works necessary for salvation. Thus grace becomes merely the basis which makes possible the real thing. The real thing is the meritorious works; they are the key to the process of justification. For it is by works that we see whether the grace lent to us is actualized and put to good use, or whether it remains instead idle capital. In the strict sense, therefore, initiatory grace is really the basis of the possibility, the indispensable condition of the real event. In relation to the merits which are normative for salvation, justification has liberating and creative power. Its position is rather like that of a means to an end.[2]

In thus characterizing Thomistic faith as a "means to an end," we should not forget, of course, that this is an exaggerated formulation because in Thomism grace is in some sense final as well as primary. For what man merits is grace in its quality as an end, as ultimate "goal." Between the two, however, merits have a decisive position, since they can challenge and even block the way from primary grace to ultimate grace.

In Rome's assigning of a key position to works, the Apology sees not only an infringement upon the exclusiveness of Jesus Christ, but also a threatened perpetuation of the assaults of doubt [*Anfechtung*] which Luther sought to overcome. "If those who are regenerated are supposed later to believe that they will be accepted because they have kept the law, *how can our conscience be sure that it pleases God*, since we never satisfy the law?"[3] If good works occupy the

[2] The same point is made by the Apology in another connection (IV, 109) where it opposes the Thomistic concept of "faith fashioned by love" [*fides caritate formata*] on the ground that here it is not faith *itself* which mediates justification but the *love* which expresses itself in works. In relation to this love faith is only a liberating power and thus only a means to the real end. *BC* 122-123.

[3] Apology, IV, 164; *BC* 129 [italics by H. T.].

key position in the process of justification, then the assurance of our being accepted and justified by God (the "sure conscience" [*conscientia certa*]) is continually threatened. For this assurance depends in turn on the assurance that we have fully kept the Law, an assurance that can never be definitive and unequivocal. To the degree that the decisive phase in the process of justification passes into the hands of men, there is always instability, and hence assaults of doubt.

Over against this view of justification, the point of the Lutheran doctrine can be stated precisely in terms of perspective: Man is enabled to look away from himself, not only from his works and virtues but even from his faith.[4] He is empowered to cling wholly to the objectivity of the divine promise of forgiveness. Otherwise man is simply "turned in upon himself," and this *incurvitas in se* is a devilish circle even when it purports to be pious and godly and relies upon faith as a *habitus*.[5] It affords no peace. Luther trod the bitter way of this *incurvitas* to the end in his struggles in the cloister.

The critical point in the Roman Catholic scheme of justification is best expressed in relation to Luther's doctrine of Law and Gospel. For Roman Catholicism makes the fundamental mistake of not differentiating between these two elements. It bases justification decisively on—or at least regards it as conditioned by—the fulfilling of the Law. Hence the Gospel is not liberation from the Law. It is merely God's ratifying response to the keeping of the Law. Thus the Law's sphere of jurisdiction is not breached but continues instead to embrace the conditions under which alone the Gospel can become operative. Servile fear is thus perpetuated, and with it the afflicted and uncertain conscience, the *conscientia incerta*. Actually, the comfort and assurance of the act of justification depend on the objectivity and absolutely unconditioned character of the divine promise. But these can be expressed and confirmed theologically only when Law and Gospel are strictly differentiated and carefully set forth in their dialectical opposition to one another. To relate them to one another positively is at once to lay conditions on the divine work of grace, and to deliver up the conscience to the devilish circle of the *incurvitas*.

---

[4] See the Luther quotation documented above at p. 16, n. 2.

[5] See Luther's *Treatise on Good Works* (1520), (*PE* 1, 190). [*Habitus* has reference to a quality, state, or condition independent of works.—Ed.]

The result is a depressing inversion of perspective and the necessity for anxiety-ridden efforts at self-control. The same objection to Thomas recurs again and again in the confessional writings. It is directed against the characterization of grace and faith as bases on which, in the name of the Law, the merits necessary to salvation are then built. The mistake here, says the Apology, is to think that faith is merely "the start of justification or a preparation for justification [the idea of faith as basis]. Then it would not be faith, but the works that follow [this faith], by which we would become acceptable to God."[6]

This Roman conception of the doctrine of justification constantly casts its shadow upon evangelical theology as well, and even upon interpretations of Luther's own teaching. Thus Karl Holl[7] expounds Luther's doctrine as follows. Acceptance into fellowship with God is only the condition for raising up man and stripping off his sinful nature; it is not itself, however, the actual raising and stripping off. The goal which God pursues in justification is renewal. Thus Holl distinguishes two acts in the process of justification. The first is justification in the narrower sense, whereby justification is presented as a "way" or "disposition." The second is the real "goal," i.e., renovation. In other words, in justification God anticipates the final judgment on man. He sees proleptically what man will become once the foundation has been laid in the form of justification.

The only difference between Luther and Scholasticism, as Holl sees it, is that for Luther the new image of man in the state of justification is exclusively God's work. But this naturally does not exclude—it rather includes—the fact that man as thus embraced by the divine action has to be active himself, and that his salvation is consequently linked with this activity, his merits, his taking seriously of the divine imperative. Here too, then, the decisive point is man's obedience vis-à-vis God's purpose to be gracious to him. Hence the indicative, whereby God pronounces me righteous and receives me into fellowship as his child, is de-emphasized in favor of the imperative. The objectivity of the divine promise of grace threatens to be

[6] Apology, IV, 71; *BC* 116. This is the secret of Rome's "faith fashioned by love" [*fides caritate formata*].

[7] See Karl Holl, "Die Rechtfertigungslehre in Luthers Vorlesung über den Römerbrief mit besonderer Rücksicht auf die Frage der Heilsgewissheit," *Gesammelte Aufsätze zur Kirchengeschichte*, I (Tübingen, 1923), 119, 123, 128.

obscured by the subjectivity of the keeping of the Law and by the signs of renovation. There is thus the danger that the disquietude of the afflicted conscience will be made permanent.

For Holl too, therefore, faith is a mere beginning of justification, i.e., a "part" to which additions must be made. Here again there is the threat that the *sola fide* will be dissolved, and that the Reformation understanding of faith will be destroyed at its decisive point. Ethics again becomes an adjunct to the event of justification, whereas actually it can have as its content only the fulfillment of this event, its other side.

We thus conclude that there are certain forms in which the imperative, and hence ethics itself, may enjoy a relative autonomy. But in these forms the relation between Law and Gospel is falsified, the *sola fide* shattered, and the assurance of salvation replaced by permanent assaults of doubt.

One-sided Emphasis on the Indicative

What happens when particular emphasis is laid on the indicative? We may also formulate the problem of this section thus: What happens when we emphasize merely the automatism of works, that automatic element which precludes our viewing the renewed state of the justified from the standpoint of the "new obedience" he has to render?

Classic theological examples of this approach are to be found, for example, in Andreas Osiander (1498-1552) and even more so in the modern holiness movements. The holiness movements push to the extreme certain tendencies found in Osiander and thereby completely and radically abandon any connection with Luther's doctrine of justification. Osiander himself maintained such a connection; he thought of himself as a consistent disciple of Luther and regarded Melanchthon as an apostate from the true Lutheran position.

Osiander too begins with Luther's *sola fide*. But he gives the doctrine a mystical turn inasmuch as he teaches an essential indwelling of Christ by faith, and thus presents a sharp antithesis to Melanchthon's theory of imputation. His point is that the business of justification is not exhausted simply in the forgiveness of God, nor in that automatic process whereby the stone lying in the sun becomes warm (in this process the sun always remains something outside the stone!). Rather there is a real indwelling of the righteousness of Christ in us, an ontic appropriation into our very being, and it takes place through faith. The practical concern of Osiander here is to ward off the danger implicit in the imputation theory, namely, that of making the existence of the justified run off in two directions. For on a purely forensic view of the act of justification, there is always the risk of thinking that the possession of righteousness by faith is still possible even when the conduct of the Christian bears no relation to it whatsoever, when there continues in the Christian instead a rich burgeoning of sin. It is for this reason that Osiander, in his famous "second Königsberg disputation," warns against setting one's mind at ease on the ground of the remission of sins. God cannot regard the justified man as righteous if his life displays an absolute incongruity between the righteousness promised and the righteousness actually on hand. In this sense Osiander can make the extreme statement in his *Ein gut Unterricht* (1524) that "only the works which the Holy Ghost works in us count in the presence of God." Justification in this sense is a being *made* righteous [*justum effici*] in virtue of that real indwelling of Christ. The term *efficere* [to make] thus involves the sharpest possible antithesis to Melanchthon's *imputare* [to impute], and Osiander believes that his view is true to Luther.

At any rate we have in Osiander and Melanchthon two interpretations of Luther's doctrine which have become more and more exclusive—at least so far as theoretical reflection on them is concerned—as a result of the bitter disputes among their successors. In the indwelling of Christ by faith Osiander sees already the beginning of the *justum effici*. God's purview embraces the entire process from beginning to end. If he pronounces man righteous now, it is because he also sees the end of the process and takes it into account in his judging.[8] However, Osiander does not make it finally clear that this process of renewal has its *root* in the divine work of remission. The idea that the divine forgiveness is based, so to speak, on a divine prognosis, on an anticipatory vision of how the whole process will turn out, obscures the fact that forgiveness is an unconditional act which is itself the condition of the total renewal. On the other hand, Melanchthon "was not able to make it clear in his *doctrine* of justification that faith in forgiveness is necessarily accompanied by

[8] Cf. what was said above regarding Karl Holl on p. 77.

the renovating indwelling of Christ. Thus justification and renewal are sundered in him; justification becomes purely forensic." [9] It was because Luther's doctrine had been thus distorted and truncated that Osiander was driven into opposition and finally pushed to the other extreme.

It would, of course, be an over-simplification to say that phenomenologically Osiander did not see the actual incongruity between the divine righteousness imparted to us and the righteousness we in fact display. He can actually portray this incongruity in such eloquent terms that sometimes it is difficult to discern any divergence from Melanchthon at all. On occasion Osiander expressly insists that the sinful will is not uprooted by the implanting of the new see, that while we have the mind of Christ we do not follow it after the flesh. [10] In such a formulation Luther's *simul justus et peccator* would seem to be fully safeguarded.

Nevertheless, for all the similarity of utterance there is a radical difference. For the opposition of the sinful flesh to the divine righteousness does not here imply that God constantly grants us afresh his promise of grace in spite of the persistence of our concrete sins. It implies rather that the righteousness which has been mediated through the indwelling of Christ, and which has become, as it were, our state or condition [*habitus*], is from the very first complete and perfect in itself, but that it draws us only gradually and "progressively" into that completeness and perfection. It is not a matter of man, who is to be characterized in his totality as a sinner in keeping with Luther's view of the "whole man" [*totus homo*], [11] being less than totally and unconditionally justified, as if there persisted a radical contradiction between the *peccator in re* and the *justus in spe*. For Osiander, the problem is rather one of inadequate synchronization between the righteousness which is essentially present and posited in us on the one hand, and our total interpenetration by it on the other.

The imperative, which here too can still be accepted as a postulate

[9] Fritz Blanke, "Osiander," in *Die Religion in Geschichte und Gegenwart* (2nd ed.; Tübingen: Mohr, 1930), IV, 818.

[10] See especially Osiander's *De Unico Mediatore Jesu Christo Et Justificatione Fidei. Confessio Andreae Osiandri* (Regiomonte, 1551, Oct. 24).

[11] Cf. Erdmann Schott, *Fleisch und Geist nach Luthers Lehre unter besonderer Berücksichtigung des Begriffs "totus homo"* (Leipzig, 1928).

of sanctification, thus has no longer the character of a real imperative. At best it consists in the demand to allow the seed implanted within us to develop. Herein lies the essential difference from Luther: the *alienum*—the fact that God's righteousness is something outside of me [*extra me*]—is replaced, or at least decisively restricted, by an immanent substantiality.

Osiander attempts to avert the dangerous proximity of these theses to the Thomistic doctrine of *habitus*. He expressly and repeatedly observes that faith is not to be understood as being a virtue which, as *fides caritate formata*, then effects our righteousness and so constitutes a claim to merit. Rather, faith is our righteousness only to the degree that it includes Christ and implies his dwelling in us. Our righteousness is thereby grounded in an immanent substantiality to be sure, but this immanence is understood, as it were, only in the sense of a statement of location: God's righteousness has entered into us and been made our own inwardly. Although it has thus entered into us, however, it is not a "part" of us. It retains so to speak the character of an alien body, of a seed planted in us which remains distinct from the earth in which it is planted. In contrast to Thomism, that which is produced when we are made righteous (the *effectus* in the *justum effici*) does not consist in meritorious works which we ourselves have done. It lies rather in the righteousness of Christ himself, apprehended by faith, implanted in us, and made our own.

I mention this distinction from Thomism in order that Osiander's view of justification may not appear to be too crude and one-sided. There is always a danger of over-simplification when a theological system is examined from a particular standpoint, as in our investigation of the absolutizing of the indicative in Osiander. Yet we have seen to what degree this absolutizing is actually present. In Osiander, the fact that the justifying God stands over against me, and the external character (the *extra me*) of his righteousness, are both de-emphasized in favor of a quality inherent in man. His concern is with the unfolding of this indicatively posited quality, with the coming into being of what I am, that I let the seed implanted within me develop. Thus the absolutizing of the indicative and the absolutizing of the imperative both lead to the same result, namely, to a quality immanent in man, and to the relative dissolution of the partnership with the divine Thou as the Person confronting man. In Thomism this reversion to man may be seen in the fact that the key position in justification is assigned to meritorious good works. The result is necessarily a particular emphasis on the imperative, which requires these good works of us and which of necessity stirs into action, as it were, the disposition granted to us by grace. This explains the legalistic tendency in Roman Catholic practice. In Osiander, on the other hand, the reversion to man is displayed in the doctrine of the mystical indwelling of Christ which takes place in faith, and in the doctrine of a true and essential quality of righteousness which has been implanted in man as a result of that indwelling. This leads to the absolutizing of the indicative, i.e., to the postulation of a natural unfolding of the seed thus implanted, after the fashion of natural causality.

Thus whenever the imperative is isolated from the indicative or the indicative from the imperative, the absolutizing in either case leads to autonomy of the ego, to its separation from the fellowship between God and man, to its release from all connection with the *alienum*. When the imperative is isolated, or relatively so, man is thrown back on his own resources for the attainment of salvation and the *solus Christus* is robbed of its exclusiveness. The final result in secularism is the self-creation of the ego and the freeing of ethics from theological control. The mark of an existence thus based on the imperative is uncertainty and servile fear.

When, on the other hand, the indicative of justification is isolated, as in Osiander, the result is an autarchical form of the ego, to which a *habitus* is imparted and which is thus referred, not primarily to the historically present Christ, but to itself as the mystical tabernacle of Christ's presence. Faith is thus driven to seek constant reassurance from perfectionistic experiences. The mark of an existence thus essentially determined by the indicative is assurance, and the constant suppression of the actual state of affairs whereby alone this assurance may be attained. Retrospectively, these aberrations show us once again how decisive it is, if evangelical ethics is to have a proper foundation, for us to demonstrate that in the Bible a connection exists between indicative and imperative, and that we should also analyze carefully the precise nature of that connection. The connection can be demonstrated only within the sphere of a biblical anthropology, where the concept of the person is such that man can be described only in terms of his relationship to the God present in Christ.

The Twofold Significance of the Imperative

Our review of the necessary material, though brief, has been sufficient to clarify at decisive points the problems involved in the two motives, fruits on the one hand and the imperative on the other. These problems are identical with those of the new obedience generally, and thus include the key question of evangelical ethics as a whole. For evangelical ethics stands or falls with the way in which it defines the significance of works, of conduct, and hence of existence as such. This is why we must seek to be utterly precise in our definitions, since every inch put behind us on the drawing boards here in our discussion of basic principles will mean many miles saved later when we come to the specifics of ethics and ethical programs.

As I see it, one of the acknowledged difficulties in relating the indicative and the imperative in Paul, and in transcending the mere assertion of a logical antinomy, rests on the fact that the concept of "imperative" has been taken too much as a lump sum instead of being differentiated into essential nuances in its usage.

Exact analysis of what Paul might subsume under the concept "imperative" necessarily shows us that there are two quite different meanings, each involving a distinctive relation to the imparted gift of new existence. In order to point up these diverse meanings we would recall the most essential of our previous conclusions, namely, that good fruits are to be regarded not as products of religious subjectivity but as operations of the Holy Spirit. Here again we see a peculiar feature of biblical anthropology, i.e., that it is impossible to view man as in any sense an isolated cause or independent subject of anything, e.g., good works. The biblical concept "man" is rather always a relational concept. For man always stands in some relation to the Holy Spirit, whether his attitude toward Him be positive or negative. This relational character of the human ego is theologically safeguarded in a decisive way by the fact that Reformation theology defines the Holy Spirit not as a psychical fluid but as a Person, the third Person of the Trinity, thereby pointedly making the Holy Spirit the object of man's attitude, the one to whom man stands in relation. It can be shown with all necessary clarity—and we would show this now—that in Paul the imperatives refer to man's relationship to the Spirit, but in two very different ways.

The Demand to Decide for the Spirit

On the one hand, the task is imposed upon me (I am under the "imperative" obligation) of relating myself to the Spirit. In other words, I have the task of deciding whether "Spirit" [$\pi\nu\epsilon\hat{v}\mu\alpha$] or "flesh" [$\sigma\acute{a}\rho\xi$] will be the power which dominates me. It is "imperatively" required that I decide, in the sense in which Jesus spoke of this decision: "No one can serve two masters . . . You cannot serve God and mammon" (Matt. 6:24; Luke 16:13). The summons to decision has always an imperative significance. That is true, even though the making of this decision takes place under the ultimate proviso that a decision has been made concerning us, and that it is the predestinating action of God which fixes the sphere within which our deciding is possible in the first place. This is why from man's point of view the doctrine of predestination too always has the "appearance" of an imperative. The invitation always involves at the same time a command to accept it. The guests invited to the royal marriage feast have no control over the fact that the king prepares a feast much less over the fact that they are invited to it. But now that the meal has been prepared and they have been invited, the invitation comes to them as an imperative, as a question which demands decision (Matt. 22:1-14).

In this first form, then, the imperative does not imply a requirement that I do "good works" (that would be to fuse the Law incomprehensibly into that automatic process whereby good works take place as a matter of course). Rather, the imperative here refers exclusively to the fact that I am set into a particular relationship to the Spirit, a relationship by virtue of which that automatic process itself is first set in motion.

This first imperative relates, as it were, to the attainment of the right starting point, on the basis of which, once it has been attained, everything else will follow naturally of itself. In this first form, then, the imperative does not involve the intervention of a legal demand thrust into the automatic process as a disruptive alien factor, a factor which in some sense wrecks the operation. On the contrary, the imperative and its fulfillment are rather a presupposition enabling the automatic process to function in the first place. To this extent the opposition between indicative and imperative ceases already to have the character of an antinomy.

It is only in these terms that one can understand the demand of Paul: "You also must consider yourselves dead to sin, and alive to God in Christ Jesus" (Rom. 6:11). The imperative does not refer to the dying. Over this we have no control, since Jesus Christ has died for us and we only receive the gift of his dying and are drawn into it. The object of the imperative is that we should take this death into account, take it seriously, and thus make the gift become a gift in which we participate.

Basically, then, the demand is always that we should let God the Spirit work to bring about the mortification, the transformation. As Luther puts it, God has laid over us the mantle of divine righteousness as an adornment and a robe of honor, but our feet still peep out from under the mantle and the devil takes pleasure in bravely nipping these exposed extremities. However, Luther does not draw the conclusion that we should forcefully go after the monster and in so doing become as active as possible. He argues rather that we must pull our feet in under the mantle and thus allow God himself to protect us.[12] This putting of the feet under the mantle, this readiness to let God the Spirit work for us, and to subordinate ourselves to him, this is the real fulfillment of the imperative.

In other words, the crucial thing—and the whole point of this first form of the imperative—is that we should drink from the right source. If we do this, there need be no concern about the physiological working out of the organic process which follows once we have taken food and drink. This is an automatic biological process which cannot be moralized, which is not under the influence of the will or of imperatives. I can influence the process only insofar as I determine the source from which I drink. And even this is an unguarded statement, as we shall see in a moment. The command relates exclusively to selection of the right source: "put on the Lord Jesus Christ" (Rom. 13:14), "by the Spirit put to death . . ." (Rom. 8:13). Hence there can be no collision between the command and

12 Cf. WA 39^I, 519-521.

the automatic process. They relate to one another in the same way as the imperative, "We should fear and love God . . ." relates to the consecutive clause, ". . . so that we do not steal, commit adultery, etc."

The Demand to Renounce Whatever Hinders the Spirit

Now there is also a second form of the imperative in which the relation to the Spirit is again not just a self-chosen act which we can perform arbitrarily. The demand, "Work out your own salvation with fear and trembling," stands over against, indeed is interwoven into, the further statement, "For it is God who works in you both the willing and the achieving" (Phil. 2:12-13). Hence it is not left to our own initiative, action, or capacity to come to the right source and to open our mouths. This too is a matter of our calling by the Spirit. But the fact that it is God who works, and not we ourselves, requires closer investigation. Why are we "not able" to do it?

Naturally the first answer is that we are victims of the sinful *incurvitas in nos,* and that what we seek is consequently not that Source at all but always ourselves. When we seek God what we find is an idol, which is ultimately a God fashioned according to our own image and representing only our love of self. When we seek eternal felicity, what we attain to is only a sublimated pride in having saved ourselves, such as is expressed in Pharisaism. When we seek self-perfection, what we end up with is only the sublimated self-adoration of the superman.

These things inevitably follow, and this is why for the Reformers grace does not simply "link up" with human nature, not even in the sense of correcting its present condition. On the contrary, there must necessarily be the miracle of a "new creation," a total reconstruction which is not a mere alteration. Man's co-operation is quite impossible, since his works can only express but never overcome the given condition. Works can only intensify but never dissolve the *incurvitas,* as may be seen again in Pharisaism. To this degree the imperative has no place here. For the imperative is a call for action. But action is quite incapable of effecting the transition to the new man. Action simply throws me back upon the old man: "Wretched man that I am! Who will deliver me from this body of death?" (Rom. 7:24).

Nevertheless, in the paraenetic passages of the New Testament we constantly find imperatives which refer to the contesting of this situation of the old ego. How are we to interpret these passages, other than in terms of a dreadful theological inconsistency?

The state of hopeless *incurvitas* is expressed in specific acts or attitudes which, for Paul, are concrete hindrances to the activity of the Spirit and obstacles to his working. There are certain impediments, certain "prohibitory things" [*prohibitiva*], which at times can be specifically named. This is not to suggest that once these obstacles are removed the way to heaven is opened. Theologically, the question cannot be put in this positive form. Thus one may not conclude from the existence of these negative *prohibitiva* that their removal would elicit grounds for positively grasping salvation.

On the contrary, we can only speak of these *prohibitiva* negatively and say that there are certain conditions under which, in principle, the work of the Holy Spirit cannot take place. More precisely, there are certain conditions and attitudes with which the Holy Spirit cannot in any case co-exist under the same roof in the same ego. But the action of man might well have some influence on these special conditions and on their removal, and therefore it could make sense to have an imperative which would demand such action.

What are these *prohibitiva* according to the New Testament? We shall restrict our enumeration of them to those which are decisive.

The first is fornication (πορνεία). "Be sure of this, that no immoral or impure man, or one who is covetous (that is, an idolator) has any inheritance in the kingdom of Christ and of God" (Eph. 5:5). According to this passage, fornication has prohibitory significance because it implies bondage to idols. The primary point at issue is not just the transgression of a specific commandment, as in such "comparatively" casuistical transgressions as stealing or lying. Though the breaking of any commandment involves, to be sure, an offense against the First Commandment, it is particularly obvious in the case of fornication that we are here setting ourselves flagrantly under another lordship. And though no action of ours can free us from this spurious bondage—to avoid fornication is after all not to find emancipation from bondage to the flesh but only to drive the flesh to other forms of expression—we are nevertheless enjoined to avoid its flagrant expression. As Luther might say, in the flagrant expression we would be giving strong consent.

At this point the question arises why this worship of false gods, this subjection to alien rule should come to expression particularly in fornication. Paul discusses this ethical distinctiveness of fornication in I Corinthians 6:9 ff.: "Do you not know that the unrighteous will not inherit the kingdom of God? (which means that the kingdom cannot come to him who by his consent to sin in effect protests against the kingdom of God) . . . neither the immoral, nor idolaters, nor adulterers, nor homosexuals will inherit the kingdom of God."

Here we are told clearly that certain things prohibit the renovating activity of the kingdom of God. Their prohibitory character certainly does not consist in the fact that the sinful acts as such exclude me in principle from salvation. For I fall victim to them even in the state of justification. Moreover Paul tells us expressly that "all things are lawful for me" (I Cor. 6:12). He thus indicates that in faith we are free from the Law, that in love and by virtue of the authority inhering in the obedience of faith we may follow paths that are contrary to the letter of the Law.[13]

On this basis it would in principle be an open question whether there might not be freedom for adultery or for certain sexual actions if they were done out of love. As it is possible to steal out of love— we need only think of the things that happened in Germany during the lean years just after the Second World War—so one might ask whether there could not be cases of sexual surrender which take place sacrificially and by means of which a man might possibly be saved, cases which to all external appearances seem to have the form of adultery. As a "question," at any rate, this possibility is not to be quashed immediately. I will not attempt here to construct or recount instances in which this question might become concrete. People responsible for the care of souls meet such cases constantly, and the theme has not infrequently been treated in literature.[14]

As we have said, we will refrain from giving specific examples and simply assert the possibility that that which concretely diverges from the letter of the Law may *not* necessarily have of itself the prohibitory significance of which we are speaking. Otherwise there could be no place for the *simul* (*justus et peccator*) in Christian life. Prohibitory

[13] Cf. the freedom of Jesus in relation to the commandment regarding the Sabbath (Mark 2:23-28; 3:1-6).

[14] A well-known example is that of Sonia in Dostoevski's *Crime and Punishment,* who out of love becomes a prostitute. See below pp. 635 ff.

significance arises rather out of something quite specific in connection with these acts, namely, the idolatrous bondage into which they betray me. Paul alludes to this in the same verse where he speaks of all things being lawful when he says that I am free only for that which does not rob me of my freedom, i.e., subject me to a new and anti-godly power (ἐξουσία, I Cor. 6:12).

This bondage to an alien and anti-godly power can express itself in *all* the different forms of passion and indulgence, even in such harmless things as smoking or dependence on a hot-water bottle. Both these come under the category of freedom and therefore of the lawful (ἔξεστιν). But both may imply bondage to a way of life, e.g., that of middle-class comfort, which I am not ready to give up for the sake of the Gospel and obedience, and which I thus prefer to God. If asceticism has any valid point at all, it can only be the fact that it enables me to sit loose to life and thereby makes me ready for new ventures in obedience. Even such indisputably worthwhile things as "goods and kindred" can assume the role of the enslaver (ἐξουσιάζειν). In every way of life there lurks the possibility of its becoming a *prohibitivum*. This is true even of those which are "most moral," as may be seen for example in the case of the rich young ruler (Mark 10:17 ff.). He had observed all the commandments from his youth, but was still brought to ruin by wealth and a high standard of living.

Now Paul maintains that in certain instances the "possibility" of becoming a *prohibitivum* is in fact a sheer "necessity" because in the nature of the case the thing is such that in principle and, so to speak, by a kind of natural causality it impedes the coming of God's kingdom. Such necessity obtains, for example, in the case of fornication, which always places under idolatrous bondage and hence cannot in principle belong to the sphere where "all things are lawful." Here there is no flexibility, just a hard either-or. This is brought out clearly in the ensuing verses (I Cor. 6:16-17): "Do you not know that he who joins himself to a prostitute becomes one body (ἐν σῶμα) with her? . . . But he who is united to the Lord becomes one spirit (ἐν πνεῦμα) with him." Accordingly there is in this situation no *simul:* Either we are related to Christ in the relationship of spirit, or we are related to the harlot in the relationship of flesh. Here then is a case of a *prohibitivum* which is so by necessity.

Paul goes on to state why fornication is such a special sin: "Every other sin which a man commits is outside the body; but the immoral man sins against his own body" (v. 18). Other sins can be, as it were, "outside" me. In relation to them it may be true that "I live—yet not 'I.' " In other words, the wrong I do as a Christian, in that state in which I am characterized by the "Christ in me," is in some sense outside my Christ-ego, so that when I do it I am in a sense not myself. As Luther puts it, my sin as a Christian is the remnant of the serpent skin which I have long since sloughed off. Since sin no longer "possesses" me, I "am" no longer identical with sin. Sin is an "it," something which still clings to me and continues to work, but is no longer "I." For "I" am in Christ, not in sin. "Who shall bring any charge against God's elect? God is here who justifies. Who is to condemn? Christ is here" (Rom. 8:33-34). This means that when I am summoned to judgment and charged with specific sins, Christ answers in my stead, as it were, "Here!" He stands in my place. I am no longer "I." My identity with myself is broken.

Now this is precisely the possibility which no longer exists as long as I am in bondage to fornication. For fornication always means that I am subjected to an alien power. This is one sin which I cannot keep at a distance by saying it takes place "outside" me. On the contrary, when I commit it, I am right there, fully involved.

The so-called "glass of water theory" of sex is out of the question. I simply cannot say: It is not I but merely the reaction of my glands and hormones; it is just something I allow to happen to my body, as if *I* were not this *body*. But I *am* this body! I am personally involved as a participant. Hence I am qualitatively changed by it. This is why prostitution, more than any other sin, leads a person to the very limits of his humanity. Hence it is characteristic that Paul should here use the term "body" [σῶμα] rather than "flesh" [σάρξ]. Σῶμα aways refers to me, the ego, insofar as "I" exist physically. Σάρξ on the other hand is something to which I am no longer tied, inasmuch as I need not necessarily identify myself with it. We also recall the distinction between living "in the flesh" [ἐν σαρκί] and "according to the flesh" [κατὰ σάρκα].

In saying all this we are referring of course not to individual lapses but to fornication as a way of life to which those who practice it give their consent, and which is thus in their thinking legitimate (whether in terms of the libertarian "glass of water" theory or of some perverted concept of freedom which may even be called "evangelical"), or as something to which men give themselves against their better judgment, thus according it not theoretical but *de facto* assent. We mention this in order to avoid the impression that every sexual offence is a particularly aggravated case of the sinful act which separates us from God and necessarily excludes us from forgiveness. What we are saying is simply that there can be no *simul* between claiming forgiveness on the one hand, and willfully and deliberately engaging in fornication on the other. Fornication is indeed a most important example—but only an example!—of the prohibitory character of sinful attachments.

To sum up, we maintain that fornication is a *prohibitivum* of the first rank because it implies, not the mere possibility, but the necessity of my being enslaved by an alien power. It occupies this relatively unique position because it can never take place "outside" me, and "I" consequently cannot isolate myself from it. It obviously cannot be compared to Luther's "serpent skin," nor can it possibly be included in the sphere of what is "lawful" (ἔξεστιν). Hence it cannot be linked by any *simul* with the kingdom of God, the lordship of the Spirit. The two are mutually exclusive.

A second *prohibitivum* is that of "sitting at the table of demons." In this connection too there characteristically arises the question of what is lawful (I Cor. 10:23), the question whether we have here a sphere of freedom, an adiaphoron. Is there here merely the possibility or is there rather the unavoidable necessity of a *prohibitivum?* Here again Paul decides that what is involved is a case of necessity.

He discusses the point in the context of the Lord's Supper and of the eating of pagan sacrifices: "You cannot drink the cup of the Lord and the cup of demons. You cannot partake of the table of the Lord and the table of demons" (I Cor. 10:21). In his rendering Luther interpolates the word *zugleich* [at the same time] which, while it is not in the Greek original, is wholly in line with the sense of the text. The *simul justus et peccator* cannot refer to this twofold attachment to Christ on the one hand and demons on the other. For this reason the eating of meat sacrificed to idols—if one believes in these idols— is a real *prohibitivum*. II Corinthians 6:14-16 represents another instance where the *simul* is excluded because of the clear-cut either-or.

In Titus 1:16 there is a similar reference to obstructive attitudes, particularly deeds or attachments which make the knowledge of God impossible. Moreover, there are forms of action which do not merely hamper the growth of faith but actually destroy faith. Denial of the duties we owe to the community of believers is an example (cf. I Cor. 5:5-7).

The third great group of *prohibitiva* is sharply defined; it consists in "denial of the Lord." If I deny him, he and the angels will deny me at the last day (Matt. 10:33; Luke 12:9; II Tim. 2:12). Faith and the denial of faith are incompatible. No *simul* can embrace them both.

If it is true that there are certain works or attitudes which necessarily exclude and prohibit faith, the Spirit, and the kingdom, then it is also true that *any* action, however moral or faithful to the Law, can take on this prohibitory significance. If we keep this fact clearly in view, what we have said thus far cannot lead to casuistry. The prohibitory significance of works also explains why, despite the fact that the sinner is justified, there is still a judgment according to works.[15]

We may summarize the nature of these *prohibitiva* as follows. In regard to the first meaning of the imperative we said that the main thing is to drink from the right source, and that we are summoned to turn to the Spirit because once the right basis has been established the automatic process will then be set in motion. So in regard to the *prohibitiva*, we may now say: Whoever keeps his mouth closed is unable to drink from this right source. This happens—we keep our mouths closed—when we disqualify ourselves from faith by our fornication, idol-feasts, denials, or other impedimenta.

To state the matter even more pointedly, we may say that the New Testament knows no particular disposition *for* faith. All are called, even harlots and publicans, so long as they are poor in spirit and take no pleasure in the blossoming of their sins, so long as they do not assent to sin. There is, however, a specific disposition *against*

---

[15] Cf. the Formula of Concord, Solid Declaration, III, 41: "[It is impossible that] on occasion true faith could coexist and survive . . . with a wicked intention." *BC* 546. As surely as meritorious works do not have to precede faith in order to bring the believer into a right "disposition," so surely do faith and wicked intention mutually exclude one another—and so surely does wicked intention constitute a *prohibitivum*.

faith, a disposition with which the new man, the new creation, can in no case coexist (cf. I Pet. 3:7; John 7:17; 18:37b).

We are now able to understand the significance of this second form of the imperative which carries with it the demand that we should fight against these *prohibitiva*. As we have said, the result of a successful conflict here cannot be that actual cause is given for the emergence of the new man, but only that the indispensable precondition for that emergence is not sabotaged, the condition without which the new man remains embryonic.

Here the relation between indicative and imperative is at once clarified. At any rate, it loses the character of direct antinomy. The purpose of the imperative is not to intrude upon the automatic process and so declare that justification of itself is incapable of producing the "new creation." On the contrary, the imperative is rather a demand that we should attain to that starting point where the automatic process goes into operation. And this is done in two ways, both of which involve an actual demand. First, the imperative requires us to drink from the right source, and second, it commands us not to keep our mouths shut (through embracing the *prohibitiva*).

7.

Law and Gospel as Constant Partners

We are now in a position, by way of systematic review and development of previous insights, to bring out the significance which the Law of God continues to have for the justified, and therefore for those who stand under the Gospel.[1] For the sake of completeness we shall first summarize briefly the problem of indicative and imperative.

Within its significance as an imperative the Law has two functions which must be differentiated. First, the imperative has the function of calling for our decision with respect to the Gospel. It makes the gift [*Gabe*] into a task [*Aufgabe*], the invitation [*Angebot*] into a command [*Gebot*]. For one may refuse a gift or invitation. We expressed this first significance of the command in the figure that we must drink from the right source. Second, the imperative relates to the renunciation of *prohibitiva*. Though it cannot of itself give us a disposition for faith, it does draw our attention to certain dispositions against faith.

The Distinction between Law and Gospel[2]

In the state of justification the Law has to bear witness to our existence as *de facto* sinners [*peccatores in re*]. By keeping this fact before us, the Law protects us from the illusion that we are "perfect Christians" who no longer live at the point where we must continually begin again [*incipere*], needing the forgiveness of Jesus Christ each day afresh.

In opposition to such an illusory conception we point to our crying out of the depths, our walking through the valley of the shadow of death, the cry: "Lord, I believe; help my unbelief" (Mark 9:24),

---

[1] On the *usus legis in renatis* see John Calvin *Institutes of the Christian Religion* 2. 7. 4, and Wilfried Joest, *Gesetz und Freiheit* (2nd ed.; Göttingen, 1956).

[2] The remainder of this chapter was first published in *Festschrift zum 80. Geburtstag von Th. Wurm* (Stuttgart, 1948) and in my *Theologie der Anfechtung* (Tübingen, 1949).

the unresolved tension between the *justus* and the *peccator*, the constant assaults of doubt and temptation in face of which we must believe, the permanent presence of the old Adam who must be put to death anew each day. And we insist that all this is and remains the mode of our existence as Christians so long as we walk by faith and not by sight. There is no healing of this wound, unless we are to fall victim to the most seductive and dangerous illusions concerning ourselves, unless the miracle of grace is to become something routine and automatic, and supernature "second nature" to us. Hence the Law has the function of being "gauze in the wound," to prevent it from false healing.

We have now reached the point where we may collate the insights which have led Lutheran theology to lay decisive stress on the strict differentiation of Law and Gospel, indeed to make the radical distinction between them the criterion by which to test the legitimacy of a theology.[3]

First, it is a mistake to understand the Law as Gospel, for this is to confuse the two spheres of meaning rather than keeping them distinct. This is what would happen if we were to see in the very fact of the Law's existence a guarantee that it can also be fulfilled—in the sense of Kant's axiom, "You can, because you ought"—and were to regard the keeping of the Law as identical with the winning back of fellowship with God. The result in such a case would be false security (*securitas*). Failure to measure up would be readily admitted, but it would be regarded simply as a quantitative failure to reach the norm, not as a qualitative expression of the fact that fellowship with God is broken. But as regards fellowship with God, the real question is not the quantitative one of the different degrees to which it may exist, and the fervency of it, but the purely qualitative one: Does the fellowship exist or does it not?

If it is believed, however, that the Law of God can be fulfilled, and peace with God thereby attained—if, in other words, the approach is quantitative, the *securitas* which inevitably results can only understand the Gospel in a sense which is restricted in two possible ways.

---

[3] It is not by accident that this problem of Law and Gospel is the point where Karl Barth first comes into conflict with Luther's basic theological presuppositions. See Karl Barth, *Community, State, and Church: Three Essays* (Garden City, New York: Doubleday, 1960), and Hermann Diem, *Karl Barths Kritik am deutschen Luthertum* (Zollikon-Zürich, 1947), esp. pp. 32 ff.

Either the Gospel is thought to be a form of divine assistance granted to empower us the better to fulfill the Law (as in Wilhelm Herrmann), or the Gospel is regarded as an authority which grants us forgiveness for that wherein we have failed to meet the norm. This *securitas* implies a corresponding depreciation of the Gospel. Theologically, the majesty and pre-eminence of the Gospel can be assured only if it is not confused with the Law.

The same is true also if the opposite course is taken and the Gospel is for its part understood as Law. This is what would happen, for example, if the Gospel were thought to require an "imitation of Christ," and thus to be "good news" only insofar as I fulfill this *lex evangelica*. In this case the Gospel would again cease to be an expression of the free and unconditional grace of God, since it would be linked with certain presuppositions, with certain acts of obedience and sanctification. It would no longer protect me against that torturing and enslaving inversion of outlook which must arise the moment I have to go on taking stock of myself and at the same time running my life on the basis of whether I have fulfilled the required conditions, whether I have met the demands, and whether I may consequently (!) apply to myself the promise of the Gospel. Here it can no longer be a matter of "grace alone," but only of explicit "co-operation" and "conditions."

If the confusion of Law and Gospel first mentioned leads to false security (*securitas*) this second form leads to assaults of doubt [*Anfechtung*], to uncertainty of faith, and to a stress on mastery of self. In place of the glad and sure acceptance of a filial relationship which is granted to me *sola gratia*, the best that this second form can offer me is a sublimated form of servile fear which forces me constantly to produce evidence that I am—not a son but—a good servant who may remind God of the reciprocal obligations of the contract and of His covenant faithfulness. The profoundest mystery of justification, namely, that God protects me against myself and the accusation of my own heart and conscience, that he is *Deus defensor* against the *cor accusator*, is quite inexplicable along these lines. Instead of receiving the comfort and assurance afforded me in the miracle of divine mercy, I am condemned to permanent assaults of doubt.

Second, only a strict distinction between Law and Gospel can produce the kind of theological reflection in which both holiness and love are appropriately expressed and extolled in thought as essential and determinative aspects of the divine majesty. It is vital that these two aspects should be strictly maintained theologically if love is to be seen for what it really is, namely, a miracle, something which cannot possibly be postulated, the deed which God performs in the miracle of the incarnation.

If this tension between holiness and love, judgment and grace, is not adequately expressed in theological reflection, indeed, if it is weakened even in the slightest degree, love will appear to be *the* nature of God. But this quickly gives rise to an optimistic Christian world view in which the supreme value is "the good Lord," the divine goodness. From here it is only a step to the cynical observation of Heinrich Heine that it is the function of God—"that's his business!"— to forgive, to be good to us, and hence to let us enjoy with relative unconcern the blossoming of our sins.

If we take from the concept of God this dualistic tendency which sustains it, the result is a pseudo-Christian monism. It is of the essence of monism of every kind that sin—and judgment too—at once acquires a teleological reference. Sin and judgment "make sense" only to the extent that they may be linked with a happy ending or even directly reduced to a common denominator with goodness and with grace. This is why Hegel's monistic philosophy of spirit, in which evil is only transitional, is deeply unchristian. This is also why the Lutheran confessions reject the doctrine of apocatastasis, which teaches a monistic consummation and so plays down the character of existence as a situation calling for decision.

It could also be said that what is at issue here is the problem of the historicity of revelation. When we said a moment ago that all monistic tendencies, such as the inclination to a one-sided proclamation of the love of God, really lead to a pseudo-Christian "world view," we were expressing already the implied threat to this historicity. For a world view has as its content certain truths which claim to be of timeless value. To put it even more precisely, these truths possess only such existence as accrues to them by "valuation." What is involved is always a timeless and static idea, such as that of "the good Lord." The love of God, on the contrary, as that love is proclaimed and enacted in the salvation history [*Heilsgeschichte*] set forth in the Bible, is a miracle—a miracle that conceivably may not have happened but did as a matter of fact actually take place. The love of God is not something men "value," which exists because and insofar as it is "valued." It is something which simply "is," by virtue of an almighty and authoritative act. In his concept of God Luther makes this point through constant reference to the fact of conflict in God, the fact that God had to wring from himself the resolve to love, that it cost him something in the sacrifice of his Son. What is thus felt on the Barthian side to be a threat to the unity of God in consequence of this tension does not rest on mere speculation. It rests rather on Luther's express soteriological concern, namely, to consider, to praise, and to express theologically the *miracle* of divine love, its utter inexplicability, the fact that while it may not have happened it did as a matter of fact take place.

For Luther, this tension in God expresses itself also on the level of faith. Luther constantly brings the act of faith itself into explicit tension by his refusal ever to regard faith as a kind of *habitus,* an allotted attribute of existence. Luther understands faith rather as movement, as a flight from God the judge, the *Deus accusator,* to the God who is gracious to me in Christ, a flight from the hidden God [*Deus absconditus*] to the revealed God [*Deus revelatus*].

Just as justification is present only in act, and therefore only as a constantly ongoing justification (in face of the permanent *peccator in re*), and just as I am continually pulled beyond the range of the magnetic field of Adam which encompasses me, so faith is constantly referred to that in face of which we must believe, namely, the assaults of doubt, the possibility of despair, the threat posed by the power of sin poised on the horizon of my existence. Only when this tension remains, i.e., only when justification as an ongoing act is not done away by justification as a given and perfect state or *habitus,* only then does Christ remain the mediator. And only then do we remain dependent upon and oriented to the historicity of the miracle which happened then, the incarnation, and the miracle which happens now, its appropriation by and application to me.

Barth on Gospel and Law

It is thus understandable that the dissolution of this tension which undergirds our faith should necessarily lead to unhistorical thinking.

It is understandable that where a theological system contains—however covertly—certain tendencies toward a lack of historicity, they will press for the removal of this tension, and thereby betray themselves.

This seems to us to be the case in Barth's theology, where Gospel and Law are regarded as two sides of the same thing. When they are thus viewed as correlatives, the tension between them is eliminated. In the words of Barth's famous definition, "Thus, we can certainly make the general and comprehensive statement that the Law is nothing else than the necessary form of the Gospel, whose content is grace." [4] As the form of the Gospel, the Law thus conceals the Gospel. The Law does not stand over against the Gospel as something which accuses and destroys us, before which we cannot stand. On the contrary, when we reach the point of confessing our failure, we are already raised up by that which the Law is really intended to be, namely, the promise that "you shall be" what is demanded.[5] Hence judgment and destruction are no longer total. Carried to its extreme by Barth, this means, "The very fact that God speaks to us, that, under all circumstances, is, in itself grace." [6] But is this really so? Is it really grace when Adam is asked, "Where are you?" Out of the dreadful judgment, devoid of grace, which falls on Adam, salvation comes through the grace which God wrings from himself, painfully, as demonstrated at Golgotha. But is not the miracle precisely this? And is this not quite different from Barth's correspondence of form and content in terms of conformity to Law? Otherwise, was not Adam's hiding a case of blind panic? In other words, cannot God's speech to man be something terrible, something that plunges us into a hopeless impasse? Who in that case may speak of "grace" without doing violence to God's judgments, and without taking from grace the very thing which makes it grace, namely, the element of miracle, the inexplicable, that which conforms to no law? Who would dare to call judgment "grace," and understand God's speech in every case as grace? "If you were blind, you would have no guilt" (John 9:41); if you were deaf, you would likewise have no guilt. But now that you must in fact see, now that you must in fact hear the Law, you are without excuse and the judgments of God break upon you. What

[4] See Karl Barth, *op. cit.*, p. 80.
[5] *Ibid.*, pp. 78, 82.
[6] *Ibid.*, p. 72.

else is this but an indication that when our eyes encounter only night, and our ears only silence, this could be because God is graciously sparing us? How then can we ever describe God's speech "in itself" as grace? We can do so only if we weaken the antithesis between Law and Gospel, for judgment then loses its dark aspect and whispers that we will have a happy ending.

Such a doctrine of Law and Gospel introduces an abstract monism into theology and makes it into a philosophical world view of grace: "The Word of God [is] . . . properly and ultimately grace: free, sovereign grace, God's grace, which therefore can also mean being Law, which also means judgment, death, and hell, but grace and nothing else."[7] Even hell has its place in this philosophy of grace. When will the logical consequence of this monism be drawn, namely, the forthright confession and solemn proclamation of the universal consummation of all things, the ἀποκατάστασις πάντων?

If what we have said is true, namely, that a monistic weakening of the antithesis between Law and Gospel leads to the timeless and static "idea" (the idea of love, the idea of grace, etc.), then it should be possible to trace such consequences in Barth's theology since all the monistic presuppositions are there present.

In fact, it has long since been pointed out, and with some frequency, that in Barth's theology, with its tendency towards abstract timelessness, there is no movement of saving history from the Old Testament to the New. The two Testaments are rather related after the manner of two concentric circles having their foci at a single center.[8] This timeless interrelating of the Testaments, which has been most profoundly and comprehensively criticized by Oscar Cullmann,[9] is even plainer in Barth's followers than in Barth himself. This is especially true of Wilhelm Vischer[10] and Hans Hellbardt.[11]

The understanding of the Old Testament within the Barthian school reveals particularly clearly the timeless and non-historical character of its exegesis. Vischer and Hellbardt are concerned with cer-

[7] *Ibid.*

[8] See Karl Barth, *Church Dogmatics,* ed. G. W. Bromiley and T. F. Torrance (Edinburgh: T. & T. Clark, 1936——), I², 70 ff.

[9] See Oscar Cullmann, *Christ and Time,* trans. Floyd V. Filson (Philadelphia: Westminster, 1950); revised edition 1964.

[10] See Wilhelm Vischer, *The Witness of the Old Testament to Christ,* trans. A. B. Crabtree (London: Lutterworth, 1949).

[11] See Hans Hellbardt, "Abrahams Lüge," *Theologische Existenz heute,* No. 42.

tain theological presuppositions of exegesis which we shall now briefly review. In so doing we must constantly keep in mind the central purpose of this review, namely, to show that this kind of timeless understanding is closely linked with a weakening of the antithesis between Law and Gospel, or, as we have also shown, with the negation of historicity in the name of a pseudotheological monism.

The Old Testament, for the Barthian school, is the church's source of revelation. It does not have its origin in the history of religion, though ("according to the flesh") that is where it is located. Because its place is there, it is susceptible to confusion with others, and so partakes of interchangeability, the form of a servant. As source of revelation the Old Testament, it is held, can be understood only when the revelation in Jesus Christ—for there is no other—is made the interpretative center of the Old Testament. When the matter is thus expressed in such a general way, it may appear that what is involved is simply a familiar theological thesis, comparable to that involved in an Old Testament exegesis conceived in terms of "prophecy and fulfillment." Yet careful note should be taken of the special nuance that the reference here is to the "christological center" of the Old Testament.

Before considering this nuance, however, we must indicate another concern in Vischer's exegesis. He seems to suggest that the disastrous gap between scholarly study and preaching can be closed only through a strictly christocentric understanding of the Old Testament. At any rate the dualism between the professor's podium and the preacher's pulpit is done away the moment the sermon's message of Christ is asserted to be at the same time the kerygma exegetically derived from the Old Testament.

This movement towards the reconciling of research and proclamation is attested externally by the fact that Vischer repeatedly and with extraordinary consistency begins with Reformation and Patristic exegesis, and in so doing does not hesitate even to adopt some of their allegorizings. The very fact that he thus ignores the time interval between the fathers and the present, and all that it has meant for the development of historical studies, is itself indicative of non-historical thinking.

There is no getting away from the fact, however, that exegesis of this sort constitutes a summons to get to the point theologically.

That point is the revelation in Jesus Christ, in other words, a historical fact. The attempt is to lay bare this historical fact by liberating it from the relativity of historicism and bringing it into direct and absolute relation to "my" existence. This relation is for its part viewed quite historically. It refers to the history of Christ "with me." I am the one who is dealt with. It is a matter of Christ's presence pure and simple, as that presence is expressed in the affinity of Christ to my existence ("here and now"), and as it is thus necessarily expressed in the corresponding affinity of all times—including Old Testament pre-Christian times—to Christ. What Kierkegaard calls the "transparency of the intervening time" between Christ and me, and what Bultmann calls the "dialogue with history," [12] is thus in the first instance historically intended, i.e., as the immediacy of a historical datum to me. The only question is whether the time continuum, within which the historical entity of Christ occurs, can be destroyed in favor of a historical continuity with "me." Is not this to jeopardize—precisely in the interests of a basically historical concern!—historicity as such?

That this is in fact the case is seen the moment we consider the relation between the Old and New Testaments implicit in this particular theological position. As we examine that relation we shall give special regard to how the time continuum, within which the Testaments arise in historical succession, is preserved. We would make four main points.

First, the dialectical school lays strong emphasis on the eschatological character of the New Testament. The manifestation of Christ is said to have fulfilled the promise, not in the sense of ending it and replacing it by the thing promised, but in the sense of making the promise full, complete, unambiguous, and confirmed, so that all blessedness is now a blessedness in hope, and all faith is Advent-faith, a waiting upon the revelation yet to come. The consequence of this radically eschatological interpretation of the New Testament is that the two Testaments are juxtaposed under the aspect of a common expectation of salvation. The apostles are aligned with the patriarchs and prophets on the common ground of their waiting. The former have not seen what the latter hoped for (as in Matt. 13:16-17); they all hoped alike. There

[12] Rudolph Bultmann, *Jesus and the Word*, trans. L. P. Smith and Erminie Huntress (New York: Scribner's, 1934), p. 12.

is no qualitative difference between the two Testaments, only a distinction in the manner in which salvation is dispensed and offered, namely, the distinction taken over from Calvin between the "mode of administration" and the "substance" of salvation.[13]

With respect to Barth's formulations,[14] it must be asked whether and in what way the message of Christ brings any redemption from the Law of Moses. For the antithesis between them, that is to say, the antithesis between Law and Gospel (though in our sense too it would be a mistake simply to equate the Old Testament with the Law) is completely effaced. In the New Testament the promise of redemption from the bondage of the Law is precisely not fulfilled; it remains a promise, and is simply confirmed. The story of Christ is nothing but a renewal, a particularly impressive demonstration and confirmation, of an already existing promise, a promise which as such had been linked from the very outset with the Law, thereby subduing the judgment character of the Law.[15]

According to Barth, the Gospel (in its quality as fulfillment) does not take the place of the Law by liberating us from the Law in its character as a curse. Instead the Gospel reveals itself as the content of the Law by confirming the promissory character of the Law. But this involves, both theoretically and practically, abolition of the temporal succession which is expressed in the once-for-all nature, the ἐφάπαξ, of the manifestation of Christ. Without this temporal succession, the ἐφάπαξ is eliminated in favor of a Docetic presence of Christ at every stage of salvation history, and in favor of a purely symbolic understanding of the historical incarnation in terms of its "demonstrative character." Christmas thus becomes a mere Advent, perhaps a particularly radiant Advent—with five candles!—but not a real Christmas involving a decisive juncture of the aeons.

---

[13] Calvin *Institutes* 2. 10. 2: "The covenant of all the fathers is so far from differing substantially from ours, that it is the very same; it only varies in the administration." John Calvin, *Institutes of the Christian Religion*, trans. John Allen (Philadelphia: Presbyterian Board of Publication, 1909), p. 386. This passage is directly quoted by Vischer, *op. cit.*, pp. 20-21.

[14] Barth, *Church Dogmatics*, I², 81 ff.

[15] Barth cites some quotations from Luther (see *ibid.*, I², 76-78) in order to prove that Luther held a similar view of the Old Testament, and that in Luther's thinking the New Testament simply brings another mode of the event of salvation. But these passages must be seen against the background of Luther's doctrine of Law and Gospel. By divorcing them from this background, Barth takes them out of their context and gives them an artificial and misleading emphasis.

Second, to state the matter sharply, there is thus an implied identity of the two Testaments. In any case there is no significant movement in salvation history. Even the schema of "prophecy and fulfillment" implicit in salvation history can no longer be applied, because in the first place it rests on a fundamental succession, and because in the second place it contains the idea of "fulfillment" whereas in this extreme "eschatologism" the most that can be said is that the "promise" comes into "full force."

What is at stake in this view is not simply the prophecy of the Messiah, not simply the covenant God of the Old Testament who is indeed the Father of Jesus Christ, but the immediate presence of Jesus Christ in the Old Testament itself. Wilhelm Vischer is still a bit restrained in this respect. He can say, for example, that the Old Testament only knows "what" Christ is, whereas the New Testament knows "who" he is. But Hans Hellbardt is more wild and radical. He unguardedly draws the logical consequences of Vischer's position and describes the angel of the Lord in the Old Testament as Christ, the manna as Christ's body and blood, the faith of Abraham as faith in Christ, etc.[16] Jesus Christ is already "there." His incarnation loses its ἐφάπαξ character. It is, as it were, only another way in which his previously existent activity continues to manifest itself. In other words, it is simply another "mode." (This is what Calvin's concept of "administration" suddenly looks like after a few more stages of development.)

With reference to our present problem of timelessness and non-historicity, we may thus say that the time continuum within which the year 1 A.D. occurs at a specific point is obliterated. "Theologically, it is of secondary importance that in point of time Jesus came into the world after the Old Testament."[17] The "concept of time" is "eliminated."[18] In place of the physical entry of Jesus Christ into history ("the Word became flesh") we thus have a historical Docetism, a ghostly incorporeality. The time continuum within which the concrete points of salvation history may be dated ceases to exist and in its place there is set a circle which at all points is equidistant from its christological center.

[16] See Hellbardt, op. cit.
[17] This is exactly how Hellbardt puts it in "Die Auslegung des Alten Testaments als theologische Disziplin," Theologische Blätter, 7/8 (1937), p. 141.
[18] Ibid., p. 142.

Third, the immediate presence of Jesus Christ in the Old Testament means necessarily that he must be found in every verse. It is no longer a question of a historical line which leads to him and which may, of course, be determined even retrospectively by him as the decisive geometric point. On the contrary, it is a question of the Old Testament being totally "filled" by Christ.

Hence Vischer believes, quite logically, that those who distinguish in the Old Testament between sub-Christian elements and elements confirmed by Christ have surrendered the Christian confession that Jesus is "the Lord" of the Old Testament.[19] Hans Hellbardt makes the similar statement that "within this revelation one cannot introduce the category of the 'more or less.'"[20] The hermeneutical principle necessarily involves squeezing the text until it proves christologically fruitful. What is sought in the text is not what it has to say to its historical situation, its own particular point on the time line. No attempt is made to distinguish the individual sources, e.g., J, E, and P in the Pentateuch, and to discern their theological significance. Instead they are all seen simply in their timeless relation to the timeless presence of Christ. Says Walther Eichrodt, in critically characterizing timeless-allegorical exegesis of the Vischer-Hellbardt variety: "They think it is not necessary first to ask what the text is seeking to say in the time of its composition and in the situation of the authors [the time line!]. Instead they already know what the text must have in view, and so can relate it to its point of reference as a similitude which has to be treated in terms of allegory and unrestricted typology."[21]

Fourth, if then the historical setting of the individual sources is eliminated and not interpreted, and if salvation history thus becomes merely the historical, allegorical illustration of a timeless idea, then the rift between the church's message and historical scholarship clearly becomes all the greater. In fact, it becomes intolerable. For Old Testament scholarship is then surrendered indeed to complete secularism, no matter how fervently and laboriously a person may try

---

[19] Vischer, *op. cit.*, p. 26.
[20] Hellbardt, "Die Auslegung . . . ," p. 136.
[21] Walther Eichrodt, "Zur Frage der theologischen Exegese des Alten Testaments," *Theologische Blätter* (1938), p. 77.

to hide the fact or smooth it over, as Emanuel Hirsch has tried to do in his brilliant and honest book.[22]

Following this criticism of the Barthian concept of history we would pause for a moment to recall the general context of our discussion. We were trying to show that the tension between Law and Gospel which Luther emphasized so passionately is the only guarantee that in theological reflection the historicity of revelation will be preserved. If this tension were to be done away, theology would be taken over by a timeless, monistic principle such as the idea of love or the idea of divine goodness. Instead of being an exegesis of salvation history theology would be changed into an optimistic-Christian philosophy or world view. Since in Barth the tension has in fact been done away, it was important for us to observe whether and how far this theoretically postulated tendency towards timelessness is actually to be found in him and in his followers.

Now that we have seen clearly the emancipation from history which has taken place in the area of Old Testament exegesis, we can put the same question with respect to the New Testament. Here too the tendency towards timelessness manifests itself with equal sharpness.

It is expressed in Barth's thesis that the end of the world does not take place in time, that the Parousia is not historical. In his own words, "Resurrection means *eternity*. . . . The moment when the last trump is sounded, when the dead shall be raised and the living shall be changed, is not the last moment of time, but is time's τέλος, its non-temporal limit and end. It comes ἐν ἀτόμῳ, says Paul, in an indivisible, non-temporal, eternal now. Is it yesterday, tomorrow, today? Is it ever? Is it never? In each case we may answer Yes and No. . . . The resurrection of Christ, or his second coming, which is the same thing, is not a historical event; the historians may reassure themselves —unless, of course, they prefer to let it destroy their assurance—that our concern *here* is with an event which, though it is the only real happening *in* is not a real happening *of* history." [23]

"Do not our ears burn when we hear this? Will there never be an end to all our ceaseless talk about the 'delay' of the Parousia? How

[22] Emanuel Hirsch, *Das alte Testament und die Predigt des Evangeliums* (Tübingen, 1936).

[23] Karl Barth, "Biblical Questions, Insights, and Vistas," *The Word of God and the Word of Man,* trans. Douglas Horton (Boston: Pilgrim, 1928), pp. 89-90.

can the coming of that which doth not 'enter in' ever be 'delayed'? The End of which the New Testament speaks is no temporal event, no legendary 'destruction' of the world; it has nothing to do with any historical, or 'telluric,' or cosmic catastrophe. The end of which the New Testament speaks is really the End; so utterly the End, that in the measuring of nearness or distance our nineteen hundred years are not merely of little, but of no importance. . . . Knowing that here every word can be 'only' a parable, . . . [who] shall be able to set us at rest with this 'only'? Who shall persuade us to transform our expectation of the End . . . into the expectation of a coarse and brutal spectacle? . . . What 'delays' its coming is not the Parousia, but our awakening." [24] One notes the same lack of historicity in Barth's exegesis when he treats of the resurrection of Christ.[25] All this seems to underscore our assertion that the monistic tendency in Barth's theology, which finds expression in his obliteration of the antithesis between Law and Gospel, leads to timelessness, the elimination of salvation history, and hence a philosophical world view.

Now it might be objected that we have selected rather one-sided quotations, namely, those deriving from an epoch in Barth's thinking which he himself has long since transcended. For Barth himself has admitted that since the appearance of *Rechtfertigung und Recht* in 1937 [26] he no longer maintains the transcendence of God in terms of absolute diastasis, which would in fact involve a certain lack of relation to history. He has presumably abandoned this Platonic, Neo-Kantian conception in favor of a more synthetic trend which, on the basis of the Second Article, carries with it an explicitly positive inter-relating of God and the world. However, it is not our purpose here to interpret the extraordinarily sharp curves in the theological development of Barth, nor to consider the even sharper twists of his followers in their strained efforts to follow him. We would simply note that we are aware of these shifts, and in our theology we seek to take them into account.

---

[24] Karl Barth commenting on Romans 13:11-12 in his *The Epistle to the Romans*, trans. Edwyn C. Hoskins (London: Oxford University Press, 1933), p. 500.

[25] Karl Barth, *The Resurrection of the Dead*, trans. H. J. Stenning (New York: Revell, 1933).

[26] See the translation of this text by G. Ronald Howe, under the title "Church and State" in Karl Barth, *Three Essays*.

For the present we shall have to be satisfied simply to assert that the monistic and timeless tendency has remained unaltered through all the phases of Barth's theological development. It has simply shifted its center of gravity within the system to other loci (from the earlier problems of *ante Christum-post Christum* and of eschatology to the newer problem of Law and Gospel). It has also changed its indicators. The earlier diastatic phase worked out the antithesis of time and eternity, of immanence and transcendence, primarily from the standpoint of judgment, the No of God, death, and the absolute void. The newer synthetic phase proceeds from the standpoint of a harmony of Law and Gospel to express the Yes of God to the world and its forms, e.g., the state; and it does this in the name of the love of God, or—in terms more polemical but also more precise—in the name of the idea of love. In both cases what is involved is an enforced unity whereby all the dualistic tendencies in theology are necessarily subsumed under a common denominator. The only difference is that the negative indicator of the *monon*, the idea of judgment, has now changed into a positive one, the idea of love. Through the various phases of development one thing thus remains the same, namely, the monistic timelessness.

The monistic tendencies in Barth's theology, and the resultant trend towards the non-historical, e.g., in the purely tangential relation between time and eternity, are intensified in the *Church Dogmatics;* rather, their ultimate theological foundation is here more clearly revealed. Symptoms of the non-historical detected in earlier works are manifest here as well. Among them we may mention, for example, the idea that there is no real movement of history from Old Testament to New. The reason for this lies in the structure of Barth's Christology, which does not follow the plain example of the New Testament in commencing with the event of the incarnation (λόγος ἔνσαρκος) and then reflecting on the pre-existence of Christ (λόγος ἄσαρκος), but instead reverses the procedure, taking the pre-existent Christ as its starting point and working out from there by logical deduction.[27]

It is hard to say whether Barth's doctrine of predestination (to which we must now turn) is determined by this christological starting point, or whether the opposite is true, that Barth's conception of

[27] On this point see Gerhard Gloege, "Zur Prädestinationslehre Karl Barths," *Kerygma und Dogma,* IV (1956), 193 ff., esp. 209 ff.

election, in which "the electing God and elected man [meet] . . . in the name of Jesus Christ," [28] necessarily implies starting with the pre-existent Christ.

However that may be, the principle of Christ's pre-existence not only involves the thesis of Christ's presence in the Old Testament but also carries with it certain other consequences: The incarnation is no new event, it is not the inauguration of a new covenant; it cannot be the historical juncture of the aeons, but simply recapitulates, clarifies, and reveals events which, as enacted and completed facts, belong to the perfect tense of Old Testament salvation history, or even to the pluperfect tense of pre-temporality. "The *Word of God*, which has come so near in the act of the divine mercy [performed on Israel] . . . , is identical with the *Word of faith* proclaimed by the apostles of the Church . . . the apostolic message of God's mercy actualized in Jesus Christ does *not* speak of any *new revelation* . . . the *one old* revelation of God in which Israel participates is as such the message which is proclaimed by the apostles." [29] The incarnation, the entry of Jesus Christ into history, is thus one among many melodies in the already established theme of the age-old mercy of God. It is, as it were, one of the forms or directions in which God's primal decree has been implemented. What we have here is simply the play of waves over a timeless (or primal) deep—a movement which in its restlessness bears a resemblance to the waves of history.

It need hardly be shown that in contrast the New Testament (e.g., Hebrews 1:1 f.; 7:27; 9:12; 10:10) emphasizes the uniqueness, the once-for-allness, the newness and particularity of the Christ-event. A. A. van Ruler holds that the New Testament brings no new and independent revelatory event as compared with the Old Testament, that it is only a "commentary" on the true Holy Scripture which is the Old Testament.[30] His thesis, of course, is an extreme view which differs from that of Barth in many details. However, it is the logical outcome of starting at Barth's starting point. It may be that van Ruler is off the track as far as the Calvin-Barth tradition is concerned; still, even a derailed car enables us to plot its previous course. The point of comparison between the views of Barth and van Ruler lies in the one

---

[28] Barth, *Church Dogmatics*, II², 59.

[29] *Ibid.*, p. 247 [italics by H. T.].

[30] A. A. van Ruler, *Die christliche Kirche und das Alte Testament* (1955).

determinative thesis that the fact of salvation is "decreed" from the very beginning and that the New Testament essentially provides us only with exposition, "commentary." In other words, the New Testament carries with it noetic but not ontic-historical progress. (The fact that van Ruler has no christological doctrine of pre-existence but interprets Christ simply as an "interposed emergency measure" is not to be overlooked. This is where the two theologians differ, but the distinction plays no decisive role so far as the historical problem is concerned.)

Thus far, in accordance with the theme of the chapter, we have traced back the trend towards non-historicity in Barth essentially to his relaxation of the tension between Law and Gospel. But the ultimate basis for this trend is to be found in the above-mentioned principle of Christ's pre-existence. This it is that leads in the first place to the setting aside of the Law-Gospel dialectic. In order to see what is involved here, we must turn briefly to Barth's doctrine of predestination.[31]

"The election of grace is the sum of the Gospel . . . , the Gospel *in nuce.*" [32] It has this significance primarily because in his doctrine of election Barth does not speak of a *Deus absconditus,* a darkness outside and alongside the *Deus revelatus* which can unsettle us and make of predestination a "horrible decree." Instead election is said to proclaim Jesus Christ. God and Jesus Christ dare not be sundered, he claims, as has been done in Calvin's doctrine.[33] "If we would know what election is, what it is to be elected by God, then we must look away from all others, and excluding all side-glances or secondary thoughts we must look only upon the name of Jesus Christ." [34]

This concentrating of the doctrine of election in Christology is carried so far that Christ stands as representative of the universally electing or saving God on the one hand, and of elected man on the other. In him God is he who elects and he who is elected, both in one.[35] Since Christ represents all humanity, election is universal and all-inclusive. Now the concept of non-election, i.e., reprobation, necessarily corresponds to that of election, and so it may be asked

---

[31] *Church Dogmatics,* II², 3-506.
[32] *Ibid.,* pp. 13-14.
[33] *Ibid.,* p. 111.
[34] *Ibid.,* pp. 58-59.
[35] *Ibid.,* p. 59.

whether Barth, by arguing that all humanity, man as such, is elected in Christ, and by thus positing a universalism of salvation, has not closed the path to this corresponding concept of reprobation.

The solution which he offers, and which he tries to support by means of the traditional doctrine of double predestination (used however in a radically different sense), is quite astonishing: Reprobation does not refer to a definite number of men, to the mass out of which God does his electing. (How could that even be possible given Barth's christological presupposition?) Rather God rejects *himself* by means of the incarnation of his Son. Reprobation is taken up unto God himself. Election and reprobation are in the strict sense no longer a drama between God and man. They are an intratrinitarian occurrence. In Jesus Christ God elects and rejects himself.

"In the election of Jesus Christ which is the eternal will of God, God has ascribed to man the former, election, salvation and life; and to himself He has ascribed the latter, reprobation, perdition and death." [36] The two belong together. For if in Jesus Christ God has "elected Himself as man's Friend and Partner," if he has "elected fellowship with man for Himself," [37] this means that God has voluntarily placed himself in jeopardy. This is true already as regards his relation to man in his primal state, i.e., to Adam before the fall: "What a risk God ran when He willed to take up the cause of created man even in his original righteousness, when He constituted Himself his God and ordained Himself to solidarity with him!" [38] This "risk God ran" becomes a calamity that has befallen Him when through the incarnation of Christ He enters into solidarity with fallen man when "it is the lost son of man who is the partner of the electing God in this covenant." [39] "For if God Himself became man, this man, what else can this mean but that He declared Himself guilty of the contradiction against Himself in which man was involved; that He submitted Himself to the law of creation by which such a contradiction could be accompanied only by loss and destruction; that He made Himself the object of the wrath and judgment to which man had brought himself; that he took upon Himself the rejection which

36 *Ibid.*, p. 163.
37 *Ibid.*
38 *Ibid.*, pp. 163-164.
39 *Ibid.*, p. 164.

man had deserved; that He tasted Himself the damnation, death and hell which ought to have been the portion of fallen man." [40]

Since grace thus becomes a universal Yes (inasmuch as it no longer involves a "calling out," an ex-ception), there is strictly speaking no counterpart to it. It has no vis-à-vis. This is true in two senses: there is no longer any such thing as a sphere in which grace refuses to operate, thereby manifesting itself in terms of its "choosing" and "culling out," and there is no real resistance to the call of grace on man's part. God's Yes becomes an all-inclusive Yes. For the lost Son of Man is the representative of all humanity. There thus disappears not only the dualism between Law and Gospel but also the dualism between the God who seeks man and the man who rebels against God. For man as such, not as believer but simply as man, is found. He is freed from his lostness, because God has taken man's perdition upon himself. The resultant monistic tendency finds expression in the christological construction of the doctrine of predestination, for strictly speaking God confronts only himself, and the event of salvation is reduced to a monologue which God holds with himself in his character as Father and Son, as representative of deity and representative of humanity. God is he who elects and he who is rejected, both in one.

The resultant dissolution of history is apparent at the following three points.

First, since Jesus Christ as "elected man" from eternity [41] implies our election, and since election is therewith grounded on the pre-existent "God-Man . . . who is as such the eternal basis of the whole divine election," [42] the incarnation is robbed of its character as event. It is no longer the "juncture of the aeons." It no longer denotes God's disruptive intervention in that process wherein the Law acts as a curse. The incarnation ceases to be the miracle of Christmas, a new work of God. At most it can be only a significatory demonstration of the eternal will of God. As such, it is not an ontic event but only a means of noetic clarification. What takes place is not an event but only the explication of an event. Its character as event is reduced to the noetic sphere. (We may recall in this connection that for Barth baptism has

[40] Ibid.
[41] Ibid., p. 103.
[42] Ibid., p. 110.

no ontic significance as a sacramental effecting of the state of salvation, but only noetic significance as the sign of a state already present and effected by other means.)

There can be no doubt that, for all his deviations in detail with respect to the absolute decree and double predestination, Barth is moving along the same lines as Calvin. For Calvin too the new covenant does not really introduce anything new. It simply means that there has been "more clearly and explicitly revealed the grace of the future life." [43] The distinction between the Old Testament and the New Testament is similar to that between "the shadow" and "the body." [44] The new covenant does not involve a historical turning point. It simply brings a noetic clarification of the original decree of atonement laid down in the pre-existent Christ. To take the pre-existent God-man as the starting point of Christology is to rob the incarnation of its ontic character.

Second, since God ascribes to himself man's reprobation, and thereby addresses to man his pre-temporal Yes, it becomes quite impossible to regard evil, rebelliousness, man's urge to be his own master, as elements of tension in the history between God and man. But this history itself is therewith threatened. To put it as sharply as possible, this history is in danger of becoming a pantheistic movement, which again has the character of a divine monologue.

Evil is in fact—and what a strange deformation of the views of Augustine and the Scholastics this is!—to be described only in terms of privation: "The possibility of existence which evil can have is only that of the *impossible,* the reality of existence only that of the *unreal,* the autonomous power only that of *impotence.*" [45] Hence evil is "nothingness." In this respect it is like the chaos before creation, which has no potency of its own, but simply crops up within a theological statement about creation [Gen. 1:2] as a mere potentiality, a possibility which God "ignored and left behind," which he rejected, and which is therefore definitively superseded, obsolete, past. [46] Hence evil cannot be spoken of as something real. It can be described only in negative terms. It is "what God does *not* will. It lives only by the

---

[43] Calvin *Institutes* 2. 11. 1; p. 405.

[44] *Ibid.* 2. 11. 4; p. 408.

[45] Barth, *Church Dogmatics,* II², 170 [italics by H. T.].

[46] *Ibid.,* III¹, 108-109.

fact that it is that which God does *not* will. But it does live by this fact. For not only what God wills, but what He does not will, is potent, and must have a real correspondence. What really corresponds to that which God does not will is nothingness." [47]

Thus Barth does not deny to evil "a real dimension." [48] Nevertheless evil cannot be regarded as ontically autonomous. It appears, so to speak, only as the reflection of a reality; it exists only insofar as God passes over it and wills it *not*. It is only this curiously farfetched and concocted notion—that even the non-willing of God is "potent"— which keeps nothingness from being submerged in the "merely nothing" and elevates it instead to a form of being, albeit in the negative mode. And as evil is raised to "a real dimension" and saved, as it were, from being "merely nothing" only by its negative relation to the divine non-willing, so it is raised to a sham existence, not by itself, but by human error, and hence by way of falsehood. [49] In the light of Jesus Christ, however, "there is no sense in which it can be affirmed that nothingness has any objective existence, that it continues except for our still blinded eyes, that it is still to be feared, that it still counts as a cogent factor, that it still has a future, that it still implies a threat and possesses destructive power." [50]

The loss of a history with God and its replacement by a primal perfect finds expression not merely in the robbing of the incarnation of its character as event but above all in the dissipating of eschatological expectation. For if we have to use the perfect tense here, if the world is no longer subject to evil, if the Christian can lightly disregard evil as something past, as "a wasp without a sting," then it is illegitimate "to think of [nothingness] as if real deliverance and release from it were still an event of the future." [51] There is nothing more to expect because everything has already taken place. There can be nothing yet to be consummated because we are already participants in the consummation.

Over against this view Gerhard Gloege has assembled a wealth of New Testament materials to show that in the Synoptic Gospels, Paul, and the Gospel of John the believing community is still very much

[47] *Ibid.*, III[3], 352 [italics by H. T.].
[48] *Ibid.*
[49] *Ibid.*, pp. 350-351.
[50] *Ibid.*, p. 363.
[51] *Ibid.*, p. 364.

exposed to assaults of doubt and temptation, that it is warned against the powers of seduction, and that in the provisional state of a faith still assaulted by doubts it awaits a consummation in which alone the definitive victory over evil will take place and our provisional believing be overcome by an ultimate seeing.[52] The church does its waiting in an ongoing history; it lives in the imperfect tense, and is assaulted by the reality of evil. Such knowledge as it possesses is not the conviction that evil has become nothingness, that the church now lives in the perfect tense of victory, but only the assurance that the promise is true: evil can no longer gain power over us, we will overcome its assaults.[53]

Third, we hesitate to take up a concept which has become almost a slogan. Nonetheless we must say that to establish the event of salvation on a primal perfect is thereby to deprive the event of historicity. Nothing remains but a play of waves over a timeless deep. Gone is the tension-packed commerce between God and the world (which nonetheless still has its "ruler of this world"!). All that remains is a mere monologue of God with himself. The result is that everything is subject—and here finally is the "slogan"—to a dominant *Christomonism*.

The resultant non-historicity threatens to pass over ultimately into a cyclic movement of salvation which, as the antithesis of a directed and progressive time line, is a symbol of the non-historical. This cyclic character would be seen in the fact that Barth would have to teach universalism, as the fitting expression of God's universal, pre-temporal Yes to humanity.

As a matter of fact Barth is constantly on the fringe of this very doctrine. On the one hand he says, "We cannot . . . make the *open* number of those who are elect in Jesus Christ into a *closed* number to which all other men are opposed as if they were rejected." [54] This thesis seems to contradict the twofold outcome of history and to demand a universal monistic salvation. Yet Barth tries to evade this conclusion by arguing that we are also not to "equate their number with the totality of all men." This is impossible because "in Jesus

---

[52] Gloege, *op. cit.*, pp. 247 ff.
[53] On the dissolution of history by this "perfect tense" see Regin Prenter, "Glauben und Erkennen bei K. Barth," *Kerygma und Dogma*, III (1956), 176 ff., and esp. 190-191.
[54] *Church Dogmatics*, II[2], 422 [italics by H. T.].

Christ we have to do with the living and personal and therefore the free will of God in relation to the world and every man." [55] For "nowhere does the New Testament say that the *world* [as such] is saved, nor can we say that it is without doing violence to the New Testament. We can say only that the election of Jesus Christ has taken place on behalf of the world, i.e., in order that there may be this event in and to the world through Him." [56]

Now the reason Barth cites here is exceedingly strange, for it positively rejects a systematically necessary conclusion on the ground that the New Testament itself does not contain this conclusion. It seems to me that Barth should have felt compelled to ask *why* the New Testament does not actually teach this doctrine. Could it not be that the different eschatological conclusions found in the New Testament rest on premises quite different from those of Barth's theology, especially his premise of the pre-existent God-man. A theological starting point can be judged in terms of its theological consequences—even more so in terms of the consequences suppressed by means of that *deus ex machina* called absence of biblical support.

The second argument brought by Barth against the universalist conclusion is equally non-systematic. It is the reference to the dubious theological company (Origen, Schleiermacher!) into which one falls by championing such a doctrine. This too is a secondary and heteronomous argument. In any case it is not clear, if one follows the thrust of his teaching, how Barth can escape the doctrine of apocatastasis. On the other hand, it is understandable why Barth does all he can to avoid this conclusion—though that avoidance takes more the form of a simple desire to steer clear of it than of a reasoned argument for doing so. The reason is that to expose the universalist thesis would be to confess openly the non-historical cycle. It would be an admission that the history of salvation does not consist of events and turning points and divine resolves, but involves only some noetically significant demonstrations of primal facts, and hence is an intratrinitarian circle.

What "takes place," as Barth sees it, is not an ontic occurrence in the strict sense, but only the occurrence of demonstrations, of indicatory and noetic processes. At this point at least there is a remark-

[55] *Ibid.*
[56] *Ibid.*, p. 423 [italics by H. T.].

able parallel to Bultmann's understanding of salvation history, where again one cannot say that "something takes place," but only that "self-consciousness arises." [57]

Luther on Law and Gospel

We thus maintain the basic Lutheran concern for a sharp distinction between Law and Gospel. Our concern in the first place is to safeguard the miracle in the Gospel whereby God overcomes himself, and his love saves us from the threat of his holiness. Our concern in the second place is to safeguard the historicity of revelation, which would simply be dissolved away into the timelessness of an idea, and therefore of a philosophical world view, were it not for this miracle, this contingent act, datable in the time line by virtue of the Word's having become flesh. Our concern in the third place is to see to it that neither Law nor Gospel is weakened and robbed of its true character by the attempt to establish a teleological relationship between them, between God's love and God's holiness. The Law is robbed of its true character when it can no longer imply total judgment but points instead to the telos of grace, when it no longer occasions man's dying. And grace is weakened when it no longer raises us from the dead (since indeed we have not died), when it is no longer "unconditional" but is linked in various ways with the fulfillment of the Law (since it merely empowers us to fulfill the Law, or forgives us for that wherein we have fallen short of fulfilling it). It is on these grounds that Luther's theology rejects any teleological relationship, any correlation, between Law and Gospel.

God is the author of the Law and the author of the Gospel, to be sure. He is both in one. Yet Luther rejects the attempt to establish in the name of this personal unity of the one God any ultimate point of reference or *monon* in which this dualism too would be dissolved. Luther knows only that we must believe in this unity without having seen it, without even being able to conceive of it. Now it goes without saying that Luther was aware of the problem of the unity of God, and how that unity is threatened by the Law-Gospel antithesis. He had no need to be alerted to that problem by modern theological

<hr>

[57] Cf. my essay "Reflections on Bultmann's Hermeneutic," trans. John Macquarrie, *The Expository Times*, LXVII (February-March, 1956), 154-157 and 175-177.

speculation on matters of identity. Indeed it was he who in his "table talk" drew attention to the continuing nature of the problem involved in the Law-Gospel antithesis: "There is no man alive on earth who can distinguish between Law and Gospel. We think we can when we listen to sermons, but we are way off. Only the Holy Spirit can do it. Even the man Christ could not do it, on the Mount, which is why the angel had to comfort him. Though he was a learned doctor, come from heaven, still he had to be strengthened by the angel. I should have thought I could do it because I have written so much about it, but when it comes right down to it I realize that I am way, way off. Only God the most holy therefore must and should be our teacher." [58]

This means that in theological reflection the distinction between Law and Gospel does not admit of any conceptual perfection and completeness, *perficere,* to use the terminology of Luther's doctrine of justification yet to be discussed; it is always rather a matter of intention and striving [*intentio*], of beginning again from scratch [*incipere*]. We may add that this tension is in principle irresolvable because Luther knows of no underlying unity which could "remove" it. The bracket enclosing the *Deus absconditus* and the *Deus revelatus,* the God of judgment and the God of grace, the author of predestination and the author of the Gospel, is not visible to us. Indeed it is something we must not see, or even remotely hope to perceive. The Word of God that has to do with us, the *verbum Dei quoad nos*—which alone is the subject of theological discussion—never encounters us except in this "dualistic" unfolding. And the concomitant tension is never overcome except in the flight for refuge, in the movement of faith, in the overcoming of God by God. [59]

The mystery of Christ's humanity comes to expression precisely in the fact that he subjected himself to this tension of the *verbum Dei quoad nos.* In Gethsemane, and in his question "Why?" on the cross, Jesus was a Christ under assault by doubt [*angefochtener Christus*]. He did not "understand" the judgment meted out to him vicariously. He did not "understand" the *Deus absconditus* in his

---

[58] *WA,* TR 2, No. 1234.

[59] See the Luther citations in Karl Holl, *Gesammelte Aufsätze zur Kirchengeschichte,* I (Tübingen, 1923), 77, n. 3: ". . . when God's word contravenes God's word" (*WA* 14, 299); ". . . press through to God in spite of God" (*WA* 19, 223).

hidden love. This means he did not "understand" the resolution of the tension between holiness and love, judgment and grace, Law and Gospel. It had to be attested to him by the angel. For the angel whom Luther, in the "table talk" just quoted, expressly introduces as a sign of the problem of Law and Gospel, does not explain away the tension in terms of a principle of unity; he simply "witnesses" to Jesus that the Father is behind that tension, and demands that Jesus cling to this testimony. For Jesus too the tension is overcome only in the flight for refuge, only in the movement from the decision of God taken against him to the decision of God taken for him, as attested by the angel. And when Jesus undergoes the assault of doubt once again upon the cross, and gives utterance to it in his question "Why?" the same flight or movement may here too be clearly discerned. For in this very question Jesus is addressing God. He is fleeing to God. Indeed in this very cry of dereliction Jesus is already saying, "Father, into thy hands I commit my spirit" (Luke 23:46). This Lucan word must really be construed as an exposition of the cry of dereliction. For it expresses in the moment of complete victory that terror before the awful God of judgment which first found expression in the question "Why?"—a question which in the very personal form of its address, however, already pointed to the goal of victory in "flight."

The unity of Law and Gospel can never be more than an object of the movement of faith, with the emphasis falling equally on both "movement" and "faith." The negative counterpart of this form of unity would be some given principle of unity. Here movement would be dissolved in the timelessness of something static and given, and faith would be dissolved in seeing, specifically in insight into a particular principle. Only the Holy Spirit, however, really "knows" the tension, and consequently also the unity; and for him it is no principle. Hence we have the unity only as we have the Holy Spirit, i.e., only as we believe in him and therefore do not really "have" him. If we try to make this unity a theological theme in the direct sense, we are really seeking the God who is beyond Law and Gospel, God "as such" [an sich], as he is in himself. The result is an illegitimate distinction between the *verbum Dei extrinsece* and the *verbum Dei intrinsece,* between that which is the Word of God intrinsically and that which is the Word of God extrinsically, the Word of God as it is

119

in itself [*intrinsece*] and the Word of God unfolding to meet us as Law and Gospel [*extrinsece*].

Calvin on Gospel and Law

On this whole question of Law and Gospel, Barth is undoubtedly following Calvin rather than Luther. For Calvin repeatedly takes the unity of God as the starting point in his deliberations on Law and Gospel. Consequently, he sees the two not in tension, but in a harmonious and complementary relation to one another.

This is illustrated in Calvin's interrelating of the Old Testament and the New, in which we have at least the germ of what we have called the unhistorical approach of Karl Barth and Wilhelm Vischer. For Calvin, there is basically only one covenant in many varied forms. These variants are rooted in a common unity, and they complement one another in representing this unity. In the same sense, Law and Gospel are simply modifications of a basic common factor from which both can be deduced. Expressed in this conception is that philosophical background—the tendency to think in terms of a principle and to draw logical conclusions from it—which Lutheranism has always found to be a weakness in Calvinism.

With respect to this criticism of Calvin, and to similar criticism of Barth, it might be objected that one should not forget that theological goal or intention of the act of thought which lies beyond the philosophical medium of reflection in which the act is enclosed; certainly one should not confuse the goal or intention with the medium of its implementation. It is in fact true of Calvin that the philosophical principle of his systematics is constantly overcome, broken down, and driven into fruitful inconsistency by the fact that he is at every point an exegete. He always has at hand a corrective drawn from the history of salvation. This control of abstract systematics by exegesis may be observed particularly well in his doctrine of predestination. Nevertheless, one cannot desist from criticizing the philosophical medium just because there stands behind it a true intention, because this medium, e.g., the principle of unity in the doctrine of Law and Gospel, repeatedly comes to the fore only in the pupils of the master, as is the case apparently in such disciples of Calvin as Karl Barth and especially his followers.

Calvin is speaking in terms of this principle of unity when he says,

"Christ, therefore, now declares, that his doctrine [i.e., the Gospel] is so far from being at variance with the law, that it agrees perfectly with the law and the prophets, and not only so, but brings the complete fulfillment [*solidum complementum*] of them." [60] There is thus a real "agreement [*consensus*] of law and Gospel." [61] This consensus, this complementary relation, is expressed as follows in respect to the Old and New Testaments: "God had, indeed, promised a new covenant at the coming of Christ; but had, at the same time, showed, that it would not be different from the first, but that, on the contrary, its design was, to give a perpetual sanction [*sanciret*] to the covenant which he had made, from the beginning, with his own people." [62] The Gospel can never mean that the Law is made imperative: "For since the rule of a devout and holy life is eternal, it must, therefore, be unchangeable, even as there is indeed only one single and ongoing righteousness of God." [63] This is what we earlier called the unifying principle which is beyond Law and Gospel, existing in each.

Writes Calvin, "Let us therefore learn to maintain inviolable this sacred tie between the law and the Gospel, which many improperly attempt to break. For it contributes not a little to confirm the authority of the Gospel, when we learn, that it is nothing else than a fulfillment [*complementum*] of the law: so that both, with one consent [*mutua concordia*], declare God to be their Author." [64] Here it is not, as in Luther, the Holy Spirit alone who knows the ultimate relation between Law and Gospel. Here this relation is open rather to human perception, for there is a teleological connection between them, so that they are complementary the one to the other. The Gospel can therefore of necessity be distinguished from the Law—and the Old Testament from the New—only quantitatively, not qualitatively. For the one is only "a literal doctrine," the other is "a spiritual doctrine"; the one is only "engraven on tables of stone," the other is "inscribed on the heart." [65]

[60] John Calvin commenting on Matthew 5:17 in his *Commentary on a Harmony of the Evangelists, Matthew, Mark, and Luke*, trans. William Pringle (3 vol.; Grand Rapids, Michigan: Eerdmans, 1949), I, 275.
[61] *Ibid.*
[62] *Ibid.*, p. 277.
[63] Cf. *Ibid.*
[64] *Ibid.*, p. 278.
[65] Calvin *Institutes* 2. 11. 7; p. 411.

Even when Calvin quotes passages from Scripture which tell us that the Law preaches death and condemnation while the Gospel preaches life and righteousness, and that the Law is abolished while the Gospel remains,[66] he changes the obvious qualitative distinction between Law and Gospel into a quantitative one. For he does not allow us to interpret these passages as if the work of the Law had been purely negative, "as if [it had been] . . . without any beneficial result."[67] On the contrary, the effect of the Law is wholly productive and positive in that it leads to "disquietude of conscience," and thus causes the people of the old covenant to flee already "for refuge to the Gospel."[68] Thus the Law did not plunge the people of the Old Testament into despair. It rather produced in them a fruitful unrest which made them open to the Gospel; it had a demonstrable teleological relation to the Gospel, and was actually the shadow of the Gospel cast backwards into the Old Testament. The Law was therefore salvatory insofar as it carried with it a positive disposition for the Gospel.

This is what enables Calvin to recast the qualitative distinction between death and life, condemnation and righteousness, into a quantitative one. For this fruitful "death" and productive "condemnation," while they have, to be sure, an evangelical content, do not yet, merely with this content, attain to the fullness of the Gospel, which is now fully displayed and which is all comfort and grace. The Law-Gospel antithesis is thus to be understood as being quite relative, not unconditional. Only "by way of comparison" [*per comparationem*] is it presented in the extreme form of death and life; for only thus can the Gospel's full "plenitude of grace" [*affluentia gratiae*] in "comparison" with the Law be shown.[69] The Law is therefore related to the Gospel as partial grace is related to total grace. What was once a distinction has now become a mere "difference"; the contradiction has become a mere dissimilarity [*dissimilitudo*].[70] The qualitative relation is now a quantitative one:

[66] *Ibid.*
[67] *Ibid.* 2. 11. 8; p. 412.
[68] *Ibid.* 2. 11. 9; p. 413.
[69] *Ibid.* 2. 11. 8; p. 412.
[70] *Ibid.* 2. 11. 7; p. 411. The chief passage in which this transition occurs reads as follows in context (the italics are mine): "This difference between the 'letter' and the 'spirit' is not to be understood as if the Lord had given his law

perspicuous, calculable, and teleologically determined.

The fact that this could actually happen is connected with Calvin's concern for the unity of the covenant in the covenants, the unity of righteousness, and the unity and personal union of the author of the Law with the author of the Gospel, which constitutes the starting point of his thinking. In this connection it is neither necessary nor possible to play off the divergent organizational structure of the *Institutes*—it does *not* begin formally with an analysis of the concept of God—against the starting point we have ascribed to Calvin. The true starting point does not have to be expressed in the organizational structure of the system by formally being made the subject of the first chapter. The formal schema of the concept of God into which all the loci are integrated is to be seen rather from case to case, from locus to locus.

Nevertheless, we must admittedly be on our guard here too against artificial constructions. Calvin's theology is certainly more and other than a philosophical treatise cloaked in a Christian camouflage. In fact, if we interpret Calvin along the lines indicated we will see that his ideas are like the "resultants" of a parallelogram of forces, a parallelogram determined by his philosophical efforts on the one side and by his exegesis of Holy Scripture on the other. The very fact that we distinguish between the philosophical medium which provides Calvin's structural principle of organization, and the content of Holy Scripture which he aims to present, must surely indicate that our purpose here is not at all to set up some kind of a fundamental and unbridgeable antithesis between Calvin and Luther, or to buttress any kind of rigid confessionalism. Nonetheless, we would in conclusion sum up as follows the differences between Calvin and Luther with respect to the relation between Law and Gospel.

Luther versus Calvin on Law and Gospel

Calvin co-ordinates Law and Gospel from the standpoint of the one God who remains identical behind the two modes of operation. Consequently he finds in the modes themselves only forms of "administration" which are "secondary" and which cannot constitute the

to the Jews without any beneficial result, without one of them being converted to him; but it is used *by way of comparison*, to display the plentitude of grace with which the same Legislator, assuming *as it were* [!] a new character, has honoured the preaching of the gospel." *Ibid.* 2. 11. 8; p. 412.

THEOLOGICAL ETHICS – FOUNDATIONS

true point of departure. He thus begins with their identity, with the personal union of the author of both Law and Gospel. Whether overtly or covertly, this identity is always his point of orientation. Hence Calvin views Law and Gospel in their unity, teleologically. Luther, on the other hand, begins with the factual distinction between Law and Gospel. In other words, he begins with the Gospel as the miracle which has actually taken place to rescue us from the judgment of the Law. In order to preserve this character of the Gospel as miracle Luther is forced to resist any attempt to establish a teleological relation between the two. Since we cannot know this teleological relation between Law and Gospel, knowledge of the divine identity which stands behind them is also denied us. That knowledge belongs only to the Holy Spirit.

The result is that in the Calvinist tradition there is an unhistorical trend which leads finally to the extreme of relating Old and New Testaments in the fashion of two concentric circles. On the Lutheran side, however, the accent plainly rests on the historicity of revelation. This concern finds particularly pointed expression in Melanchthon's well-known statement from his *Loci Communes* of 1521: "To know Christ is to know his benefits." Here "Christ in himself," i.e., the Christ repeatedly represented in the static and timeless schema of the doctrine of the two natures, gives way to salvation history. In this connection the history of his "benefits" is to be understood not only as the history of the years 1-30, but also as the history of the church which after all is his body, indeed as the history of Christ "for me." In the categories of Martin Kähler, Christ does not stand behind this history [*Geschichte*] as a "Christ in himself" (whether as a metaphysical entity or as the "historical [*historischer*] Jesus"); he exists in it. Christ is his own history [*Geschichte*]. To express this fact theologically is really the purpose of that not wholly felicitous phrase with which Lutheranism has described the incarnation, God's becoming history: *finitum capax infiniti* [the finite can encompass the infinite].

The fact that the Law continues to have significance even for the justified—a fact expressed by the Lutheran doctrine of Law and Gospel and by the formula *simul justus et peccator*—assures that existence, as existence *coram Deo,* is in fact a history. This historical existence "is" only as it is constantly "achieved." And it is "achieved" only in the incessant movement from judgment to grace, from the assaults of doubt to comfort, from the despairing contemplation of existence (from the human perspective of *coram se ipso*) to the confident contemplation of existence (from the divine perspective of *coram Deo*), from the *cor accusator* to the *Deus defensor*.

Where there is a teleological understanding of Law and Gospel, or on the other hand, an isolation of the Gospel from the Law, existence is in danger of becoming unhistorical. This threat to the historicity of existence appears where the element of decision is played down and existence is understood monistically. It appears too where existence is understood in terms of an entelechial self-unfolding of the mystically construed "essential righteousness" of Christ (Osiander). The true function of the Law with respect to the Gospel is to serve as "gauze in the wound." There is no healing of this wound. It remains open in order that we will return continually to the doctor, in order that we will pray each day, "Forgive us our trespasses." For as we do so the history with God is "achieved"—a history which has the character of event and is replete with decisions.

8.

The Continuing Pedagogic Significance of the Law for Believers

That the Law continues to serve an educational function even in the justified is obvious in light of the fact that our Christianity is never something complete and finished but is constantly in process of becoming. To be sure, the Gospel frees us from the dominion and curse of the Law, and to that extent the Christian who has wholly "crept into Christ" [1] does not need the Law any longer. He really is like the stone lying in the sun, which need not be commanded to become warm. Nevertheless, the question arises as to who is that kind of a Christian. For whom is Christ in fact the exclusive basis of existence? Are we not always on the *way* to this, but never at the *goal?* By putting the question this way, by asking "to what extent" Jesus Christ is the determinative basis of our life, it becomes clear that the Christian state has a quantitative as well as a qualitative aspect.

The qualitative consists in the fact that as justified persons we are accepted into fellowship with God, and total righteousness is thus ascribed to us. Now we are either in this fellowship or outside it. There is no third possibility, e.g., of grades of acceptance or of levels of approximation to this fellowship. This kind of thinking in terms of levels or degrees always ceases the moment we think in terms of personal fellowship with God. It begins again only when we think ontologically, thus postulating different levels of perfection in which it is possible to achieve different grades of approximation to God through different degrees of imparted grace and attained merit.

How different things are when we think in terms of personal fellowship may be seen from the story of the Canaanite woman in Matthew 15:21-28. She stands only on the threshhold of faith; her

[1] See p. 16, n. 2 above.

faith is still in its very infancy [*incipere*]. She makes no explicit confession of faith but simply cries for help and puts her trust in Christ despite the assaults which test her to the breaking point. Confident that he can help her, she is thus wholly in the state of righteousness: Jesus calls her faith "great." A similar case is that of the official of Capernaum, who at his first encounter with Christ is also accepted through faith (John 4:47 ff.). Hence the state of justification is, from God's standpoint, perfect from the very first. It cannot be improved or transcended. It qualifies a man definitively and absolutely.

The Quantitative Aspect of Justification

Nevertheless, there is also a quantitative way of viewing the state of justification. If we are not to become entangled in hopeless contradictions, we must of course insist that this is in the strict sense only a "way of viewing" it. It is a matter of perspective. The quantitative perspective arises when I turn again from that which God has done to me and for me absolutely, and look "back" upon myself. For then there arises the question of what must accordingly happen to me and in me, i.e., of what has not happened yet. The question is: How am I, as the one who does the living of my life, to express in my decisions this fact which has taken place with regard to me? And this poses the question of whether the qualitative perfect tense of my justification, my acceptance into fellowship with God, does not imply at the same time a struggle, a change, an advancing or progressing on my part. In other words, does not the perfect tense of my having been justified [*justificatum esse*] imply a being justified more and more [*magis et magis justificari*]?

Before we pursue this quantitative aspect of the problem of justification, and the continuing pedagogic function of the Law which it entails, I should like to guard against the possible misunderstanding that what is involved here is a consideration of man *coram se ipso*, that inverted viewpoint from which justification has freed us by turning our gaze away from ourselves and toward the author of the Gospel. If I am not mistaken, we must rather distinguish, with Luther, between three types of perspective within justification.

First, we look to God as the object of our faith, to the God who justifies us in Christ. Here it is a matter of our looking *away* from ourselves and our own faith as we "creep into Christ."

Second, we look *at* ourselves. This is the true introversion to which the Law leads us in its "accusing role" [*usus elenchticus*]. The Law leads us into the depths of perplexity and despair. Under the eye of the divine Judge, we are confronted by the persistent *peccator in re*, by the continuance of the old Adam. Here even faith, as a "psychical act," a "work performed," cannot help us; it crumbles in our hands. Here we stand before the *cor accusator*, before absolute uncertainty. Seeing ourselves from this standpoint we do not even know "whether" we believe.

There is, however, a third perspective which applies in the present case, i.e., in our quantitative consideration of the state of justification. I now turn my gaze away from the object of faith—where it was previously stuck!—and turn it *back* upon myself as standing before the justifying God [*coram Deo justificante*], in constant relationship to him. This second look at myself, however, does not plunge me into despair. For I can no longer abstract myself even for a moment from the fact that before God I am justified. This second look drives me then not to despair but to repentance. Unlike despair, repentance begins with the fact that, while I am totally inadequate *in re* for the justification ascribed to me, this inadequacy—because I am justified! —no longer means damnation for me but only hope and command. The "Woe to me!" turns into an imperative, specifically the imperative associated with the indicative ("nothing can separate me any longer from the love of God"). This third perspective thus places me at what for me is not the end, but the beginning, the point from which I have to advance.

"Advance" is really the key word suggested by this quantitative consideration of the Christian state. This way of viewing the matter has reference to progress in our Christianity, to the process of maturing spiritually. Thus it has reference negatively to the fact that I for my part have to apply to and fulfill in myself the death which I have died in Christ. I have to struggle to free myself from all that stands in contradiction to this death. I have to live a life of self-mortification. My repetition or re-enactment of that which God has already done to me and for me is called "repentance" [*Busse*]. Positively, the progress, the maturation process consists in my allowing the resurrection of Christ, into which I have been drawn, to take place in me—in the form of a new life and through my own affirmation and acceptance of it. My letting this happen is called "sanctification."

To sanctify oneself is to "put oneself at God's free disposal, to be used where and how and when he wills," after he has first laid hold and taken charge of us.[2] Penitence and sanctification thus imply that, as one who was and is the object of the divine mercy, I am also summoned now as a subject to grasp and actualize that mercy, or better, to let it happen to me and in me.

When we put it thus, it is plain that the quantitative way of viewing the Christian state is to be understood only from the angle of the imperative which is associated with the indicative. This imperative simply contains the command that I countersign God's already signed pledge that he will be to me a gracious God. Repentance and sanctification are really in the strict sense only re-enactment, repetition. There can be seeking, struggling, and dying only when I am already found, only when I am already overcome, only when the death of Jesus Christ has already become effective for me. I can make a meaningful beginning only when God has ended with me. Growth in sonship can take place only after I am a son. Seeking begins only after I am found; then it is that I am really called to seek. This seeking is what causes me to advance and progress, to find more and more [*magis et magis inveniri*], to be justified more and more [*magis et magis justificari*]. It brings me even deeper—again in the *quantitative* sense!—into death, into aversion from my former life. "He [God] was therefore found and not sought. But once found, he wants to be still further [note the quantitative expression!] sought for and again and again to be found [*magis inveniri*]. We find him when we are converted to him from our sins and we seek him as we persevere in this conversion." [3]

---

[2] Fritz Rienecker, *Praktisches Handkommentar zum Lukas-Evangelium* (Giessen, 1930), p. 479.

[3] Luther's commentary on Romans 3:28 (*LCC* 15, 120). Cf. the other notable contexts in which Luther speaks of progress: "As long as we live on this earth we cannot possess God but must seek him . . . and ever keep seeking him." (*LCC* 15, 91); "There arises in [the spiritual man] a greater and greater hatred against that which assails him" (*LCC* 15, 190). See also *LCC* 15, 108, 112, 113. The growth of which Luther speaks, of course, has reference not to the righteousness imputed to me (*justitia imputativa*) but to the righteousness which is actually operative in me (*justitia formalis* or *effectiva*). Quantitative observation of *growth* in the latter is possible only on the ground of the qualitative *fact* of the former (see *LCC* 15, 19; cf. WA 39I, 443-444).

If, however, there really is this progress, this summons to advance, then the obvious question is whether our talk about the justified having no further need of the Law is not merely a dealing in abstractions. The abstraction would lie in the fact that in our existence as pilgrims this kind of justification simply does not exist. We only move towards it. We for our part stand at the very beginning, in the stage of *incipere,* though God's work on us and for us is finished and we may possess the intended fellowship with him. He who has fully crept into Christ no longer needs the Law. But have we done this? Have we fully crept into Christ? Can it be said, then, that the abiding unfitness of our *de facto* condition has been truly ended and overcome?

This point is made by Luther in his Third Disputation against the Antinomians. The argument runs as follows: "To those who are freed from the Law, the Law is not to be applied. The church is freed from the Law. Therefore the Law is not to be preached [to the church]. Paul says [in I Tim. 1:9] that 'the Law is not laid down for the just.'" To which Luther's pointed response is: *"To the extent that they are that"* [*In quantum sunt tales*].[4] In other words, Luther concedes that in the Christian state of liberation from the Law, the Law has no further business; it is not to be interjected there anew. But he can understand this state and this liberation quantitatively: "To the extent that we are that" the Law has nothing more to do with us. This means that the liberation is not something complete and perfect. As surely as we must insist that the believer is radically and totally liberated from the Law, so surely must we also put the question of whether we are such believers, to what "extent" [*in quantum*] we are, and whether it is practical to throw away our crutches before we can really walk.

If our Christian state is not something complete and perfect— though this is precisely not the case—we would have to agree that it makes sense to have flashing red lights on both sides of the path we have to tread, as reminders that it runs through alien territory and is flanked by deep abysses. We are redeemed children of God, but there are unredeemed areas in us. Perhaps we are not aware of the extent to which a particular business activity, a certain kind of behavior toward our neighbor, or some sexual attitude, represents a

4 *WA* 39^{I}, 528 [italics by H. T.].

wandering off the given path. Perhaps I do not realize that the various strands of my highly complex ego tend to pull apart from one another in quite different directions, according as I am motivated by thought, feeling, will, sex, and the demands of my calling, with some parts striking out on their own or failing to keep up with the rest, with some parts hurrying on in advance and others tarrying behind. Hence it is no real contradiction, as one might at first suppose, to say that we are redeemed children of God but that there are unredeemed areas in us.

The contradiction is only apparent. It appears because the two statements are made from two different perspectives. When I say that we are redeemed children of God, I am speaking of God's gracious acceptance and of the fellowship with him which is already effected; this is the *justitia imputativa*, the imputed righteousness. But when I speak of the unredeemed areas in me I am thinking of the fact that I for my part—looking back from God's action to my own action, to my re-enactment of his deed, my operative righteousness [*justitia effectiva*]—am still in *process* of relating all spheres of my life, all sectors of my ego, to this reality. My conversion may have been such that I was delivered from some particular bondage. This bond, then, was my point of crisis, the place of my emancipation. At this point my life is now plainly in line with the divine redemption. But how do I stand at all the other points? Do I not continue to live a disintegrated existence, going off in many different directions in the various departments of my ego? Do I not continue to live in self-contradiction, in what Jesus called "hypocrisy"? For ὑπόκρισις does not mean simply a cunning, sophisticated, and conscious pretence. It does mean that various spheres within me follow different laws, or at least that they do not all obey the Law of God. Hence it is that I pray, fast, and serve God, without realizing that even as I do so another part of me is looking at men and paying service to them, so that I am objectively in self-contradiction (cf. Matt. 6:5, 16; 7:5; 22:18; 24:51; Luke 13:15).

The event of sonship is one thing, the relating of the individual spheres of life and the ego to that sonship is quite another. The flashing lights of the Law on both sides of the path serve as a reminder of that fact. Or, to change the metaphor, the Law is a kind of sheep dog whose purpose is to recall the members of the flock to the path of the shepherd. Now it is the shepherd who does the leading, not the dog. It is not the dog but the shepherd who is the center and focal point of the flock, the one to whom the sheep know themselves to be related. Moreover, the flock wants to follow the shepherd. It is aware that here is its place of safety. What the sheep do not realize is that this or that appealing tuft of grass a little distance from the path is actually a temptation to stray away from the path, that it stands, as it were, in contradiction to the will of the shepherd. For this reason the existence of the sheep dog is a reminder of the way of the shepherd. It is a reminder of what the sheep really ought to do and in fact want to do, and of the fact that this uniformity of will and duty is imperiled by the divergent and disintegrating character of their instincts, in this case their desire for safety on the one hand and for food on the other. The dog reminds them that their desire for food, or the sheer laziness which makes them lag behind, or the solicitude which prompts a mother sheep to dawdle with her young, that all these different spheres in what is on the whole the incomparably simpler psyche of the sheep are related to the way of the shepherd. The sheep are reminded that in following the shepherd they will find satisfaction for all these individual spheres, but that if these spheres lead them to go off on their own and pursue an autonomous existence they will inevitably come to grief and perish utterly. This is why the sheep dog tries to prevent the disintegration. He is there because it is realized that there are instinct complexes and that it is therefore a task in itself to relate these complexes to the one goal of following the shepherd.

To state the matter differently, the Christian life is a pilgrimage, and this fact poses a problem because during the pilgrimage there is the danger of breaking away, of disintegration and dispersal. Such disintegration derives from the complex character of the ego. It is this complexity which makes possible the existence of unredeemed spheres even within sonship. Consequently, the Law is necessary in the Christian life to remind us that *all* spheres are to be related to our sonship, and that as pilgrims we are in a permanent state of objective ὑπόκρισις which continually demands a constantly reiterated mortification and conversion.

It was in this sense that Luther spoke of the abiding significance of the Law for the imperfect Christian, the pilgrim,[5] though to be sure his statements are rather mild and restrained, and twentieth-century theologians have to address their age very differently. Luther was understandably reserved on this point for he was writing with Romanism and its nomistic doctrine of sanctification in view. He could, of course, break his reserve with astonishing intensity in his later disputations against the Antinomians, whose views represented a position which was the direct opposite of Roman Catholicism. Our twentieth-century situation, however, is quite different. In the "age of discontinuity" (Max Piccard) and cleavage of consciousness, in a secular age which proclaims the autonomy of individual spheres of life and in which it seems—but only seems—that a man can be a Christian in one separate sector (*modo Christiano*), a business man (*modo oeconomico*) in another, and a politician (*modo politico*) in still another, in such an age the flashing red lights are needful. Indeed the work of the sheep dog in rounding up and holding together is far more necessary now than it was in the Middle Ages, which did still enjoy at least a relative unity of life.

Law as a Loving Reminder

We may sum up as follows. The Law retains its significance because, from the human standpoint, the Christian state is always incomplete and in process of becoming. The Law serves to relate the individual spheres and stages of existence to the fact of justification. It is thus a countermeasure to combat the constant disintegration of Christian life and its objective hypocrisy. It is a constant reminder of the real theme of our justified existence.[6]

It is evident that in this respect too the Law has lost its pre-Christian character as one of the "powers" [ἐξουσία]: It is no longer the devouring wolf. Instead it helps to preserve intact the flock's connection with the shepherd who has already accepted the sheep as his own. Hence it always has reference to the work which God

[5] See Luther's *A Treatise on Good Works* (1520), PE 1, 199-200, where the "four kinds of men" (actually only three of the four) ought to be interpreted—as in the case of Jesus' four soils (Matt. 13:3-23)—not in terms of various distinct groups of Christians but in terms of various possible states of the one Christian heart.

[6] Cf. the reasons given by The Formula of Concord for the continuance of the Law in the article on "The Third Function of the Law." Solid Declaration, VI, 2-3 and 7-9; BC 564-565.

has completed in us and for us: "The Law is to be retained in the third place in order that the saints may know what works God requires for the exercise of obedience to him."[7] Here pointed expression is given to the 'fact that the Law refers back to the completed act of God. For the "obedience"—that which God's will intends —is already present. It is present and given in the very fact that we have "yielded ourselves" to God as obedient slaves (Rom. 6:16), and in the fact that we cannot do other than "press on to make it my own, because Christ Jesus has made me his own" (Phil. 3:12).

To this degree the sustaining, nourishing, and perfecting Law no longer has a normative significance, in the sense of leading me as if it were the shepherd instead of simply the dog. It has rather a regulative significance in that it reminds me of the way in which my given obedience—to which I could never be forced by any law—may be exercised. It points me to specific areas in which the question of obedience is acute. It breaks up the one light into the different colors of the spectrum, as it were, so that I may recognize in the different spheres (symbolized here by the different colors) the one beam which illumines and guides my life. The Law does not *make* the new man, but it does *exercise* him and shows him the full range of relationships in which his newness is relevant.

In regard to this function of the Law, by the way, Calvin makes some surprisingly similar statements concerning the Law's "didactic

[7] See WA 39I, 485. This citation is not in the most important manuscripts of Luther's Second Disputation against the Antinomians, and it may well derive from Melanchthon. Nevertheless it seems to us to be true to Luther's intention, and only a nuance of meaning is involved if it is ascribed to Melanchthon instead. Werner Elert (see above p. 55, n. 4) may be right in charging that this quotation was lifted almost verbally from Melanchthon's *Loci Communes VI, De usu legis divinae* (*Corpus Reformatorum* 21, 406) and put on the lips of Luther by later editors. Still Elert's charge of "blatant falsification" is exaggerated. The opposing scholars quoted by Elert ground their assertion of a threefold use of the Law in Luther on a broader basis than merely this single passage and the other one adduced from the Church Postil of 1522 (WA 10I, 456). See Gustav Kawerau, "*Antinomistische Streitigkeiten,*" in *Realencyklopädie für protestantische Theologie und Kirche,* I, 588; Reinhold Seeberg, *Lehrbuch der Dogmengeschichte* (4 vols. in 5) IV1, *Die Lehre Luthers* (2nd ed.; Leipzig: Deichert, 1917), p. 207; Friedrich Loofs, *Leitfaden zum Studium der Dogmengeschichte* (4th ed.; Halle: Neimeyer, 1906), p. 861, n. 1; Franz Frank, *Die Theologie der Conkordienformel* (4 vols., 1848-65), II, 389. [Most German and Scandinavian scholars of the twentieth-century Luther Renaissance now deny any such "third (i.e., didactic) use of the Law" in Luther. See Gerhard Ebeling, "On the Doctrine of the *Triplex Usus Legis* in the Theology of the Reformation," in *Word and Faith,* trans. James W. Leitch (Philadelphia: Fortress, 1963), pp. 62-78.–Ed.].

role" for believers, the *tertius usus legis*. For Calvin, in contrast to Luther, this third use of the Law is indeed "the principal one, . . . more nearly connected with the proper end of [the Law]. . . .[For the justified it is] an excellent instrument to give them, from day to day [here again comes the quantitative way of viewing it!] a better and more certain understanding of the divine will to which they aspire, and to confirm them in the knowledge of it." [8] Here again there is that backward reference. The Law refers back to an effected indicative, namely, our aspiring towards the will of God. This aspiring is not produced by the Law, only unfolded—and that through the unfolding of the will of God to which it has reference. In this respect, and *only* in this respect, there are "advances": "No man has already acquired so much wisdom, that he could not by the daily instruction of the law make new advances into a purer knowledge of the Divine will." [9] Even though the saints "with whatever alacrity . . . labour for the righteousness of God according to the Spirit," still it is true on the other hand that "they are always burdened with the indolence of the flesh, which prevents their proceeding with due promptitude." [10] For the spiritual man the alacrity, the promptitude, the spontaneity of obedience are given. Yet Calvin too sees areas of the ego which are not immediately caught up in this movement, even though the whole man is the object of justification. The Law must therefore make clear that those other areas too are intended, that they too are related to the theme of justification. Calvin groups these other areas under the rubric *caro*, "flesh." This is something Luther would not do, because for him "Spirit" and "flesh" are not parts of man; each is rather man in his totality—the *totus homo*—only seen from a different perspective. The difference in terminology, however, need not and should not prevent us from seeing in Luther and Calvin a similar definition of the function of the Law in the life of the justified, even though the presuppositions of Calvin's thinking cause him to accentuate this "third" function far more heavily than Luther.

Against this mode of retaining the Law in the life of believers there cannot be played off the demand of Paul to the Galatians that they should stand fast in the freedom for which Christ had set them free

[8] Calvin *Institutes* 2. 7. 12; p. 323.
[9] *Ibid.*
[10] *Ibid.*

THEOLOGICAL ETHICS – FOUNDATIONS

and that they should not submit again to a yoke of slavery to the Law (Gal. 5:1). In this passage Paul is dealing with an issue which vexed the Galatian congregation, namely, whether the Old Testament Law is still binding for believers in Christ. To the degree that the Law represents the written code which kills, it is to be set over against the Spirit who gives life, and over against the freedom which he bestows (II Cor. 3:6, 17). For the law of the Spirit, namely, the life in Christ Jesus, has freed us from the Law of sin and death (Rom. 8:2). This is why, according to Paul, there is no Law laid down "for the just" but only "for the lawless and disobedient." In the same context Paul even connects the Law with an entire catalogue of vices (I Tim. 1:9-10) in order to show the Law is designed to oppose the opponents of God.

As we have said, one may not conclude from these verses that the Law is abolished altogether for believers, as though they no longer had to respect it as a binding authority. The verse which immediately precedes (I Tim. 1:8) gives us better instruction: "The Law is good, if any one uses it lawfully." Paul speaks in the same vein in I Corinthians 9:19 ff. where he is emphasizing his freedom: he is not merely outside the Law to those who are outside the Law, but also a Jew under the Law to Jews who are under the Law. Herein may be seen how the Christian—in this case the apostle himself—acquires a relation to the Law which is wholly new but by no means merely negative, since one result of his freedom is a wholly new regard for the Law. The relation between Law and freedom is by no means purely antagonistic. In his new freedom the Christian is oriented to the Law in a new way.

That the apostle becomes a Jew under the Law to Jews who are under the Law is something we can hardly dismiss as a merely tactical device, as pastoral strategy which in terms of what is actually involved was not seriously intended. On the contrary, Paul here regards the Law as a true authority. But he has a new relation to it, different from what he had "before." He stands in relation to the Law no longer as a servant but as a free man. He still respects it as a declaration of the will of God. But this declaration expresses as a demand what in his new creaturehood has now become the object of love and willing fulfillment. For the old—the old that enslaved me!—has passed away (II Cor. 5:17), it has died with Christ (Rom.

6:7 ff.). We no longer live under the sign of the fact that the old must be enslaved and oppressed. Instead we have become children who can say, "Dear Father" (Gal. 4:6-7; cf. Rom. 8:17; 9:7; Gal. 3:7; 4:28).

Thus the Law itself is always the same; its authority too is unchanged throughout. Only my relation to it is different; I am no longer a servant but a son. I am no longer an opponent of a law which confronts me as the Law of the "Lord." I am rather one who in free love actually wills what his Father wills.

The same holds true also in the second case, where Paul is "outside the Law to those who are outside the Law." Here too his freedom is not purely antagonistic toward the Law which confronts it. For he is "not without law to God but under the law of Christ" (ἔννομος Χριστοῦ, I Cor. 9:21). This is why that ἔννομος Χριστοῦ is the most felicitous and precise designation imaginable to elucidate the substance of what is meant by "not being myself under the law" (I Cor. 9:20) and yet at the same time "not being without law toward God" (v. 21). For the "ἔν-νομος" shows us that under Christ the Law is fulfilled but not done away. It is simply deprived of its dominion. It claims us, but by way of a different orientation, as those for whom Christ has fulfilled the Law and who are now subordinated to his authority in perfect freedom.

The thing we object to is that people throw out the baby with the bath. We are not to regard the authority of the Law as done away simply because there has been a change in our relation to it. As Luther says, it is nonsensical to command the sun to shine, or to command that two and three add up to five.[11] Similarly, the Law is "the eternal, irrevocable, unchangeable will of God."[12] It remains the same whether man subjects himself to it willingly or unwillingly, whether he fulfills the Law in free self-determination or sighs under the compulsion of the Law. It goes without saying that one does not command the sun to shine, or the stone lying in the sun to get warm. These things just happen of themselves. Yet as they take place a "law" is fulfilled, in this case a law of nature. The voluntariness with which the sun shines without having to be commanded is not a freedom antagonistic to the law; it is a fulfillment of this law.

[11] WA 17ᴵ, 130.
[12] WA 45, 149.

There are thus two relations to the Law, and two ways in which the Law is valid. There is first the Law which drives us, which judges and accuses us; from this Law we are delivered. Then there is also the Law to whose fulfillment we are impelled by the Holy Spirit, as those for whom Jesus Christ has already fulfilled the Law and who now live in a new bondage (Rom. 6:16).

In face of this Law—which is valid but no longer condemns—there arises the question which we have answered already in our discussion of the quantitative problem but which still stands in need of final theological foundation. This is the question of whether, in the new creaturehood of the justified, we are already truly and totally subject to the Law and do fulfill all its demands. To put it another way: Does the ἔννομος Χριστοῦ imply of itself an orientation to and acquiescence in the totality of the divine will? This is doubtless the case insofar as, in the state of justification, we do "believe" and to that extent do right by God, thus fulfilling the First Commandment. Beyond that nothing can be demanded of us "additionally." At the same time it is still an open question whether the *whole* of our existence in all its dimensions is actually related to our believing, i.e., whether our faith is also actualized in all the spheres of our existence.

This question must necessarily arise, if for no other reason than the fact that the First Commandment does not stand alone: there is a second table of the Law as well. The relation between the two tables of the Law would be misunderstood if it were thought that the second is merely an "addition" to the first, i.e., that we have in it "supplementary" demands. The second table is rather exposition of the first from the standpoint of the plurality of life's relationships. To "let God be the Lord" implies such and such for my relation to parents, spouse, neighbor, etc. The second table thus contains regulations for the execution of the first. It raises with us the question of the actualization of the First Commandment in the various relationships of life.

It is in this sense that, for the justified, special significance attaches to the fact that the Law of God is not simply done away, but retains its validity. The justified man must inquire of the particular laws [νόμοι] whether and how far he has really given attention to the actualization of his faith in the particular spheres of life. It is at this point, as we said, that the Law has the significance of the sheep dog.

But even in its role as a reminder [*Anamnesis*], the Law is changed. For in performing its task, in reminding us continually of things omitted and forgotten, it no longer kills us. Instead it is friendly; it helps us. After all, the sheep dog is anything but a wolf. The wolf kills; the dog tries to protect us from it.

This wholly essential function of helping, which is only very imperfectly suggested by the traditional phrase "third use of the law," loses its theological foundation if we regard Law and freedom as mutually exclusive, if we fail to understand the new relationship which faith establishes between them. For if we regard them as mutually exclusive, then the Law is simply "done away," and hence can no longer serve as a reminder in the way God's love intends. We must instead take seriously the change which grace has brought over it: The condemning judge has become a helper. It contains regulations for the execution of the First Commandment, for the execution of faith. It seeks to prevent faith from falling into self-contradiction and consequently into hypocrisy (I John 4:20; II John 9).

Law as a Comfort in Time of Doubt

We have seen already that assaults of doubt are implicit in the false perspective to which we constantly fall victim. When I consider myself *coram me ipso,* when I listen to the *cor accusator,* it becomes impossible for me to believe in my salvation. My situation seems hopeless. It seems no less hopeless when, to this natural way of looking at myself, I add the spiritual aspect. At least that is the case when I begin looking for certain spiritual phenomena in myself, for some spiritual experience or psychic process. When that happens my faith evaporates away to nothing. It may be understood in terms of psychology or environment. But it then ceases to be what it purports to be, namely, the act of God, and therefore something which has been elicited from me from without.

What follows then is a psychical reduction of faith and of its object which corresponds exactly to the reduction of the history of revelation to religious history. This happens the moment I fail to view the events of revelation as a transparency for what is wholly other, and instead stop at the outward form of the events, seeing them only in their continuity with what is fleshly. Thus the assaults of doubt arise the moment I regard the saving work of God "in us," e.g., the awakening of my faith, as something which has taken place "through us," when I lose sight of the true author and doer and allow my gaze to become fixed upon myself instead.[13]

In face of this introversion and the uncertainty based upon it, I need to be summoned to seek, not myself, but him who wills to be sought, and who has instead already been found in that faith which, as regards its psychic instability, disintegrates before my very eyes. Hence the command of God has for the justified the further significance of pointing me away from my weakness in prayer, my sense of unworthiness, to the fact that God has commanded, "*Seek* ye my face" (Ps. 27:8), and that I am to *be* strong and of good courage (Josh. 1:7, 9). Because of the subjective impossibility of prayer, the man who prays can dare to open his lips only in view of God's clear command. In Luther's judgment, he is perfectly right *coram se ipso* who insists on the subjective impossibility: "On my account this prayer would amount to nothing; but it is important because God has commanded it." [14] In this command I have to do with the very same Lord who regards my prayer as my works, indeed as me myself, "not on account of the person, but on account of his Word and the obedience accorded it." [15]

The command turns my inverted view once again and points me to the one who does the calling. God's command takes from the man who obeys it all the torturing concern about whether his actions (his praying, his believing) are meaningful and warranted, for the simple reason that the one issuing the command himself assumes all accountability. This is why the command means liberation from the assaults of doubt which spring from introversion of outlook and which make all spiritual activity appear to be meaningless or silly and hence unwarranted.

Luther regards the Law as so great a help in time of doubt that he cleaves to the First Commandment even when Christ himself threatens to disappear. When the whole theology of the cross [*theologia crucis*], when that strange and hidden detour of God by

---

[13] In *The Babylonian Captivity of the Church* Luther writes, "Faith is a work of God, not man, . . . The other works he works through us and with our help, but this one alone he works in us and without our help." *LW* 36, 62.

[14] Luther, The Large Catechism, III, 13; *BC* 422.

[15] *Ibid.*, III, 16; *loc. cit.*

way of the cross threatens to break down and to lose all meaning, Luther clings to the fact that this long way around was taken, not because Luther had proved its necessity theologically or in any other way, but exclusively because God so *willed* it. Hence the final court to which I can appeal in time of doubt, when all meaning and support fail, is simply the will of God as it confronts me in the First Commandment.[16]

Law as a Servant of Love in the Political Sphere

We have already pointed out that the Law's "political function" [*usus politicus*] does not belong to the complex of questions relating to the indicative and imperative, since what is involved is not a specifically "Christian" *usus*, intended for the *Christian*. For the *usus politicus* is characterized by the fact that it is in force quite independently of any personal relation to God; as the author of the political law God is anonymous, or at least is capable of remaining anonymous. Even the civil or criminal code is in force whether I know the legislator personally or only indirectly, and whether I approve or disapprove of his intentions. I am subject to his order and power only externally, but to that extent I am indeed his subject. In exactly the same way, I come within the *usus politicus* of the Law whether I be heathen, Turk, or atheist. For human existence is impossible unless life in society is ordered and organized—whether or not the question of the author of the order is raised, whether or not I stand in a positive relation to him, and however I may conceive of him. Hence the anonymity of the author of the political law necessarily makes this aspect of the Law into a universal order binding upon all men, an order which is "neutral," as it were, and does not depend upon existence being Christian, i.e., upon justification.

It is along these lines that we are to understand the well-known words of Luther [17] in which he says that there are flourishing political states outside Christendom too, and that if it came to a choice he would prefer a wicked but wise prince to a good, i.e., Christian, but

---

16 For the appropriate citations from Luther see Paul Althaus, "Gottes Gottheit als Sinn der Rechtfertigungslehre Luthers," in *Lutherjahrbuch* (1931) or in *Theologische Aufsätze* (Gütersloh, 1935) II, 1 ff.; also Karl Holl, "Was verstand Luther unter Religion?" in *Gesammelte Aufsätze zur Kirchengeschichte*, I (Tübingen, 1923), 73, n. 5.

17 See *LW* 9, 19.

stupid one. "It is not required [by God] that a person be Christian to be a magistrate. Hence it is not necessary that the emperor be some holy person. It is not required that he be Christian in order to rule. All that is necessary is that the emperor be guided by reason. This is why the Lord God has preserved the kingdom of the Turks." [18]

The normative authority for the technical ordering of political affairs is reason. It is true that with our present knowledge of blatantly secular systems, e.g., in totalitarian states, we have to realize far more clearly than did Luther—we shall go into this matter in detail when we come to political ethics in our next volume—that this universality of the *usus politicus* relates only to the technical problems of organization and government. It would be a fatal error to suppose that the state should not merely be organized in conformity with reason but should also be based on reason. For reason itself stands in need of a foundation; without it, it is a free and dangerous vagabond.

For, as Luther knows well (even though he failed to take it into account in the field of political ethics), reason is a purely instrumental thing which supplies our aspiring and fearful existence with the moral alibis and *ex post facto* rationalizations which serve its purpose. If existence is ordered aright, reason discharges its ministry. But if existence is not ordered aright, reason becomes a harlot, a spinster who throws herself at everyone and fulfills his desires. It can be hired as a dictator hires writers to do editorials, or even scientists to do research, which will support him with rationalizations and retrospectively vindicate his will on intellectual and moral grounds.

In this sense every aggressive and ideological state has also tried to justify itself spiritually and morally, and has thus made "instrumental" use of reason. For it is of the essence of the ideologies produced by reason that within them evil is not wrong. On the contrary, in the very nature of the case evil is assigned a certain vocation, and thus sanctioned, in the name of the highest value proper to that ideology. If the nation is the highest value, then "whatever serves the nation" is good, even though murder committed in the name of the nation was traditionally regarded as wrong. The slogan, "My country right or wrong," is accepted in this sense. Thus liquidation of the mentally ill or of the middle-class intellectuals can never be wrong within the sphere of those ideologies which demand it or, more

[18] WA 27, 418.

precisely, which prepare the ground for it ideologically and from the very outset sanction it. In this sense the "rational" ideologies are able to establish the polarity of good and evil in whatever way is agreeable to and innocuous for themselves. That is, they can adapt this polarity, making it amendable to the very same "interest" which has autocratically set itself up to put reason to work for itself, employing it as the intellectual defender of its own purposes.

This is why Lutheran Orthodoxy rightly rejected the *usus normativus* of reason and proclaimed instead its *usus organicus,* i.e., its function to serve, to listen, and to obey. To be sure, the *usus organicus,* which expresses what we affirmed to be the functional significance of reason, can be a virtue only if reason stands in the service of the authority of God, i.e., of revelation. If instead it serves the authority of another "interest"—and it must always be "serving" something!—then it becomes, illegitimately, a normative authority. This is why we believe we must go even further than Orthodoxy's doctrine of reason. We cannot simply regard the *usus normativus* and the *usus organicus* as co-ordinate but alternate possibilities. We must rather insist that reason always has the function of serving, and thus always involves in some mode the *usus organicus.* If it serves itself, i.e., if reason becomes the defender of man and of his purposes, it necessarily claims an illegitimate normativity. If on the other hand reason serves God, then it really serves. Hence we would do better to distinguish between the *usus oboediens* and the *usus normativus* (*autocraticus*) as specific instances of reason's one constant function, the *usus organicus,* the function of serving.

That these distinctions must always be worked out in regard to the use and position of reason is due to the fact that reason shares in the fall of man and hence is subject to abuse and perversion. Theologically, therefore, one cannot understand by reason a mere timeless a priori capacity. Reason must be regarded rather as something that is completely wrapped up in the history of man. Hence it is necessary to distinguish between original, fallen, and regenerate reason. The *usus normativus* can be understood only in the light of the fall, for in the *usus normativus* reason takes unto itself the destiny of fallen man; it disobediently transgresses its limits, seeks self-determination, and violates the majesty of God. What Kant means epistemologically by his critique of transcendental reason, i.e., by the critique of reason which transgresses its limits and becomes dogmatic metaphysics, is only a shadow of that actual historical process which we have in view when we speak of the "fall of reason" and of the resultant transgression of limits and self-absolutization.

The way in which reason is integrated into the order of existence depends entirely on whether it is to be understood as autonomous reason or as reason under the authority of God. One may thus assume that this position of reason will work itself out in the sphere of its political use. That it actually does so is plainly shown by historical experience, and especially by the political history of secularism.

While it is true that certain social and other laws operative in political life partake of a kind of universal rationality, still it may be that very different intentions are actually being pursued in the various regulatory and organizational—all uniformly rational—laws. This is in fact the case; these laws simply have the character of means or instruments. The decisive accent falls on the intention, not on the means to realize it. The result is that only in the sphere in which it is a "means"—in which the point at issue is the technique of exercising power, the mechanics of operating the government and preserving the form of rule and organization—only in this sphere does the *usus politicus* possess a universally "rational" character in terms of which it is the same among Christians, Turks, and other groups.

The picture changes, however, the moment we consider the intention present in the Law's "political function." For this is very different among Christians on the one hand and pagans on the other. One has only to consider that the determinative intention behind political action may include such diverse factors as the desire for lebensraum, the concept of the freedom of the individual, or imperialism. But if this is so, then the *usus politicus* of the Law necessarily has a distinctive meaning when it is referred to Christian existence. Inasmuch as the Christian community lives not apart from others but precisely with the political configurations which embrace all men, it too is subject like everyone else to the *usus politicus*. This subjection, however, takes place in a very distinctive way, not externally different, as if there were specifically "Christian" states or "Christian" economies, but inwardly different, in terms of intention.

It is in this sense that we must investigate the *usus politicus* of the Law, that Law which continues to be valid for the justified. We

144

do this, however, only in the sense indicated. Any other way of treating the *usus politicus*, i.e., as a Christian monopoly, will be off limits, since it can lead only to the concept of theocracy and to various forms of clericalism. At this stage in our inquiry we need only indicate briefly the specifically Christian significance of the Law's political function, since we shall be treating the matter fully later in our next volume. For the Christian there is expressed in the *usus politicus* a definite relation to the author of the Law and to the neighbor to whom we are "politically" related. It is of this definite relation that we are to speak, even though intrinsically the *usus politicus* is equally valid for both Christians and pagans, and even though the author of the Law may remain anonymous, as we have seen.

The relation to God to which the *usus politicus* points consists in the fact that God in his patience gives order even to disordered man, who since the fall has turned against God and thus brought himself into jeopardy. God gives man this order so as to preserve him from self-destruction and to secure for him the sheer possibility of physical existence. Christians thus understand the political order in light of the love of God, and they acknowledge it with gratitude. For Christians the political order is decisively characterized by the fact that the author of the Law is not anonymous, at least for them. On the contrary, in the political order he makes himself known as the Emmanuel. He shows himself to be the one who by his Word upholds all things, including political reality, orienting them to the goal of his salvation history.[19]

The relation to the neighbor which is, for the Christian, implicit in the *usus politicus* is determined by the fact that such a relation exists at all. That is what political order is all about, that is its intention, the ordering of the relationship between man and man. Politics, therefore, does not imply a structure which is an end in itself, one which is concerned basically with the construction of an ideal world order. The relation to the neighbor which thus determines the political order has as its norm the "law of love."[20] This law alone must be the goal of all laws.[21] Luther says well that even

[19] See the fifth thesis of the Barmen theological declaration of 1934 in Arthur C. Cochrane, *The Church's Confession under Hitler* (Philadelphia: Westminster, 1962), p. 241.

[20] See *LW* 45, 98 and *PE* 5, 69.

[21] *WA* 15, 691.

the power to judge and punish is not given to the prince "because of his beautiful, yellow hair."[22] Political power derives rather from the Fourth Commandment, i.e., from paternal authority, and is thus grounded in love, the love of the heavenly and the earthly father.[23] The twofold form of the command to love, however, shows that love for the neighbor can take place only in conjunction with love for God. Hence from this angle too the *usus politicus* is for the Christian linked with the author of the Law and the Gospel.

Even in this final form, therefore, the Law is for the justified a constant and necessary reminder—a true *anamnesis*—of the one who encounters him in justification as the gracious God and who beyond that still wills to be gracious to him also in the political use of the Law. For herein God secures for man his time in this aeon. He gives him the basis of physical existence, without which human life would collapse. He bestows upon him one of those possibilities which are essential if man is to be able to show his love and thereby express his thanksgiving and praise.

Thus in the political form of the Law, we see again the over-all significance which the Law in all its forms has for the justified. It reminds him, with respect to all the spheres of life embraced by the commandments, in what ways and in how many ways the "new obedience" is to be exercised. The Law has the task of introducing this obedience into all the contexts of life, so that there will not be any unredeemed spheres whose relation to the God of judgment and grace remains unrecognized or is disregarded, which necessarily leads to "disintegration" and "hypocrisy." Among these contexts is also the political sphere. It too could become one of those unredeemed spheres which the Christian does not relate to the fact of justification but abandons instead to autonomy. (It would not be difficult to recount from recent history dreadful illustrations of the "unredeemed" and autonomous sphere of the political.) Hence this sphere too could lead to hypocrisy, this sphere too is in need of cleansing. The task of the *usus politicus* for the Christian is to remind us of this fact.

[22] *PE* 5, 60.
[23] Luther, The Large Catechism, I, 141-142, 150; *BC* 384-386. In thus grounding the order of the state on the personal I-Thou relation, Luther differs from Melanchthon, whose concept of society contains an idealistic Utopianism (see p. 9 above).

THE COMMAND OF THE CREATOR AND
THE LAW FOR THE FALLEN WORLD

9.

The Command of Creation and the
Image of God in Man

All the forms and uses of the Law thus far considered have one thing in common: the Law always relates to fallen man (who though called to redemption is still fallen man). The Law has to expose his sin and (like "gauze in the wound") keep it exposed. It has to call him out of his perplexity and doubt, and therefore out of a particular actualization of his fallen state. It has continuing validity on the assumption that there are "unredeemed spheres" even in the justified, so that the justified finds himself to be advancing and maturing but not yet perfected. The doctrine of the Law must always be viewed against the background of the fall. The dark foil of this fallen aeon, which flees from God and blatantly defies him but is nonetheless called back by him, must be considered at every point when we speak of the Law of God. Whatever we say theologically concerning the nature of the Law must take its bearings from the Pauline statement that the Law was "added because of transgressions" (Gal. 3:19).

From this it follows that one cannot understand the Law simply as "the" will of God. The Law is rather God's will as it pertains to us [quoad nos]. It is his will as it appears in the refracted light of our particular situation. Perhaps one might even say that it is the will of God as altered by the fallen world.

God's Original Will and the Law in Salvation History

If biblical evidence is sought for this "refraction" of the divine *nomos* through the medium of this aeon, reference may be made to the way the command [Gebot] of God in creation (Gen. 1:28) is changed into the law [Gesetz] of the Noachic covenant which has reference to the fall (Gen. 9:1 ff.). In the latter the original command is still recognizable, but it has undergone remarkable modifications.

In Genesis 1 man's ruling position in the cosmos is regarded more as a hierarchic position in the order of creation, which man "has" purely by virtue of the divine appointment. Within the Noachic covenant, however, this position is subject to the law of conflict and terror which is operative in the fallen aeon; it is linked with "fear and dread" (Gen. 9:2) and related to guilt and atonement (9:6). When God ordains capital punishment, one may appropriately say that here the original will of God is not expressed in the Law. What is expressed is rather the will of God as it is altered by sin, or better, the will of God which adapts itself, which does not withdraw from fallen man in rigid self-assertion but instead pursues him and makes itself known to him in a form which takes into account the disrupted possibilities of the fallen world. If it did not do this, the will of God could only mean death for the fallen world.

Biblically, the same point is made in another connection. In the name of God, Moses can command or allow that "a certificate of divorce" be given (Matt. 19:7; Mark 10:4). It is expressly emphasized, however, that this is only because of "hardness of heart" (Mark 10:5). Hence the Mosaic directive is to be distinguished from the true will of God. What is here proclaimed as Law is not the original "command of creation." For "from the beginning of creation" (Mark 10:6; Matt. 19:8b) it was not so. According to the true command given at the beginning of creation, husband and wife are rather thrown back upon one another. But if God were to maintain unswervingly the original command, if in his grace he did not reach out to us in the particular circumstances of our fallen state —in this case if he did not take up the matter of divorce—an intolerable situation would arise, and the unmodified command of creation would cause real misery.

Theologically, therefore, it is important to view the Law of God for this fallen world in terms of this constitutive relationship to sin. When Galations 3:19 says that the Law was "added because of transgressions," or when Romans 5:20 says that the Law "came in to increase the trespass," we are compelled to conclude that the "Law" of God must be distinguished from the "will" of God, the former being a specific instance of the latter, i.e., the "Law" is God's "will" as modified by the fallen world.

To avoid misunderstanding, we may note expressly that this modification is not to be regarded as man's work, in the sense that man through the fall has forced God to abandon his original plan. On the contrary, God and God alone is to be regarded as the subject of this modification. The change is a sign of his gracious condescension. It is a sign that God goes after the prodigal son all the way to the far country, in order that he might not lose him but reach him. We must see to it—as the Bible repeatedly does—that the son remains aware of this approach, that he always views the Law which prevails in the far country (this does not mean simply whatever is the law of the land there!) against the background of the true Law which obtains in the father's house. We must see to it that he relativizes the former in the name of the latter, which is the real Law. In the Bible this is done in Jesus' reference to the Law at the beginning of creation. It is done in the Sermon on the Mount. It is done in Paul's relating of the Law to transgressions. It is done by the very fact that the first chapter of the Bible is there at all.

The "coming in" of the Law (Rom. 5:20), the fact that it is historically and in terms of its very substance connected with sin, can thus be understood only if the Law is not immediately equated with the will of God. For the form in which God's will operates is always different. God's will is actualized not only in the command of creation, not only in the Law for fallen man, but also in the declaration of his will for the coming kingdom of God, as that is prefigured in the anticipated eschatology of the Sermon on the Mount. The form in which God's will is expressed changes continually. If, then, we are to know the secret of the divine Law, we must see in it one of these forms of expression of the divine will. We must understand it both in terms of its function in salvation history and in terms of its particular locus in salvation history.

It would be quite erroneous to try to understand this Law in terms of the "abstract truth" it contains or to ascribe to it the kind of "timeless validity" expressed, for example, in the categorical imperative. The moment we do this the Decalogue becomes "natural" law and the axis of a corresponding system of "natural" law. It becomes a moral idea which moves in the void and is no longer rooted in history. The existence of a system of natural law always indicates a crisis in the concept of history. Behind the historical phenomenon, behind the positive law which changes, there is sought a constant factor, a timeless Platonic idea of law, an abstract norm of the moral. To be sure, this negative judgment is not the only criticism which theology has to make of natural law; if that were the case it would imply a thoroughgoing committal of theology to positivism. Nevertheless, behind every attempt to interpret the Law of God simply as natural law there is this attack on history. History becomes mere illustration or mythical adornment of that which is timelessly valid.

If our theological ethics is to treat natural law critically, as it must, the ultimate foundation of such criticism must be laid at this stage in our discussion. This foundation consists in demonstrating that the Law of God has significance for a particular epoch within salvation history, and that it is thus restricted to a specific phase of this history, namely, the interim between the fall and the last judgment. With respect to the Law there is no such thing as a continuum stretching from the beginning of creation across this interim to the eschaton; there is no such thing as a legal norm which could be neutral or indifferent toward differences in time, and hence itself timeless.

Now the fact that the Law has significance only for a particular historical epoch can be perceived only if we set the Law of one epoch, that of the fallen world, over against the Law of another "epoch," the original state. Only thus is it possible also to see clearly the distinction which we made above between the "proper" and the "improper" will of God. As we can know sin itself only against the background of the unblemished creation—which is why there can be no doctrine of sin without a corresponding doctrine of original innocence —so we can know the Law which refers to sin only against the background of the pure command of creation. The very existence of salvation history and the fact that the Law is rooted in it are confirmed only by this correlation with what was primal. It is this correlation alone which also gives ultimate justification to any differentiation of theological ethics from its philosophical stepsister, and from all ideologies of natural law. Everything we have previously said about this differentiation must also be seen in this light.

To show this correlation we must first describe man's original image of God [imago Dei] because the image can be understood only in terms of the divine command in creation and because the nature of the command is portrayed in its human reflection. As the

Law of God in the fallen world corresponds to man as sinner, and to the man who in his sin is justified, so the command of creation corresponds to the *imago Dei*. And as fallen man can be understood only in his relation to the broken (really fragmented!) *imago Dei*, so the Law which has "come in" (Rom. 5:20), and which has reference to this aeon, can be understood only in the light of the original command. As the Law in the fallen aeon tells us who we are, so the original command tells us whence we come. We can be known as we are, however, only in terms of the fact that we come, and from whence we come. For this reason, the two lines of approach are inseparably related.

Hence discussion of the *imago Dei* belongs in theological ethics. For ethics speaks of what man "ought" to be, and that can be done only on the basis of what man "is." [1] And what man "is" can be understood only if we establish whence he "comes."

Once we have considered the original *imago Dei*, we shall be free to recognize "man in revolt," and in the light of this man to understand the Law of the fallen world as an alteration of the original will of God. In this connection we shall then speak of the nature of original sin and of the problem of freedom and the bondage of the will.

We would now introduce our teaching on the *imago Dei* with certain theses which we shall then proceed to develop. The command of God the creator and the corresponding obedience of man his creature together describe the fellowship with God which we call the divine likeness. We therefore understand by the command of God the creator that command in which God hands over to his creature the life of man in all its fullness, and lays upon him the task of living that life in fellowship with God. This command insures that according to the Christian view of creation the human creature was made to be a person, "man." Faith in creation sets me in a personal I-Thou relationship to God, not primarily into an I-it relationship to the created world. Christian thinking about creation is personalistic. Moreover, the doctrine of the *imago Dei* does not describe a given attribute or property [*habitus*] of man which may be demonstrated in detail. It has reference rather to the alien dignity which man

---

[1] See our discussion of the relation between what is and what ought to be in terms of the relation between indicative and imperative on pp. 51-93 above.

possesses by way of his divine prototype [*Urbild*], that original which is present in Christ alone. The *imago Dei* therefore, in keeping with the reciprocal interrelation of indicative and imperative, is not simply a gift; it is also the task implicit in the gift. In the concept "destiny" both these factors are brought into juxtaposition. Since man cannot fulfill his destiny in his own strength, the divine likeness is to be attained only in the gift of the Holy Spirit.

The Imago Dei as Task and as Destiny

The divine likeness—the fact that man according to Genesis 1:26 is created after the image of God and is thus particularly singled out from among the various forms of creation—is not only a gift but a task as well. In this respect it shares the common meaning and purpose of creatureliness. That the divine likeness is understood in this way in the Bible may be seen among other things from the fact— we shall develop the point more fully later—that real use of the concept is always made in connection with the phenomenon of the indicative and imperative. The image *has been* created by God (indicative), but we are now also to *put on* the "new man" who is renewed on the model of this image (Col. 3:10). The image is both gift and task. It is to be noted that precisely in Colossians the verse which refers to the image is surrounded by nothing but imperatives (Col. 3:5 ff., 12 ff.; cf. also Eph. 4:24, where we must again "put on" [imperative] the man created after God [indicative]). If we recall that in the New Testament the divine likeness necessarily finds its norm in the image of God in Jesus Christ—and hence is emphatically integrated into the order of redemption as well as that of creation— then the constant relation to the phenomenon of the indicative and imperative will become perfectly clear (cf. Rom. 8:29 f.; II Cor. 3:17 f.).

In other words, the creator is here viewed, not merely as the prototype from which human existence derives, but also as the model after which it should be patterned and to which it should conform. Divine likeness is not just that *from* which I come. It is also that *to* which—through the gift and claim of God—I go. With respect to the divine likeness being that "from which" I come, we would miss the point of this coming if we were to understand it only in terms of historical derivation, if we were to have in view only the time of that primal period in Paradise—and thus bring ourselves into conflict with the scientific findings of paleontology. Similarly, we would miss the point of the "whither" if we were to understand it only as the content of a prognosis and not as the content of a promise which claims me. On the contrary, the "whence" especially is to be understood in terms not so much of history as of the substance of the matter itself (though to be sure in this instance history and substance are all wrapped up together, and must be if we are not to fall victim to spiritualizing symbolism). The "whence" is to be understood in terms of the matter itself in this sense, that I was intact and whole when I came from the hands of God, and that my *de facto* existence as a sinner is therefore my own responsibility.

Just as the divine likeness cannot be viewed aesthetically (from the safe distance of non-commitment), so it cannot be viewed historically, in the sense of a pedigree in black and white which settles the matter and can safely be stored with my other valuable papers. On the contrary, the divine likeness constitutes a disturbing goad. For because of it the Law of God is able to address me with its claim. I do not learn of the divine likeness through a historical account that "once it was so." That would only lead to sentimentality along the lines of the "good old days." On the contrary, I learn of the divine likeness by being addressed with a claim which I cannot meet. The content of the message concerning the *imago*—the fact that as gift it is also task and Law—determines also the form in which that message may be perceived. For it is by the fact of having it granted to me once again that I come to see what it was I lost; in the *imago* I perceive also the "Law." Hence I learn of the divine likeness both in the promise and in the acute judgment, both in Gospel and in Law.

It is in this sense that, on the basis of the *imago,* Jesus addresses us in the Sermon on the Mount: We are to be perfect as our Father in heaven is perfect (Matt. 5:48). Only against this background, i.e., only in the light of this understanding of the perfection of the divine likeness, can we understand the Sermon's radicalization of the Mosaic Law. The divine likeness is not just something of which we are told. It is something on the basis of which a claim is laid upon us. It is something which—in the strange ambivalence of imperative and future which marks this saying of Jesus—encounters me as Law and promise, as judgment and grace.

Thus the message of the *imago Dei* addresses me on the basis of my real but forfeited existence in order that precisely on account of this message I may turn from my "past" (that from "whence" I come). It addresses me in order to give me my future, which is promised to me anew in this very message. If we can call the original state of the *imago Dei* "original righteousness" [*justitia originalis*] then we can designate the image of the "new man" given me in Christ as "final righteousness" [*justitia finalis*]. This concept of what is "final" expresses not only the character of the *imago* message as a claim, but also the fact that in Christ what is involved is not simply a restoring, but a transcending of the original—in any case something "new"—comparable to the rejoicing over the returning prodigal which exceeded that over the son who remained at home (Luke 15:7, 22-24, 31 f.), or to the grace which, where sin increased, "abounded all the more" (Rom. 5:20).

The fact that the divine likeness implies a "task" and fixes a "goal" suggests something concerning its real nature. It suggests that the divine likeness is exclusively a teleological and not an ontological concept. In other words, what we have here is a state of relation and not a state of being. We shall now consider successively those two conceptions, the teleological and the ontological.

The Teleological View of the Imago Dei

According to the creation stories in Genesis the teleological reference of the divine likeness is in two directions, first towards the fellow creature, and second toward the creator. We shall deal first with the relation to the fellow creature.

In Genesis 1:26 the divine likeness is perceived in the peculiar and pre-eminent position accorded man in relation to the rest of creation. According to the Priestly code there is a crescendo of creation with a clear break after the creation of the beasts. At this point there is inserted a divine monologue which prepares the way for the extraordinary thing which now follows, the creation of man. Here the distinctive feature of man as compared with the rest of creation is that he appears as the goal of all the rest, not only in the quantitative sense that he is the apex of the whole creation but also qualitatively inasmuch as he alone receives the divine likeness and so has an orientation which is "above." The effect of this special position in the cosmos is that man is to rule over all the rest of creation (Gen. 1:26, 28).

The fact that man is the goal of creation is expressed more precisely in two different nuances, first in God's resolve in the divine monologue to create man with this goal in view (v. 26), and second in God's command to man that he should now voluntarily bring the plan to realization (v. 28). The goal is thus built into man indicatively and imposed upon him imperatively at one and the same time. This setting of the goal in two ways, which for man implies a twofold pre-eminence and distinction as compared with the rest of creation, is what really constitutes the nature of the divine likeness. For the nature of a thing is always to be "defined" in terms of its telos or goal. (*Finis* is at the very center of the word "definition.") The nature of the creation is such that it serves, that it is ordered and directed by that which is above it. In this sense the creation is "referred to man and needs his dominion as an ordering principle. That man should be furnished with the divine likeness for this purpose is logical enough, for even earthly rulers, when they cannot be present in person, usually set up their images as signs of majesty." [2] 'As creation receives its order from man, so man receives his order from God.[3]

Thus all creatures are related to God. For some the relationship is direct, for others indirect. These various relations describe the original hierarchical order of the world.

It is characteristic that the various references to the divine likeness in Genesis (1:26; 5:1; 9:6; cf. also Wis. 2:23; Sir. 17:3) do not give us statements of ontological content but restrict themselves to these hierarchical relations, to the position of man in the total cosmos. This is true even of a verse like Genesis 5:1, where the concept of the divine likeness is, as it were, a preamble to the first genealogy in the Bible, the one that ends with Noah (5:32). If we recall that genealogies are repeatedly introduced into the salvation history—and thus are

---

[2] Gerhard von Rad, εἰκών, in Gerhard Kittel, *Theological Dictionary of the New Testament*, edited and translated by Geoffrey W. Bromiley (Grand Rapids, Michigan: Eerdmans, 1964——), II, 392.

[3] See the paragraph transcribed by Rörer in Luther's *Lectures on Genesis*, WA 42, 46, n. 1. [The paragraph is from an edition not used in *LW* 1.—Ed.]

determined by a teleological interest in salvation rather than a chronological interest—we will recognize that here too the significance of the *imago* concept is in terms of relation. The genealogical data simply set forth in a certain form the history of that which has happened in respect of God's plan for man, both in the sense of man's non-realization of the plan and in the sense of a divine action which despite man's disorder and guilt still sticks to the plan. God sticks to his plan because the existence of the image, despite the fact that it is laid upon man as a task, is not dependent upon man's actualizing it through his obedience to the command of creation. If it were, then the image would indeed be completely shattered with the disobedience of the fall. The image, however, is something which God has undertaken to project and to create—this was the first of our two nuances —and therefore the image cannot be done away even though the form of its existence constitutes a difficult theological problem the moment the divine plan and man's attitude towards it conflict.

In commenting on Genesis 1:26, Luther distinguishes between a "private" and a "public" likeness.[4] This distinction corresponds roughly to the one we made between the ontic and the relational concepts of the *imago*. For by "private" likeness, Luther has in view man in himself, in his constituent elements, without external relations, as seen from within [*Eigenschaften*] rather than from without [*Aussenschaften*], e.g., in respect of his righteousness and goodness. Luther believes that Paul teaches the loss of this "content" of the divine likeness through the fall. On the other hand, Moses is held to speak of the "public" likeness revealed in relation to the world and expressed in man's ruling position. This formal[5] likeness, the likeness viewed in terms of its outward reference, remains intact: ". . . in our image."

---

[4] [See *WA* 42, 51, where lines 1-12 are from an edition of Luther's *Lectures on Genesis* not used in *LW* 1.—Ed.]

[5] To avoid misunderstanding I may point out that I am using the word "formal" in a different sense from that of Emil Brunner. (See *Man in Revolt*, trans. Olive Wyon [Philadelphia: Westminster, 1947], pp. 499 ff., esp. pp. 512 ff.) What Brunner mostly has in view are formal qualities of man such as freedom, responsibility, the gift of reason, and he sees in these factors relics of the original state. I am using the term, however, in the Scholastic sense of *forma* as the reason for a thing, its point and purpose. This is why I use the term in speaking of the position of man in the totality, his relation to the creator and to his fellow creatures.

The divine likeness is thus a relational entity because it is manifested in man's ruling position vis-à-vis the rest of creation, or better, because it *consists* in this manifestation, in this exercise of dominion and lordship. The attempt to differentiate the essence of the image from its manifestation, and therefore to understand man's ruling position of lordship only as a result of the true properties of the image (reason, will, freedom, etc.), has no foundation in the Bible and betrays a Platonic mode of thinking. The image of God consists in the manifestation, for it is of the very essence of a picture —that is its point!—to "effect" something, for example, in the person who looks at it; it "consists" in this effect, not in the variety of colors. The *imago Dei* does not consist apart from its specific operation. Luther's repeated insistence that God is always *actuosus,* in action, holds true also of God's image. The same approach is to be seen in Melanchthon's view that the nature of Christ is to be seen in his benefits, in his operation in salvation history, rather than in his metaphysical attributes. It is an approach which opposes differentiation between capacity or ability on the one hand, i.e., an attribute which enables us to do something, and activity on the other, i.e., this attribute in operation. We have to see the nature of God, not in his attributes [*Eigenschaften*], but in his outward relations [*Aussenschaften*], in what he does with us, in his relation to us, in his being "Emmanuel" [God with us]. The image of God in man is to be similarly defined. It is not a constituent capacity inherent in man but a relational entity, namely, man's ruling function vis-à-vis the other creatures.

To be sure, this is only one aspect of the teleological reference. There is, of course, another side to the matter as well. For the *imago Dei* consists also in a relation to God. In terms of the matter itself it would be more fitting to speak of this aspect first, since logically it precedes man's dominion over creation. But in adopting the present sequence we have been swayed by the fact that the creaturely aspect is given primary emphasis in the Bible. This sequence also gives stronger safeguards against speculation about the image in terms of constituent qualities or attributes.

The relation to God which man possesses as *imago* is again one of strict pre-eminence as compared with the rest of creation. It is so in the following directions. In the first place, according to the

Yahwist, God imparts to man the breath of divine life and thus gives him a share in God which sets man in an intermediate position between God and earthly matter, a position not granted to the rest of creation (Gen. 2:7). In I Corinthians 15:45 Paul alludes to the fact that there is here an advance intimation of man's resurrection body and therefore of the state of fellowship between man and God consummated in Christ. For the ψυχὴ ζῶσα, the "living being" which is all that the first Adam is in contrast with the πνεῦμα ζωοποιοῦν, the "life giving spirit" of the second Adam, this ψυχὴ ζῶσα comes into being—to use the language of the Septuagint—through the inbreathing of the πνοὴ ζωῆς. Hence, as originally projected, the first Adam too is already one who participates in the spirit [πνεῦμα] of God and shares in his life [ζωή]. From this standpoint too, then, the divine image expresses, not constituent qualities of man, but an external relation of fellowship. Precisely from this standpoint it is clear that the *imago Dei* in its quality as gift implies also a task, namely, to actualize this fellowship, to countersign it from the human side.

Now there is a second side to the Godward aspect of the *imago's* teleological reference, in which this is expressed even better. In both the Priestly code (Gen. 1:28) and the Yahwist (Gen. 2:16) man is addressed in the form of a command. In the one case man is to act as God's vicegerent in taking over the lordship of creation; in the other he is forbidden to eat of the "tree of the knowledge of good and evil"—on pain of death. Through God's address to him man is thus brought into relation with God, but at the same time specifically kept at a distance from God. God reserves for himself a zone of majesty which man must respect, thus maintaining "merely" a divine likeness and not pretending to some kind of equality with God. Here it is suggested—again as a prerogative accorded only to man of all the creatures—that man's special situation in creation is characterized also by special danger, a danger which is peculiar to man alone. Man can transgress the command and thereby fall victim to death, he can alter his quality as divine likeness in the most dangerous manner—but only alter it, not destroy it. For it is in the very nature of the image itself to expose man to this peril, to allow him but also to burden him with the making of decisions which are unknown to the beast and which therefore cannot bring the beast to ruin. Failure is something specifically human; it belongs specifically to the divine likeness.

To the question whether man has lost certain ontic attributes we cannot, then, give either a positive or a negative answer, since the question is wrongly put. For what is involved is always either the maintenance or the breach of a certain relationship. But even a broken relationship is still a relationship. The relation of protest, the mode of negation, is still a prerogative of man which the other creatures do not possess even as a possibility. The *imago Dei* is thus a *character indelebilis*, in both the positive and the negative mode.

Hence the crucial point is to affirm this quality of the image as relation. Luther defines it thus when he describes the image of God as "perfect knowledge of God, supreme love for God, eternal life, eternal bliss, eternal security." [6] In the same context he rejects any definition of the image in terms of content, since this would logically force us to deduce that the devil is also God's image.

Hence it is the fellowship with God as both a gift and a task which constitutes the *imago Dei*. This fellowship finds expression in the twofold fact that God endows man with his breath, his $\pi\nu\epsilon\hat{\upsilon}\mu\alpha$, and that he addresses man, setting him over against himself as a Thou. The rest of creation, like everything that is perishable, is also a likeness of God (Ps. 104). But there is a profound difference precisely in the form of the similitude. These other creatures receive their life from God unconsciously. God's creativity and providence are poured out in the plenitude of creaturely life, but without involving any question, appeal, problem, or possibility of failure. Unconscious nature, whether organic or inorganic, has to do only with the breath of God; more precisely, it is simply the object of this breath (Ps. 104:30). Man on the other hand has to do with God's Word. Herein lies his distinction, his dignity, and his burden.

The biblical account of creation and the later references to it in Genesis say nothing about the constituent qualities of the divine likeness. Instead they restrict themselves to the afore-mentioned relations, without throwing any light on the inherent qualities necessary to fulfill them. This means that there is something ineffable about the *imago Dei*. It cannot be made the subject of concrete statements. Now this ineffability has nothing whatever to do with the

[6] See p. 155, n. 3 above.

methodological difficulties of theological speech; these we can and should try to solve. The ineffability of the *imago Dei* is in the strict sense rooted in the very nature of the matter itself. As Gerhard von Rad has rightly pointed out, this veiling of the concrete is understandable, since ultimately it is "the ineffability of the divine being" which in turn "draws a veil of mystery over the Godward side of man (in his quality as *imago* of this God)." [7]

The Ontological View of the Imago Dei

We have now said enough to demonstrate the impossibility of defining the divine likeness in an ontic sense. Yet we cannot leave this negative aspect of the matter without explicitly mentioning certain ontic features on which there may be, and indeed continually has been, reflection. One might refer, for example, to certain qualities which appear to afford man in the first place the possibility of exercising dominion over creation or of making the requisite decision with respect to God, qualities such as personality, freedom, responsibility, conscience, dignity, or the free exercise of the moral disposition. Now there is nothing of this whatsoever in Genesis, and we believe we have shown that this silence is more than just an inadvertent omission, that there are actually sound reasons for it both soteriologically and in the nature of the matter itself.

At best, one might still be inclined to think that Genesis sees in man's corporeality a particular mark of the divine likeness. But von Rad has rightly shown that the distinction between body and soul is an unwarranted intrusion into the text, and that emphasis on corporeality arises only in protest against an illegitimate spiritualizing. The genealogy in Genesis 5 may be interpreted only in this sense. For the reference to the physical progeny of Adam is important in that the *imago* character thereby applies to his whole race and is not regarded simply as an idea to which the first man alone stood in a unique and unrepeatable relation. In the same sense the *imago* has a physical representation (Gen. 9:6): one can sin against it by murder, i.e., by physical extirpation. Nevertheless, any attempt to

---

[7] Gerhard von Rad, *loc. cit.* Cf. Luther's statement: "Therefore when we speak about that image, we are speaking about something unknown. Not only have we had no experience of it, but we continually experience the opposite; and so we hear nothing except bare words." *LW* 1, 63.

go beyond this, and to make corporeality a specific ontic feature of the *imago Dei*, leads inevitably to speculation. It is in this sense that one must understand and evaluate the view of Thomas Aquinas that the human body symbolically expresses by its erect stance the glory of the soul's divine likeness, and that the body thus possesses a divine likeness *per modum vestigii.*[8]

The most profound objection against an ontic-qualitative description of the *imago* is doubtless the one raised by Luther when he points out, as we have already noted, that the devil possesses these constituent *imago* qualities to an even higher degree than man. To be sure, Luther is also quite uninhibited in making ontic statements as well. Thus the original divine likeness consists for him in perfect "knowledge of" and "belief in" God, and in the fact that Adam lived "a life that was wholly godly; that is, he was without the fear of death or of any other danger, and was content with God's favor." Adam could thus "speak with the serpent without any fear, as we do with a lamb or a dog."[9] Luther can go almost to the point of fantasy in this depiction of the ontic marks of the image. Not only does he say that Adam's "intellect was the clearest, his memory was the best, and his will was the most straightforward . . . also those most beautiful and superb qualities of body and of all the limbs," prior to the fall. He also describes with dramatic and glorious naïveté the superiority of man to the animal kingdom: "I am fully convinced that before Adam's sin his eyes were so sharp and clear that they surpassed those of the lynx and eagle. He was stronger than the lions and the bears, whose strength is very great, and he handled them the way we handle puppies."[10] Indeed, Luther can go so far as to say that the earth itself was more beautiful, radiant, and fruitful in this primitive state: "I hold that before sin the sun was brighter, the water purer, the trees more fruitful, and the fields more fertile."[11]

One should not attach too much theological significance to these humorous and carefree imaginings. For, as we have seen, all these

[8] "The very shape of the body represents the image of God in the soul by way of a trace." *Summa Theologica,* part 1, ques. 93, art. 6, reply obj. 3; *The "Summa Theologica" of St. Thomas Aquinas,* trans. Fathers of the English Dominican Province (2nd rev. ed.; London: Burns Oates & Washbourne Ltd., 1922), p. 295.

[9] *LW* 1, 62-63.

[10] *LW* 1, 62.

[11] *LW* 1, 64.

ontic qualities are decisively determined and characterized by the distinctive mark of obedient submission to God in love, knowledge, and faith. While reference is actually made to certain lofty ontic qualities, their significance is relativized by the framework of the original hierarchical relationship in which man is subordinated to God, i.e., by that which Luther describes as the *imago publica*. The crucial importance of this question of the distinguishing mark or token becomes manifest when it is said that the devil himself possesses these ontic qualities to an even higher degree: "If these powers [namely the ontic qualities] are the image of God, it will also follow that Satan was created according to the image of God, since he surely has these natural endowments, such as memory and a very superior intellect and a most determined will, to a far higher degree than we have them." [12]

Luther thus brings his doctrine of the *imago* within the framework of his doctrine of justification. For here too it is not our works or qualities which in themselves constitute our value before God; it is rather faith alone, i.e., the mark or sign, the attitude towards God in the name of which those works and qualities are exercised. If divorce were ever to take place under this sign, it would be no sin; apart from this sign, i.e., in unbelief, even worship of God is idolatry. [13] Even the demons know that God exists—that is, they have the ontic gift of *cognitio*—and shudder (Jas. 2:19). It is at this point that we see the significance of the fact that, notwithstanding Luther's flights of fancy in their depiction, the *imago Dei* for him is not to be understood in terms of constituent qualities but exclusively as a relation. It is at this point that we see the continuous line running from the doctrine of creation to the doctrine of justification, from the image as originally given to the image as given again in the new creation [καινὴ κτίσις].

If we keep in mind the fact that the real intent of Luther's teaching on the primitive state lies in the direction of a personal relationship, then his occasional remarks about the remnants of the image, the "relics of creation," will fall into their proper place. For example, Luther says, "We are not so thoroughly inclined toward evil that

---

[12] *LW* 1, 61.

[13] See the Theses of 1520 on the question whether works contribute to justification, esp. Nos. 1, 10, and 11; *WA* 7, 231 f.

there is not left to us a portion which is affected toward the good." (He is thinking above all—as he says here—of the conscience, the "*synteresis*.")[14] Here, as in certain of Luther's statements concerning natural theology, what we have are relics of Scholastic thinking which cannot be reconciled with his real theological intent as we have described it. One must be on guard precisely at this point against quotations torn from their context in the very heart of Luther's theology.[15]

The ontic constituents of the divine likeness thus have only instrumental significance. They cannot be regarded independently of their purpose. They can be described only as means which either accord or conflict with this goal. If they conflict with it, this does not imply —despite Luther's fantasies—a quantitative diminution of the ontic qualities. It does mean that they are qualitatively transformed into their very opposite: the being which after God was most highly endowed becomes Satan in a Luciferian fall. Man too experiences by his fall, not a quantitative loss of reason in the sense that he sinks to the level of the irrational creatures, but a qualitative alteration in the sense that reason now serves only to make him "even more brutish than the beasts." The higher reason is, the more dangerous it is.

If we have pressed our rejection of the ontic character of the *imago* to the limit, we may state in conclusion that the *imago* is thus not a residual condition, a property in the sense of a *habitus*, but something which actively happens in the moment. It has the quality of an event. This eventful quality consists in the fact that the relation to God which constitutes the *imago* must be constantly realized afresh. Fellowship always exists only as it is achieved, only as it is realized. God must posit it; man must realize it. The continuous aspect of the relationship lies exclusively in the divine promise, in God's loving will, which causes him to create an image and stick by it through every breach of relationship. The momentary aspect lies in the fact that the maintenance of the image in turn depends on man's obedience, on man's consenting to it and entering into it. In other words, the continuous aspect is the indicative of the divine promise, the

[14] Martin Luther, *Lectures on Romans*, translated and edited by Wilhelm Pauck (Philadelphia: Westminster, 1961); LCC 15, 88 [see also LCC 15, 24, n. 46.—Ed.].

[15] On the hermeneutic considerations which underlie my objection to such an interpretive method see my *Kirche und Öffentlichkeit* (2nd ed.; Tübingen, 1948), pp. 68 f.

momentary aspect is man's attitude of obedience or disobedience in face of this indicative.

To safeguard the momentary and personal aspect, and therefore to express the relational character of the *imago*, one must accept the extreme formulation: "The divine address constitutes the person." We put it thus in conscious opposition to the antithetical formulation that mere "addressability" is what constitutes the person, that the person is the object of a "contacting action" [*Anknüpfung*]. To adopt the latter formulation is to move in the ontic sphere of thought which draws attention to certain ontic factors, to certain "relics" of the created person which, as intact nature, form the basis for the grace given in the Word. Such an understanding of person is unbiblical; it derives rather from Greek and Aristotelian origins.

This conception is impossible, not merely in terms of the concept of person, but also in terms of the Word. For the person who has turned from God is not susceptible of being addressed by the Word in the sense that his hearing is his own contribution or co-operation. "The unspiritual man does not receive the gifts of the Spirit of God" (I Cor. 2:14). There is no such thing as a capacity for hearing that corresponds to the fact that man is addressed. The fact that the divine majesty is objectively presented in the works of creation (Rom. 1:18 ff.) by no means implies a corresponding capacity in man to lay hold of this presentation and, as it were, put it to fruitful effect. It is quite possible for God's revelation to be objectively given while it is subjectively obstructed. Hence the revealing of revelation takes place only in the miracle of the Holy Spirit.

This argument is strange and alien to post-Kantian thinking. According to Kantian epistemology, the object of experience—which here corresponds to the content of God's revelation in creation (Romans 1) or even to his revelation in the Word—is identical with an objective capacity to experience; experience is a transcendental function of the subject doing the experiencing. Over against this view it must be said that revelation is by no means identical with revelation-to-*me*. Revelation can imply, and insofar as I look at myself it *must* imply, a being-hidden-from-me. This "being-hidden" of course is such that it roots not in the nature of revelation but in my own nature, which turns in upon itself, and which stands under the wrath of God because of its blindness and deafness.

Thus in the modern doctrine of man's "addressability," theology has not merely made common cause with certain elements in the Aristotelian and Thomistic doctrine of nature, as did Protestant Orthodoxy with its doctrine of "general revelation" and "relics of the primal state." It has also adopted Kantian epistemology and its doctrine of the subject-object relation in experience.

That this subject-object relation simply does not exist as regards revelation—let us come right out and say it, as regards "religious experience"—that there is no such attribute or epistemological quality as "addressability," we would assert as forcefully as possible by our statement that it is the divine address which constitutes the person as *imago Dei*. If that be true then the human person cannot be demonstrated in his given ontic qualities because he exists only in God, witness Paul's pregnant phrase about being "in Christ," i.e., participating in the divine likeness of the divine Son.

The divine likeness rests on the fact that God remembers man, in spite of the corruption which prevents our identifying certain relics of the *imago* and from them reconstructing the *imago* itself. The genitive in *imago Dei* is a subjective genitive. What is involved is "our image" as that image consists in God's remembrance of us. What is at issue is the image which God has of us, the image of the prodigal which the Father "retains." He keeps him in remembrance as "his" son, even though the sonship is corrupted to the point of extreme depravity, even though those who know the son are forced to ask, on the basis of "experience," on the basis of what they see in him ontically, "What? He is supposed to be_____'s son? It is hardly believable." But the Father keeps the true image in his heart, and only here is it truly carried. But here too it is hidden. For what is carried in the heart cannot be projected outwards, nor truly seen from without. Hence the *imago Dei*—man!—is the object of faith and not of knowledge. Man really exists only in the consciousness of God. Hence man is present to me only as God himself is present, namely, in faith.

It would certainly be inappropriate at this point to protest that the sonship does not merely lie in the remembrance of the Father but is also ontic in a biological sense. We simply cannot reflect on the indelible character of such a biological sonship, since in the parable of the prodigal son the sonship is viewed not from the standpoint of biological existence but from that of the "worth" implicit in a relationship: "I am no longer *worthy* to be called your son" (Luke 15:19). It is self-evident to the returning prodigal that his biological sonship remains intact. But he knows equally well that he cannot appeal to his biological nature, that it is inadequate for restoring the father-son relationship. He knows that what matters is the filial relation, the required "worth" which is either recovered or forfeited. He *has* forfeited it, and so no longer possesses any demonstrable qualities of sonship. He is wholly and utterly thrown back on the father's remembering him, on the image that the father has kept of him, which is certainly not the image which he offers of himself *de facto*. The biological standpoint is relativized by the fact that God is able "from these stones to raise up children" to Abraham, even as the Gentiles can be adopted in place of Israel (Rom. 11:17 ff.).

This interpretation of the *imago Dei* brings us up against a theological fact which is significant from two angles. First, theological ethics cannot accept any anthropology in the usual sense, whereby man is treated as a phenomenologically accessible model who may be known in terms of his qualities. For what he really is—the sign, as it were, before the brackets of his existence—is not apparent. It is kept and hidden in the heart of God as an alien factor. What is phenomenologically accessible in man as a model leads only to astonishment over the fact that God remembers him at all: "What is man that thou art mindful of him?" (Ps. 8:4). *In re*, i.e., phenomenologically, there is nothing in man to justify this remembrance. In fact God does not remember man because of what he "is." On the contrary, man "is" what he is because God remembers him, because God has this image of him. If we understand this remembrance as love, we can find the most precise description of the whole situation in Luther's definition of the divine love: "The love of God does not find, but creates, that which is pleasing to it. The love of man comes into being through that which is pleasing to it." [16]

Second, we have here at the same time a change or shift of subject as in the concept of the righteousness of God. Here too what is involved is a subjective genitive. It is the justifying grace, usually called "active righteousness" [*justitia activa*]. Since man exists only in relation to this justifying God, the righteousness can also be

[16] *Heidelberg Disputation* (1518), Thesis 28. *LW* 31, 41.

attributed to *man*. But it is not an imparted, ontic quality in the sense of infused grace or a *habitus*. Rather it denotes in the strict sense the "righteousness which avails before God," i.e., the righteousness which avails only before him because it is the image of us which he wills to bear in his heart. Hence it can be described only as a relation: "with reference to," not "in and of itself." It is not a *justitia in se*, but a *justitia quoad Deum*.

This raises again a question on which we have briefly touched already, namely, whether the image thus defined can be lost. There can *not* in fact be any such loss; the divine likeness constitutes a *character indelebilis*.[17] There is, of course, a negative mode of the divine likeness. But this negative mode of the *imago*, being merely a mode, still bears witness to the existence of the *imago*, and is therefore different in principle from the *imago's* non-existence or its ceasing to be.[18] As both Pascal and Kierkegaard have pointed out in keen formulations, it is the very misery of fallen man which reflects his greatness. "For who is unhappy at not being a king, except a deposed king?"[19] "All these miseries prove man's greatness. They are the miseries of a great lord, of a deposed king."[20]

Nor is it merely the loss of a positive relation to God as such which points to this greatness, i.e., by the fact that such loss presupposes the prior state. Testimony is also borne to the prior greatness by the fact that man can *reflect* upon this loss, that he can be *addressed* on the basis of it. Herein is manifest his continuing function as a responsible subject, his uninterrupted obligation to reply. Hence his dignity as the image continues vis-à-vis the rest of creation. "The greatness of man is great in that he knows himself to be miserable. A tree does not know itself to be miserable."[21]

Kierkegaard makes the same point in his discussion of despair.[22]

---

[17] At this point we can only concur with Barth in his criticism of the Reformers; their occasional references to the loss of the image do betray Scholastic influence. Cf. *Church Dogmatics*, ed. G. W. Bromiley and T. F. Torrance (Edinburgh: T. & T. Clark, 1936——), III[1], 192-193.

[18] Brunner seems to have something similar in view when he speaks of the "destruction" of the image in *Man in Revolt*, *op. cit.*, p. 105.

[19] Blaise Pascal, *Pensées*, trans. W. F. Trotter, introduction by T. S. Eliot (New York: Dutton, 1948), No. 409.

[20] *Ibid.*, No. 398.

[21] *Ibid.*, No. 397.

[22] Soren Kierkegaard, *The Sickness unto Death*, trans. Walter Lowrie (Princeton: Princeton University Press, 1941).

For despair points to the primal perversion of man's relationship to his source as this is expressed either in the despairing will not to be oneself or in the despairing will to be oneself.[23] In this "possibility" man has of relating himself to himself is to be found his "spirit being," and hence his infinite advantage over the beasts—in terms of our present train of thought we would say his dignity as the *imago*. When Kierkegaard says that regarded in a purely dialectical way the ability to despair expresses both an advantage and a drawback for man,[24] he is alluding to the very quality in the image which we have called its indelible character, namely, that even in its negative mode, in the inception of despair, it reflects the distinctive position of dignity proper to the *imago*.

Kierkegaard makes the transition from the positive to the negative mode as follows: "If it is an advantage to be *able* to be this or that, it is a still greater advantage to *be* such a thing. That is to say, being is related to the ability to be as an ascent. In the case of despair, on the contrary, being is related to the ability to be as a fall. Infinite as is the advantage of the possibility of despairing [the positive mode], just so great is the measure of the fall into actual despair [the negative mode]."[25]

When we seriously try to see the image in these two existent forms, the positive mode and the negative mode, it becomes evident that we have cut the ground from under the feet of those who would inquire about the permanence of ontic factors. There is no longer any point in demonstrating the existence of abiding factors which change only in mode, since it is precisely the mode which is here not an accident but the really decisive and sustaining factor.

To bring out the fact that the negative mode is an abiding mark of dignity, we should perhaps mention that even the power of Satan is subject to it. For the idea that Satan is a fallen angel, and the idea that the devil's work is a blasphemous imitation of the action of God, both rest on the insight that in the existence and nature of the devil it is a *divine* quality that is expressed, only in the form of its direct opposite. The greatness of a fall is always measured by the height of its starting point. It is a basic concept of biblical

23 *Ibid.*, p. 17.
24 *Ibid.*, pp. 19-20.
25 *Ibid.*, p. 20.

demonology that all devilish perversions presuppose a particular qualification. The most disruptive ideas are thought to have always had at first a lofty rank and a kind of paradisal innocence (think, for example, of the ideas of equality, liberty, and fraternity). Similarly man's endowment with reason, with a special dignity as *imago*, serves only, as we have seen, to make him "even more brutish than the beasts."

This indelible character, maintained even in the negative mode of the *imago Dei*, may be finally stated as follows. Man cannot get rid of his humanity. He cannot dehumanize himself. If he could this would imply his ability actually to reach the bestial sphere beyond good and evil which he seeks to attain. But this is exactly what he cannot do. Even if he be untrue to himself, man remains in his misery the deposed king. He can thus be addressed on the basis of his humanity, as in Romans 1:18 ff. he is addressed on the basis of his creatureliness even though that creatureliness is repeatedly "suppressed by wickedness." It may be shown that the pastoral conversations of Jesus constantly revolve around this problem; he addresses harlots and publicans on the basis of their origin and destiny. The negative mode of humanity implies no loss of humanity. It implies no plunge into amorality, into the non-accountability which is sought. What we call "inhumanity" is precisely guilt, guilt in respect of the humanity which, as both gift and task, continues to exist even in the negative mode. Even in the negative mode man still remains responsible.

Only in this form is there any demonstrable continuance of the *imago Dei* from the human standpoint. To adapt Kierkegaard's word about despair, we may say that in despair man necessarily remains himself, no matter how great the corruption or perversion. Even the demons stand under this necessity. This is why, when they are confronted by Jesus Christ, they are always thrown back upon themselves; they have no option but to declare themselves (Mark 9:20; 3:11 *et al.*).

The divine likeness is thus, in the first place, that which man *in re* no longer is, but on the basis of which he is nonetheless addressed. It is that lost state in which he was when he first received his being from the hands of God, and which he cannot escape—inasmuch as he is constantly being addressed on the basis of it—though it is in the negative mode that he must necessarily live it out.

However, the divine likeness is, in the second place, that which is to be received afresh in Christ as a quality *in spe*. It is thus something which is given to man, and has to be given again. Consequently, it expresses not man's own immanent and ontic dignity, but that alien dignity which is grounded in and by him who does the giving.

10.

The Christological Character of the Imago Dei

This character of the *imago Dei* as an *alienum*, something alien, is supremely brought out by the fact that as a *proprium*, as a true ontic possession, an attribute in the strict sense, it is ascribed solely and exclusively to Jesus Christ. It is ascribed to him as a *proprium*, not merely in the sense that in him alone it has remained intact, but above all in the sense that it is present only in him. In the absolute sense Jesus Christ is the only man. More precisely, he is the only man who fulfills humanity; he does not possess humanity merely in the negative mode, as an unrealized possibility. We can say this, of course, only if we at once add the safeguard that "humanity" cannot here be understood as an a priori concept expressing a knowledge of man enjoyed prior to and apart from Jesus Christ. If it were so understood, then Jesus Christ would be understood as merely fulfilling, or having to fulfill, an idea of humanity deriving from our own sovereignly creative consciousness. Our thinking must take the very opposite course. We must first learn from Christ and perceive in him—*ecce homo!*—what man is. We must first learn from his divine likeness wherein the divine likeness of man consists. For man's divine likeness is fulfilled only in Christ, in our participating in his divine likeness.

The New Testament expresses this in many different ways. Christ is the image of the invisible God (Col. 1:15; II Cor. 4:4)—so much so that our own divine likeness finds its fulfillment in this, that we are "conformed to the image [εἰκόνος] of his Son" (Rom. 8:29). For the "image" of Christ is the primal, the original image. As we are conformed to it, we are "perfect, as our heavenly Father is perfect" (Matt. 5:48), and thus attain ourselves to the true form of the εἰκών. When it is said that we are to bear the "image of the man of heaven," what is meant is the "image" of Jesus Christ, and that means the "image of God" present in him. For it is here that we come to participate in the "last Adam," who is the "life-giving spirit," and

171

who is therefore Christ, the representation of God himself (I Cor. 15:45, 49).

The recurrent expression "in Christ" also points to the fact that all those whose existence is grounded in Jesus Christ participate in his divine likeness; they are taken up into it, as it were. Inasmuch as they are, one might say, "in Adam," they belong to the old fallen creation; they are therefore the image of this "old Adam." Inasmuch as they are "in Christ," however, they are a "new creation" (II Cor. 5:17); they are therefore the image of the new Adam. That is to say, they are fashioned in accordance with the original, Jesus Christ himself, in whom God is, and in whom all discord is resolved (II Cor. 5:19).

We must now endeavor to develop more precisely this basic thought that, as those who are "in Christ," we share in his divine likeness. Only thus can we bring out in sharpest relief the alien character of man's divine likeness. We shall seek first to document more precisely the fact of this participation, and then to define its nature.

The Fact of Our Participation in the Divine Likeness of Christ

As regards the fact of our participation in the divine likeness of Christ, it is important to note first its pronounced eschatological character (especially in I Cor. 15:49: "we shall bear the image of the man of heaven"). This raises the question whether the divine likeness is not here taken to be essentially future. Is it not the content of a promise, with no reality as yet? But if so, can man be addressed already here and now on the basis of it? Or is his present condition a matter of no consequence inasmuch as the participation in Christ's divine likeness cannot in any case be a present reality? Is man therefore to be abandoned to a laissez faire policy? Does Luther's thesis that man is "sinful in fact, righteous in hope" [peccator in re, justus in spe] imply for the doctrine of the divine likeness that the disciple will never be an "image" of God "in re," that he will rather maintain the character of a disturbed, confusing, and broken reflection, never advancing to the point where righteousness becomes a characteristic property of his very being?

But would this not have serious consequences for ethics? Does such a capitulation still allow for any concrete goals of action and instruction? What place is left for Christian pedagogy? Must not all pedagogy be anointed with a drop of humanistic oil, by providing it with certain humanly attainable goals or ideals?

At this point we are confronted by a host of questions, though actually the question we confront is always the same one, the one that always sets before us the "smooth" solutions to the problem offered by Roman Catholics and by the humanists. We have already faced this problem in relation to the doctrine of justification, and especially the idea of progress. We shall constantly face it again in practical ethics. Yet we cannot refrain from pointing out that it is also acute in relation to the eschatological character of the *imago*.

A vital insight into the solution of the problem of how future and present interrelate in this promised *imago* is offered by Paul in his debate with Judaism in II Corinthians 3:7-18. Paul reminds us in the course of his argument that Moses veiled his face after he had met God on Sinai (Exod. 34:29-35). As Paul understands the Exodus story, the reason for this veiling is twofold. First, the glory of God, as it appears in connection with the Law of God, has a menacing and terrifying quality. Hence those who are near Moses cannot bear the radiance of the divine majesty (cf. Isa. 6:5; Exod. 33:20). Second, the radiance diminishes, however, the longer Moses is among men, which is why it is restored with every new saving encounter. Thus the second motive for the concealment probably lies in the fact that Moses wishes to hide the diminution of the glory lest his authority should be impaired by it.

Neither of these reasons, neither the terror of the glory nor the diminution of its radiance, applies in the case of the new covenant. The glory of God has lost its consuming and accusing character (Luke 9:32; John 1:14; 11:40; 17:24; Acts 7:55; Rom. 8:33 f. *et al.*), because it now appears "in the face of Jesus Christ," i.e., in gracious condescension. Moreover, what is involved here is a lasting and definitive glory which does not fade like that displayed in the limited interim period of the Law (II Cor. 3:13). It is something continuous, something which has abiding relevance to every moment of our living and dying (Rom. 14:7 f.; II Cor. 5:9).

Why can the glory of God be borne now, i.e., with the dawn of a new aeon? Why does it now illumine rather than consume? Whereas the Law of God brought servitude, both through the service of obedience which it entailed and through the guilt and servile fear into which it plunged us, freedom prevails under Jesus Christ. There is the free, confident, and glad intercourse of the child with its father, who does not demand that certain conditions be fulfilled before he will be near us and we may be near him (II Cor. 3:17). This is why the transformation into the "likeness of God" may begin here and now. Inasmuch as we stand before God "with unveiled face" (II Cor. 3:18), we reflect the same likeness and are changed from glory to glory.

In this same passage, which sums up our problem as in a compend, a number of things are said concerning the new *imago Dei,* to which we will now turn for further insight into its present and future character. To begin with, it is characteristic of the image given and promised in the new covenant that it implies a relation of immediacy to God. This immediacy is metaphorically expressed by the removal of the isolating veil. It is posited by the fact of Christ, since it is in him that bondage through guilt and fear is replaced by the freedom of sonship.

As a consequence one may say that the marring of the image consists in the destruction of the original immediacy of child to Father or creature to creator. The Law had the task of continually reminding us of this destruction, of constantly addressing us on the basis of it, and in so addressing us of bringing it out of latency into unmistakable and terrifying clarity. The loss of this immediacy promptly produces certain concrete symptoms, the existence of which is known to all men (even to non-Christians) but the meaning of which can be understood only theologically, i.e., only in terms of their actually being symptoms of a particular loss of a particular immediacy.

For fallen man, i.e., man fallen from this immediacy, is no longer related directly to God. Instead he interposes between himself and God various protective intermediaries, entities which are attuned to himself (Rom. 1:18) and thus humanly oriented. These intermediaries, these creatures misrepresented as gods, for their part are characterized by the fact that they no longer display the creator. "Nature" is a sphere of creation, but in the pantheism or atheism of natural science it has in fact lost its character as a transparency of the creator. Indeed, it has become a symbol of the self-contained, non-transcendent finitude which bears its own religious or a-religious meaning within itself. As an intermediary, "nature," instead of disclosing the creator in a decisive way, functions to isolate creator and creature from one another. The same phenomenon may be observed in connection with the history of the secularized philosophies: the individual creaturely forms lose their "significative" character, their ability to point beyond themselves, and become instead the sign of a self-sufficient immanence which desires to be self-sufficient and which finds its security in this desire.

This is the secret of these intermediaries. Man does not by any means turn to creaturely intermediaries in order to find in them transparent symbols of the deity of God. On the contrary, it is precisely because he must here encounter the transparency of God, because he must here encounter the plain manifestation of God (φανερόν, Rom. 1:19), namely, his invisible "power and deity," that man makes use of the medium in which this deity appears to him in order to silence the eloquent sign and rob it of its significance. Indeed, he causes it to point in the opposite direction, to the self-sufficient and self-enclosed world, to his world. Thus the symbol becomes a protective and non-transparent insulating layer.

But then man has to go even further, for the reality of God can never be contested in a purely negative form. Merely to isolate God, however, would be such pure negation. The "anti" is not enough; man has to set up an anti-"thesis." This consists in the fact that the creaturely form is interpreted not merely in terms of its isolating function, but beyond that as a medium in which man actually represents himself, so that instead of an *imago Dei* we have an *imago hominis*. As we have seen, man understands the creaturely form as "his world." It is his magnified ego. For a cogent illustration of the anthropocentrism in terms of which creation is here viewed one has only to recall how among the left-wing Hegelians the Hegelian "world-spirit" irresistibly reverts to man as an attribute of man, and how there thus occurs a reversal of the original hierarchy of absolute and finite spirit. Then there is the even more crass but easily overlooked example to be found in the race concept of National Socialism.

It may thus be seen quite easily how with the loss of immediacy towards God a concealing veil is sought, and how this veil may be manufactured of many different materials and in many different colors, even in the form of creaturely entities which are robbed of their transparency and made into an insulating layer. Here may be seen man's *incurvitas in se,* which is no less a true curving in upon himself simply because there is an apparent outward orientation to such concrete elements of the world as spirit or bios. For all these are only extended symbols of the self.

The immediacy thus lost is restored in Jesus Christ, in spite of the incomparably greater glory. It is a glory which we are far less able to bear than Moses was (II Cor. 3:7 ff.); nevertheless, we are able to bear it because we no longer live in bondage through fear and guilt but are free children of our Father. Hence, in the strict sense, it is not a matter of our simply being "able to bear it." It is rather a matter of the fulfillment, the joy (παρρησία, I Cor. 3:12) of the Christian state itself. It is indeed a matter of our being "trans-figured" into that which is our true destiny. "Transfigured" [*verkleret*] is Luther's apt rendering of μεταμορφούμεθα in II Corinthians 3:18, where he also speaks of our "reflecting" [*spiegelt sich*] the glory of the Lord.[1]

Because in Christ we are thus "transfigured" and the immediacy of the parent-child relationship is restored, we are then enabled to "reflect" the glory of God—but only in the sense that it is God who is actually reflecting himself in us. For what is always involved in the reflection of a mirror—the analogy provides its own limitation—is merely borrowed light. Man can never boast that his character as an "emitter of light" is his own, a kind of residual property inherent in himself [*habitus*]. He cannot claim that as a person, as one who possesses and uses reason and freedom, he is in his very nature a composite of divine qualities. He cannot boast that his acts and achievements bespeak his status as the divine likeness. Paul does once take this course, but only in passing and for pedagogic reasons, to demonstrate to others his own credibility. And in so doing he explicitly calls it folly. It is really not necessary. Indeed, it ought not to be. It is not in keeping with the reality at issue. He has to be forced to do it (II Cor. 11:16). Hence all that he says along these lines is expressly parenthetical. It is an alien body intruded into the true text of his message. For in distinction and in opposition to this folly which causes him to point to himself and his own achievements, the

[1] See Luther's German New Testament of 1522. *WA, DB* 7, 146.

real point, if we take the metaphor of the reflection seriously, can only be that the light does not enter into him in the way it enters into phosphorescent material which can absorb it and then continue to give it off even after the original source of light has been withdrawn. Man can never look to the measure of his own enrichment with grace and holiness in the sense of a *habitus*, nor ascertain the degree to which he is saturated with divine working materials, as is somehow always done in the case of the doctrine of infused grace [*gratia infusa*].

On the contrary, if the source of light is withdrawn, the mirror immediately becomes dark. For it possesses only borrowed light, a light which is borrowed *ad hoc* and moment by moment. Never for a single moment does it have a light which is its own, appropriated by it as an "attribute" or property.

We emphasize this point so strongly because here the fact of the *alienum* recurs again in the sphere of the *imago* doctrine. In the last resort the attribute of the "image" is not an attribute of man. It is an attribute of the glory which stands over against him. Or, as we put it earlier, the prodigal does not remain the son because he has maintained his qualities of sonship, or because his father could still be "seen" in him—to use again the terminology of the mirror figure—even in the far country. He remains the son only because of another who has maintained his qualities of fatherhood even during the period of exile and estrangement. What is at issue here are the qualities, not of the mirror, but of the one casting his reflection in it. Here, and here alone, is the reason why the Pauline and Reformation tradition views the concept "righteousness of God" so ambivalently, a fact which Roman Catholic thought finds so difficult to understand. On the one hand, it is the righteousness "of God himself" (subjective genitive) and therefore a quality of God; yet on the other hand it is also the righteousness which "avails before God" (objective genitive) and therefore—we say it with tongue in cheek—a kind of quality of man.

We now understand why this concept of God's righteousness has to be ambivalent. It simply cannot be understood at all if the bearer of righteousness is regarded alone, in isolation. For this righteousness refers to the relation between two partners, namely, God and man. More precisely, it refers to the fact that this relationship is in good order, to the fact of fellowship, and consequently to the one who establishes the fellowship. But we have seen already that it is God alone who restores the relation to its proper order. Fellowship with the prodigal is maintained only in the heart of the father. It is the fatherliness which sustains the sonship, and not vice versa. But we must be even more precise. In the giving of the only-begotten Son (John 3:16) there takes place something whereby God actualizes that which otherwise remains hidden in him; he restores the original order and immediacy of the parent-child relationship.

When we put it thus, it becomes crystal clear why we cannot isolate the two partners and affirm righteousness to be an immanent quality of both. In this light too we can see that it is completely beside the point to ask whether justification is a mere forensic declaration, the "as if" of the divine pardon with no corresponding qualitative presuppositions in man, or whether on the contrary it is the recognition of a previously imparted essential righteousness, an infused quality. This question, which has played so fateful a role in dogmatic history, is a problem only if God and man are illegitimately isolated from one another and solipsistically considered apart. But to consider them thus is to get off at once on a false track. Such a consideration leads to the doctrine of qualities [Eigenschaften] or attributes of character over against which we have already set the concept of outward relations [Aussenschaften]. God's revelation in Jesus Christ has made it clear that God and man must be considered exclusively in terms of their relation to one another. Man exists as the imago Dei only insofar as he is oriented with respect to God either toward God or away from him. What man could conceivably call his qualities are simply the qualities of this relation, and Luther has rightly said that in the last analysis only two "attributes" remain, faith and unbelief. This fact is the logical consequence of man's being what he is, a relational entity.

The fact that in Jesus Christ God too is present only in relation to man is naturally not to be interpreted in a mystical sense. It is not as if we had here a reciprocal relationship in which each one posited the other, so that God could not live for a single instant unless I did. That God wills to exist only as Emmanuel, as God for me [Deus pro me], as a relational entity, is not due to any superior ontic order which embraces both God and man and thus is imposed on God. It is due exclusively to God's gracious condescension and therefore to the initiative of God himself, who in Jesus Christ wills to pour himself out, withholding nothing.

Consequently there are, in the last analysis, as Luther consistently emphasizes, only two attributes in God too: love and wrath, Gospel and Law. Even from a purely etymological standpoint this statement of course indicates a crisis in the whole concept of attributes. For love and wrath, Gospel and Law, do not express static, immanent constituents of the character of God. They express rather God's active being *ad extra*, in relation to something. They bespeak an activation which can find expression only within the framework of the partnership between God and man, an activity which can be expressed only in terms of the problem of fellowship between the two, namely, the problem of the breach and the restoration of this fellowship.

In this sense it has rightly been pointed out that God's is not merely a passive or distributive righteousness [*justitia passiva et distributiva*] in the sense of a juridical declaration, but an active righteousness [*justitia activa*] which becomes operative within the God-man fellowship. In the same sense the righteousness which "avails before God," the righteousness which is within the province of man, is not an attribute in the sense that man is its subject, the one who creates it (even if only to the degree that he produces it as a merit with the help of prevenient and co-operating grace); instead it is truly and properly the grace which approaches man and comes to him, in virtue of which God renews the broken fellowship.

These statements enshrine the essential impulses which drove the Reformers to protest vigorously against the Scholastic doctrine of infused grace. For in the theory of the *gratia infusa* the grace of God ceases to be exclusively an attribute of the divine-human relationship, a characterization of the divine action upon man, as it was in the question which precipitated the Reformation: how are we to find a gracious God? In the doctrine of infusion the problem is instead how grace as a divine attribute becomes a human attribute. To return to our metaphor, what is involved is the enriching of man with luminous phosphorescent material. The mirror is no longer turned exclusively to the downward ray of the divine act of salvation, as in the Reformation and even in Protestant Orthodoxy's semi-recatholicizing of the Reformation.

To sum up, the *imago Dei* depends on the reflecting of alien light, a process which is always under the control of the glory of God which casts the reflection. As *imago*, man is not luminous phosphorus; he is merely a mirror. In other words, he actualizes the image by being —and only to the degree that he is—a disciple. He actualizes it by living in fellowship with Christ and by being bound to Christ as a member of his body. In Christ, the glory of God is present. Hence man is the *imago Dei* only in the sense that he is "in Christ." For this reason the verse which is so decisive for our problem (II Cor. 3:18) is re-enforced by the earlier verse (II Cor. 3:5) in which Paul expresses the fact that his "sufficiency," i.e., the fact of his being equipped to bear spiritual office (v. 6), does not consist in his own excellence or virtue but is wholly "from God" and therefore excludes all boasting (II Cor. 11:16; cf. 4:7).

We thus reaffirm our original thesis that by reason of its christological background the image of God consists in an alien dignity. It is not an attribute of man himself [*Eigenschaft*], but an attribute of the relationship in which he stands [*Aussenschaft*]. In connection with the concepts of grace and justification we thus come upon two very different doctrines of the *imago Dei* which we shall have to consider in greater detail later in the chapter devoted to our debate with Roman Catholicism (chapter 11).

This christological description of the *imago Dei* carries with it the answer to our original question about the extent to which the divine likeness of the new man created after Christ is not merely future but also present. For even now we already stand in the light of the divine glory "with unveiled face" insofar as we have Christ. Here are present and future, fulfillment and promise, in one. What is now *sub tecto crucis* [under cover of the cross] will then be manifest in glory. Applied to the figure of the mirror, the *sub tecto crucis* implies that the mirror itself is an "earthen vessel" (II Cor. 4:7) which has no power or quality of its own, whose light is simply the reflection of something else. It implies too that the reflective function, the very existence of the image, is constantly threatened by whatever comes between the mirror and the light to disrupt the relationship, to intrude and to deflect and dispel the light, namely, the assaults of doubt [*Anfechtung*]. These assaults still determine the present form of the image, since this form is bound to faith and faith for its part is still under attack and is constantly summoned to believe nonetheless, despite the assaults of doubt.

The form of the image is fulfilled, however, in sight, in a seeing which comes only from love. If it is true that I shall then know even as also I am known by God, then I dare not, even in this fulfillment of the eschaton, ignore the relation to God. Indeed this eschatological existence can be described only as a relationship of reciprocal knowing, and hence as ultimate co-ordination and mutuality. In it, however, I am more and other than merely a reflecting mirror. My knowing here becomes analogous to that of God. The analogy of being [*analogia entis*] which arises here—and only here—means that I have myself become a bearer of this knowledge, that I am the one who does the knowing, that it is now properly ascribed to me as an "attribute," and that nothing can any longer intervene between the light and its reflection. For the luminosity has become a property with which I have now been endowed.

Hence the presence of the *imago Dei* which is given me here and now, that *imago* which is represented in the figure of a mirror, has always the character of a promise. It is a kind of earnest money, a "guarantee" (II Cor. 1:22; 5:5; Eph. 1:14). For we are constantly in process of being changed (in the sense of the μεταμορφοῦσθαι of II Cor. 3:18). The image of glory seeks ever clearer delineation in us. The stone lying in the sun necessarily becomes warm.

Now this is not to be misunderstood as if what were involved were some kind of advance in sanctification, in the sense of my being enriched with holiness in such a way that I could actually watch the level of holiness in me rise. There may be such growth. In fact, there certainly is. But it is something which definitely cannot be watched and measured. And there is no such thing as concluding from the observation of such growth, by watching the increase of faith and the increasing harvest of good fruits, that faith actually exists, or better, that faith's object actually exists.

False developments of this kind are all rooted in an illegitimate inversion of outlook, in the *incurvitas in nos* whereby we are turned in upon ourselves. This *incurvitas* is itself theologically conditioned by the fact that we understand the *imago Dei* as a quality of man, and that we consequently understand justification and sanctification as the infusion and development of this quality. In the Reformation sense, however, justification is the direct opposite. It liberates us from this inversion of outlook and fixes our glance on the object of faith. For the Reformation break-through accomplished by Luther consists supremely in the fact that he no longer understood grace as a medicine whose working, once it has been administered, has then to be observed, with all the oscillation between disillusionment and despair which such observation must entail. Luther learned to believe, not in himself as a product of the medicine of grace, but exclusively in the "gracious God." The basis of his salvation was not he himself as the image restored through his church's means of grace; it was rather the light of the gracious God, the light which the believer reflects.

This involves no denial of progress, of growth, of increase in justification [*magis et magis justificari*]. It simply disqualifies such development as an object of observation, as a possible "outlook." For such progress and growth is, as it were, one of those things which in passing are added to us "as well" (Matt. 6:33; Luke 12:31). Our "seeking" must have as its object something wholly different. Whoever confuses the purpose with the by-product—identification of the object of our seeking will quickly disclose such confusion—is practicing idolatry. For idolatry is nothing but the attempt to ground oneself in what is created rather than in the one who creates, in the products rather than in the one who produces them.

Here, too, Luther offers us incisive theological safeguards. Though he does not always use the term in a consciously programmatic way, he does as a matter of fact protect the *imago Dei* from being misunderstood in terms of a quality—and Christianity from being consequently misunderstood in terms of the growth of this quality. In so doing, he also provides safeguards against inversion of outlook. At this point, we need only mention the theological means by which he does this, since the theme of the book will compel us to refer to them again and again.

In the first place, by making faith the basis of the restored image, by making it the thing that moves the mirror into the light, Luther guards faith against the misconception that it is a human virtue. He does this by attempting to describe the subject of faith as an unextended mathematical point, a *punctum mathematicum* [2] which ac-

2 See WA 40II, 527, line 9; cf. LW 12, 239. Cf. also WA 40I, 21, line 12 and LW 13, 128.

cordingly is not to be construed as a physical theatre where certain actions occur which may be observed and evaluated—and in both cases treated apart.

This does not mean, of course, that Luther denies that faith is a psychical accomplishment, a work, an experience. He does not deny that it may as such vary in degrees of intensity. It goes without saying that as a psychical act faith is a work of man, just like any other work. The very fact that faith is commanded, and that it is thus an act of obedience, makes it a work, an accomplishment of the human subject. Indeed, it is the "chief work," not to be set "apart from all works of the other virtues." [3] Consequently, faith can be described as "a work that man must do." [4] Luther even goes so far as to play off the significance of faith as a "work" against its misunderstanding as a *"habitus."* [5]

These statements, however, say nothing about the justifying significance of faith. For there cannot be the slightest doubt that faith in its quality as a work does not justify. It simply lays hold of its object, namely, him who alone can justify. [6] Hence faith can sometimes be admonished not to believe in itself, not to be introverted or avid for experience; it must cling to its *object.* [7] Thus faith is not a quality of the *imago;* faith actually constitutes the *imago* in its significance as a mirror. For the mirror does not absorb the light. It shines only by way of reflection, i.e., only in relation to the object which casts the reflection.

A second safeguard of this thought by Luther is found in the manner in which he speaks of progress in the Christian state, of being made more and more righteous [*magis et magis justum effici*]. Progress (*proficere*) comes only by constantly going back to the beginning, by starting, as it were, all over again. [8] But this beginning again means to stand ever anew before the prototype; it means constantly turning

---

[3] *PE* 1, 189-190.

[4] *LW* 23, 23.

[5] See p. 76, n. 5 above.

[6] *"Non . . . justificat [sc. fides] ut opus . . . ',,* Paul Drews (ed.), *Disputationen Dr. Martin Luthers* (Göttingen: Vandenhoeck and Ruprecht, 1895), p. 705.

[7] See the Luther quotation documented above at p. 16, n. 2.

[8] "To go forward [*proficere*] is nothing other than constantly to begin again [*incipere*]. And to begin without going forward is to go backward [*deficere*]." *WA* 4, 350; cf. *LCC* 15, 91, n. 4.

as a mirror to the source of light. Progress cannot mean that we regard the foundation as already laid—in the form of infused qualities —and then build upon it. On the contrary, this foundation must continually be laid afresh [*incipere*].

In this connection it is characteristic that Luther does not point his explanation of the commandments in the direction of growth and progress, or some kind of doctrine of perfection. With each new commandment he emphasizes anew that we should "fear and love God." This fearing and loving is the initiatory act with which our Christianity begins. But it is also its fulfillment. There is nothing above or beyond it. Whatever follows can only be a completion of this beginning, just as there is nothing above or beyond baptism since whatever follows it can only be a further creeping into baptism. Nevertheless, this is not just marking time. It involves real progress. A state of faith in which there were no such increase or development [*proficere*] would call in question the genuineness of the beginning [*incipere*]. For it would imply the impossible phenomenon of a faith without works, of justification without sanctification, of lying in the sun but remaining cold. And in the sphere of doctrine it would imply the infantile and hence impossible existence of a believer who to the very end could not tolerate solid food but drank milk only (Heb. 5:12-14).

To sum up, we may say that the renewal of the *imago Dei* consists in the fact that as believers we are "in Christ," and that we thus participate in the divine likeness of Christ. This participation takes place now, as surely as we are here and now incorporated into his body as members. Yet it contains at the same time a promise for the future, the promise that we shall one day be "changed" with the end in view that we may know even as also we are known. Hence the image restored "in Christ" exists not as a quality of man but as a relatedness to Christ the prototype. Nor is it strictly an "image"; it is rather a "mirror." It does not live by the *proprium* of what it possesses, but by the *alienum* of what Christ is for it.

This, then, suffices as a description of the fact of our participation in the divine likeness of Christ. We must now attempt a more precise delineation of the christological form of this participation.

The Form of Our Participation in the Divine Likeness of Christ

We intentionally open this discussion by criticizing a concept which in our view defines and explicates this participation erroneously. We refer to the concept of *imitatio*. The "imitation" of Christ simply means the effort of a disciple in following him to approximate as closely as possible—to imitate—the example of Christ himself. It implies, as it were, a participation in him based on the believer's effort to achieve similarity of life and conduct.

At much the same time as painters took their sacred figures out of their golden heaven and implicated them in life here below— they were enabled to do this by the discovery of perspective—a similar return to this-worldliness took place in Christology. This occurred with the portrayal of Christ as above all a model or example for the religious and ethical existence which was also brought to light in the new this-worldly perspective. In this connection we must not think only of Thomas a Kempis and his *Imitation of Christ,* although his was the first prominent attempt in this direction. Behind his work there still lies, of course, the older Christology, in the light of which alone his *Imitatio* can be properly understood. With the growth of this-worldliness or "secularization," however, the "example" idea became more and more independent until, in the nineteenth century, the figure of Christ became no more than an illustration, or better, an incarnation, of ethical postulates. Imitation for its part came to denote the ethical training through which we strive to approximate more fully those postulates—"by way of Christ," of course, which is to say stirred and animated "by him."

The piety of "imitation" has thus a legalistic aspect. Accordingly, one may justly suspect that even before it came out into the open with a program of its own such piety was already latent in dogmatic history. Indeed it was necessarily at work wherever there was special emphasis on the Law, wherever men thought in terms of human co-operation and of man's capacity for such co-operation (as did, for example, Pelagius).

Now when it is held that Christ is an example, and that discipleship consists in imitation, what does this do to the concept of our participation in Christ, and of the *imago* which corresponds to this participation?

On the basis of this *imitatio* piety, the participation of the Christian

185

in Christ can be thought of only if man stands on the same level as Christ, with only a quantitative distance but not an infinite qualitative difference between them. For after all I can imitate the example only of someone to whom I am basically akin. Hence the piety of imitation has a tendency either to reduce Christ to the human level or to exalt man in his capacity for Christ-likeness. In fact this is the first reason why the imitation of Christ cannot be a means to participate in the divine likeness of Christ, because it exalts man and abases Christ instead of exalting Christ and abasing man.

A second reason why imitation cannot possibly be "the first step" is that it makes of Christ a Law, whereas he is supposed to free us from the curse of the Law. In certain forms of the piety of imitation the whole problem of obedience or disobedience to the Law is transferred to Christ, and discipleship accordingly can consist only in the failure adequately to measure up to his standards. This would clearly mean a dissolution of that basic distinction between Law and Gospel which is constitutive of all theology.

For on the one hand it makes the Gospel into Law. Ordering replaces giving. Conditions of operation are imposed. There must be some minimal approximation to the normative model of Christ. The Gospel ceases to be comfort and consolation in time of perplexity and doubt. Instead it exercises the function of Law precisely by plunging me into perplexity and doubt, and keeping me there. The resultant attitude is that of despair.

Conversely, the Law intimated in the demand for imitation is for its part made into Gospel. The promise that we shall be like Christ is attached to the fulfilling of the Law. The attitude displayed in this Pelagian attempt to exalt man is that of pride [hybris]. Nor is it any accident that despair and pride are the two attitudes which result from every legalistic approach, wherever Law and Gospel are confused.

The piety of imitation is thus wrong in principle. Luther has decisively corrected it, however, by describing Christ, not as an example, but as "the exemplar," the original prototype or model. Christ is the prototype [Urbild] for Luther in a twofold sense. He represents God's saving activity, and he represents man's state of salvation.[9]

[9] On the concept of "exemplar" in Luther, see H. Thimme, *Christi Bedeutung für Luthers Glauben* (1933), pp. 18 f., and Ernst Wolf, *Peregrinatio* (Munich:

What does it mean that Christ represents the saving activity of God? In the New Testament Jesus Christ is not the "teacher" of divine righteousness as von Harnack thought when he made his famous statement that "the Gospel, as Jesus proclaimed it, has to do with the Father only and not with the Son." [10] Christ is rather the prototype of divine righteousness. That is, he does not merely give information and instruction—concerning God's mercy toward the anxious, disturbed, and guilty, concerning the coming of the kingdom of God and the outpouring of the Spirit—and then impersonally withdraw behind his own statements as artists, scholars, and teachers must do. On the contrary, he himself *is* everything that he proclaims. He does not merely know and state the truth; he is the truth (John 14:6). Nor are his disciples characterized primarily by the fact that they know and have comprehended the truth imparted by their Lord, like good pupils, like the initiates of mystery cults. Rather they are characterized by the fact that they *are* of the truth (John 18:37), and thus have apprehended the truth. The fact that Jesus is this saving truth, that in his own person he represents the event in which this truth is actualized, is what makes him the prototype. And having mentioned what this means for the disciples who live from and of this truth, we have thereby already indicated wherein participation in Jesus Christ the prototype, the image, consists.

This is brought out even more plainly when we consider the further fact that Christ is the prototype of man's state of salvation. This is most succinctly expressed in the statement in Ephesians 2:14 that "he is our peace." Here again it must be emphasized that he does not merely give us peace, just as he does not merely impart truth. He thus discharges a different function from that of the priest who reconciles and pacifies an angry deity in the sacrificial cultus,

Kaiser, 1954), pp. 72 ff. Luther derives his concept of the prototype [*Urbild*] wholly from the concrete history of salvation. By it he means (a) the action of God and, conversely, (b) the man toward whom God acts. Schleiermacher, on the contrary, following Hegel's concept of the absolute, uses the same term to refer to the "idea" of a thing which exists first and which is then realized or actualized, e.g., Christ fulfills the idea of man, Christianity fulfills the idea of religion. We must be careful not to be misled by later usage into making of Luther's concept of the exemplar or *Urbild* a philosophical term to be understood in the light of a climate which is quite foreign to its spirit.

[10] Adolf von Harnack, *What is Christianity?* trans. Thomas Bailey Saunders (2nd rev. ed.; New York: G. P. Putnam's Sons, 1903), p. 154.

or from that of the psychotherapist who is concerned to replace unrest and anxiety by a feeling of psychical balance which may be called peace. On the contrary, Christ himself *is* at peace with the Father. Hence, if we are "in Christ," as the preceding verse puts it (Eph. 2:13), we are set in the same peace, i.e., we stand at the same place where Jesus Christ stands in relation to the Father. This very phrase "in relation to the Father" expresses clearly the difference between this kind of peace and that which is understood as a psychical *habitus*. We recall the poetic description of the latter:

Of what he ought to be each person has his own conceit,
And till the goal is realized his peace is incomplete.

This kind of psychical peace is the only one that can be attained or even imagined so long as the conception of man is ultimately solipsistic. For in such a case the only real question is how man can be brought into harmony, and so be "at peace" with himself. Peace is here the congruence between what man is ideally and what he is in actuality, between what he was destined to be and what he has turned out to be. In Goethe's terms, it is the realization of the "entelechy," of the "daemon," which is impelled neither by accident [*tyche*] nor by necessity [*anangke*] but by itself, and can therefore to some extent be pacified or composed, put "at peace."

As secularization has advanced, this solipsistic conception has come to prevail more and more until it has finally issued in the icy solitariness of existential philosophy. In the light of such a conception, the peace mediated through Christ can be understood only as a kind of psychical infection, accomplished through the "inner life of Jesus" working by way of example. In any case it is understood as a psychical process, a psychical quality. It is no longer interpreted "metaphysically," in the way Ephesians understands this peace and its impartation. For in Ephesians it is really understood as peace between two powers, between God and man, and not as peace between two parts of the inner self. In other words, peace is not a matter of "domestic policy," the peace of the self with itself. It is rather a matter of "foreign policy," a state of peace with God. Here too we could repeat: peace is not a psychical quality of man himself but an attribute of the relationship in which he stands.

Only in these terms can we understand the ontological language of Ephesians when it says that Christ *is* our peace. He stands in the breach which has opened up between God and us. His coming, his history, the simple fact of his existence, *is* peace. Only in this light can we come to grips with the further thought which Herder had in view. For it is only on the basis of this objective peace with God that harmony can be attained between what man is supposed to be and what he actually is. Then—but only then!—could something be said about how "peace of mind" can arise through this inner harmony of the self with itself. In the present context, however, our concern is only with the fact that in this statement of Ephesians Jesus Christ is regarded as the prototype of our state of salvation. Since he is peace between God and man (Luke 2:14), we are *in* this peace insofar as we are "in him" (ἐν Χριστῷ).

That Christ is the prototype of a new creaturehood, namely, of my existence as a Christian, can also be described in a different way, as may be seen from Galatians 2:20: "I live; yet not I, but Christ liveth in me: and the life which I now live in the flesh I live by the faith of the Son of God." Here Paul gives a comprehensive description of my participation in the life of Christ, or, as one might also say, of my being taken up into his prototypical existence. We would expound the meaning of what is involved here by considering two relations suggested by this saying.

First, "the life which I now live in the flesh, I live by the faith of the Son of God." It may be assumed that here, as, for example, in Romans 3:22, 26, the genitive "of the Son" is to be understood as a subjective rather than an objective genitive. Hence the meaning of the phrase is not that I have the new life through faith *in* Jesus Christ—though in terms of the matter itself and quite apart from the signification here intended this statement is indeed true—but rather that I have the new life through and in the fact that Jesus Christ believes, so that here he is thus taken to be the prototype of my faith. He, Jesus Christ, lives as he gives himself in unreserved trust to his Father. If I am "in Christ," I participate in his life, and the point where I stand is thus the very point where he so believes. The truth is the same as in the statements that I die and rise again with Christ, and that my life too, along with his, during the period before the Parousia "is hid . . . in God" (Col. 3:1-4). The believing, dying, and rising of Jesus Christ are in this sense prototypical. As facts of salvation history they do two things. They make my salvation, my peace, possible. And they thereby make of me a participant who believes and dies and rises again with Christ; they cause me to be fashioned in accordance with the model, Jesus Christ.

This participation of the Christian in Christ has been referred to since Martin Kähler as "inclusive representation." Exclusive representation is comparable to that involved in ordinary division of labor, whereby one person, for example, runs the train on behalf of all those who ride in it. Where deputyship of a *personal* nature is involved the situation is quite different. The soldier who does battle on behalf of his country, the public servant who takes over for all the people the tasks of running risks and making decisions, the theologian who bears the burden of doubt and perplexity on behalf of the entire church, does so not in order to "exclude" the people entrusted to his care from all similar concern and activity but precisely in order to "include" them by way of having the people share with their representatives in understanding and undergirding the work. Such vicarious action on the personal level is intended to awaken those who merely slumber and vegetate, and call them to responsibility. It is "inclusive," drawing them into the life and action of their representatives. The suffering of the master "for" his people is not to dispense them from suffering; quite the contrary (Matt. 10:25). This concept makes clear how the prototypical believing, dying, and rising of Christ, and his peace with God, are matters into which we are drawn and in which we participate. His vicarious representation is "inclusive," as is in the last analysis his significance as *imago Dei*. His *alienum* lays claim to my *proprium* and takes it up into itself; yet the "inclusion" is never such that actual "representation" is no longer involved, as if my *proprium* were my own attribute or *habitus*. His *alienum* too does not remain aloof. In fact, *proprium* and *alienum* are no longer what they were; each is converted into the other. As Luther says: "When it is necessary to discuss Christian righteousness, the person must be completely rejected [one must look away from the sphere of the *proprium*, the personal attribute]. For if I pay attention to the person, or speak of the person, then, whether intentionally or unintentionally on my part, the person becomes a doer of works who is subject to the Law. But here [instead] Christ and my conscience must become one body, so that

190

nothing remains in my sight but Christ, crucified and risen. But if Christ is put aside and I look only at myself, then I am done for. For then this thought immediately comes to mind: 'Christ is in heaven, and you are on earth. How are you now going to reach Him?' " [11] To turn my gaze on myself is to end up "subject to the Law," stuck with nothing but my own *proprium*. Then Christ's *alienum*, his righteousness, remains aloof, beyond me; it is exclusive rather than inclusive, and I do not participate in the Son's *imago Dei*.

The Galatians passage inevitably speaks of yet a second relation implicit in my participation in Christ through being drawn into him. If it is true that I live "in Christ," it is thereby true in the second place that Christ also lives "in me": I live, to be sure; yet not I, but Christ lives in me. By now it ought to be clear to us that here too it will not suffice to understand the matter psychologically, as if there took place in me an alteration of my inherent disposition or quality [*habitus*], a magical transformation, as it were, in consequence of the reception of the deity into my personal being. If we were to adopt such a view, we should be guilty of that illegitimate inversion of outlook which isolates us from relation to Christ and sets us in an abstract *in se*: ". . . if Christ is put aside and I look only at myself [*in me*]." In fact, however, we are persons only "in relation." Luther's *in me* has perhaps a seductive sound to modern ears, but what is at issue here is again not a property of man in himself but an attribute of the relationship in which he stands. What is said is simply that my connection with Jesus Christ, my incorporation into the body of which he is the head, implies that I may ascribe all his qualities to me, that I may understand his *alienum* as my *proprium*, i.e., that I may relate myself inclusively rather than exclusively to his vicarious representation.

By the qualities of Christ we are to understand—again in contrast to the humanistic virtues—the qualities of his relationship to the Father. By healing the breach in my relationship to God, by taking my guilt to himself and bearing it away, he sets me in precisely this relationship to his Father. Hence it comes about that I can claim his qualities as my own: his faith is my faith, his righteousness my righteousness, his peace my peace. And for this very reason all my qualities now apply conversely to him: he is my guilt, my death, my

[11] *LW* 26, 166.

unrighteousness. "Faith couples Christ and me more intimately than a husband is coupled to his wife . . . faith is no idle quality." [12] All the attributes one can ascribe to the Christian are predicable not of qualities inherent in the person himself but of the "communication" with Christ. "A sure trust and firm acceptance in the heart . . . [faith] takes hold of Christ in such a way that Christ is . . . the One who is present in the faith itself." [13] The concept of faith poses the greatest challenge to the concept of attributes, since it speaks pointedly of a relationship.

This insight that all the qualities of Christ, all the qualities of his relation to the Father, may be ascribed to me, is simply expressed in the familiar stanza of Zinzendorf's hymn:

> Jesus, Thy blood and righteousness
> My beauty are, my glorious dress:
> Midst flaming worlds, in these arrayed,
> With joy shall I lift up my head.

Luther deals constantly with this theme, and he can even speak of a "conformity" to the Son's *imago Dei* which is constitutive of the Christian.[14] It is precisely because Christ accepts "conformity" with our sin, our dereliction, our perplexity and doubt,[15] it is precisely *because* he lived a human life and "offered up prayers and supplications, with loud cries and tears, to him who was able to save him from death" (Heb. 5:7, cf. also 2:18) that we are "conformed" to his divine likeness and he takes us up into his life (cf. I Pet. 4:13; 2:21-25; II Cor. 1:5).

We can sum up as follows our insights into Christ as divine likeness and prototype, and into our conformity with him. Since the *imago Dei* is not to be defined ontologically in terms of certain demonstrated qualities, but as a relationship between God and man, it cannot be lost, just as man could not forfeit his humanity even if he wished to do so. The *imago* can only pass into the negative

---

[12] *LW* 26, 169.

[13] *LW* 26, 129.

[14] Cf. Ernst Wolf, "Die Christusverkündigung bei Luther," in *Peregrinatio* (Munich, 1954); and Ernst Wolf, *Staupitz und Luther* (Leipzig, 1927), pp. 204 ff.; Walther von Loewenich, *Luthers Theologia crucis* (2nd ed.; Munich, 1933), pp. 133 ff.

[15] ". . . that Christ was to have been a sinner." *WA* 27, 109.

mode. Hence the *imago* is not "nothing." But neither is it merely something "left over," something which has survived as a relic of the original endowment at creation. On the contrary, the image is really present, but only in that negative mode which implies negation of the original fellowship with God, a negation however which is still a prerogative distinguishing man from the beasts. Man's very failure to attain the telos for which he was destined is part of the prerogative of him who was created in the divine image. For even in the negative mode he bears on his forehead the mark of his original nobility, and his misery is always that of a "deposed king."

The *imago Dei* in man is therefore not an immanent quality, nor (subsequent to the fall) the relic of a quality, a leftover from our first creation which provides a foundation on which to rebuild. Since it is rather a relation to God, it can attain to the positive mode only in the one who is "our peace," i.e., the prototype of our unfallen position: only "in Christ" (cf. II Cor. 4:4; Col. 1:15; Rom. 8:29; II Cor 3:18; Col. 3:10; Eph. 4:24). This opens up the question how then there can be participation in this, the Son's unimpaired image of God. Do we attain to such participation by the fact that he is our example and that we become his imitators? We have seen that this form of imitation involves a confusion of Law and Gospel, and that this way is therefore closed. We can attain to participation not because he is our "example" but because he is our "exemplar." He is the original prototype both of God's saving activity and of the reality of our own salvation.

I must first be in a right relation to Christ in his quality as exemplar before I can follow him as example. I must first be set in the sun before I can become warm. I must first know that I am loved before I can love in return. I must first have attained to the *incipere,* i.e., I must first be set as a mirror in front of the original, before I can accomplish the *proficere,* i.e., before the original can become increasingly transfigured in me, before there can be the μεταμορφοῦσθαι.

To adopt the terminology of Luther, it is not that we imitate Christ in order to be as he is, i.e., to participate in his divine likeness. It is rather that his qualities count as ours, his *alienum* becomes our *proprium,* his fate becomes our fate (Matt. 10:25) because in the state of faith we *are* as he is. For the fact that he is an example is "the least important thing about Christ, in respect of which he is like other saints." [16] As an example, "Christ is of no more help to you than some other saint. His life remains his own and does not as yet contribute anything to you. In short this mode [of understanding Christ as simply an example] does not make Christians but only hypocrites." [17] "It is not the imitation that makes sons; it is sonship that makes imitators." [18] It is our being taken up into the sonship of Christ, into his original and likeness, which first makes it possible for us to be copies, and not vice versa.

Hence Reformation doctrine does not understand the image of God as an ontic relic of creation, which is to be described as a quality proper to man, a quality whose various elements (reason, conscience, free will, etc.) may be brought out diagnostically—or even "existentially"—and understood and used as a natural substructure for grace. On the contrary, since it is only a particular relationship to God, the *imago Dei* is imparted to me in its positive mode as an "alien righteousness" [*justitia aliena*] in Christ, both eschatologically and in the present. The divine likeness of man in this sense is present only in the prototype, in Jesus Christ, and is therefore imparted to me only "in Christ." In its positive mode the *imago Dei* is nothing other than faith.

[16] *WA* 15, 396, line 18.
[17] *LW* 35, 119.
[18] *LW* 27, 263.

11.

The Evangelical-Roman Catholic Debate on Creation and the Original State

Everything that we have said till now about the *imago Dei* forces us into debate with Roman Catholic theology, for on this matter we have radically abandoned its normative intentions. Such a debate ought to be pursued from two main standpoints.

First, we have been led to assert that the *imago Dei* exists only as a personal relation (in both the positive and the negative mode), and that (in respect of its positive mode) it consists in participation in the alien righteousness [*justitia aliena*] of Christ. This naturally raises the question of the legitimacy of this assertion in face of the Roman Catholic statement that the *imago Dei*, however it may be described in detail, is always an ontic quality of man, and that it thus belongs to him as a *proprium*, being his either through creation or by grace. We shall pursue this question in the present chapter under the rubric "Creation and the Original State."

Second, it goes without saying that these divergent doctrines of the *imago Dei* must profoundly influence also the doctrine of justification. As already indicated, justification as viewed in terms of the Reformation doctrine of the image has its center of gravity in an event outside of man, namely, in the *alienum* of the righteousness of Christ. Hence it necessarily looks away from man's own merit or worth, his own *proprium*, no matter whether this *proprium* be his by creation or as a superadded gift of grace. Justification viewed in terms of the Roman Catholic doctrine of the image, however, has its thrust in a very different direction. Here the image is a quality of man, and justifying grace must therefore always imply an alteration of this quality, whether in the sense of "increasing" it or "correcting" the distortions of it due to sin, or "restoring" it to its original perfection. The significance ascribed to merits in justification underscores this understanding. For the possibility of earning merits depends on certain qualities of man which for their part are produced by grace; it depends on certain infused virtues.[1] This attempt of Roman Catholic theology to find a basis for the process of justification in the *proprium* of man (*habitus,* merits, etc.) as well as in the *alienum,* will be considered in the next chapter.

The ethical relevance of these problems hardly needs demonstration. For upon our decision in respect of these two questions depends our whole teaching on, for example, conscience and natural law. Is conscience a relic of the original *imago* of creation? Does the *imago* thus persist in and beyond the fall in this form, the form of conscience? Is natural law another sign that nature has remained intact, at least in certain spheres? And beyond that what is the character of man's action in the state of justification? Is it still to be described as *de facto* sin [*peccatum in re*], or is it really renewed in the sense of there being a renewed *habitus*? Does it indicate a certain *proprium* of man, a *proprium* that has been renewed by the infusion of grace?

But the deepest and most provocative question, the question which casts doubt upon the whole enterprise of Reformation ethics, arises at another point and may be stated as follows. If we reject an ontic definition of the image in terms of content and speak only of the relationship to Christ, we are ultimately confronted by the alternative of standing either "in Christ" or *outside* him, either *in* faith or *outside* of faith (Rom. 14:23). But in this case Christian doctrine seems to have nothing to say to those outside the Christian faith; it appears to offer them no binding norms, not even the divine commandments. For those outside the faith are all included in the one body of sin. There is then no point in the philosophical quest for truth, for everything is falsehood. There is no point in quickening the conscience, for everything is sin. Indeed, there is no point in taking seriously the commands of God—even if only as natural law—for whatever was done in obedience to them would be nothing but deceptive vices.[2] Would not this imply a most dreadful abandonment of the world at large? Would it not make of Christianity an esoteric club cut off from all communication with the outside world?

---

[1] See Chapter 7 of the *Decree on Justification* by the Council of Trent (1545-1563), Session VI, in Henry Denzinger, *The Sources of Catholic Dogma,* trans. Roy J. Deferrari (30th ed.; St. Louis: Herder, 1957), No. 800.

[2] See above p. 59, n. 10.

Here we are indeed confronted by a problem which goes to the very foundations, a problem which is ultimately posed for us by Roman Catholicism. For Roman Catholicism itself does possess directives which have universal validity for both Christians and pagans, inasmuch as it recognizes a universal, ontologically defined *imago* quality present in all men. But Roman Catholicism too is confronted by a basic question: Can this *imago* quality be established in any other way than through an analogy of being [*analogia entis*], that is to say, through a reduction or even an elimination of the *justitia aliena* of Christ, which means the surrender of the *sola gratia* and of the corresponding *sola fide?*

At this point we seem to be caught on the horns of a real dilemma. We have arrived at what is a key problem in any discussion of the foundations of ethics. The decisive premises for its solution have already been worked out. But we have yet to consider the Roman Catholic solutions, and in confrontation with them to test our own premises.

The Roman Catholic View of Nature and Supernature

Roman Catholic thinking is profoundly ontological, Reformation thinking profoundly personalistic. To think ontologically is to see being and its forms from two angles, first, as elementary constituents to be demonstrated in their qualitative particularity, e.g., reason, conscience, sensuality, body, and spirit; and second, with reference to the ordered relationship in which these elementary parts stand to one another.

Where men think ontologically they postulate certain constant factors which are simply given facts; these are thought to be quite independent of the question "Christian or non-Christian, faith or unbelief." What reason is, what conscience is, this can be determined objectively, apart from any ideological consideration, since what is at issue is a permanent ontic factor. Such factors are to be found in what we called the elementary constituents.

Where men think ontologically they also postulate certain constant factors which are posited or set down, namely, those ordered connections through which the various elementary constituents are related to one another. The relation may be very disordered—this is in fact

197

the case since man's fall into sin—but it has a validity nonetheless which is objective and again quite permanent, independent of any ideological considerations and therefore quite independent of faith or unbelief. This ordered connection is constantly posited.

To think personalistically, on the other hand, is to see all the realities of human life exclusively in terms of the personal relatedness of God and man, or more precisely, in terms of the fellowship between God and man given in Christ. Since this fellowship is given only in faith, personalistic thinking also means to see all these realities exclusively from the standpoint of faith. It means to see them under the sway of the alternative that they are characterized either by the believer's existence in faith and are therefore "justified," or by unbelief and are therefore "sin" (Rom. 14:23). But in this case there would be no constant factor, no continuum, no neutral base of the ego, and consequently no common ground (how could there be?) between Christians and non-Christians. At any rate, these are consequences worth thinking about. We do not hesitate to draw them here very sharply because we want to point up the gravity of the ethical problem ahead.

One of the most difficult problems of theological interpretation is to ascertain whether in Roman Catholicism this ontological way of thinking comes first—perhaps as a heritage from Aristotle—and as such predetermines the doctrine of the *imago,* its ontic elements and the ordered relationship among them, or whether instead the substance of the doctrine comes first—perhaps as a heritage from Scripture or from tradition—and for its part requires this ontological method as its most appropriate form of thought and expression. The problem is further complicated by the difficulty in deciding whether this ontological doctrine of the original state determines the structure of the doctrine of justification—whereby justification must effect a correction of the image fragmented in the fall and a restoration of integrity—or whether instead the Roman Catholic doctrine of the original state is conceived in terms of, or as a kind of projection back from, justification. In respect of certain parts of the doctrine, e.g., the differentiation between natural gifts and supernatural gifts [*dona naturalia* and *dona supernaturalia*], the second interpretation is correct. For this differentiation is something which cannot be proved from the biblical doctrine of the original state, as even some

Roman Catholics concede.[3] It obviously follows the outlines of a doctrine of grace which was developed first, without reference to the original state.

On the other hand, what we have called the personalistic tendency of Reformation thinking necessarily raises the question whether the biblical message of salvation is not here subjugated to an "ism," specifically "personalism," and whether in the strait jacket of such an "ism" the fullness of the biblical statements is not thereby curtailed. At any rate, we cannot justify our doctrine of the *imago Dei* simply on the grounds of personalistic thinking as such, especially since the Bible itself is not lacking in ontological references. Rather we can only use personalistic thinking *ad hoc*. Its function can only be that of serving [*usus organicus*] in given instances the specific biblical doctrines of salvation. At the same time we must always be prepared to let it too be broken and remolded as may be necessary, even though it may seem to us to be particularly appropriate and to be present especially in the writings of Paul.

Ontological statements concerning the *imago Dei* are to be found already in the early church fathers, who consistently regard reason and freedom as ontically distinctive features of man. The spiritual nature of man thereby described is therefore similar in "species," so to speak, to the nature of God. It is characterized by the immateriality of cognition and volition. This spiritual nature of man implies also his superiority to his fellow creatures, as superiority which expresses itself according to Genesis 1:26 in the form of his having dominion over the rest of creation. For many of the early fathers, however, man's dominion over the rest of creation does not derive from his spiritual nature, as though the dominion first arose as a consequence of his character as the *imago*. Instead they link man's dominion directly with the analogy to the divine ruler and to his omnipotence, though there is no question but that even for them man's dominion is an ontic feature of the divine likeness—there is no debate about that.

What is debated among the fathers, however, is the question whether and how far the human body is an ontic component of the divine likeness. It is a question suggested particularly by man's

[3] See Diekamp, *Katholische Dogmatik nach den Grundsätzen des heilsgen Thomas* (Münster, 1939), II, 122.

upright posture, which points to his dominical status. However that may be, we can undoubtedly say it has always been part of the Roman Catholic concept of nature that nature has within it ontic elements of likeness, phenomena evincing a parallelism between divine existence and human. This is clear already from the etymology of the term, as Roman Catholic theology has pointed out. For as the Latin *natura* comes from *nasci,* and the Greek φύσις from φύειν and φύεσθαι, the reference in both cases is said to be to the origin by conception and birth. Now what is mediated by birth is something similar—like begets like—and therefore that which is born is like that which gives it birth. Hence the concept of nature is said to be particularly well adapted to express the analogy between creator and creature of which we are presently speaking. In particular the concept of nature is thought to bring out especially well certain elements in this analogy which we would now enumerate.

First, by the "nature" of a thing or being is meant the qualities of the species, that which connects it with other things or beings of the same kind. The term does not refer to individual peculiarities of the thing or being in question. For after all birth takes place within a species, and involves primarily and supremely the transmission of the collective qualities of the species. Thus it is of the nature of the hen for example to be "henlike," i.e., to cluck, to lay eggs, to hatch them out. If it should happen—and zoology has occasionally come across such a case—that a hen lays eggs but then pecks them to pieces instead of hatching them out, such a hen would not be "henlike." It would be "deranged" in the sense that it had moved outside the natural order of its species. It would not have the qualities of the being "hen." It would not be in conformity with its own "nature." "Nature" comprises the qualities of the species, and in so doing denotes the "essence" of the thing which resides in the species. Individuality therefore is not particularly "natural." It can in fact be quite abnormal, eccentric, and unnatural. Hence the concept "nature" is to be differentiated not only from supernature which is above it, as we shall see later, but also from the non-conforming individual which is below it.

Writes M. J. Scheeben, "To the natural . . . the antithesis strictly speaking is not the supernatural but the non-natural, i.e., that which does not belong to nature or proceed from it or correspond to it. We must bear this firmly in mind when we attempt to conceive of nature in its broadest sense as that which is in general the essence of things so far as their activity is concerned. Here nature is regarded as an abstract, as essence itself, which as such does not include that which after all determines what it will be in any given being, namely, individuality. Whatever determines nature only in individual beings does not really belong to nature—insofar as nature is held to be identical with essence—and is therefore not 'natural' at all, though it is natural insofar as it is closely tied up with nature and is not higher than nature." [4]

Second, the concept "nature" always implies an entity which imparts its own character by giving birth, and also an entity in which the impartation of character takes place as that entity is born. At this point one may clearly discern the indisputable influence of Aristotle's understanding of the relationship between form and matter (εἶδος and ὕλη), according to which the former impresses on the latter its character, its "form." "Hence . . . nature as a concrete and collective term denotes primarily the totality of all beings living in matter, . . . the world of material things, which are subject to the antithesis between what gives birth and what is born, as also to the antithesis between what is animated or at least determined matter and what is animating or determining form: in contrast to the purely spiritual world, in which, by virtue of its simplicity, there is no place either for matter or for the antitheses which it entails." [5]

We thus arrive at Scheeben's definition: "Nature is . . . the quality of a substance as it arises from and is thus also determined by the essence of a thing, and its disposition to live, act, and move in accordance with goals appropriate to that essence." [6] In this definition three characteristics of nature are particularly important: (1) Nature is essence as defined by qualities of species. (2) Nature thus constitutes a substance which is not variable, which in particular cannot be modified by the contingencies of individuality. It is that which in any given thing is unchanging and persistent despite all other changes. It is that which cannot be missing in a thing if it is to be

[4] M. J. Scheeben, "Natur und Gnade," in *Gesammelte Schriften* (Freiburg, 1941), I, 18-19.

[5] *Ibid.*, p. 16.

[6] *Ibid.*, p. 31.

a thing of this kind at all—in contrast to that which within this species may vary and change. For this reason—and here is where theologically the greatest weight belongs—it is also indestructible, maintaining its continuity intact even in face of the powers of sin and grace. (3) For all that, however, it is not yet something complete and perfect. In Scheeben's terminology, it is a "disposition" [Anlage], a potentiality or possibility. That is, it points beyond itself to a fulfillment which has yet to be imparted to it.

It was Irenaeus who introduced into the *imago* doctrine a new feature which was worked out later and came to have far-reaching importance for the doctrines of sin and justification. This was the distinction between the original disposition of man and his destiny, between his "natural endowment" and his "supernatural goal." To his natural endowment belongs his corporeality as well as reason and freedom. But his supernatural destiny goes far beyond this. It consists in the incorruptibility ($\dot{\alpha}\phi\theta\alpha\rho\sigma\dot{\iota}\alpha$) and perfection ($\tau\epsilon\lambda\epsilon\iota\acuteo\tau\eta s$) which constitute the vision of God. To the question whether or why God did not give man this perfection from the very first, as part of his natural endowment, Irenaeus answers that as an infant ($\nu\acute{\eta}\pi\iota os$) Adam was not at first able to bear this strong meat.[7] Hence, along with the original disposition, that which transcends the purely natural disposition of the original *imago* is also given to man as his future destiny: "We have not been made gods from the beginning, but at first merely men, then at length gods."[8]

Irenaeus sought to find exegetical support for this distinction between original natural endowment and supernatural destiny by arguing that the two terms for image in Genesis 1:26 are not synonymous, *imago* ($\epsilon\dot{\iota}\kappa\acute{\omega}\nu$) representing the original natural disposition (endowment with reason, etc.) and *similitudo* ($\dot{o}\mu o\acute{\iota}\omega\mu\alpha$) representing the supernatural destiny. Although more recent Roman Catholic theology has long since abandoned this exegesis, the substance of the distinction has remained. In fact, it has exercised a basic influence on the whole theological system. In Thomism, as we shall see, the doctrines of creation, sin, and grace would be quite unthinkable apart from this distinction. But first let us examine more closely the

[7] Irenaeus *Against Heresies* iv. 38.

[8] *Ibid.* iv. 38. 4; *The Anti-Nicene Fathers*, ed. Alexander Roberts and James Donaldson (Buffalo: Christian Literature Publishing Company, 1885), I, 522.

distinction between these two concepts in Irenaeus himself.

The *imago* ($\epsilon\dot{\iota}\kappa\acute{\omega}\nu$), as natural endowment, is on the one hand actualized in the body, and is thus a neutral thing.[9] No less neutral, so far as the matter of man's attaining to or losing his higher destiny is concerned, is his spiritual endowment, i.e., the fact that he is "endowed with reason, and in this respect like to God, having been made free in his will, and with power over himself." [10]

The *similitudo*, on the other hand, consists in the combining of the naturally endowed soul [*anima*] with the Spirit of God,[11] so that man allows himself in free obedience to be led and fashioned by this Spirit even to the point of complete deification. Hence the *similitudo* is not perfect from the outset. It is dependent on man's voluntarily seizing and actualizing his final destiny. It comes into being only through man's participation in the Spirit of God. This is why the *similitudo* has a supernatural character. It is nature perfected by the divine intervention, an intervention to be sure which must be seized and appropriated by man in complete freedom.

It is very important that we understand this distinction between natural endowment and supernatural destiny, whereby the former is something firmly given and in substance indestructible whereas the latter is something variable which can be won or lost. For it is through this distinction, first made by Irenaeus, that there has arisen that distinction so basic for later Roman Catholicism between nature and supernature, between nature and grace, the distinction which allows for the *imago's* having an explicitly ontological character which continues intact through its impairment by sin and its restoration by grace. For with this distinction two ontic spheres are marked off in man. One of these, the sphere of nature, is constant. It cannot be altered in substance. The other, however, the sphere of supernature, is variable, i.e., it is added or not added, and it can be lost.

How are these two spheres related? God adds to the ontic

9 "If the Spirit be wanting to the soul, he who is such is indeed of an animal nature, and being left carnal, shall be *an imperfect being, possessing indeed the image* [of God] *in his formation* [*in plasmate*], but not receiving the similitude through the Spirit; and thus is this being imperfect." Irenaeus *Against Heresies* v. 6. 1; *op. cit.*, I, 532 [italics by H. T.]

10 *Ibid.* iv. 4. 3; *op. cit.*, I, 466. Cf. *ibid.* iv. 37. 1: "God made man a free [agent] from the beginning, possessing his own power" (*op. cit.*, I, 518).

11 See note 9 above. Cf. *ibid.* v. 12. 2 and v. 16. 2.

qualities of the original image, i.e., to reason, freedom, etc., a supernatural endowment. He does so really by way of "addition." This supernatural endowment, which may be called the *similitudo,* takes two forms.

In the first place, it consists in "original righteousness" [*justitia originalis*], i.e., in the fact that man is holy and just before God, that he is God's child and an heir of heaven, that he shares in the divine Spirit. Whereas Irenaeus understood this supernatural endowment in terms of a goal set before man, to the attainment of which man had to be led, later Roman Catholic theology regards it as a gift already bestowed upon man in his primal state. In virtue of this endowment Adam is said to have possessed already that which Christ has won for us—or won again after Adam had lost it. (This very statement of the matter suggests how, within the framework of ontological thought, sin and redemption take on very strongly, at least in form, the character of addition, subtraction, and new addition.)

In the second place, the supernatural endowment consists in the "gifts of integrity" [*dona integritatis*]. These are not an addition to nature, in the sense of an extension of nature into the supernatural sphere. Rather they work within the sphere of nature itself, imparting order to it and thereby enabling its individual ontic elements to fulfill their function. This is why they are sometimes called "praeternatural gifts," because while they are additional to the natural endowment they do not point to something beyond it but refer rather to that which is within it. The significance of these praeternatural gifts for the ontology of the divine likeness is seen best in respect of the most important of them, namely, the "gift of rectitude" [*donum rectitudinis*]. Aquinas claims that the rectitude of divinely endowed man was that in him "the lower powers were subjected to the higher, and the higher nature was made so as not to be impeded by the lower." [12]

In other words, if the ontic elements of the "natural" image are to "function" at all and work as they were intended to do, then there must be added to them an order [*ordo*] which will relate them to one

---

[12] *Summa Theologica,* part 1, ques. 94, art. 1; *The "Summa Theologica" of St. Thomas Aquinas,* trans. Fathers of the English Dominican Province (2nd rev. ed.; London: Burnes Oates & Washbourne Ltd., 1922), p. 307.

another in a way that is in keeping with that intention. Without this ordered relation—or after its destruction—the natural ontic elements remain intact, but instead of constituting an ordered and "meaningful" mosaic, they are like a chaotic jumble of individual fragments. In this sense the creation of man can be conceived in two stages. In the first there are fashioned out of complete chaos the elementary parts (reason, freedom, sensuality); these are individually ordered within themselves but they are not related to one another. They thus constitute a secondary chaos, in which the qualitative particularities which are individually ordered within themselves are all at odds with one another. If we attempt to conceive of that which Scholasticism calls nature or the image, and to do so without considering this praeternatural ordered connection which has been added, the *rectitudo*, what we end up with is this secondary chaos. Consequently the gift of rectitude has as its function the ordering of nature within itself; it is to bring nature to itself, i.e., to its intended destiny.

The order of rectitude thereby implied might be called "the hierarchy of elements," i.e., the relationship of reciprocal superordination and subordination which derives from the rank assigned to each of the individual elements as its destiny. To establish this hierarchy ontologically it is necessary to begin from above, i.e., with God himself. The superior power must in every case be given its proper rank if the inferior elements are to be ordered aright and if there is thus to be a hierarchy of rectitude.

This means concretely that reason can function properly, in a manner consistent with the intended order, only if it does not regard itself as autonomous, which means in this case, only if it subordinates itself to God and does not claim to be the topmost authority in the sphere of nature. Again, sensuality and the other lesser powers of the soul can function meaningfully only if they subordinate themselves to reason and allow themselves to be guided by it. Finally, the body is granted the function intended for it only if it subjects itself to the soul.[13]

This whole ordered arrangement is upset at once if we do not begin with the two greater powers and their right relation to one

---

[13] "This rectitude consisted in [man's] reason being subject to God, the lower powers to reason, and the body to the soul." *Ibid.*, part 1, ques. 95, art. 1; *op. cit.*, p. 317.

another, i.e., if reason, instead of subjecting itself to God tries to be autonomous and to create its own religious and philosophical views. For when this happens, the rest of the hierarchy below this level does not remain unaffected. On the contrary, the whole order is disrupted. Reason is then no longer determined from above (by God). Instead it is determined from below, by the "lower powers," and there thus arises, for example, a vitalism and its corresponding philosophical expressions.

On the other hand, if reason remains in its right place in the *ordo* hierarchy, there is the possibility of true rationality, i.e., of sound philosophy. For the reason which is then operative is a reason unmodified by the lower powers. In the sense of Roman Catholic usage, then, "Christian philosophy" is not a philosophy which has to borrow from supernatural (revealed) dogmas, and which would thus in heteronomous fashion be an ancillary of theology. It is rather a philosophy which remains within the sphere of nature, that nature which has its rightful and intended place within the *rectitudo* hierarchy and is thus brought to itself.

The moment the highest of the normative norms [*norma normans*] loses its authoritative position, there arise incessant perversions and rebellions among the inferior members of the hierarchy. For "the first subjection was the cause of both the second and the third." [14] As long as concupiscence, for example, remains ensconced within this *ordo,* as a kind of lowest member in the hierarchy, it is not an "inordinate appetite"; it is without sin. [15] Whether or not this ontic element of the original natural endowment actually becomes sin, depends entirely on whether or not it is integrated into the *rectitudo.* It will always remain an element of the *imago;* the question is whether it will also serve as an element of the *similitudo.*

We have thus clearly exhibited the decisive duality in Roman Catholic anthropology and its doctrine of the divine likeness. The sphere of the imago includes nature. In terms of its substance this nature is static and self-contained; it cannot be augmented or diminished, improved or destroyed. It cannot be destroyed because it consists of an accumulation of ontic parts, each of which is in itself

[14] *Ibid.*

[15] See, for example, the *Decree on Original Sin* by the Council of Trent (1545-1563), Session V, in Denzinger, *op. cit.,* No. 792.

unalterable. If this nature is to point or lead beyond itself, it cannot do so of itself; there must be a new creative act. There must be the impartation of the supernatural gifts which lead from the natural *imago* to the supernatural *similitudo*.

What takes place, then, at the fall? Negatively, it may first be said that the fall does not alter nature. It cannot affect the *imago* qualities of man. It can only undo what God had done in adding the supernatural gifts. Only in respect of these gifts do we move in that variable sphere which can be affected and altered by sin and redemption. The subtraction of supernatural gifts which takes place at the fall works itself out in various ways which would be delineated in a detailed dogmatics. Within the limitations of our present ethical concern it will suffice to concentrate on the single fact that among these gifts is the *rectitudo* in virtue of which the individual elements of nature are brought into an ordered relationship such that nature is enabled to fulfill its function. The subtraction implied in the fall means primarily that this ordered connection disappears, that reason sets itself above God, the lower powers set themselves above reason, and the body sets itself above the soul.[16]

Each of those rebelling elements remains individually intact for they all belong to nature, which is unalterable in substance. But as regards their relation to one another they relapse into chaos. What was an ordered and meaningful mosaic in consequence of the gracious gift of rectitude, now becomes a bizarre medley of unrelated stones.

Implicit in this conception of the relationship between nature and supernature are certain evident postulates which will determine how a doctrine of redemption would accordingly be constructed. It would be constructed in such a way that grace would link up with certain natural elements which are already present and intact. Grace would first bring them into proper order, and second carry them forward into the supernatural sphere.[17] This is what is meant by the well-known statement, "Grace does not destroy nature but perfects it"

[16] In other words, original righteousness [*justitia originalis*] consists—among other things—in the harmonious ordering of the natural elements, and original sin [*peccatum originale*] consists in the dissolution of this order. See Thomas, *Summa Theologica*, part 1[II], ques. 85, art. 3; *op. cit.* (New York: Benziger, 1915), p. 44; and part 1[II], ques. 82, art. 1; *ibid.* (1915), p. 412.

[17] See the Dogmatic Constitution concerning the Catholic Faith (Chapter 2, on Revelation) of the Vatican Council (1869-70), Session III, in Denzinger, *op. cit.*, Nos. 1785, 1786.

[*gratia non destruit naturam, sed perficit*]. The perfecting here includes both qualitative ordering in the sense of rectitude and quantitative extension to a supernatural end. There may be seen at this point both the analogy of being [*analogia entis*] and the integration of nature and grace.

It would admittedly be one-sided to say that ontological standpoints alone are involved in this conception. Roman Catholic thinking is a combination of opposites. One-sidedness would be inconsistent with its very character. The very fact that the whole *rectitudo* hierarchy depends ultimately on the relation to God injects a personal element. On the other hand, within this relation there are found the purely ontic components of nature. The result theologically is the question whether it suffices to say that what we have here is a combination or even synthesis of ontological and personalistic thinking, or whether we must instead insist that the personal relation is overthrown in principle the moment ontological statements are made at any one point, i.e, whenever at any one point there are set apart spheres of being which are neutral with respect to the personal relation.

An answer to this question from the standpoint of the Pauline and Reformation understanding of the matter would clearly have to be in terms of the second alternative. Paul said that "whatsoever is not of faith is sin," and the moment his "whatsoever" is limited in the slightest degree, e.g., by the setting apart of certain ontic spheres which are neutral as regards faith, the *sola gratia* and *sola fide* are abandoned in principle. Here it is a question of all or nothing. These dire consequences are to be seen in the Roman Catholic doctrine of grace even more plainly than in the doctrine of the *imago*. For it displays a similar combination of ontological and personalistic understanding (grace is regarded as both a material medicament and a divine disposition), and it enables us to see clearly that in this combination the ontological side predominates and the personal element is increasingly absorbed.

Two things here are of particular interest and importance from the ethical standpoint. The first is the fact that there is a constant ontic sphere which is neutral in respect of sin and grace. The very existence of such a sphere means that the catastrophe which overtook man at the fall could not possibly be total, affecting the entire man. This necessarily restricts the Pauline statement that "whatsoever is not of faith is sin" (Rom. 14:23).[18] The second important thing is the fact that the justification of the sinner does not mean a wholly "new creation" (II Cor. 5:17), and does not rest exclusively on the *alienum* of the righteousness of Christ. Instead the redeeming action of God links up with the *proprium* of man, i.e., with the natural elements which have remained intact; it simply joins itself to them by way of "addition."

The extraordinary consequences of this conception for ethics may be seen in two directions.

First, because of the persistent ontic *proprium* of his nature, man is capable of co-operating in the work of salvation. This thesis is not altered in the slightest by the insight of the Synod of Orange (529) that grace precedes any activity of this kind, so that the initiative throughout remains with God.[19] This insight certainly goes beyond what cheap Protestant polemics generally regard as the Roman doctrine of justification. But in the last analysis it can mean only that prevenient grace actualizes nature's existent disposition for grace, and the capacity for co-operation latent in this disposition. Divine grace gives, as it were, the initial push which makes possible man's autonomous participation, his "co-operation" in the attainment of salvation by virtue of which he is able to win "merits."[20]

This ontological point of departure in Roman Catholic thinking about the *imago Dei* in man thus has important implications for the entire doctrine of salvation, and leads directly to the Roman Catholic —Reformation controversy. The fact that there is this ontic remnant implies that in addition to the *justitia aliena* of Christ there is also the *proprium* which, after its actualization through grace, has a significance all its own. Thus there arises a partnership between God and man, and a certain co-ordination of the modes of operation,

18 Cf. Luther's exposition of this verse in the 1520 Theses on the question whether works contribute to justification, WA 7, 231, esp. Theses Nos. 10 and 11.

19 See Canons 3-5 on Grace of the Second Council of Orange, Denzinger, *op. cit.*, Nos. 176-178. Cf. also the emphasis given to prevenient grace "without any existing merits" on man's part—apparently in reaction to Luther—in Chapter 5 of the *Decree on Justification* of the Council of Trent (1543-1549), in Denzinger, *op. cit.*, No. 797.

20 The Apology of the Augsburg Confession regards this disposition, this idea of "initial grace," as destructive of the *sola gratia* and the *solo Christo*. See Apology, IV, 162-163 and IV, 17; BC 129, 109.

namely, of grace on the one hand and merits on the other. Beyond that, the sharp distinction between Law and Gospel is thereby eliminated, and the sequence of justification and sanctification is blurred. In other words, the doctrine of the *imago Dei* contains the seeds of the Roman Catholic doctrine of justification against which the Reformation protested with its passionate emphasis on the *sola gratia, sola scriptura,* and *sola fide* [the *particula exclusiva*].

The consequences of this ontological approach are to be seen in yet another direction. Because the *imago* is a basic neutral component of man's nature, two things are possible. It is possible to discern in all men an identical ego-core to which appeal may be made. And it is possible by virtue of the harmonious agreement of nature and grace, image and similitude, not only to declare the right order in which the various components of the ego are to be related to one another, but also to define the right order which, as superpersonal nexus, has to unite the individual egos, i.e., the individual bearers of nature, to one another.

On the basis of the first of these two possibilities Roman Catholic moral theology regards conscience as an objective component of nature, to which appeal may be made. And if conscience is admittedly capable of error this is because, as an element of nature, it has fallen away from the order of rectitude. But it may be brought back into the *rectitudo;* this means finally that it must be subjected (*recta subjectio*) to the *norma normans,* i.e., to "God" as represented by the church.

On the basis of the second of these two possibilities Roman Catholic moral theology recognizes a natural law which can grasp and reconstruct an order of being that is valid for all men, and not merely for Christians. This order of being is valid for all because it has to do with the order of the natural elements which remain identical and intact in all men; this is its ontic side. Moreover, this order can be demonstrated by theology because theology deals with nature as it is restored to recititude by grace; this is its noetic side. In Roman Catholicism there is thus a moral teaching which is objectively based and which claims universal validity. Beyond that, Roman Catholicism has also a political, economic, and cultural ethics which rests on the evident axioms of natural law, and which in this sense claims to be unquestionably perspicuous and authoritative even for non-Christians.

We need hardly point out what extraordinary vistas are hereby opened up for the Roman Catholic mode of world conquest, of cultural, political, economic, and social action. We are certainly confronted by a clear and consistent chain of argument which leads from subtle distinctions between the original *imago* and *similitudo* to a conception of theological ethics and finally to the concrete programs and practices of Roman Catholicism in the sphere of public action. Neither the programmatic guidelines of the papal encyclicals nor the directions taken by certain Catholic political parties are conceivable without that original distinction.

The Evangelical View of Man Before God

Before we turn to the Reformation criticism of the Roman Catholic conception, it would be well for us to point out that this conception was by no means universally or consistently abandoned in Reformation theology, where more or less distinct relics of it continued to be found.

That this is hardly true, however, of the Lutheran confessions—apart from certain terminological echoes such as the use of words like "substance" and "nature" [21]—is clear from the fact that in them the primal state of man is simply described as original righteousness. Within this *justitia originalis* no distinction is made between *imago* and *similitudo;* they are regarded as synonyms. The concept of original righteousness is definitely not understood in terms of ontic qualities, neither as the "balanced physical constitution" of Adam involving a general harmony of all his physical and spiritual powers,[22] nor as his obedience to the second table of the Law as this is made possible by such ontic integrity. On the contrary, original righteousness is conceived here wholly and exclusively in terms of a per-

[21] As we noted earlier, similar echoes, and sometimes naïve, ontic descriptions of the first estate are to be found even in Luther, who uses such terms as "nature" and "essence" in connection with original sin, comparing sin's corruption of nature with leprosy's poisoning of the flesh and even referring to sin as belonging to man's essence (*LW* 1, 165-166, 113-114). These echoes were of fateful significance later, for as Orthodoxy became increasingly ontological (especially in Flacius and his doctrine of original sin) it could appeal, at least formally, to these passages in Luther.

[22] Apology, II, 17; *BC* 102.

sonal relation, and is manifested in the fulfillment of the first table of the Law, i.e., in fear, love, and faith towards God.[23]

Since the divine likeness is thus understood exclusively in terms of relationship, specifically as a relationship to God in the sense of the First Commandment, sin cannot be regarded as a subtraction from or a diminution of the original divine likeness. Sin is rather a radical breach, a transition to the diametrically opposed "negative mode." Original sin "means that human nature has completely fallen." [24] The Reformation thesis that the divine likeness is completely forfeited at the fall is, as we have seen,[25] understandable in this light. But it is a deduction which clearly goes beyond the biblical evidence. The Bible nowhere speaks of any such loss. This is why we referred in the nature of the case to the divine likeness in the "negative mode."

The concept of the *imago Dei* did not, of course, maintain this unequivocally personal form in the history of Protestant theology. The ontic features of man's existence in the primal state, which were not denied in the Apology, but not emphasized either, soon came to the forefront again and took on constitutive significance. With them it was only natural that the concept of the original *habitus* should also recur. These ontic characteristics—the terminology leaves us in no doubt as to the intellectual tradition from which they derive— include the perfection of knowledge,[26] free will and its holiness which is directed to the good, the perfect harmony and extraordinary agreement of the superior and inferior powers,[27] physical immortality, unrestricted dominion over all creatures, and perfect bliss of both body and soul in paradise.

This Orthodox ontologism of man's primal state is fully as unequivocal as that of Scholasticism. The only difference is that it does not have the distinctive safeguards which Scholasticism found in its differentiation between the *imago* and the *similitudo*, between nature and supernature. This differentiation enabled Scholasticism to limit, as it were, the area devastated by the fall, and to salvage from the

[23] Apology, II, 16, 18; cf. Formula of Concord, Solid Declaration, I, 10.
[24] *LW* 1, 114.
[25] See pp. 167-170 above.
[26] *Lux seu perfectio aliqua habitualis* (!) *intellectus* in J. W. Baier, *sapientia originalis* in Polanus.
[27] This harmony, which was the praeternatural gift of *rectitudo* in Scholasticism, appears in Hollaz as a characteristic of the divine likeness given in nature.

disaster certain remnants. As we have seen, this limitation meant that only the supernatural gifts were subtracted while the natural state itself remained intact.

Protestant Orthodoxy is here in a very different and far more precarious position. For it inherits from the Reformation its theological concern to understand sin as a radical encroachment which affects the totality of man. Once we understand man consistently, in terms of his "relation to God," as Luther particularly did, we can no longer think in terms of gradations but only in terms of a clear cut either-or. The Orthodox understanding of the *imago* is complicated by the fact that it has to translate into a very different framework of thought statements which were originally framed within this schema of personalistic thinking. It has to translate them into an ontological thought system. But here the radical aspects necessarily take on a very different appearance. Indeed, they inevitably suffer serious distortion.

For instance, if original righteousness consists in a specific ontic structure, i.e., in the ontic characteristics mentioned above, its radical disruption by the fall must be understood also as an ontic disruption. The disruption can no longer be understood as a breach of fellowship and therefore as a transition to the negative mode of the divine likeness (in which mode, however, the likeness itself continues to exist). It must rather be interpreted as the outright removal of these ontic features themselves. Logically there is no other possibility; this must inevitably be the consequence, the only possible consequence. But since the very essence of man consists in these features, since what is involved here is his very nature, without which it is impossible to conceive of man at all, the fall means nothing less than the self-surrender, indeed the annihilation of man.

One reason for this difficulty is that the radicalism of the Reformation statements cannot possibly be translated from personal into ontological categories; the moment that is attempted, it takes on a very different—and misleading—meaning. A second reason is that in the transition from the one schema to the other no use can be made of safeguards which are present only within the ontological schema; I refer specifically to the safeguard implicit in the distinction between natural and supernatural gifts. But—and this is the third drawback— Protestant Orthodoxy cannot claim this safeguard if it is to be true to its original theological concern, i.e., if it is to follow the Reformation tradition in its radical understanding of the fall; for the very point of the safeguard built in by Scholasticism was to *de*-radicalize sin and to keep free of sin a neutral substance which would be the base of the human ego.

Thus Orthodox thinking steers with assurance into what is bound to be a fatal impasse. It tries to expound a Reformation concern by means of Scholastic methods, and this is quite impossible. There is genuine drama in the attempts made by Orthodoxy to free itself from this impasse. As I see it, certain essential developments in Protestant dogmatics cannot be understood at all unless we regard this dilemma and the efforts made to overcome it as the ultimate motivation and the implicit theme of the whole theological movement.

How in fact did men seek to resolve the difficulty? There were two possibilities. The first merely made the dilemma worse and led to an open rupture in the doctrine of justification. The second sought to save the situation by inconsistency and thus hardly deserves to be called a thoughtful effort at all.

The first of these two possibilities is found in Flacius, and specifically in his doctrine of original sin as the "substance" of man. The crisis in this conception becomes acute when we come to the doctrine of the regenerate man, whom Flacius can describe only as the bearer of two antithetical substances whose strife can be decided only by the help of divine grace, which must ever be actualized anew. This amounts to a complete surrender of the Reformation doctrine that "sin does not reign." It even means that Scholasticism's optimism about nature has been replaced, if you will, by an oppressive pessimism about sin and corruption which completely annuls the message of a "new creation." When the radicalism of Luther's doctrine of sin is thus translated into ontological terms, it leads to a doctrine of two antagonistic "substances" which inevitably destroys from within the whole Reformation understanding of justification.

The second attempt of Orthodoxy's two major attempts to escape the impasse took the form of simple inconsistency. Flacius was consistent to the last. He was so consistent, at least so far as his dogmatic work was concerned, that he did not shy away from espousing the mythological monstrosity of having two substances in man. There are indications that he was one of those theologians, fortunately not

214

too rare, who are better (which in this case means less consistent) than their theology, and who know more of justification as "Christians" than they do as "theologians." Theologically, however, Flacius was alarmingly consistent, once he had been enticed onto the ontological battlefield by his opponent, Victorin Strigel.

It seems that other theologians, Johann Gerhard for example, realizing or at least dimly suspecting to what fatal end consistency would lead, deliberately sought to be inconsistent. It was in fact quite logical to draw from the impossible combination of a personal and an ontological *imago* doctrine the consequences Flacius drew. Once man has fallen away from his primal fellowship with God, sin by logical necessity becomes the true substance of the ego, so that even with respect to the ontic particularities man becomes the very opposite of what he was—he loses his identity.

The inconsistency of other schools of Orthodoxy in respect of the ontological presupposition is that they assume the preservation of certain "remnants" of the primal state. The ontological presupposition is this, that original righteousness consists not in a superadded gift as in Scholasticism, a gift which is subtracted at the fall while nature itself remains nonetheless intact; original righteousness instead is now completely and in its totality identified with man's natural endowment. The theological concern expressed in this presupposition was prophylactic. The desire was to interdict from the very outset, already in the doctrine of the primal state, any possibility of dividing man into two sectors, a natural and a supernatural. It was rightly seen that to neutralize the natural part of man by such a division would be to restrict the fatal consequences of sin. By such an approach sin could always be limited to only one part of man, and then the redemption in Christ would also be particular rather than total in scope; it would mean restoration of the shattered "part," by way of a new "addition."

The conclusion to be drawn from this presupposition should have been that there is no such thing as a relic of creation, there is no neutral element of the ego which survives the fall, no *natura* which remains intact. However, Gerhard and others refuse to draw this conclusion. Instead they speak once more of the survival of "certain tiny remnants of the divine image." [28]

[28] Johann Gerhard, *Loci Theologici* (Berlin, 1885), I, 4.

In other words, at the very point where things should have become quite serious, where what is at stake is the real purpose of correcting the Scholastic doctrine of man's primal state, at that very point there is suddenly a cheerful return to the Scholastic way. There is a cautious staking out with all sorts of little flags of that area where the relic of nature had been levelled, and the site is vigorously rebuilt with houses called "humanity." For when it is asked wherein the relic of nature consists, the answer is now that it consists in "man's rationality and volition," and in his "natural knowledge of God and of the Law." [29] In respect of these remnants, therefore, the divine likeness is not destroyed. There has again been only a "subtraction," and a remainder is left.

The fact that what is said about these remnants is said with restraint, almost with tongue in cheek—the particles are ever so slight!—does not alter the direction in which this theology is necessarily moving. The movement is away from the original personalistic starting point and in the direction of quantitative statements.[30] Paul Althaus rightly points out that where this happens a quantitative orientation must likewise follow in the doctrine of justification: Christ no longer brings back to the Father's house a prodigal who is completely lost and who has no claims. Instead he simply reverses the subtraction by adding once again to the surviving relic those particles which had been lost. In these terms justification is no more a total event than is the fall.

We thus see to what confusion the "unnatural" combination of personalistic and ontological thinking leads. On the basis of the Reformation understanding of Scripture there was first a radical view of man's primal state and of the fall, but this radical view was then whittled down ontologically. So it is only on the basis of Scripture that there is once again a summons to reject this weakened conception, and obedience to this summons involves a severe inconsistency with respect to the starting point of the whole system. For there can be no doubt that it was a scriptural summons which Ger-

---

[29] Man's conscience is also said to remain alive and to belong to that relic of nature which is the divine image. *Ibid.*

[30] "Logically, the work of Christ, the restoration of man as the image of God, would also have to be understood in quantitative terms, and hence as partial rather than total." Paul Althaus, *Die christliche Wahrheit* (Gütersloh: Bertelsmann, 1948), II, 96.

hard thought he was obeying. He believed that he was brought back onto the right track especially by Romans 2:15,[31] although what he was in fact brought back to was this unfortunate business of the relics of the image, which is wholly out of keeping with the reality of the situation.

By thus pursuing the Scholastic doctrine of the *imago Dei* and bringing out its incompatibility with the personalistic understanding of the Reformation, we have implicitly worked out already the essential points which must be made in any criticism of the Roman Catholic doctrine of the *imago*. Our only remaining task, therefore, is to give a systematic review of these findings.

The essential Reformation objection to the Roman Catholic conception of the image must be based, as it is in Luther, on the fact that ontological definition of the *imago* leads to a neutralizing of the state of nature—unless we would commit ourselves to the daring and impossible enterprise of Flacius. But this neutralizing implies that the essential ontic qualities of man (freedom and rationality) are defined without reference to the fellowship with God. The criticism of the Reformers on this point is best expressed in the statement of Luther that on this view the devil himself must be fashioned in God's image, since these qualities of the rational soul are even stronger in him.[32]

The logical consequence of this neutralizing is that the fellowship with God, and hence also faith, is not part of the true and proper nature of man as God created him, for if it were, man would have to be measured constantly and exclusively against the norm of his origin. Faith and fellowship with God are rather among the "super-added gifts" or "accidents."

Accordingly, sin is not an event which impinges upon the essence of man. Instead it is understood simply as a subtraction of the accidents. The theology of redemption is in turn determined by this view of sin inasmuch as justifying grace (justification) remains for its part in the sphere of "accidents" and can have the significance only of adding anew the gifts once added but subsequently lost.

The Reformers must of course concede to Roman Catholicism that the image of God, indeed the humanity of man in general, has

[31] Gerhard, *op. cit.*, II, 115.
[32] See the Luther quotation documented above at p. 162, n. 12.

an ontic "side," just as faith has its psychology and believing thus involves real psychical and even physiological processes. There can be no doubt on this score. However, there is on the other hand every reason to doubt, indeed to reject, the idea that man or his faith is in some way characterized by these ontic factors. On the contrary, he is characterized solely by what he receives from the alien factor of divine grace. The characterization can thus be expressed only in personal categories. There is here then no place for any attempts to go beyond the assertion of this one, exclusive sign of human nature. It would be inappropriate to attempt to investigate the ontic base of the ego where this characterization is supposed to take place, to ask concerning the ontic manner and fashion in which this characterization is thought to occur, e.g., by the impartation of a *habitus* or by demonstrable changes of the ego in consequence of the gift of grace. All such attempts are foredoomed to failure.

For the inversion of outlook, by means of which alone this ontological analysis can proceed, compels us to see man in illegitimate isolation, as man "in himself," and in so doing to dissociate him from his personhood and consequently to overlook the true theme of his existence. At first glance it may seem to be a relatively innocuous matter to set alongside the alien character which is given man by his personal fellowship with God statistical data concerning his ontic process. It soon becomes apparent, however, that it is impossible to stop with the mere statistics of these processes; instead the *proprium* goes on to assume at once a normative rank.

For example, statements are made—and the autonomy of ontological thinking demands that such statements be made—specifying wherein the human partner in the fellowship ontically consists, i.e., "who" it is with whom God enters into fellowship. But this question of the "who" can only imply the assertion of an ontic substratum which exists as the *humanum* prior to and apart from the fellowship with God, and to which the fellowship consequently has reference. Ontological thinking cannot avoid putting this kind of question. But this ineluctability entails the further necessity of postulating a neutral factor that remains identical and intact at all times, through all the various phases of God's salvatory and non-salvatory dealings with man. This would be something which is there when God releases man from his creative hands, which is there again when God summons this being to fellowship with himself, which is still there when this fellowship is forfeited through sin, and which is there again when the fellowship is restored in redemption. The autonomy implicit in this line of thought demands that the *natura* concept be conceived. "Nature" stands for that substratum which always remains identical and intact, beyond both the fellowship with God and the breach of that fellowship, and which must therefore be regarded as essentially neutral.

The logic at work here is as follows. The first stage involves the consideration of ontic qualities alongside the personal relation. The second stage involves what is already implicit in the first, namely, the postulating of an abstract ego which is to be understood neutrally. The third stage involves what is already implicit in the second, namely, the transition from a statistical interest in these ontic qualities to a consideration of them as normative. This involves a restriction of the sphere of sin and grace, inasmuch as they only amount to a subtraction from or addition to that constant neutral substratum which always remains the same and cannot be changed qualitatively.

One indication of the profound influence which the ontological aspect of the Roman Catholic doctrine of the *imago* has had, and of the extent to which it has resulted in a neutralization of the ontic qualities of man, may be seen in the fact, not only that nature as such is regarded as neutral, but also that the obvious conflict within man's nature between flesh (matter) and spirit, instead of being connected with the rent in creation caused by sin, is interpreted in purely dynamic terms within the framework of the Aristotelian teaching on form and matter, and hence is regarded as neutral. Now if our thesis is not to be misunderstood, it must be realized that what Thomistic theology connects with sin is the dissolution of the *rectitudo*, i.e., the triumph of flesh over spirit, but not the actual conflict between them. The existence of concupiscence, which is what repeatedly precipitates that conflict, is itself metaethical; it constitutes, as it were, the last not completely formed relic of matter. This relic must necessarily remain, for a complete forming of matter by spirit would be a contradiction; it would involve the complete absorption of matter by form. Hence the tension between flesh and spirit never can and never should be resolved. And while it is perhaps painful for us to have to bear this relic of earth, the relic itself is not evil. It is an ontic-dynamic phenomenon.

As Scheeben says, "In itself, spirit does not have the power to remove that tension [the tension in which it stands with matter, with the *inferiora*], and thus to concentrate all nature in itself in a higher unity. On the other hand, matter for its part will not allow itself to be in some sense suppressed in its spontaneous development. Consequently there is in the nature of man as such neither the power nor the necessity for resolving the tension through the establishment of that unity." [33]

This thesis is actually a necessary consequence of the ontological approach; "nature," which is itself a neutral factor, necessarily consists in an intrinsically neutral interplay of forces between its component parts. On the other hand, the moment we regard man exclusively and wholly as one who exists *coram Deo*, this neutralizing ceases. In the Sermon on the Mount, where this "existence in relation" is set forth in the most radical terms, it is clear that concupiscence as such, as well as the tension which it creates for man, is itself a sign that man has already fallen (Matt. 5:28). What seems to be only a dynamic play of forces when viewed from the standpoint of man, i.e., *coram se ipso*, is *coram Deo* a sign of division, a breach in the ready spontaneity and immediate readiness of man's love for God, a symptom that instead of his whole heart being fixed on God only a part of man can be brought into this Godward relation.[34]

To sum up the dilemma of Roman Catholic ontologism in a single sentence, we may fitly say that it involves an illegitimate concept of the human *proprium*, i.e., its schema constantly forces it to a consideration of man "in himself" and of man's qualities "in themselves."

This criticism must be pressed and substantiated in connection with the Roman Catholic doctrine of justification. Here too we must keep in view the nature and concept of the *proprium*. For it is immediately apparent that the treatment of this concept is the crucial turning point in the whole development of Christian ethics. This concept of the *proprium* determines how human action will be

[33] Scheeben, *op. cit.*, p. 39.
[34] Cf. Karl Holl, "Der Neubau der Sittlichkeit," *Gesammelte Aufsätze zur Kirchengeschichte,* I (Tübingen, 1923), 158-159.

understood, e.g., whether it will be construed as possessing the rank of "merits," co-operation with grace, an expression of free will, or as being in some other way related either to man or to God as its author.

12.

The Evangelical-Roman Catholic Debate on Justification and Grace

We may venture to say that as regards the beginnings of justification, the "first act," there is a measure of agreement between the Reformation and the Roman Catholic understandings, inasmuch as in both God is regarded as the ultimate author of justification.[1] The real question is—more accurately *man* is really asked—whether and how far he is prepared to acknowledge and maintain this primacy of God as the acting subject.

The Evangelical View of the Human Person in Relationship

The main thesis of Reformation theology, which in Luther as well as Calvin finds its most impressive buttress in the doctrine of predestination, is to the effect that at no stage in the event of justification do *I* ever become the subject instead of God. We shall try to clarify this thesis, which is also to be understood as an antithesis to the Roman Catholic doctrine of justification, by means of a quotation which is central in Luther's thinking and typical of the debate.[2] We shall divide the passage into separate sections with an accompanying commentary.

Luther says, "You are [*es*] righteous through steadfast love and mercy." The crucial point here is the understanding of the *es*, of the ontology of righteousness. To what extent is this a statement concerning the *de facto* condition or quality of the justified man? Luther continues: "This [i.e., this being righteous] is not [a matter of] my disposition [*habitus*] or a quality of my heart, but something for-

[1] See, for example, Canons 3 and 5 on *Grace* of the Council of Orange II (529) in Henry Denzinger, *The Sources of Catholic Dogma* (30th ed.; St. Louis: Herder, 1957), Nos. 176 and 178; and Chapter 5 of the *Decree on Justification* of the Council of Trent in *ibid.*, No. 797.

[2] The quotation is from Rörer's notes on Luther's 1532 lecture on Psalm 51:2 as given in *WA* 40II, 353, lines 3 ff. Cf. *LW* 12, 328, which is based on Veit Dietrich's printed version of the same lecture.

eign [*extrinsecum*], namely, the steadfast love of God." It is at first a matter for surprise that the *es*, which seems to denote a statement concerning the condition of man, should in reality be a statement concerning God, indeed should be coupled with a conscious and systematic attempt to guard against any possible misunderstanding in terms of quality. One might even dare to express Luther's intention here by means of the equation: the righteousness of man (which was meant by the *es*) = the mercy of God. The righteousness of the human person can be defined only in terms of the divine person, never as a quality of man alone.

It is as if we were reminding a highly gifted but proud person: "Your gift? That is not you yourself; it is God's gift in you." We must repeatedly remind ourselves of this distinction, for ontological thinking is so natural to us that we constantly forget it. We forget all too easily that which the very term itself declares so plainly, namely, that a "gifted" person is not one who stands alone but in relation to the Giver. This is why even a statement which on the face of it seems to refer to man alone (*es praeditus* = "you are gifted") can refer essentially to God, since man is the "subject" only in a derived sense and God is the true subject. At any rate, this sheds light on what otherwise seems to be logically such a difficult problem, namely, the extent to which the "righteousness of faith" can at the same time also be called the "righteousness of God." This identification poses difficulties, particularly logical difficulties, only from the standpoint of our modern categories of thought, which are shaped by solipsism and grounded in ontology.

We may add another illustration to supplement the one just given. A father says to his child, "You are my precious child." If the child had to expound the statement, and had the necessary reflective ability to do so, he would not really see in it the extolling of some quality of his own. He would realize that the glad (the "evangelical") part of the statement is not what is said in it concerning "myself," but what is said in it concerning "my father," namely, that "I am loved by him." More precisely, then, the father's statement should be, not "You are precious [*lieb*]," but "You are loved [*geliebt*]," with the implication "I am the one who loves you."

Even more exact analysis would disclose that there is a sublime distinction between the statements "You are a good child," and

"You are a precious child." The word "precious" is by no means merely a more tender and heartfelt expression for the sober moral judgment that a child is good. Instead it carries with it the implication that "your goodness derives from the fact that you have discerned my love." The goodness is, so to speak, of a responsive variety. It is an attitude released by "my love," a form of being within the fellowship which exists "between us." Thus here too a statement couched in the fashion of Luther's *es* in the second person ("You are precious") can be understood only if the predicate quietly exchanges its subject, or better, only if the indicated quality is referred not to the human person but to a relationship between persons in which God is subject.

Finally as a third illustration, we may recall the way in which the Bible writes history. It records events, to be sure, which take place on the horizontal plane of the world: people are born, they sin and are righteous, they fall sick and get well, they wage war and conclude peace, and finally they die and are gathered to their fathers. In reality, however, the reference is not to what actually transpires on the horizontal level insofar as this is technically susceptible of narration. The true reference is rather to the author of the history, the one who is active alike in creating and destroying, in grace and in judgment. The human subjects here decked out with predicates are only symbols for that other who is the real subject at work. Everything that is said can be rightly interpreted, therefore, only if we keep in mind the relationship between God and man, and the fact that in this relationship God is always the subject.

The Luther quotation which we are presently expounding then goes on to deal with the question of whether and how far there is contained in this statement concerning God a statement concerning myself as well, or, more precisely, whether and how far I myself am here specifically the subject of a statement: ". . . for we know that our sin has been forgiven and that we live in his great love and abundant mercy." This answer to the question thus leads us directly into the same train of thought we developed when we were discussing the significance of the *imago Dei* in terms of the figure of a mirror. Even that "specific" statement concerning "myself" cannot be abstracted from the relationship "in" which we live, as we saw earlier in connection with the ἐν Χριστῷ.[3]

Our existence, then, is characterized by the fact that either we are right with God or we are not. There is no third possibility. By comparison all other distinctions pale into insignificance, e.g., the distinction between clergy and laity, or differences in gifts and talents, or in possessions, or even in moral achievement as between publicans and Pharisees. Here again the critical importance of faith is apparent. Since I am right with God if I believe, and not right if I disbelieve, everything that proceeds from faith is right, and everything that proceeds from unbelief is sin.

At this point we may pause to reflect critically on a question posed for us precisely by the confrontation with Roman Catholic theology. If "I" am defined simply as a factor in a relationship, i.e., as the object either of God's wrath or of his grace, then it might well be asked whether this does not violate my personhood. Does not the suspicion arise that man may be understood merely as an "object" or even an "effect"? And is not this to do away altogether with that which constitutes man, in distinction from the beasts, as the *imago*? It is perhaps against this very view that Roman Catholicism intended to build in certain safeguards, using its doctrine of infusion, for example, to put man on his own feet, i.e., to throw him back on his own working and initiative, on himself as an acting subject. There is a representation of this in Michelangelo's fresco of the inbreathing of the divine breath on the ceiling of the Sistine Chapel in Rome. God gives life to Adam with his finger. But as he does so, Adam's leg is already drawn up, and the very next moment, by virtue of the divine quickening, Adam will be able to stand on his feet and be a self in his own right.

We are here confronted by a problem which is central to the Reformation doctrine of justification, a problem which could lead to

[3] "I take grace in the proper sense," writes Luther in his treatise *Against Latomus* (1521), "as the favor of God—not a quality of the soul, as is taught by our more recent writers. This grace truly produces peace of heart until finally a man is healed from his corruption and feels he has a gracious God." *LW* 32, 227. In his 1519 commentary on Psalm 1:2 Luther says: "Here 'delight' [in the Law of the Lord] stands, first of all, neither for ability [*potentia*] nor for the indolent habit [*habitus*] which was introduced from Aristotle by the new theologians in order to subvert the understanding of the Scriptures, nor for the action [*actus*] out of which, as they say, that ability or habit proceeds. All human nature does not have this delight, but it ᵕust necessarily come from heaven. For human nature is intent and inclined to evil, . . . The Law of the Lord is truly good, holy, and just. Then it follows that the desire of man is the opposite of the Law." *LW* 14, 295.

the doctrine's own undoing. For it is quite conceivable that the personalistic concern of the Reformation doctrine, as distinct from the material-ontological concern of Roman Catholicism, should have the very opposite effect and lead to a dissolution of the personal aspect. This could be the latent danger in the Reformation's stress on the utter passivity of man, i.e., on the purely relational significance of the ego, its significance as merely an object.

To see how Reformation theology ought to react to this objection we may recall once again the concept of "progress" in Luther's teaching. This concept has repeatedly helped us already to vital insights concerning, for example, the lasting significance of the Law in the state of justification, the interrelation of justification and sanctification, etc. It can also render us important service in clarifying the concept of "person" in this kind of relational thinking. It can do so for this reason: It is simply impossible for us to conceive of personal life as being in the static equilibrium or rest of a particular relationship—whether that relationship be right or not right—and beyond that, within either one of these two possibilities, not susceptible of further modification or qualification. We can conceive of personal life only as being in movement, and therefore as progressing and advancing. But this in turn implies self-development. And self-development for its part implies the existence of an independent, or at least a relatively independent, entity which does the developing. Hence the question of how Luther conceives of progress will necessarily afford us insights into the ontic character of the person, his being as subject, and hence also into the way in which we may guard against the possibility of the person becoming a mere "object" or "effect."

What does Luther mean in this respect when he says that "to go forward [*proficere*] is nothing other than constantly to begin again [*semper incipere*]"? [4] We have already seen that he intended for one thing to counteract the idea that what is involved here is simply the self-unfolding of a seed of sanctification implanted within us. In the Christian state there is rather a constant referring back to the starting point, i.e., to our fear and love of God, to the fact that we allow his grace to be addressed to us. Thus the *incipere* implies the turning away from self and the turning—which is an ever renewed and never outmoded turning—to the *alienum* of the divine act which is and

[4] See p. 183, n. 8 above.

remains the beginning, the middle, and the end of the state of justification.

At this point there thus arises for us the question whether any further self-development in man is to be rejected. If it were, then the ontic existence of the person, his being as subject, would also be denied, and the schema of personalistic thinking would be carried to such an extreme that the person as such would be done away.

There can be no doubt that Luther has no intention of rejecting such self-development. This is clear already from the numerous references, especially in his early writings, where he speaks of being "justified more and more," of being "found more righteous," of being "made more and more righteous," of being "moved from good to better." [5] The passive form of the predicates of course shows us unmistakably that, whatever else may be said about that which here takes place in man, this development can never be referred to man as an isolated subject. There is here a clear and unambiguous reference to progress and to the matter of development in the state of justification, and hence to something that really takes place here in the ontic sense. There is no suggestion of an abstract state of rest. Nonetheless, everything depends on giving to this event its proper theological significance. This can best be expressed by saying that the progress which here transpires in the ego has the significance only of a by-product, of something which is added "as well" when we seek "first" the kingdom of God. Man's real task, however, is to attend to the "production process" of the main product. A technician or chemist who fixes his attention on the process whereby the by-product is developed will inevitably upset the over-all program which issues in the main product.

That we are here expressing a central concern of Luther's theology may be seen from the image mentioned earlier of the stone which, when it lies in the sun, becomes warm of itself. It does not have to be commanded to do so. With this image Luther is not simply attacking the use of the Law which is designed to revalidate the imperative and which expresses an inadmissible lack of confidence in the sun and its efficacy. On the contrary, Luther uses the image above all to dispute the lofty status accorded to the Law in the life of believers.

[5] See above p. 129, n. 3, and p. 183, n. 8; see also Luther's commentary on Romans 12:2 in *LCC* 15, 321-322.

If the Law is given equal status with the Gospel as a co-operating factor, it turns man's gaze away from contact with the sun, from fellowship with God, and therefore from the absolutely decisive process, and fixes it instead upon a secondary event, i.e., upon that which, in consequence of this contact, takes place in me, or ought to take place, and which I must help along by my own action.

From this angle too we thus achieve an insight which has already suggested itself to us earlier: As there is no denying an ontic structure of the *imago Dei* in man, so there is no denying an ontic happening in the human subject. This question of the ontic, however, is by no means the decisive problem. Far more decisive is the question of the noetic. The real question is not whether there is this ontic structure of the *imago*, this ontic development, but whether this is something men may know, whether we should be directing our attention to it, or whether the shift of attention thereby demanded will not cause us to miss the real thing which is here happening to me, and to which I am summoned. It is again a matter of perspective.

One cannot insist strongly enough that this, and this alone, is the real controversy with Roman Catholic theology.[6] It is a mistake to make of the protest against ontological thinking a denial of the ontic facts as such. The sole point at issue is our attitude to these facts, the place which they occupy in theological understanding, and especially the question whether we accord them the rank of a major theological concern. How important this question is, we have seen already in our illustrations from the physical world. For the way in which we approach a theological topic is decisive for how it will appear to us and what we make of it.

It is thus essential to learn from the passive verbs in the quotations adduced from Luther that, while something does take place in us, while there is progress and development, it can never be "observed" as something which is an end in itself. We must rather be content simply to affirm in passing, implicitly, that which happens within us, even as we go about the business of insisting that it is God who is at work in us. We can speak of the *proprium* of man, as it were, only cryptically, as we speak of the *alienum* of God. We do not deny the existence of this *proprium*, or of development within the human

---

[6] At this point I am happy to record my substantial agreement with Rudolf Hermann, *Luthers These 'Gerecht und Sünder zugleich'* (Gütersloh, 1930).

subject, not even in the ontic sense. What we do deny is a particular way of regarding the matter and speaking of it which involves far more than merely a variation in perspective.

The *proprium* really exists and man really is a subject, though only insofar as both the *proprium* and the subjecthood are constantly posited by God through his speaking to us. A necessary confirmation of their existence is to be seen in the fact that there is a certain continuity about the ego of believers on the horizontal level. The ego undoubtedly could not be regarded as a subject if it arose each moment by new divine positing, by a continually repeated creation out of nothing [*creatio ex nihilo*], as if its existence were like that of a scene on the motion picture screen, which comes into being anew twenty times every second, with each successive picture on the film. Yet the important point so far as this continuity of the ego is concerned is that, while the ontic being is there, it is not something that has its existence in itself and can be regarded in isolation. On the contrary, it is to be seen as something whose point and purpose is determined by its fellowship with God.

This means that "today, when we hear his voice" we are not such as have been newly called out of paganism or unbelief and made to stand, as it were, for the first time, as if that "today" were really a new day of creation, not preceded by a yesterday which also lay in the sphere of this creation, this καινὴ κτίσις. On the contrary, when we hear his voice "today," it confirms and strengthens what it said yesterday, that wherein it called us. Moreover, when we hear his voice today, it confirms and strengthens also that which it said long before yesterday to the patriarchs and prophets, to Abraham and Isaac and Jacob, that wherein it called all of them. God's summons in the present always assumes the existence of the church and of my own Christianity as given, and hence also the continuity of both. The body of Christ and the fact of a man's membership in it are both objectively given. They do not exist only in the flickering, filmlike manner of a continually renewed—almost with frenzied speed—*creatio ex nihilo*. They have a continuous existence on the horizontal level, not only a momentary existence at the point where vertical and horizontal intersect.

As we have said, this horizontal continuum obviously has an ontic side. Only here too it is evident that the ontic side of the continuum can never be the basis of the subjecthood of the church and of the believer. In this sense the ontic side is not something that can be "observed." It is rather something that is added "as well," in passing. The moment we try to view the continuity from its ontic angle, we begin to detect, for example, in church history, in its epochs and individual events, demonstrable traces of the kingdom of God, and we are tempted to make of it a proof of God or of the divine foundation of the church (this is what has actually happened in Roman Catholicism). Reasoning in terms of the practical syllogism, we also begin to make of the experiences, leadings, and achievements of the individual believer a basis for believing (this is what Roman Catholicism does by and large with its adoration of saints, and necessarily so).

For this reason it is essential that we regard the continuity of the Christian life as a continuity of fellowship with God, as a continuity which is bestowed, bestowed ever anew. To be sure, stating the case in this way could possibly suggest that there is a certain instability about this continuum. For if it must continually be given afresh, this obviously implies, as the reverse side of the coin, the negative fact that it can one day be withheld, that when that happens it ceases, and as a result we exist after all only momentarily, at a given point of time. In such a case the continuity which existed in the past, e.g., that involving Abraham, Isaac, and Jacob, would also be represented only as a momentary bringing together of calls which were in fact "discontinuous" events.

Logically, this negative aspect ought indeed to be postulated. It would not be in keeping with the facts, however, to give it an equal place alongside the other aspect, namely, that of continuity. For the negative aspect arises only when the matter is viewed in a way which is not consonant with the point at issue. The "punctual" view of fellowship between God and man, in terms of existence in the moment at the point of intersection between vertical and horizontal, arises only if we consider the ontic structure of the human subject, i.e., only if we consider the fellowship—and the faith which corresponds to it— in relation to what is going on in the inner ego. There is, for example, the matter of the decision which must be made ever anew, and hence may go unmade. There are the assaults of perplexity and doubt which must constantly be overcome, but which may therefore also triumph over me. And there is the matter of the believing which is daily required of me, and which may not only be affirmed but also denied. When the matter is viewed only in this way, when I look only at myself [*in me ipsum*], the continuity of the ego is dissolved— and with it the continuity of the ego as a "new creation"—into a sum of discontinuous points.

What we described as discontinuity from the epistemological angle is called in existential-theological terms despair, and in terms of perspective *peccator in re*. To consider myself *coram me ipso* is to forfeit my identity, i.e., my identity with the righteousness of Christ. It is to forfeit my continuity, i.e., the fact that I exist by the faithfulness of God. This view throws me back upon my own action, and upon the elusive "moment" on which I cannot count. It takes from me the comfort of the sonship which was promised to me in eternity, assigned even before the foundation of the world (Eph. 1:4), and will be sustained at the last judgment (John 5:24). The continuity of the ego may be seen only when we view the ego from the standpoint of God, who posits it as an ego called to fellowship with himself. More precisely, this continuity is grounded in the promise of God and in the faithfulness with which he stands by his promise. Thus the continuity of the human person is the continuity of God's faithfulness. It is along these lines that we are to understand the saying of Paul that he who began a good work in us will bring it to completion (Phil. 1:6). The beginning and the completion of our fellowship, and hence also its continuity, lie in the promise of God.

It is thus clear why we cannot speak of the discontinuity of this fellowship, e.g., of the "punctual" character of our decision, as the "negative side" of the same coin, and hence of equal standing with the other side. To postulate the negative aspect in this way is false. And we can show and prove precisely why it is false; hence we are not being illogical. The negative postulate is false because in the first place it proceeds on the abstract assumption that God and man are two equal partners, so that the relationship between them may be viewed equally well in terms of either. It is false in the second place because it proceeds on the abstract assumption that God can be "defined" in terms of the concept "person," and that he can thus be understood as the author of decisions which are always open, as the maker of a decision on which we cannot count. On the basis of these two misconceptions, namely, that the two partners are equal and that we cannot count on God, the discontinuity or "punctual" nature of the "new creation" has to be accepted.

On the other hand, if the relationship between God and man is seen in terms of man and his ontic circumstances, and if we then accept discontinuity, we are adopting a radically false perspective, that of the *incurvitas*. This perspective is false, as we have said, because the co-ordination of God and man which underlies it must be ruled out as an illegitimate logical abstraction. Again, if we seek to define God in terms of the concept "person"—instead of learning what "person" has to mean in this context from the self-disclosure of God in Christ—then God becomes one who makes decisions on which we cannot count, and so himself establishes the "punctual" nature of the new creation. But in his self-disclosure he has manifested himself as the "faithful God" who stands by his gracious calling. Luther defines the continuity of divine-human fellowship, and therewith implicitly the continuity of the human person, along lines such as these when he says that "wherever God speaks, and with whomsoever he speaks, whether he speaks in wrath or grace, that person is surely immortal [*immortalis*]. For the person [*persona*] of God who speaks, and the Word, show that we are the kind of creatures with whom God wills to speak infinitely [*immortaliter*] and to all eternity." [7]

Here in connection with the borderline concept of immortality it is clearly pointed out that the continuity of the human person survives even death. In highly characteristic manner, however, the definition of continuity given here avoids the perspective of ontologism, which regards the continuity as given and assured by an indestructible "immortal" substance of the soul. It is highly characteristic that the continuity consists instead in a history with God, a history which because of God's faithfulness can never cease. So long as I keep this faithfulness in view, and do not stand on my own ontic *proprium,* my continuity on the horizontal plane is given. And if there is a continuity of church history, from the patriarchs and prophets to the believing community of the last day, this is because God "is who he is," i.e., he was faithful, is faithful, and will be faithful, and because there always has been and will be a believing community (and this in turn is grounded in the promise that the gates of hell shall not

[7] *WA* 43, 481.

prevail against this community, i.e., they shall not overcome its faith).

The definition of person in terms of fellowship with God, in terms of the *alienum* of his righteousness, implies no destruction of the human person. We have shown this by examining the various threats to personal being which seem to be posed by this "being-in-relation." Consequently we cannot accept the Roman Catholic argument that, if the human person is to be preserved, it must be regarded as the bearer of ontic qualities, e.g., of an infused *habitus*, that it must have a *proprium* of its own and not be understood solely as "being-in-relation." The Reformation doctrine of man rests instead on the fact that God, who establishes divine-human fellowship, remains its subject, that his grace which calls man really remains his grace, that it does not become a *habitus* of man by way of infusion, as Roman Catholic theology asserts with respect to the widest spheres of grace, e.g., the comprehensive complex of created grace [*gratia creata*].

Before we examine this transforming of grace from a quality of God into an attribute of man in Roman Catholic theology, we must first buttress our own position. We must first show that in the Reformation view of the person the *alienum* of God remains intact and is not transformed into the *proprium* of man. And second we must show that on the basis of this *alienum* the personal character of man is upheld, so that we cannot accept the Roman Catholic thesis that in their theology they are protecting this person (i.e., for the sake of man's humanity) against its supposed "punctual" evaporation by the Reformers.

At the beginning of this chapter we ventured the statement that in the "first act" of justification there is a measure of agreement between the Roman Catholic and the Evangelical positions inasmuch as in both God is regarded as the ultimate author of justification. We have seen that in Thomistic theology, and already at the Synod of Orange, grace has an unequivocally initiatory rank. For this reason the real question in the interconfessional debate is whether the position of grace as the divine author has been unequivocally maintained in Roman Catholic theology. We then tried to show, first of all from Luther, what is involved in thus maintaining that grace is the initiatory subject. In other words, the real problems which divide the confessions arise only in the "second act" of the process of justification, where it is a matter of maintaining the position of grace as subject.

The Roman Catholic View of Depersonalized Grace

We have already pointed out that in Roman Catholic theology grace does not remain subject; it cannot do so because of the ontic relationship of nature and supernature. For since nature is the stable, intact, and absolutely continuous part of man, the righting of fallen man can consist only in the restoration of this nature's *rectitudo* and in the addition of the superadded gifts which nature had lost. In terms of the concepts *alienum* and *proprium* which are so fruitful in polemical theology, the righting of fallen man is the righting of his *proprium*. This may be seen at once in the more precise definitions of the Roman Catholic concept of grace, which are significant in two directions. (It will not now be difficult for us to see these two directions in their autonomous connection with the schema of nature and grace.)

In the first place there is the Roman Catholic distinction between grace as something God is and grace as something God produces. Grace on the one hand is a quality of God, namely, his disposition, attitude, and action towards us (*gratia gratis dans*); in this case grace is in some sense "God himself." On the other hand grace is also a working or an effect which God produces; it is that gift of substance and of power which dissociates itself from God and enters upon a history of its own, as it were, in the man who receives it. Naturally there is here an ideally close connection between the doctor and the medicine. Nevertheless, there is also a certain analogy between this "working" of God's grace and the workings of man to the degree that, when a man acts, and in the action moves outside himself, he necessarily allows his act to enter upon a history of its own. The word spoken by a politician, for example, or the book written by a poet, has its own history once it is released. It cannot be recalled. It goes forth, and there is no one to restrain it. It has in some measure taken on a life of its own, and is no longer identical with its author.

If we draw out this human analogy, we notice at once that it has its limitations. For even if God's grace has the significance of such an *effectus,* if it becomes a working which is distinct from God himself, God nonetheless remains its master. Grace is not conceivable as an *effectus* unless the man in whom it works, in whom it "takes effect" as a medicine, is related to its author. To this extent one can speak only of a relative dissociation of grace from God himself. Of this dissociation, however, one must necessarily speak.

The degree to which one can and must do this may be seen from the way in which the sacraments are said to be effective in virtue of the work which has been worked [*ex opere operato*], without any need for the personal relation of faith, indeed, quite independently of it. There is only a faint recollection of this personal relation, namely, in the fact that the recipient of grace dare not place any obstacle (*obex*) in the way,[8] and that the dispenser must have "the intention of doing what the church does."[9] Hence, apart from the grace which is "God himself," the grace that freely gives (*gratia gratis dans* or *gratia increata*), we must also speak of a grace which is its "effect," the grace freely given (*gratia gratis data* or even *gratia creata*). And it is precisely with respect to this second form of grace that we must inquire how far it is integrated into a process which makes of the divine quality a human attribute, a human virtue, such as sanctity, for example.

This leads us to the second of the two directions in which we said the Roman Catholic definition of grace becomes significant in that it makes grace a *proprium* of man. This may best be seen in the matter of the supernatural virtues of faith and hope which are said to be infused together with the grace of justification. (We cannot pursue at this point the difference of opinion regarding the question whether the infusion of faith and hope may precede the infusion of love and grace even in the initial justification.)[10] For faith and hope, being essentially forms of the *gratia gratis data*, i.e., of grace as "effect," are present in the first instance only potentially and latently. In this respect their status may be compared with that of concupiscence. For they too constitute, as it were, a kind of *fomes*, a combustible material

---

8 See Canons 5-9 on *The Sacraments in General* of the Council of Trent (1545-1563), Session VII, in Denzinger, *op. cit.*, Nos. 848-852.

9 See the bull of Eugenius IV entitled *"Exultate Deo"* (November 22, 1439) in Denzinger, *op. cit.*, No. 695; cf. Canon 11 on *The Sacraments in General* of the Council of Trent (1545-1563), Session VII, in *ibid.*, No. 854. See also Part II of the Roman Catechism on "The Sacraments in General" where the unworthiness of the minister and the validity of the sacraments is discussed; *Catechism of the Council of Trent for Parish Priests*, trans. John A. McHugh and Charles J. Callan (New York: Wagner, 1962), p. 155.

10 See Diekamp, *Katholische dogmatik nach den Grundsätzen des heiligen Thomas* (Münster, 1939), II, 535 ff.

which must first be set alight if it is to fulfill its true function. Unless they are set alight, they remain neutral, indeed are even "dead." Faith is first realized and first springs to life by becoming a "faith fashioned by love." Until that happens it is just some quiescent quality in the heart [*aliqua otiosa qualitas in corde*], a latent *habitus*. That it is activated by love means that to achieve its true stature it needs the activity of man—even though this human activity is of course made possible only by grace.

Thus man's *proprium* must be brought into play if faith is to come alive and, as a living faith, help to effect justification. Only a faith which in love has come alive produces the merits which for their part (in conjunction with the grace which underlies them) can bring about justification.

The way in which merits are structured into the process of justification shows the role that is assigned to the *proprium* of man. For the concept of "merit" can be meaningful only in connection with that which is "man's own," that which has a certain independence vis-à-vis him who then has to reward these merits—even though the recognition of them as meritorious is in itself traceable back to grace.[11]

When we consider the emphatic and unreserved attempt made by the Reformers to head off such consolidation of the *proprium,* the *persona in se,* and to think only in terms of this "person in relation," we realize that we must not allow ourselves to be irritated by the way Roman Catholic teaching constantly refers the *proprium* back to the operative grace (though this is done less and less directly as the process of grace continues). All we can do is point out that this consolidation of the *proprium* has in fact taken place.

Within the schema of ontological thought, Roman Catholic theology has done everything possible to see to it that the *proprium* of man is constantly made to refer—both backward and forward—to grace. For even in respect of its clearest manifestation, in the matter of merits, the *proprium* refers back to the grace which releases it and brings it to fulfillment. The *proprium* is also referred forward to its true goal, namely, the linking of the restored *natura propria* to supernature. Within the ontological framework Thomistic theology is therefore really and truly a consistent theology of grace. Grace stands both at the beginning and at the end. Nevertheless, the ontological

---

[11] Cf. Thomas, *Summa Theologica,* part 1II, ques. 114, art. 1.

schema is quite unable to maintain consistently the *alienum* of divine grace, its permanent character as subject.

To be sure, this is not altogether clear within the ontological thought schema itself. It is for this reason that the Roman Catholic who operates within this schema is so hard to talk to; he constantly feels that he is being misunderstood. The grace of God always remains subject, but that this is so can be seen only in the personalistic mode of thinking which takes "grace" to mean exclusively the "gracious God," i.e., a quality of God himself, and which thinks of man as the one upon whom this grace has come, and who thus lies like a stone in this sun.

The point we are trying to make here is simply this, that the interconfessional debate must begin with the scheme of thought itself; ultimately the problem is one of theological epistemology. Within the ontological framework it is impossible to argue against the Roman Catholic doctrine of grace. Here, as we have said, everything possible is done to glorify grace without merits and to assure its position theologically. There is no place here for cheap Protestant polemics, and we readily admit that the consistent regard for grace which Roman Catholic theology constantly displays within its scheme of thought gives repeated evidence of the brotherhood which binds us together "in Christ," or, if you will, of the *una sancta* which embraces us all. Nevertheless, this should not prevent us from pointing out, for the sake of truth, that this evangelical intention of Roman Catholicism cannot be realized in the given framework of thought, but will always fall short of its goal. To show this, we had to proceed in such a way as first to indicate that within the framework of personalistic thinking grace is consistently maintained as an attribute of God, and then to address Roman Catholic theology on this basis. For it is only on this basis that we can really make clear what we mean by the "alien" character of divine grace.

It is at this point that the Reformation commences its criticism of the way Roman Catholicism changes grace from a divine to a human quality. In this connection we recall the objection of the Apology that according to the Roman Catholic conception the work of Christ is restricted to the *prima gratia,* and that on the basis of this "initial grace" we are enabled to "please God and merit eternal life by our keeping of the Law." [12] Against this view it must be emphasized that "Christ does not stop being the Mediator after our renewal." [13] The Apology thus sees in the Roman Catholic conception a cessation of divine grace, or, more precisely, the cessation of grace as a quality of God (as this is expressed in the mediatorial quality of Christ) and the beginning of grace as an attribute of man, a merit-engendering virtue.

We can see clearly at this point that what takes place *in* man, and not merely what happens *to* him, has become the object of theological observation. It is again evident, as was clear already with respect to the concept of the *imago Dei* in man's original state, that the ontic element in the human ego pushes itself into the foreground. Then the theme of theology is not just the relation between God and man; on the contrary, theology then includes as an independent concern a treatment of "anthropology." Here is where the fault lies.

The crack, or better, the cracks are themselves produced by the belt of tension which necessarily arises where men attempt to combine ontological and personalistic thinking. The greatest strain and the most evident rupture are undoubtedly to be found at the place where grace ceases to be a divine attribute and becomes an effect distinct from the divine attribute from which it emanates, i.e., where grace ceases to be a personal relation to man (the "gracious God") and becomes something which is ontically infused into man and which is thus present in man, demonstrably present. For it is precisely this distinction between the gracious God and the grace given [*gratia data*] which is the starting point of the distinctively "Roman" development of the doctrine of grace. To put it epigrammatically and therefore with tongue in cheek, what men want is not primarily God himself, but "divine powers which may become human virtues and qualities" (von Harnack). At this point where grace "visibly" passes into man in accordance with certain well defined practices, e.g., sacramental operation, it ceases to be exclusively a subject and becomes a material object, "medicine." This materialization expresses itself in a variety of ways, of which we will mention only a few for purposes of illustration.

One involves the multiplicity of the modes of grace listed in every

[12] Apology, IV, 162; *BC* 129.
[13] *Ibid.*

Roman Catholic dogmatics.[14] These modes are graphically presented in a complicated and frequently intricate arrangement; there is actual grace and habitual grace, operating grace and co-operating grace, antecedent grace and subsequent grace, etc. In personalistic thinking there can of course be only one grace, namely, the divine disposition towards me. But in ontological thinking, in which the effects of grace may be ontically demonstrated and hence are indicative of ontic substrata of grace, there is necessarily a plurality of modes of grace. Again epigrammatically, it might be said that the point at issue in the Reformation concept of grace is the disposition of the doctor, whereas Roman Catholic theology (although it includes this, at least in intention) is at the same time concerned about the inventory of the drug store, about the innumerable medicaments of grace which are to be administered (infused), admittedly on the instructions and with the collaboration of the doctor.

The materialization of grace expresses itself in yet a second way, in the figures of speech employed in relation to the Roman Catholic practices of grace. Thus Engelbert Krebs compares the sacraments with veins in which the life blood of grace flows. He speaks of the sacramental graces constantly welling up to seek an entry into the souls of men.[15] To be sure, this is a metaphor. But it is not the kind of metaphor one finds in Reformation theology or even in the most dramatic of evangelistic addresses. The use of analogies from medicine is so common that we need hardly mention it.

It goes without saying that the materialization is also obviously expressed in those spheres of thought which are linked with the *opus operatum*, the "work worked," although again one should not say this without recognizing the specific theological intention which lies behind this attempt to establish the medicine of grace as ontically independent. For the *opus operatum* is designed to express the objective element, that which transcends the human person and its character as subject.[16] Indeed, one might almost say that it is designed to withdraw the sacramental event from the sphere of influence of the human *proprium*—a sphere which ontologism itself established.

[14] [See, for example, the Systematic Index (Xa - Xf) in Denzinger, *op. cit.*—Ed.]

[15] See Engelbert Krebs, *Dogma und Leben* (Paderborn, 1925), II, 66 ff.

[16] *Ibid.*, II, 81 ff.

Here again we see the tragic self-contradiction into which Roman Catholic theology is continually driven by ontologism. It attempts to use certain modes of thought to say something which not only transcends these modes of thought—this would be no reproach since it is true of every theology of Christian revelation—but which is in actual contradiction to them and hence is continually caught in the stream of these modes of thought and diverted from its original direction. There thus arises out of the objectivity that transcends the human person—which it was originally the purpose of the *opus operatum* to express—that strange, vacillating, intermediate phenomenon which is partly material and impersonal, but which is also partly linked with the character of priestly qualities and therefore with a *proprium* belonging to man as a *habitus*.

The tendency towards materialization is to be seen with particular clarity in the matter of the sacramentals, i.e., in the "ecclesiastical blessings and consecrations in which the church prays over objects of this world in order that the force of evil still operative in them may be repulsed and the glory of Christ demonstrated with power. . . ."[17] Thus it can be said in the consecrating of the oil: "May God sanctify it by his blessing, and add to it the power of his Holy Spirit."[18] With respect to the sacraments themselves, stress is likewise laid on the fact that the "matter" is consecrated, and that by the admixture of "material" grace this material "thing" receives another quality. Thus Thomas is said to believe that "Christ instituted the sacrament at his baptism by John, because for him it was at that time that water was sanctified, and so received sanctifying power."[19] Similarly Ambrose is quoted as saying that "the form and practice of baptism imply that the water is first consecrated and then the candidate goes down into it. For the moment the priest enters, he exorcises the water, then speaks over it words of invocation and prayer, thereby consecrating it."[20] In the Roman liturgy the sanctification of the water is represented by the trine immersion of the Easter candle into the waters of the font.[21]

It is obvious that this entry of grace into material things, when

[17] Michael Schmaus, *Katholische Dogmatik* (Munich, 1941), III², 69.
[18] *Ibid.*
[19] *Ibid.*, p. 75.
[20] The quotation is from Ambrose *De sacramentis* 1, 5.
[21] Schmaus, *op cit.*, pp. 78 f.

linked with the *opus operatum* teaching, necessarily gives the impression of "magical" processes. Roman Catholicism naturally does not wish to give this impression, indeed finds it quite abhorrent. Nevertheless, even if the idea of magic is a misunderstanding or, if one thinks of some of the more popular types of piety, an abuse, a kind of derailing of the load of meaning intended, still one can tell even from a derailed car the direction in which the train was moving. Even an abuse bears testimony to the use which underlies it. Naturally this is also true of certain kinds of popular piety among Protestants, which also get pretty far off the track, though these tend more in the direction of rationalism than of magic. We stress this point in order that our Roman Catholic brethren may not think we are applying this notion of the symptomatic significance of the abuse onesidely to them for polemic purposes and not also to ourselves.

From all that has been said, and not least from our final reference to the magical abuse, it may be seen that with the independent ontic establishment of the *gratia data* there is constituted the sphere of the *proprium*, an autonomous receptacle of grace, indeed a "quality" or *habitus*. This is probably clearer in respect of "things" than of "men." We think, for example, of the sacramentals and of the "matter" of the sacraments. For here there can be no appeal whatever to a personal reference, to the *gratia dans*. What we have here is a kind of theological alchemy in relation to the elements, a change of substance which is immanent even though effected from without.

In contrast, it is characteristic of Luther's theology that in respect of the "matter" of the sacraments—in the case of baptism at any rate this is quite clear—grace is not understood as a material quality in the sense of a change of substance. On the contrary, it retains its externality, its *alienum* character. Grace inheres only in the Word of God; it does not enter into the "matter" of the sacrament. The matter insofar as it is separated from the Word remains only water, "no different from that which the maid cooks with." [22]

It is very much to the point that in this context Luther cites as an illustration the commandment to honor our fathers and mothers. For parents too are not to be honored because of any quality which is theirs by nature. My parent is to be honored rather because the Word of God commands it. This commandment is "the golden chain

---

[22] See Luther's Large Catechism, IV, 22; *BC* 439.

about his neck, yes, the crown on his head, which shows me how and why I should honor this particular flesh and blood." [23] The illustration of the "golden chain" makes it clear that the element, whether the parent or the baptismal element, does not absorb grace in such a way as to make it a *habitus;* it simply has an "alien dignity." Grace is certainly bound up with the sacrament all right, but with the sacrament as a personal event, as something which takes place between the gracious God and the believing man or community. This is why elevation of the host, which is a particularly clear and extreme symptom of the ontic objectification of grace, is from this standpoint quite unthinkable. We have thus shown with all necessary clarity that it is the gift of grace, ontologically understood, which constitutes the sphere of the *proprium.*

We would turn now to the process of depersonalization which is initiated the moment grace is ontically separated from God, in order to set it forth with utter precision in the following propositions. In the first place, the grace of God in the Roman Catholic view is impersonal, not merely because as an *effectus* it has a certain autonomy in respect of its author, and not only because as the bearer of a human or material *habitus* it can become the attribute of an entity which is not God, but primarily because in the theological system as a whole—we are thinking here of Thomas—it is conceived as being in a measure present even "prior" to God. For the system of nature and supernature derives ultimately from the fact that Aristotelian ontology has taken over. Its antithesis between form and matter (εἶδος and ὕλη) precedes all Christian content. Indeed it provides the framework into which the Christian content is fitted. One might even say, it is discovered to be the most suitable container for that Christian content.

Only thus can we explain how it is possible to enunciate a doctrine of nature and supernature in the form of an ontological construction almost without making any reference to the fall. For if nature and supernature are already there as given factors, the fall can at most be only a disruption in the "inner workings" of this system. It can involve only a "dislocation," a dislocation in the form of subtraction comparable to the dislocation in the form of addition which we noted in connection with redemption. With these given factors presupposed,

[23] *Ibid.,* IV, 20; *loc. cit.*

theological thinking can never be constitutively determined by "events," by the contingent historicity of the fall and of redemption in Jesus Christ. It can never be determined by events which, by virtue of their contingency, must always transcend any system we may devise for trying to grasp them. If the system itself is to some extent already given, then the events must be fitted into it. They can only be, as it were, illustrations of an ontic order, and of a history of the world, of salvation, and of judgment which is constituted by this ontic order, a history which may thus in its basic tendencies be understood a priori, in the manner in which the "pagan" Aristotle understood it.

Personalistic thinking rests on contingency. For it relates to personal "events," e.g., to man's decision at the fall, or to God's decision to give his only Son (John 3:16). Resolves of this kind are matters of the will. They cannot be postulated. They can be known only a posteriori. They can only be attested. Ontological thinking, on the contrary, rests on regularity, a regularity which is supposed to include personal events. This regularity, e.g., the mutual flowing together of pure form and matter which underlies Aristotelian ontology, is understandable a priori. This is why there has to be, and in fact is, a proof of God in Roman Catholic theology. A clear example of this ability to postulate is the typical ontological attempt of Anselm to answer with logical stringency the question: *Cur Deus homo?*

Rooted in the same antithesis between the two modes of thought, the ontological and the personalistic, is the Thomist-Scotist controversy, in which the point at issue is whether God is subject to norms, e.g., the norm of the good. Is God encapsuled within a higher order which is itself independent of his concrete self-disclosure (Thomas), or are the norms themselves derived from the divine will and hence contingent (Scotus)?

It would be a mistake to regard Scotus, because of his opposition to Thomas, as a champion of Reformation theology in this matter, just is it is a mistake to regard Occam—the favorite example—in this light. For Duns Scotus, in contrast to the Reformers, does not base his thinking on the contingent attestation of the divine will in revelation. For him the thesis of the primacy of will over being is determined philosophically, i.e., nominalistically. Thus we are here in a very different world of thought from that introduced by the Reformation. Nevertheless, the opposition to thinking in terms of order is at any rate a "sign" that Scotus understood what was involved in these two principles of Thomist theology—the ability to postulate and the insistence on regularity—and found them to be incompatible with an essential factor in the Christian concept of God, namely, God's contingent will.

The seriousness of the threat that grace may become an impersonal law may be seen, not merely from the fact that grace can be represented independently of the particular incident of the fall, but also from the fact that Thomas can develop the essential features of his doctrine of grace without any indication of its specific character as *gratia Christi*. The only explanation is that he knows grace as a moving principle first; only later does he know Christ, in whom this moving principle is then, as it were, incarnated. Grace is thus an element outside of and in addition to Christ.

This leads us to the following polemical proposition. Protestants repeatedly accuse Roman Catholicism of rejecting the *sola gratia, sola scriptura,* and *sola fide* [the *particula exclusiva*], and of proclaiming other powers and forces as bearers of salvation along with Christ, e.g., Roman Catholicism speaks of Christ *and* the saints, Scripture *and* tradition, faith *and* works, grace *and* merit. Now according to our findings we can localize this synthetic "and" at an even deeper point, namely, in the very concept of grace and its ontic structure. The real synthesis is not between grace and merit. It is between Christ "and" grace. Grace itself is something other than Christ. To be sure, the whole concern of Roman Catholic theology is to fuse these two entities. This is a genuine concern, and we may be grateful for it. But the attempt to implement that intention is undertaken on a base where Christ and grace are regarded as being in any case not identical entities. There is thus always the theoretical possibility—which is realized concretely again and again, particularly in popular piety regarding the sacramentals, the saints, etc.—that Christ and grace will become rivals, and that in the competition between them grace will be sought apart from Christ, and thus triumph over him.[24] The new Christian wine poured into old

---

[24] As an illustration of this I might mention the common idea that Christ is an angry judge placated by Mary as the mediatrix of all graces.

Aristotelian wineskins may well burst them, but on the other hand it may itself be determined and in its character transformed by them.

What comes out here with such crystal clarity is of course a problem for every theology, including that of the Reformation; it is so by virtue of the fact of human speech. For theology, like revelation itself, uses the human speech which is at hand, with all its ideological content. Hence the ideologies with which speech is invested are always on the point of infiltrating into the revelation which they encompass, and of determining it and altering it. We need only note the extent to which Evangelical theology was shaped by Idealism in the nineteenth century, and how with respect to the Christian vocabulary today, e.g., sin, the kingdom of God, etc., the rest of the Christian message is filled with these ideological cuckoo's eggs, and is consequently misunderstood. In any case, what we are confronted with here is the most profound question which Protestant theology has to put to Roman Catholicism.

We must now consider a second aspect of the process of depersonalization initiated by the separation of grace from God. For a threat is here posed to the character of God himself as person. In the strict sense man enters upon a relationship not with God, but with his "effects," with his grace. All the earnest talk about the mystical body of Christ, that body into which we are incorporated as members and which obviously points "also" to a personal relationship, must not be allowed to disguise the fact that what we have just said strikes the true nerve of this kind of thinking. We have already pointed out that the great difficulty in debating with Roman Catholicism is that it is a system of correctives and synthetic ("and") relations. For this very reason, however, it is important not to be taken in by the apparent polarity of this thinking, with its constant tension between personalistic and ontological referents. Instead we must keep strictly to the question of what is its true focus or central point. Here we must make no concessions. We must persistently refuse to understand Roman Catholic theology as an ellipse with two foci. In terms of the Reformation question, we must insist that it is rather a circle with but one center. (Fairness seems to demand that we should enunciate clearly this principle of interpretation.)

This brings us to our most profound objection, namely, that in face of this threat of impersonality there can no longer be any theologically correct description of a genuine, i.e., personal, fellowship with God, since the author of grace withdraws behind his grace, behind a grace concerning which we may legitimately ask whether it really is "his" grace, the specific *gratia Jesu Christi.*

We are quite aware of the tremendous implications of this objection. In fact this is why we feel it incumbent to stress that our objection is to the theoretical foundations of Roman Catholic theology, not to the Christianity of individual Roman Catholics, even that of Roman Catholic theologians. It would be presumptuous, and a sin against the true brotherhood which binds Reformation theologians with individual Roman Catholic fellow Christians, if we were here to deny the possibility of their enjoying true fellowship and peace with God. It is a comfort to Protestant as well as to Roman Catholic theology to know that our Christianity does not live by theology, and that many a Christian will be preserved in the last judgment while his theology is dashed in pieces, not merely condemned by God but actually laughed out of court.

The point is, however, that this theology does not fix its attention on the heart of God, whose mirror is the Son. Instead it postulates what is ultimately an "unfathomable being" (von Harnack) which stands behind the emanations of grace. Its concern is with the emanations.

The third form of impersonality emerges when we consider the consequences of this whole conception for ethics. For not only is the Roman Catholic concept of God determined by a given ontological system into which it is integrated. Grace for its part is also integrated into a system of Law which is similarly impersonal in nature. The goal of the process of salvation, as we have seen, is the holiness of man, that condition which results from the infused *habitus* and its actualization, and which in both directions is brought about by the grace that is infused into man and made his own attribute. If we say that this holiness is the goal of salvation, then we may also say that this goal consists in the fulfilling of the Law. And grace is the instrumental means of this fulfillment. Justification is "a movement whereby the soul is moved by God from a state of sin to a state of justice." [25]

[25] Thomas, *Summa Theologica*, part 1^{II}, ques. 113, art. 6; *The "Summa Theologica of St. Thomas Aquinas,* trans. Fathers of the English Dominican Province (New York: Benzinger, 1915), p. 391.

The impersonal schema of grace is thus for its part integrated into an impersonal schema of Law which it serves. Grace establishes a *habitus* which, by its actualization with the help of the liberated and strengthened will, puts man in a position to earn merits, i.e., to fulfill the Law of God. This linking of justification and sanctification is the basis of the nomistic feature to which Reformation theology takes exception in Roman Catholicism.

We may summarize our findings as follows. First, in the Thomistic schema grace is not something God *is* but something God *gives*. By natural infusion he gives a share in his own nature. In any case, instead of speaking of the "gracious God," one has to speak of God *and* grace. And instead of speaking of the sinner as the recipient of grace, the child of God, one has to speak in terms of quality and *habitus* and holiness. It is not as if the personal element were lacking altogether; Roman Catholic thinking is always all-inclusive. But we can only say that it is "also" present, present "in" a larger scheme. And when the Reformers asserted the *particula exclusiva* against the "Papists," they were trying to express the fact that faith and grace, that Christ and the Father-child relationship given in him, cannot exist "also" and "in" a wider scheme, but can have only an exclusive mode of existence (hence the threefold *sola*). The different soundings which we have undertaken in this regard show where combinations of this kind lead.

Second, at the outset God and grace are still almost identical, although in terms of the ontology taken over from Aristotle—admittedly in somewhat modified form—this is rather a case of their having been arbitrarily "identified." Increasingly, however, grace becomes an independent natural substance, a medicine, by which God works. The containers of this medicine are Christ's treasury of merits and the sacraments. There is greater concern for the medicine than for the physician.

Third, along with this relative isolation of grace from God, i.e., along with the objectification of grace, there goes a certain diminution of trust. For complete trust is possible only when God himself is its object, and when its validity is simply the validity of God's own promise. This diminution of trust finds expression in the fact that the impartation of grace is linked to certain conditions, to merits and hence to "co-operation." The necessity for imposing these conditions is grounded in the very structure of grace, for in consequence of its ontic independence grace comes increasingly to assume a co-operative significance in the continuing process of salvation. It becomes a material instrumentality which establishes the *habitus* of man's own activity and enables him to "do it himself." Thus grace is no longer everything, and it no longer does everything. The statement "My grace is sufficient for you" (II Cor. 12:9) loses its unconditional validity. At any rate it no longer means: Let it suffice that I am gracious to you, that peace is thereby established, and that this is "my" affair and not even by implication "yours" also. When grace leads to "co-operation," it is no longer enough that Father and child should stand over against one another, or better, that the guilty child should stand before his merciful Father. Instead grace comes to the rescue of man's noetic and moral capacities; by means of certain natural infusion it makes it possible for man to stand before God in his own right. Indeed, justification depends on his attaining this independent stance. The resultant necessity to keep a constant check on oneself, i.e., to see whether the conditions are being met, leads yet again to the phenomenon of the inverted outlook, and therewith to the doubt, the waning trust, and the anxiety which are always present in legalism. Luther's conflicts in the cloister are a supreme example of this state of affairs.

Fourth, we thus perceive that on this view the position of ethics within the whole theological structure is basically altered. Our examination of the indicative and imperative has shown that ethical demands are valid only on the ground of a peace with God which has been decisively achieved already, i.e., in Christ. Thus ethics can only follow justification. In Roman Catholic theology, however, ethics occupies a key position in the actual process of salvation. Its postulates define the conditions on which justification can take place. It has no postulates arising on the basis of a justification already having taken place.

As the interrelationship of indicative and imperative, of justification and sanctification, is altered, so there is here too a basic shift in the relation between Law and Gospel. The Gospel of God's grace is no longer redemption from the Law, from constant self-questioning, from doubt, anxiety, and servile fear. On the contrary, the Law defines the conditions on which, with the help of grace, I may partake of grace. The Law is not something which has been overcome and which lies behind me. It is rather always present, and hence also future. It is the condition which the present constantly imposes on me if I am to be granted a future. It is not that which is passing away from me but that which is continually coming toward me. Reformation theology necessarily sees here a weakening of the Gospel, a weakening which results in the fact that the Law continues to share in the lordship of my life, and hence the assaults of doubt indefinitely remain.

We thus have fresh occasion to note the precision of Luther's Reformation question. Characteristically, he did not ask where he might find grace. To this he had easy enough access through the church's channels of salvation. But it was this very mode of bestowal which necessarily turned attention to the ego as the receptacle into which that grace flowed. It necessarily raised the question whether the level of the infused "liquid" was rising or stationary, and therewith whether the "conditions" were fulfilled or not. The result was the persistence of the assaults of doubt. No, Luther did not ask where he might find grace. What he did ask was where he might find a gracious God: How is it to come about that God and God alone will lay himself open to fellowship with me? Luther's question began where all ethics ended. This is why there had to be a new beginning if ethics was to exist at all.

It is this new beginning which we have been investigating and trying to reach in our discussion of the theological foundations and in our debate with Roman Catholic theology. We began the discussion by inquiring into the nature of man as the *imago Dei*. The approach one makes to this question, and the way in which the object itself is seen to dictate the approach, will determine whether one adopts either the ontological or the personalistic category of thought, indeed whether either of them will be regarded as the only possible and necessary form in which to tackle the matter. But the decision as to which category of thought to adopt will determine the whole concept of theology and ethics. It will determine how we understand grace, and how grace is related to our own action. And implicit in this determination will be the particular relation between Law and Gospel, justification and sanctification, dogmatics and ethics.

13.

Implications of the *Imago* Doctrine
for Evangelical Ethics

In the two preceding chapters we have assailed Roman Catholicism with a host of questions which go to the very foundations of its theology. Not least of all they call in question the key place it has given to ethics in its system. But while we have constantly had to draw attention to the great distance between its theology and ours we have not engaged in this debate merely for the sake of drawing distinctions and lines of demarcation. Indeed, with reference to the Christians within the Roman Catholic church we have repeatedly and expressly rejected as a misunderstanding the idea of polemics just for the sake of polemics. But even as regards the theoretical foundations of Roman Catholic thought we do not mean to suggest that polemics is an end in itself, as if it were part of the tradition and soundness of Reformation theology to feel that it is really coming to grips with things when it shows itself to be anti-Romanist. Where the debate with Roman Catholicism is genuine, its purpose can only be to allow one's own position to be challenged by that of the opponent, and then to stand up to this challenge. Thus to stand fast is required of us, not merely because the division of Christendom is sinful and the validity of continued division must constantly be re-examined, but also because in Roman Catholicism the Evangelical church is confronted by the cumulative experience and thinking of many centuries, to whose counsel it must submit and the test of whose arguments it must stand if it feels compelled to tread a different path. Before we conclude this whole discussion, therefore, we must deal once more with a question which seems to be raised particularly by Roman Catholic anthropology and which we believe poses a most urgent and serious threat to the kind of "personalistic" thinking we have espoused.

Christian Ethics and Non-Christian Morality

This question, this challenge to our own position, may be formulated as follows: Is personalistic thinking correct when it says that there is no alternative to being either in fellowship with God or out of fellowship with him? Is it correct when it maintains that there are no intermediate degrees or neutral positions between faith and unbelief, so that "whatsoever is not of faith is sin," indeed, whatsoever is of faith is righteous? [1] Thus far we have maintained the position that this is in fact correct, and we have tried to show how impossible is the ontological evasion of this clear-cut alternative. But now we must take a closer look at sensitive points in the personalism we have espoused. What these points are may be seen from the following questions.

If this alternative of faith or unbelief is the only one, if personalistic thought with all its radical implications is really correct, then is not everything outside Christ brought into the same condemnation, whether it be the piety of antiquity or Nihilism, ethical self-sacrifice or brutal and egoistic hedonism? Is Christian ethics to be reduced merely to esoteric directives for the believing community alone, while everything outside is hopelessly set on the same plane and abandoned to indifference and amorality?

If the faith-unbelief alternative is the only one, can there be any science, or philosophy, or culture, insofar as these stand outside of faith? Are not such science and philosophy as exist outside faith accursed since, even though they "subjectively" seek after truth and have a high objective ethos in this regard, there can be no such thing as a truth "outside" faith, and even if there were, it would only lead man to a hybrid kind of worldly wisdom ($\sigma o\phi ia \ \tau o\hat{v} \ \kappa \acute{o}\sigma \mu o v$)? If we are not to draw these conclusions, then we must ask whether the Christian message and Christian ethics really have some possibility of coming to an understanding with the world? Is it possible, then, that the concept "world" is something more than merely a euphemism for that mass of perdition with which Christian existence now has no affinity whatsoever? Here again at this point we see the seductive power of Rome's analogy of being [*analogia entis*] and ontological thinking. For they offer us some neutral ground between God and

---

[1] Cf. the first of Luther's Theses of 1520 on the question whether works contribute to justification, WA 7, 231, No. 1.

the world. They offer us bridges on which white flags may be waved. Can we live without these bridges?

In answering these questions, we shall first take up certain arguments on which we have touched briefly already in our previous deliberations.

In the first place, it may be recalled that there is in fact an unparalleled ambivalence about all spiritual capacities and all the acts to which they give rise, an ambivalence which finds its bluntest expression in this alternative of faith or unbelief. Luther put the argument in this way: If there is a positive and demonstrable value in natural abilities, then the devil too is the image of God. Angels and devils are distinguished not so much by objective value per se as by the sign under which they appear and from which those values cannot be separated.

However, if there is no such thing as a neutral continuum common to devils and angels, or to all the stages of human salvation and perdition, in other words if there is no such thing as that which Scholastic thinking calls "nature," does this then mean that there is in fact no such thing as a "disposition" [*dispositio*] for grace? Is there nothing in man which can be addressed? Surely Emil Brunner must ultimately be right in some way in his unwearying insistence that the Gospel does not address itself to sticks and stones, to oxen and asses, but to men. Hence the presuppositions—and here we should even have to say the natural presuppositions—are surely significant in relation to that which now happens to man in grace.

We shall have to return to this problem of the "disposition" and the "point of contact" [*Anknüpfungspunkt*] rather more extensively when we deal with the concept of conscience, and at this point we therefore cannot deal with the problem in its total range. For the moment we shall simply take up a specific part of the problem, briefly discussing the question of the disposition in connection with the fact that the Law is described as a custodian or schoolmaster (παιδαγωγός) to bring us to Christ (Gal. 3:24). In this description there is obviously ascribed to the Law the power to bring about such a disposition. Since the Law is universally valid—as surely as God himself is universally valid—and does not apply only in the sphere of the church, it prepares "man" as such for the Gospel.

But everything depends on how this preparation and disposition are to be understood. The following insights seem to be essential to a correct understanding.

First, the Law, precisely in Galatians 3, is not to be understood as an immanent natural presupposition. It is rather something given by God, given because of sin (v. 19) and in order to span the interim period which precedes the coming of the promised mediator.

Second, the Law does not create any "positive" presuppositions for the reception of the Gospel. It does not bring back to rectitude the chaotically jumbled components of nature. It does not restore order to the lower story of nature, on which the upper story of grace can then be built. On the contrary, if there be such a thing as "preparation" at all, then it is a destructive kind of preparation (Rom. 7:17). To put the antithesis to the *analogia entis* as sharply as possible, what the Law provides are really negative presuppositions in the sense of "tearing down" rather than positive foundations for a building up to completion. For what the Law does is to make it perfectly clear that we are separated from God. What really keeps us from faith is the illusory and simulated bridging of the gulf apart from Christ. This is why it is "easier" for publicans and harlots to hear than it is for Pharisees who are guilty of this make-believe. And this is why the task of the Law is to bring us into the position of the poor in spirit (Matt. 5:3; Luke 6:20), the publicans and harlots. In so doing it performs the function of gauze in the wound; it prevents premature healing and thus frees us from the illusion of health where there is no health.

Third, if it is true that there is no positive natural "disposition," and that the supposed "positive" endowments give rise to pharisaic presumption and self-confidence which are a hindrance to grace, then it is also true that we must avoid the opposite error of thinking that there is such a thing as a negative disposition, that spiritual poverty or a vacuum as such creates in us a kind of inclination to Christianity.

That we are not to think of the disposition along these lines may be seen clearly from the theology of idolatry, which proclaims as its first principle that there never is such a vacuum, but that it is always filled by false gods, or, paradoxically, by the *nihil* of Nihilism. The one who is "poor in spirit" cannot remain open and empty in his poverty. He will naturally fill it with a skeptical outlook, with a-moralism or anti-moralism, with "dangerous living," with Nihilism, or with other pseudoreligious substitutes. In other words, he will see ways of establishing himself even in his poverty, and of thus making a virtue of this particular necessity.

In relation to the Law Luther makes this clear by describing the end of the Law as "despair." Mark well, despair is the end. It cannot be regarded as a negative "presupposition," or as a "disposition." In respect of its significance as a disposition, despair is of the same order as the arrogance [*hybris*] of the Pharisee; it too keeps God at a distance. If it becomes a disposition in the true sense—and it can in fact become this—this can happen only by God's miraculously transforming it: the "natural" counterdisposition, the disposition *against,* is transformed by the miracle of "God's pleasure" [*Ubi et quando visum est Deo*],[2] i.e., by the "supernatural" operation of the Holy Spirit. Paul describes this transformed despair as "godly grief"; untransformed, natural despair, on the other hand, he calls "worldly grief" (II Cor. 7:10). The first leads to a godly end, namely, repentence; the second closes the way to this end and produces death. Here the problem of the "disposition" is put as sharply as possible. Alone, the Law produces only a despair which has the character of counterdisposition, hardening, and destruction. Only the miracle of the Spirit takes away the "counter."

Finally, from this vantage point certain statements can now be made concerning the disposition. Above all, it must be asserted that the position of man prior to faith is not a matter of indifference. To be sure, the presumption of the Pharisee and the despair of the publican both constitute a wall, and to this extent they are both a counterdisposition. Nevertheless, as the attitudes differ, so also do the ways in which they are seized upon by the miracle of the Spirit. For whereas despair may be transformed into godly sorrow and find comfort, presumption refuses to be comforted. It must first be broken and shaken out of the make-believe of illusory consolations.

This difference in mode makes it apparent why publicans and harlots are really better off than Pharisees: they are a little bit nearer, not to the Gospel of course, but to the last partition which divides them from the Gospel. This closer proximity does not give any kind of guarantee,

2 [See above p. 5, n. 1.–Ed.]

such as that possessed by the runner who knows when he enters the last lap that he will soon be breasting the tape. On the contrary, the wall at hand threatens to fall on a man; he may be struck down by it. The fact remains, however, that he is nearer this last partition. He is in a position where the wall will finally stop him—or be surmounted. Which of the two positions is actually his is something he does not decide for himself; it is decided for him by the miracle which is either waiting for him or is denied to him. Now it may be that this state of affairs takes all the fun out of the concept "disposition," although it shouldn't. After all, we are repeatedly prone (and permitted) to make human, all too human, comparisons and say that this or that complacent era needs a divine judgment, a great collapse; it needs to be tested to the very breaking point that it may thereby learn the limitations of humanity and so be set before that last partition. Using the same figure we even venture to say that great sufferings and doubts have the character of testing, of visitation intended to bring the prodigal home— and that is something we never say of the situation of self-satisfaction and of bourgeois perfectionism.

The point is that the question of disposition cannot be determined in advance, i.e., in the light of given, natural factors. It can be determined only in the light of what actually happens. This is clear already in what Luther says about the teaching [*agere*] and preaching [*docere*] of the Law, in what is a most profound analysis of this particular state of affairs: "That what we call despair is a useful thing, is due not to the Law itself but to the Holy Spirit, who makes of the Law not a robber or a devil but a custodian or schoolmaster [*paedagogus*]. Thus when the Law is taught [*agitur*] the teaching concerns the nature and power and effect of the Law, that which the Law is able to do of itself, but when the Law is preached [*docetur*] then the Gospel enters in and says: Look here, Law, are you not transgressing the limits of your own bailiwick? You are supposed to be a pedagogue, not a robber. You are able to frighten, but beware lest you kill, as you did in the case of Cain, Saul, Judas, and others. Your office here is not that of robber or devil but of pedagogue. However, this comes to pass not by the power of the Law, but by the power of the Gospel or the Holy Spirit interpreting the Law." [3] Luther's *agere* has reference

3 WA 39^I, 445.

to what the Law can effect by nature, namely, the "wordly grief," and therefore death. The *docere*, on the other hand, refers to the way in which the Law is *preached*, i.e., how it looks from the standpoint of the goal which it is to serve, how it looks when there is awareness of the promise of the Gospel's "entering in," and when therefore the "godly grief" is preached which is the goal of the Law.

Consequently, there is no objective and permanent disposition for salvation which may be known in advance. God can create such a disposition out of any situation, whether it be presumption or despair. Only retrospectively, in the light of the saving state itself, can we recognize this disposition. Hence it is better to express it by a verb rather than a noun. It is better to speak of a "disposing" by God, a "disposing" by the Holy Spirit which is accomplished when presumption is melted down into despair and despair into godly grief. Then the disposition is not something seen from the standpoint of nature; it is something prepared *ad hoc* by the Holy Spirit. Without the Holy Spirit, our distress would not have taught us to pray; it would have led us to curse. Without the Holy Spirit, the Law would not have been a custodian or schoolmaster to lead us to Christ; it would have been the pedagogue of death.

What is, as a result, forbidden is not study of the natural conscience, the psychology of what takes place inwardly prior to faith. We are not forbidden, for example, to consider Dostoevski's Raskolnikov, his illusions of grandeur, and in this sense his emptiness. What is forbidden is the attempt implicit in such study to discover therein an "existent" disposition or inclination. (What would have become of Raskolnikov without that miracle in the light of which the blessed harlot Sonia stands?) In themselves those prior conditions, including those which the Law effects "by nature," are wholly ambivalent. At any rate, this is certainly true of the only perspective available to man in and of himself, namely, that of the *coram se ipso*.

We can now move on to the second essential question which points up the real challenge to our own position: Does this ambivalence in turn mean ethical indifference on our part? In other words, is it a matter of no consequence whether I am a criminal or a virtuous idealist, whether I keep the commandments or not, whether—to use a very different illustration—healing takes place or not? Do not the commandments of God still have some affinity to wordly action outside of faith? Do they not still have some kind of objective status? But "objective" can only mean "independent of existence," as mathematical propositions are independent of the philosophical orientation of the persons who state them. Hence, if the commandments are held to persist objectively, do we not thereby again posit a universal factor in existence? And does not this mean that we are again committed to the hypostatization of "nature"? We come upon this question again and again.

Two illustrations may serve to clarify the problem, the first being from the field of medicine. If sickness and health ultimately depend on whether we are right with God (healthy) or not (sick), and if this true and ultimate health or sickness is quite independent of the corresponding psychophysical conditions with which medicine deals, then the ground would appear to be cut out from under the feet of medical practice. It could be that someone whom I as a physician restore to health is thereby made more self-confident and goes to hell, whereas another whom I cannot help, who wastes away as I stand helplessly by, takes his sickness as a visitation, the kind which brings the prodigal home, and so wins through to eternity. What then? Should the doctor heal at all? Does not the saying "I am the Lord your healer" (Exod. 15:26) exclude the service of any other physician? Is not this even clearer in II Chronicles, 16:12, where it is said of Asa, "Yet even in his disease he did not seek the Lord, but sought help from physicians"? [4]

Our second illustration is from the area of philosophy. If there is no alternative but faith or unbelief, it might be deduced that outside of faith in revelation there is nothing but Nihilism, that is, absolute negation, utter perdition, and the absolute "anti." Karl Jaspers has drawn this conclusion: "A theologian recently declared: 'It is no mere presumption on the part of the Church to say that the crucial alternative is: Christ or nihilism.'—If this were the case, there would be no philosophy. There would be on the one hand a history of philosophy, that is, a history of unbelief leading to nihilism, and on the other hand a system of concepts in the service of theology. Philosophy itself would be deprived of its heart, and where a theological atmosphere has prevailed, this has indeed been its lot. Even when works of con-

[4] Cf. Sirach 38:9, 12-15.

summate intellectual artistry have been created in such an atmosphere, they have taken their mood from an alien, nonphilosophical source, ecclesiastical religion; their independence has been borrowed and illusory, and they have not been taken quite seriously as philosophy." [5]

The deduction here drawn from the faith-unbelief alternative is that outside of faith in revelation there is not only no ethical action or meaningful medical healing, but also no genuine concern for the truth, whether in terms of general scholarship, of science, or of philosophy. Since these areas are to be described exclusively as areas of "unbelief," they come only under the sign of Nihilism.

These are two of the basic questions raised concerning our ethics. In replying to them we would refer first to two instances in the Bible where Jesus at least touches on this problem. The first is his interview with the rich young ruler (Mark 10:17-31), and the second is his discussion with the scribe concerning the commandment of love (Mark 12:28-34).

When the rich young ruler confesses that he has observed the commandments of God from his youth up, we may interpret this "observance" as ethical concern, indeed as a concern exercised outside of faith. That it takes place under a false sign, as it were, is obvious from both the beginning and the end of the pericope. The rich young ruler begins by asking how he can attain to "eternal life." In other words, his ethical endeavor is not something that takes place under the shadow of eternal life already given. It is not faith in action. It is not the fulfillment of an imperative that depends upon the indicative. On the contrary, he apparently takes the imperatives of God seriously for the very purpose of attaining to the indicative of salvation by keeping them. He has thus committed himself to an order which is basically wrong and which is bound to fail, in fact has failed. For it is directly from crisis and failure that he comes at the moment he approaches Jesus.

The conclusion of the story also shows that the perverted order is not set right. When Jesus orders him to sell all that he has, he is drawing the young ruler's attention to the fact that his actions are being performed under the wrong sign, his works are being done outside of faith. For the "one thing" lacking is not an additional law

[5] Karl Jaspers, *On Philosophy Rooted in Faith*, trans. Ralph Manheim (New York: Harper and Row, 1966), p. 1.

which must must "also" be kept. When he says "you lack one thing," Jesus is referring to the very heart of what was involved in all of the man's previous legalistic endeavor in general. The critical point is obviously that the rich young ruler, for all his ethical and religious idealism, was ultimately self-centered. For him, obedience and action were a kind of self-development and self-enrichment. They might almost be called the cultural extension of his outward standard of living.

That this was indeed his real lack may be seen from the fact that in the end he does not obey the command of Jesus. For the command to sell all that he has touches the critical point by requiring of him a certain experiment: He is to lay God and radical faith in God on one side of the scales, and his wealth, his standard of living, and hence ultimately himself as he exists in this world on the other side, and then just see which of the two is heavier. The demand of Jesus is really a demand to perform this act of weighing. And as it turns out, the side embracing himself and his existence in the world proves to be the heavier. The young man arises and goes away sorrowful, "for he had great possessions."

It is in the light of this newest failure that the man's earlier ethical endeavor—not merely his rejection of discipleship in the future—is judged. His previous efforts are exposed as a sublimated egoism which does *not* fear, love, and trust in God above all things, but serves only the self-actualization of the ego, even if in the most refined sense.

Though all this ethical endeavor is in the end negated, though it stands under the wrong sign, it is nonetheless essential to note that Jesus does recognize it: "Jesus looking upon him loved him" (Mark 10:21a). This recognition on the part of Jesus may also be seen in the fact that he enumerated the commandments for the man himself. That is to say, more precisely, that Jesus called them out, to the "outside," directing them into that existence of the rich young ruler which was beyond the pale of faith. He no doubt did this for a "Socratic" purpose. He no doubt sought to bring the rich young ruler to the "critical point" by referring to them. The important thing though is precisely this, that the commandments of God—for whatever purpose they may be used—do in fact have their significance outside of faith. For the moment we simply note this exegetical fact; we shall be considering later its theological basis.

Jesus' discussion of the commandment of love points in the same direction. The scribe, who is not a disciple of Jesus, acknowledges this commandment. In some sense, therefore, he acknowledges it outside of faith. For there is no indication of any kind that he understands love of Yahweh as love for the Father of Jesus Christ and therefore as trust in the true sense. Nevertheless, if we regard strictly the faith-unbelief alternative presently under consideration, we shall have to understand "outside of faith" really to mean that which is outside the specific faith defined by Christ. All other forms of faith and of love are "different," and hence "outside."

To grasp what is meant by this, we need only remember that the boundary lines which separate Christian communions do not run between faith and unbelief in the crude sense, but between faith and faith, whereby one particular kind of faith may be construed as "unbelief," or at least as a corruption of true faith. Thus the commandment of love and the obedience of love are not without meaning even outside the zone of true faith. To put it more precisely, here too it is significant that there is recognition of the fact that a personal love for God which claims the whole man is both more and better—we are in fact forced to used quantitative terms!—than "burnt offerings," i.e., than a "work" which merely represents me and in which I need not be totally involved.

In this sense we must not miss the particular nuance implied when the narrator says that the scribe answered "wisely" (Mark 12:34). The scribe was making a particular value judgment outside of faith. Accordingly, though the man's answer falls outside faith, Jesus does not simply negate it, even as he does not endorse it either just for being correct. Instead Jesus uses here a quantitative description: "You are not far from the kingdom of God." Hence there is such a thing as being outside, standing at the gate. But it seems to make a difference whether I am far outside or near. Apparently the keeping of the commandments, especially the commandment of love, has something to do with this difference. We thus see here the mystery of Jesus' love for the hungry and thirsty, for the seeking and the poor, even though they all stand under a wrong sign and are thus characterized by self-love.

If we inquire into the theological basis of this recognition, we shall have to say that such endeavors in obedience and love, however deficient they may be, are nonetheless manifestations of the human element, the *imago* in man, just as idolatry betrays a recollection of the creator. This is true even though in both cases—the *imago* and the creator—the search is pursued in the wrong direction. The mere fact that the relationship exists and is experienced, that the failure which occurs is a failure in respect of this relationship, shows that man is more than the beast. A purposeful action, even though its purpose be a wrong one, is better than one which is to no purpose at all. The first at least implies human existence in the negative mode; it involves that "misery of a deposed king." [6] The second implies a complete lack of human existence; it involves a relapse into an animal state beyond good and evil where one just nihilistically vegetates. It is in this sense that the publicans and harlots, the "poor in spirit," the rich young ruler and the scribe, were not "nihilists." [7]

Though the relationship which exists is lasting and indestructible, one cannot say that the proximity or remoteness which obtains within it, the movement in the direction of the kingdom of God or of Nihilism, represents a state of openness to or rejection of faith on the part of man. For the proximity, as we have shown already, is simply a proximity to that last and decisive partition; and the remoteness implies no hardening with respect to "God's pleasure," [8] since God is able "from these stones to raise up children" (Matt. 3:9; Luke 3:8). This does not give us the right, however, to treat the whole of this outside zone as if it were all on the same level, i.e., to remove utterly the very notion of gradations within it. We have no right to do that even if we must emphatically reiterate that these are only distinctions of proximity with respect to the "wall." Thinking in terms of gradations cannot be done away so long as out there beyond the pale of faith the divine likeness itself is not done away and completely destroyed. What Jaspers erroneously deduces from the faith-unbelief alternative

[6] [See above p. 167, n. 20.—Ed.]

[7] Indeed, the question arises whether there really is any such thing as Nihilism, i.e., whether there can be any consistent self-surrender of humanity to mere brutish vegetation, or whether Nihilism is not rather a borderline concept, namely, the characterization of a negative goal toward which man is oriented but at which he can never rest. For if it were possible for him to stop there, we should have to postulate the existence of a sphere in which man can no longer be addressed in terms of his humanity, his origin, and his destiny, and in which he would thus stand beyond the judgment and the wrath of God. But this would be an impossible conception from the biblical standpoint (cf. Rom. 1:18).

[8] [See above p. 254, n. 2.—Ed.]

would be true only on this assumption, though on this assumption it really would be true. Here again, therefore, it is significant that we cannot speak of such a "destruction," but only of positive and negative *modes* of the image of God.

The Validity of the Commandments Outside of Faith

This brings us now to certain important insights which indicate that the commandments are valid outside of faith. The commandments demand that there be a certain material, out of which God will form, in faith, the mosaic of our existence. God demands that there be this material, even though the mosaic may never be achieved, even though we form it instead into the image of the devil—which is exactly what happens when either despair or arrogance arises through the keeping of the commandments.

If the commandments are to be valid in this way, it is essential that we not draw from the faith-unbelief alternative the quick conclusion that the keeping of the commandments is accordingly a matter of indifference. We dare not conclude that the statements, "whatsoever is of faith is righteous," [9] and "to the pure all things are pure" (Titus 1:15), inevitably imply ethical indifference or even ethical Nihilism. Why this conclusion cannot be drawn, we have shown already in our discussion of the indicative and the imperative.

We recall that Paul specifically opposes this illegitimate dialectic in Romans 6:1 ff. and Romans 6:15. We also recall why we cannot draw from the parable of the prodigal son (Luke 15:11 ff.) the conclusion that alienation is a positive stage of development, and that evil is thus, in Hegelian terms, merely a transition. It is true that the returning prodigal is loved by his father; there is more joy in heaven over one sinner who repents than over ninety-nine righteous persons (Luke 15:7); if the dire effects of sin are great the grace of God is all the greater (Rom. 5:12 ff.). These things are true, but the reason for them does not lie in the productive power of alienation. Neither do they justify our declaring alienation and sin to be matters of indifference. They point us rather to the unfathomable mercy of God and to the secret of his inner heart. There is here no casual necessity to which God is obliged to react, as if he were subject to some superior order. Neither is it a question of there being some kind of mechanical al-

[9] See p. 251, n. 1.

ternation from the negative to the positive pole of the circuit. What is involved is the personal freedom of God.

Consequently we can never postulate a priori that the prodigal will return and be saved, as though by the dialectic of an established law. We can only confirm the fact a posteriori. And if there is here no given law, then we cannot "count on" it either. We cannot let sin increase in order to evoke even greater grace. Only when I have been saved, by being drawn into the magnetic field of the Father's heart, into forgiveness, only then can I see "in retrospect" that God has turned my greatest hurt into that which is best for me, and into the greatest joy in heaven. But then I can no longer break out into fresh sins, because I look upon these processes already from the perspective of my new slavery (Rom. 6:16-19). The transition which God effects is no longer something I can "count on," but only be grateful for; it leads not to the proclaiming of Law but to the extolling of miracle.

It goes without saying that the same is true in relation to medical healing, on which we touched in the first of our two illustrations. "The apparently logical deduction that because sickness and death can be positive they should also be desired, is never drawn. There is no summons to voluntary death. Even martyrdom is not to be sought (Acts 16:37)." [10] If sickness and death can serve the cause of my salvation, this is not to be regarded as a law of which *I* may take advantage. For that would be to regard sickness and death as the "cause" of this salvation whereas in fact it is God alone who is the cause. *God* may make use of these forces—but again only "when and where he pleases." [11] *He* may do so, but *I* cannot, and therefore I am not to say that these are indifferent matters.

It might perhaps be objected that, while it is clear to the Christian that the indicative of his salvation is linked with specific imperatives, it is not so apparent why these commandments should have a universal and objective validity even outside the Christian state. It is not so apparent why they should apply, for example, in public instruction, in the philosophical and scientific search for truth, in law, and in the matter of the relation between the generations everywhere. Yet the

[10] Eduard Schweizer, "Jesus Christus, Herr über Krankheit und Tod," *Universitas*, VI (1948), 642.

[11] [See p. 254, n. 2.—Ed.]

assertion of this general validity is undoubtedly posited along with the relevance of the commandments for the Christian.

How can the fulfillment of certain specific commandments, e.g., the paraeneses in Paul's epistles, be regarded as implementation of the obedience of faith if it is not the will of God which is expressed in them? And how can this will of God, which forbids immorality and covetousness, gluttony and foolishness (e.g., Eph. 5:3 ff.; Gal. 5:19 ff.; II Tim. 3:2 ff.), and which demands the fruits of the Spirit (love, joy, peace, patience, kindness, goodness, faithfulness, gentleness, self-control)—how can this will of God be anything other than of absolute and therefore "objective" validity? The way they mark faith's course on both the right hand and the left suggests that these negative and positive directives are apparently intended to show what man and the world are to be like if they are in accord with the will of God. They provide indicators whereby we may ascertain whether the Yes which we speak to God's will in faith is not in the next moment diluted and confounded, whether this will of God is taken seriously and affirmed in all the various dimensions of our life and not merely in the basic "sign" under which we live, whether the concrete shape of our life corresponds to the will of God. In our analysis of ambivalence we have shown already how it is possible for parts of our life to diverge from this basic assent to God's will, and that this is why a sheep dog is needed along with the shepherd.[12] But the fact that they do so indicate means that these directives really are proclaimed as the objective and universally valid will of God.

At this point, of course, one final objection might still be made, namely, that those New Testament paraeneses can be regarded as formulations of the divine will only to the degree that the particular virtues and vices of which they speak, the shapes and the misshapen distortions of our life, are evidences of the Spirit; in themselves, i.e., apart from their being rooted in either the flesh or the Spirit, they are nothing. In other words, they are always ambivalent—so runs the objection—the virtues might be symptoms of arrogance, and the vices might be indicative of a sinner who has been blessed by God and who constantly seeks refuge in grace.

Now we have already shown that works are characterized by the person who does them, and that the person is in turn characterized

[12] See pp. 130-133 above.

either by God or by self (or "the Evil One"), either by the Spirit (κατὰ πνεῦμα) or the flesh (κατὰ σάρκα). How then can the divine commandment be said to represent God's will outside the church? How can it still be "objective"? If the personal root is not determined by God's command—as it cannot be except in believers—then the person's works cannot be either. Or can they? Will there not then simply be a semblance of agreement, a semblance of peace, long-suffering, goodness, temperance, etc.? Will not this be "hypocrisy" in the strict sense of self-contradiction, the contradiction between the facade of works and the root which is the person? If the commandments of God cannot be understood in terms of what is really their "final cause," i.e., with reference to the will of God which is expressed in them, would it not then be more honest and salutary to say plainly that they have no force? To put the matter even more precisely, *if* there is no keeping of the commandments outside the church, would it not be true that the open chaos which is no longer controlled or limited by these commandments, the open warfare of all against all, the "science" which pragmatically follows all sorts of philosophical excesses, the manifest lawlessness—would not all these be a far more appropriate expression of the actual state of things than the hypocrisy which parades under the guise of external subservience to a commandment? These are the pointed questions that must be asked.

We must digress for a moment to point out that we have to see the full conceptual difficulty of the dilemma into which the Protestant, and especially the Lutheran, doctrine of justification seems to lead, if we are to understand at all the attempt, already criticized and rejected by us, to use the idea of the "orders of creation" as a means of making theological connection with those who stand beyond the pale of the church. There is one very simple thought which has led modern Lutheranism—influenced of course by Orthodoxy's teaching on general revelation—to turn to the notion of orders of creation as a way out.[13] It consists in the rather helpless, if not too plainly admitted, concession that the Second Article of the Creed, i.e., the doctrine of justification, not only does not give us of itself any such connection, but is even opposed to it.[14] But since the Second Article was regarded as un-

[13] See, for example, Paul Althaus, *Theologie der Ordnungen* (1935); *Obrigkeit und Führertum* (1936).

[14] No one in Lutheranism has ever yet hit on the strange idea pursued by Karl Barth in "The Christian Community and the Civil Community" (see Barth,

rewarding and even obstructive, there was a temptation to turn to the First Article, to get the job done by means of a construct such as the "orders of creation." To be sure, these cheap and easy solutions could be had only at the staggering price of playing off the First Article against the Second. This was done through sabotaging the Reformation doctrine of sin by setting up sinless spheres and by trying to support them with dubious quotations from Luther. The difficulty with these quotations is that while Luther certainly emphasized the totality of sin in face of the ontological restrictions of Roman Catholicism he did not draw all the necessary conclusions in relation to the concept "world" (he could hardly do that in a world which had not yet become fully secularized and been revealed in its full secular emancipation). There was thus inevitably a relapse into the doubtful concept of a natural law grounded in creation, though this Protestant form of natural law was not so carefully and precisely grounded as in Roman Catholic theology. For the latter has and has had a concept of sin, a concept of nature and grace, which gives it within its own framework both the possibility and the right to speak of natural law, whereas similar efforts in Lutheranism lead to hopeless contradictions because of its very strong and rigid doctrine of creation on the one hand and a doctrine of sin on the other which, at least in intention, is very radical in character.

To return then to the pointed questions raised above: by putting these questions we have made our problem as difficult as possible. We have made the assertion of Karl Jaspers seem almost innocuous in comparison with these formulations. We have radicalized his assertion in terms of the New Testament concept of "hypocrisy." This question of "hypocrisy" constitutes the acid test of our ethics.

The objection which we are discussing is right then in its claim that so long as the "virtues" mentioned in the New Testament paraeneses are not fruits of the Spirit, they are not real acts of obedience, a real keeping of the commandments. To be that, they would have to be an expression of the fear and love of God. But they cannot be this without the Holy Spirit. Apart from the Spirit they are simply final sublimations of self-love, as we noted with respect to the rich young ruler. To this extent the commandments cannot be kept outside faith;

Community, State, and Church: Three Essays [Garden City, New York: Doubleday, 1960], pp. 168 ff.).

indeed apart from faith it is not possible even to perceive their true meaning. More precisely, then, the question which we have to put is this: What right have we to demand in the name of God a purely external keeping of the commandments apart from faith?

For by now it must surely have become clear that the keeping of the commandments would have to be purely external where their true meaning is not even perceived. But how far can we speak of the commandments having objective and universal validity if, apart from faith, the reference is simply to externals, and if they thus end up by demonstrating and reflecting something which is false? This is the fundamental question.

Apparently what constitutes the real crisis as regards the "external" validity of the commandments is that both they themselves and the obedience paid them "demonstrate" something different. But what they demonstrate is quite ambiguous.

The works of peace, of law, of parental respect, of concern for truth, etc., which are externally demanded, and even to some extent achieved, do demonstrate in "hypocritical" fashion that the person is apparently what he ought to be. But in addition they also demonstrate, in the form of a "mask" [*Larve*], what it is that God really wants of us. Take for instance the scribe, who quite apart from Christian faith acknowledges the commandment of love, even though he does not really "know" it at all in terms of its true "final cause." He is obviously a demonstration of something more than merely that final sublimation of self-love. He demonstrates also, as in a distorted mirror, that which God really wills, namely, that we should see Him as the one who has first loved us and whom we are to love in return, and that we cannot be satisfied merely with the surrender of a divided heart, such as would be demanded by purely external obedience and sacrifice. The scribe is of course still dominated, or at least influenced, by the thought of reward; there is still a lurking egoism. Within this framework, however, he nonetheless tries to take seriously the matter of loving God, and in so doing he demonstrates that God and not self is the real theme of his obedience. This is true even though in the very next moment he turns right around and wholly distorts this intention and demonstration by laying a false stress on his own person and by thus insinuating himself into this sacrifice of love.

And what of the rich young ruler and his keeping of the commandments? He does not really keep them at all, because for him God is not the First and the Last; he obviously does not even see that this is the decisive theme of the commandments. Nevertheless, he does *seem* to keep them. It seems to *him* that he is keeping them. And he is concerned that what appears to be so should really be so. Hence he is not just a hypocrite demonstrating wholly false facts. He is at the same time one who does not seek to be his own lord, but wills to live in obedience. By his keeping of the commandments he shows also that he is concerned about the authority of God. This too he demonstrates —even though in the very next moment he draws back completely once it is made clear to him what is really implied by this authority and obedience when it comes right down to it. All this he demonstrates too, though his declination throws terrible doubt on the quality of his previous obedience.

In such a thing as obedience to parents, yes, even in idolatry, there is obviously the recollection that man does not stand at the center, but has his place within a more encompassing order in which he finds himself subjected to the authority of higher powers before which he must bow. Antiquity's "pious Greek" demonstrates this better than the emancipated man of modern mass society who acknowledges no authorities whatsoever. Hence it would be wrong to characterize the latter as "honest" in the open demonstration of his nihilistic freedom from all binding authority, and to label the pious Greek as a "hypocrite."

In the scholarly search for truth too, and in secular philosophy, there is obviously more than the mere hypocritical self-deception whereby men regard as truth that which is always at the same time also a symbol for the object of fear and hope, that which, as the object of our thought, is certainly determined at least in part by the wish which is father to the thought. For all of the researcher's most earnest efforts, there is implicit in every scholarly quest for truth something more than mere selfless acknowledgment of what is. There is implicit in his quest also a secret act of adjuration: may *this* be, rather than *that*. This is true simply because all scholarly research proceeds within the framework of a hidden or openly acknowledged world view—even if it be merely that of an announced working hypothesis or heuristic principle—and because this world view is assuredly connected with fear and hope and with that act of adjuration. But again it would be wrong to regard this "hypocrisy" as the only thing demonstrated by the search for truth. However much the seeker after truth, in his scholarly objectivity, is covertly affirming and asserting himself—in analogy with the self-love of the scribe— he is at the same time, in his concern for objectivity, moving away from self and toward the norm of truth. And by that fact he discloses that he cannot of himself dictatorially establish what truth is but that he rather stands himself under its authority; he cannot engage in intentional and tendentious lying wherever this suits his purpose but must rather submit to the truth, and bow before it. This too he demonstrates, even though his submission may be only apparent, or in part at least merely a matter of appearance.

The extent to which this second thing is actually demonstrated in science and philosophy may be seen in the case of the scholarly endeavors which took such a pragmatic turn in face of the world views of the totalitarian states. Here the fear and hope, which play some role in any act of scholarship, themselves became the dominant theme of every scholarly act, and the norm of truth was thereby solemnly stripped of its authority. This example illustrates in a negative way how the science which is, or desires to be, responsible to the truth, contains at the same time a positive demonstration of the fact that certain ultimate authorities—like the norm of truth—are not put in the hands of men, but that man is instead summoned to submission.

Obviously it is also possible to look upon natural law as a revered abode of existence furnished by man himself, something which man has accommodated to himself and which is therefore not to be equated with the commandments of God. But here too there is more than mere hypocrisy. There is also the affirmation that man is not his own legislator but has his place within a nexus wherein he is a subject. This is true no matter how plainly it may be shown that this nexus of natural law is after all not an order which for man is something given but an order which man has himself brought into being, in which—perhaps involuntarily and even in many cases unwittingly—man assumes the stance of master in the house.

From all this it is clear that the claim of the commandments to validity outside the church, i.e., the "objective" authority of the divine commandments, does not merely serve to erect a hypocritical facade.

Though the order herein proclaimed may conceal, and is undoubtedly ambiguous, it is nonetheless also a "mask" of the true will of God. There can thus be no doubt that, wherever the Christian holds a secular office as teacher, jurist, merchant, or statesman, he has not merely to keep the commandments in relation to external matters himself; he has also to press for public observance of the commandments. To put it another way, just because the "final cause" of the commandments is not to be seen here "without" is no reason for the Christian to promote chaos in law, politics, or learning in order to disclose the true state of the world. For if he did, this would be more than mere "disclosure," it would be destruction itself. It would be the breaking off in a fanatical way of that period of grace and patience which God has granted to a world oppressed by the powers of destruction. It would thus be interference in God's salvation history, a presumptuous attempt to produce by destructive human means that "godly grief" which is the work of the Holy Spirit alone. The result would be "worldly grief" and destruction. Hence the commandments of God are to be proclaimed and regarded as valid even outside the church.

Concerning this validity of the commandments outside the church we have still to add a final word in this connection. That which is falsely demonstrated by the observance of the commandments outside the church is certainly not something to be contested or done away by the promotion of chaos. On the contrary, it is the task of the church, particularly in its preaching of the Law and in its office as watchman, to show how questionable is the thing thus demonstrated. In this proclamation, in the preaching of repentance to itself and to the world, there must be heard the cry that the ax is laid to the root of the trees (Luke 3:9), that this world which vacillates between God and Satan, between obedience and hypocrisy, between longing for redemption (Rom. 8:19) and rejection of it, between the service of God and self-love, that this world lives only by the longsuffering of God, and that by his grace and patience alone the facade of this crumbling world order may at the same time be a circumscription, a surety, and—however faulty—a representation of God's true order.

This is why it is so important that the public message of the church should not consist alone in the proclamation of social, economic, and similar requirements. If that were to happen, the church would only be continuing to preach the perpetual ambiguity of the Law; it would itself be moving in the half-light of the aeon instead of bringing light. As the church constructively criticizes the various orders and reforms, it must become evident that it is after something quite different from that intended in the analogous demands of the various political parties, philosophical movements, and pressure groups. It must become apparent that what the church has in view is not some kind of pragmatic rearrangement of things for the sake of world betterment but the activity of men who have been liberated by God and summoned to fear and love. It is only when the action and organization of the world are thus carried through in the light of their existence before God and in knowledge of the "final cause" of his commandments—it is only then that the world is shaped on a basis other than the disruptive self-contradiction of hypocrisy. Only then is it shaped on the basis of an ultimate decision, a new and divinely enacted reality. By summoning to this reality, to put it quite simply, by preaching judgment and grace, by viewing both the structure and the collapse of the world within the framework of judgment and grace, by bearing witness to them expressly in these terms, the church acts publicly in the world and takes a *legitimate* outward step when it adopts a position with respect to the acute questions of public life.

For this reason one cannot speak abstractly of the "objectivity" of the commandments, i.e., of their validity even outside the church. The moment we assert a validity which may be described abstractly, we deliver these commandments over into the ineluctable half-light of the world. Then the objection which we took over from Karl Jaspers and developed is justified. Christianity can assert this public and objective validity of the divine commandments only in immediate concert with its proclamation of the message of Law and Gospel, of judgment and grace. The church which maintains the validity of the commandments outside the church has also to make clear what is the crisis of the commandments in this outside sphere. It has to point to the haziness of understanding in this sphere and to the ambiguity of the corresponding possibilities of obedience. It has to point beyond the externals to which alone the commandments can refer in this outside sphere and to speak of the internals. It has to preach those internals which, when they are apart from forgiveness, can lead only to hypocritical obedience and bring the ferment of disintegration into the world structure they intend, those internals which can serve God and neighbor in the forming of the world only as they themselves have been set in order.

The Political Function of the Law and Salvation History

We are thus at the point which the traditional concept of the "political function of the Law," the *usus politicus legis,* has in view when it proclaims the absolute validity of the commandments even outside the church. Expressed in the *usus politicus legis* is the fact that God will not send another flood, that he will not destroy the human race physically, e.g., by giving it up to the chaotic forces latent within it (in the sense of Romans 1:24).

To serve the cause of this physical preservation (in the first instance the preservation of the covenant people) is "also" one of the functions of the Law. For these laws, especially in the predominantly negative form in which they appear in the Decalogue, are indeed directed "also" against the destructive forces of sin. Thus behind the law "You shall not kill" (Exod. 20:13) there stands "also" a concern for stopping destructive blood feuds. Similarly, behind the command to honor one's parents (Exod. 20:12) there stands quite expressly the connection between preservation of authority (as between the generations) and the very possibility of historical existence (". . . that your days may be long in the land which the Lord your God gives you"); any disruption of this order disrupts existence in general. Again, behind the law "You shall not bear false witness" (Exod. 20:16) there stands "also" a concern to assure a system of law, so that the strong shall not tyrannically dominate the judicial process by means of his many children (cf. Ps. 127:5) instead of allowing right to triumph; lawlessness would in fact be an open threat to physical existence.

Thus the Law has "also" the task of safeguarding purely physical existence. One might say that this is its "incidental" task, in the sense that it is "supplementary" to its true purpose. It has this task in the same way as faith, which is likewise concerned with physical abiding and perishing: "If you will not believe, surely you shall not be established" (Isa. 7:9). We cannot, of course, make this by-product into the main theme of faith (and of the Law) in the

272

sense of turning this Isaiah text into the positive statement, "If you believe, surely you shall be established." Like faith, the Law is not a "guarantee of historical existence; but without faith (and without the Law) this existence is utterly abandoned to the transitoriness of everything earthly." [15]

It should not be overlooked, of course, that the primary reference here is to the physical protection of the covenant people, especially in the face of Gentile attack. What is involved here is particularly the safeguarding of the necessary religious, moral, and physical distinctiveness of this people. In the first instance, therefore, we may not extend this idea of the physical preservation of existence to include creation as a whole, the preservation of the whole cosmos, including the Gentiles. To this degree we do not yet have here in the strict sense an instance of the Law being valid outside the church, i.e., outside Israel. There is not yet posited here that which is ordinarily meant by the *usus politicus*.[16]

Thus when we consider the preservation of existence ensured by the Law we must do so primarily from the standpoint of the origin and goal of the covenant people. This people, and initially this people alone, stands under the "patience" of God. The patience of God for its part has in view a specific end in salvation history, namely, the redemption of the people and the coming of the Messianic kingdom. Hence the physical safeguarding of existence is never an end in itself. It is subject to a pneumatic goal which God will accomplish in his elect people. This people is thus summoned to understand the time of its physical existence, this time in which it is secure against any further flood, as the *kairos*, the "acceptable time" in which it is supposed to repent and turn to Yahweh. Physical existence is actually

[15] Walther von Loewenich, *Augustin und das christliche Geschichts-denken* (Munich, 1947), p. 8.

[16] While it goes without saying that the *usus politicus* applies also to Christians (since they too are members of the body politic) and is thus valid also within the church, it is usually thought of in the first instance with respect to "restraining the refractory," i.e., those who, living without the authority of God, threaten to fall victim to the warfare of all against all. Luther did not think primarily in terms of some sphere outside the church; he thought rather in terms of Christendom, the *corpus Christianum*. This may be why he lays so little emphasis on the *usus politicus*, although he does refer to it. With increasing secularization, however, this *usus* understandably pressed with ever greater insistence for theological attention.

the sphere of repentance and realization of salvation (II Chron. 30:9; 1:9; Deut. 32:19-47). In part it is thus determined and limited as a sphere by the way in which the people react, whether they repent or are unrepentant and disobedient. A disobedient people makes the given sphere meaningless and superfluous.

This fact is the basis of the prophetic threats of judgment and destruction, which of course intend a physical destruction, dispersion, subjection to the Gentiles, etc. The story of the sacrifice of Isaac (Gen. 22:1 ff.) is also related to this background of a physical sphere of salvation. For Yahweh's promises to Abraham in the context of salvation history depend upon the physical existence of Abraham's descendants (Gen. 17:19; 18:10-15). Hence the demand of Yahweh that Abraham should sacrifice his son is not merely a test of faith in the sense that God's command should take precedence over paternal affection. It is a test of faith in the far more radical sense that God's *will* (the command to sacrifice) is brought into conflict with God's *promise* (of a progeny which cannot be counted). Abraham's faith is shown by the fact that he leaves it to God to resolve the conflict— be it even by miracle (Matt. 3:9)—and that he himself affirms both the command and the promise. At this point the Old Testament connection between the divine history of salvation and the physical existence of the bearers of salvation is brought out with particular clarity, since it is put in the setting of a supreme test of faith.[17]

If we see that the safeguarding of physical existence is in fact related to the chosen people as those who bear salvation history, then we cannot possibly understand the preservation of existence which is "also" secured by the Law in terms of "orders of creation" or other similar concepts dreamed up by modern Lutheranism. There is here no place for the idea that there are certain basic laws which are immanent in the cosmos since creation, and which must be maintained if the world is not to perish. For the whole problem of "the Law and physical existence" is not seen here in the light of creation at all. We cannot smuggle in cosmic analyses and philosophies by shamefully maltreating the First Article of the Creed and opening it up to all kinds of ideologies. This whole complex of problems must be seen in the light, not of creation, but of salvation history and its goal:

[17] Our exposition of the story of the sacrifice of Isaac is in this respect very much like that of Luther. See *LW* 4, 91 ff.

physical existence is promised to the covenant people as the sphere in which they are to repent and await salvation.

When Karl Barth attempts a christological understanding of the state which must safeguard physical existence, he is right in this respect, that he thereby views the state in relation to its function in salvation history and in the light of its pneumatic goal.[18] The problem with his political Christology is to be found elsewhere, namely, in his attempt to provide christological foundation and explanation for every single facet of political life. This leads to the same kind of short cuts as the attempt to find a christological basis and explanation for every verse in the Old Testament. For it can result in such conceptual acrobatics as this, that since the light of God has now shone in the darkness of this world in Christ there ought therefore to be no secret diplomacy in politics,[19] or that since there is only one body of Christ and only one head, Christ himself, certain conclusions ought to be drawn regarding political democracy.[20] If we are to proceed on the basis of such formal analogies between Christology and politics, we might very well reach diametrically opposed conclusions, e.g., that in the light of the Messianic secret there ought indeed to be secret diplomacy. In fact, on this basis it would be hard to give a convincing answer to the Nazified "German Christians" who might deduce from Barth's second premise: "One *Volk*, one *Reich*, one *Führer*." [21]

We do not contest Barth's intention. By setting the question of political and physical existence in the light of the goal of salvation history he corrects the various ideologies which speak of orders of creation, and for this he deserves full credit. Our only complaint is against the way in which he goes about making these connections with salvation history and Christology. He seems to make use of an impossible christological analogy which is at least as bad as the analogy of being which is advanced in the name of the First Article. Indeed, in the last analysis he simply exchanges the doubtful rational

18 See Karl Barth, "Church and State," and especially "The Christian Community and the Civil Community," *op. cit.*

19 See Barth, *op. cit.*, p. 176.

20 *Ibid.*, p. 174.

21 See in my book *Kirche und Öffentlichkeit* (Tübingen, 1947) the note on p. 37.

perspicuity of the traditional *analogia entis* doctrine for christological mystification and artifice.

What we have, therefore, in the Law, are not "orders of creation" which are universal, rational, and have reference to the whole cosmos. What we have is rather a fact which is to be understood only in relation to salvation history, as special providence, the fact that in the name of covenant grace and covenant patience the fallen people of God receive a physical sphere of existence, a sphere which it is one of the functions of the Law "also" to protect.

Only when this is firmly established, only when a premature extension of the covenant Law to all nations and to the whole cosmos is ruled out, only then can we ask concerning the way in which the Law of God may also be generally valid outside the church and outside Israel. At this point we have thus to consider a further challenge to this general validity.

We recall that our first challenge consisted in the fact that in the outside sphere generally the Law cannot be understood in terms of its true final cause (to fear and love God). Accordingly, it can refer only to externals (the restraining of the refractory, etc.). It thereby leads to what is merely an appearance of obedience, and hence to hypocrisy. And this is what gives rise to the serious question whether the withdrawal of the Law from this external sphere, i.e., the delivering up of the world to undirected chaos, would not create a situation at once more honest and more pleasing to God. Hardly have we given a theological answer to this challenge, however, before a new challenge is posed by the fact that, at least from the Old Testament standpoint, even the reference of the Law to externals, i.e., to the preservation of physical existence, is seen to apply in the first instance only to the covenant people and to be oriented only to the goal of its particular salvation history.

When, in the light of this insight, we inquire concerning the general validity of the Law, its validity in the external sphere, we have at least to take into account this aspect in terms of salvation history. And this aspect of course gives us some vital insights into the significance of the Law of God in the external sphere.

Through the rejection of Israel, i.e., through the fact that Israel has not laid claim to the sphere of repentance and conversion granted it, this sphere (which is also physical!) has been thrown open to the

Gentiles (Rom. 11:11) and to the cosmos generally (Rom. 11:12). Humanity outside Israel is now to be grafted in, in place of the branches which were broken off (Rom. 11:17). The kingdom of God now goes out into the hedges and lanes to produce substitutes for the guests who would not come.[22] The promises made to Israel are thereby transferred to the Gentiles. The Gentiles too are now oriented to the goal of salvation history which was once marked out for Israel alone.

This means, however, that the divine commandments take on a new significance. That is, their significance is no longer confined to the covenant people alone, for the covenant is now offered to the "nations." This means in turn that the Law in its *usus politicus,* in its reference to externals, also takes on a new significance. When in this sense it creates order, prevents the war of all against all, exercises coercion, and by all this makes physical existence as such possible, this is not to be construed merely as the preservation of a certain structure given the world at creation. As previously in the case of Israel, what is involved is rather the provision of a physical basis for the sphere of repentance. It is a question of safeguarding the *kairos* which is to end with the Parousia of Christ, with the destruction of the world or with the death of the individual, i.e., with the withdrawal of the physical basis of life.[23]

The *usus politicus legis,* the universal validity of the Law, even outside the church, has its basis here and here alone, in its reference to the goal of salvation, not in an order of creation but in the order of salvation, not in the origin of the world but in the goal set for the world. In the last analysis, the preservation of humanity, and of the social structures in which alone it can exist, the preservation of both the freedom and the authority necessary to life, does not rest on a sociological or anthropological foundation. For the value of human life and of the life of humanity does not lie in itself. If we regard life only "in itself," then Nietzsche's assessment of man as the most useless vermin on the earth's crust is quite as valid as any

22 Luke 14:21-23; Matt. 22:7 ff.; cf. II Cor. 5:19, 20; Matt. 21:43; Acts 13:48, 52.

23 In this connection notice the way in which Revelation links repentance with visitations affecting the physical basis of existence, e.g., 2:16, 21; 9:20; 16:9. On the way the death of the individual limits the physical sphere of repentance, see Luke 12:20; 16:25 f.

kind of "intoxication with humanity." There is little to choose between the postulate of a solicitous preservation of *homo sapiens* and that of the destruction of vermin. It simply will not do to look at the immanent value of man because this value is not something ontically immanent. It rests in our personal relationship to the Lord who wills to have us as his children. If we exist at all, then, it is with a view to this sonship. And we are preserved by the divine commandments, not for our own sake, but for the sake of this sonship. We are "sustained" with a view to the goals of God. In these goals lies the meaning of history. Physical existence is ordered solely by this meaning, and thus is the corporal form of the *kairos*. And the *usus politicus* of the Law, which is the form of its universal validity, its objectivity, is a sign of that divine grace and patience whereby the *kairos* continues on its way and the forces of restraint (cf. II Thess. 2:6, 7) are still at work.

From a very different angle we are thus confirmed in our argument that the Law of God is not to be made inoperative outside the church in order to put an end to hypocrisy and to unmask the true state of affairs. The world dare not perish because of its actual condition, for the promises of God have gone out to it and it must have an open path amidst these promises. The world has been offered a covenant, and in the name of this covenant God's sustaining commandments are everywhere valid. Those who break them or remove them do not simply abandon vermin to destruction; they block the way to sonship and sabotage the meaning of history.

14.

The Law of the Fallen World

We have already considered the theological and ethical significance of the doctrine of man's original state. This doctrine provides the background against which "history" comes into play, the story of guilt and responsibility, of judgment and grace. It thus provides the background against which that sphere of existence arises which is the province of ethics. If, then, we are to be able to say with objective precision what sin is, we must first, or at least simultaneously, examine with equally objective precision what it means to come forth from the hands of God, i.e., to be created, innocent.

Creation and Sin

It is the concern of the doctrine of "creation out of nothing" [*creatio ex nihilo*] to set forth, buttress, and give precision to the thought that guilt was not created with the world; guilt did not enter in with creation itself but has its origin in man. We must now develop this doctrine in various directions, and particularly in respect of its ethical relevance. The doctrine of the *creatio ex nihilo* stands opposed to two other interpretations of creation which carry with them a very different attitude to guilt and therefore to ethics as well.

The first of these is the understanding of the Creator God as the "first cause" [*prima causa*]. The other is the understanding of the Creator God as the "demiurge," i.e., the architect of the world.

The understanding of God as "first cause" leads to a religion of identity and to the pantheistic equation of God and the world (*Deus sive natura*). On the one hand this may imply the removal of the dualism of good and evil altogether, for a thing cannot be held responsible if it is the mere "effect" of an external cause (this is why Luther, who like Calvin insists on both predestination and responsibility, inveighs constantly in *The Bondage of the Will* against the notion of first cause). Or it may on the other hand imply a theological blunting of that dualism on the ground that guilt is not a genuine protest against God because such a radical protest would end up out-side the sphere of the God-world identity (evil is thus understood instead as the dialectical counterpart of good inasmuch as it rep-resents a mere "transition" and is thus productive). Thus in both cases there is lacking the radicalness of the required ethical decision because evil is not understood radically as that which is truly "against" God but as a dialectical mode of being "in" God, as an attribute of God himself.

When creation is regarded as the activity of a "demiurge," or as a process of biological and telluric evolution, evil suffers a similar loss in gravity. For then it is thought to reside in the material from which the world is fashioned.

On the basis of Greek tradition there is thus a recurrent tendency in the history of Christianity to localize sin in material corporeality, especially in the sexual libido, while the spiritual spheres are sup-posedly further removed from sin. On this view man did not simply come from the hands of God. Guilt is to be attributed neither to God (which is ruled out as blasphemous) nor to man (as though he were fully responsible as the first subject). It derives instead from the material which the demiurge used in fashioning the world, and hence from without. In large measure, therefore, man's relation-ship to guilt is that of affected object rather than productive subject. Here again the radicalness of guilt is weakened.

In face of both these ways of watering down the concept of guilt, the doctrine of *creatio ex nihilo*—this is its ethical significance—insists that guilt cannot possibly be derived from elsewhere. There is no third thing besides God and me which could help to give rise to guilt. But since God cannot possibly be regarded as its author, full responsibility for it has to remain on my shoulders. Here already in the doctrine of creation, then, the question of man's responsibility is decided. Moreover, the antagonism which underlies all ethical decision ceases to be a purely inward conflict between flesh and spirit, body and soul, matter and spirit. It has reference rather to the ex-ternal relationship of man to God. For the true opponents in the struggle are the Spirit of God on the one side, and on the other the flesh as the representative of self-enclosed man trying to assert him-self against God. The terms "flesh" and "Spirit" thus have a very different content. Nor is this fact without influence on the doctrine of justification. For if the decisive antagonism does not lie in the dynamic encounter of forces within man, then there can be no such thing as a quantitative reduction of that encounter by "works," i.e., by the active triumph of the spiritual sphere over sensuality, etc. Instead there can only be a personal purification of the disrupted relationship between God and man. Here we see clearly the main lines of the Pauline and Reformation doctrine of justification.

The Freedom of Creation and the Enslaved Will

The doctrine of *creatio ex nihilo* thus marks out the original background of that reality which is the fallen world. The original world in the early morning of the day of creation is still without the shadows cast by non-transparent and isolating bodies. When we examine this background more closely, understanding it simply as background, i.e., specifically in its relation to our present concrete situation, we are forced to speak of the directness and immediacy of the original relation to God. One might use the metaphor of the closed circuit: the relation is a dialogue between God and man. No third voice can break in. The serpent has not yet begun to speak, or its voice is not yet heard. It is a life in liturgy.

Accordingly, the original freedom which distinguishes man as the *imago Dei,* and exalts him above the rest of creation, cannot be described as man's capacity to do what he pleases. It is to be defined rather as his capacity to become what he ought. To understand this correctly, we must consider the following paradox.

Freedom in face of the command of God the creator (Gen. 1:28; 2:16 f.) cannot mean that Adam and Eve are perfectly free either to obey or to fall, and that they are thus free—in terms of our formulation—to do what they please. For this original freedom (which is the prototype of all freedom, and which is hence also a judgment on our perverted and fallen form of freedom) does not involve two equal possibilities between which I could simply choose. In the strict sense, Adam has only one possibility, namely, that of doing the will of God.

That this statement is not in conflict with the warning issued in Gen. 2:17, which as "warning" seems to presuppose the possibility of the fall, is mysteriously and profoundly attested by the mention of "the tree of knowledge of good and evil." Why is the tree thus named?

Before they fall, Adam and Eve not only do no evil; they do not even know what evil is. For them evil does not yet even exist. Perhaps an illustration, one which of course has its limitations, will help us to grasp this. A young lad who is not yet physiologically developed knows nothing of the beauties and transports of *eros*. Indeed, this knowledge does not even exist until he has "known" a woman, i.e., entered into the fulfillment of sexual encounter. Similarly, one "knows" of evil only as one "does" evil, only as it leaves the sphere of mere possibility and becomes actuality. This is why "good people" often cannot understand a person who commits murder, or betrays a conviction for the sake of ambition. The name of the tree in the midst of the Garden of Eden thus means that man can know good and evil only as he takes the forbidden fruit and so "does" evil. Like God himself, whose will must be done if we are to know him (John 7:17), evil can be known only existentially.

In any case, the knowledge of good and evil, the cleavage of life into the polarity of good and evil, is certainly a typical feature of fallen man; it indicates that life has been severed from "the One." The heart now finds itself pulled in two directions, and a man's thoughts take sides and accuse first one party and then the other (Rom. 2:15). The conscience is prosecutor one moment and defense counsel the next; it is never the judge but always represents one of the two sides into which life has divided itself. The fall is characterized, not merely by the emergence of evil as a party, but also by the fact that God is thereby made into a mere "party."

We are instructed concerning this same state of affairs in Romans 7. To be torn between good and evil is in itself a mark of sin. The polarity which Idealism transformed into the sign of human dignity, the possibility of ethics and the freedom of the ethical, the substrate for the fulfillment of duty (Fichte), etc.—this polarity implies from the standpoint of the radical will of God nothing less than the presence of the opponent, the denier, on the territory of the ego, an ego which belongs to another. This is why a quantitative moral betterment, in which good takes relative precedence over evil, can never denote the fulfillment of righteousness. At best it can only be an event within the existent polarity. It can never involve more than the taking back of certain territory from an invader who, having once been allowed to enter, is still in the land.

Now the original state displays a knowledge of the good—i.e., of the will of God declared in the commandments—which is not defined by what is opposed to it, or attained through experience of its opposite. What is involved here is that *imago Dei* which is continually laid hold of anew. Nor is there any need that this continued laying hold of should be defended.

In attempting to elucidate this idea we admittedly find ourselves skirting the very limits of man's conceptual capacity, yet epistemological considerations cannot release us from making the attempt. Human thought begins to spin and swirl, and to slip into tensions and contradictions, when it is required to conceive of a good which is not constituted by the antithesis of evil. For good and evil are always relative the one to the other—have been ever since the fall, when their conflict began and our aeon got under way. This is why, since then, they can only be defined in terms of their reciprocal relationship: good is that which is not evil but contrary to evil, and vice versa.

This other world, the original world, is not accessible to human insight and knowledge because the categories of our perception are conditioned by the form in which this fallen aeon exists. To believe in an original good which does not stand in some antithesis, to believe in a good which is determined not by the fact that it is "against" evil but by the fact that it is "for" God, this can indeed be only a matter of faith, of not doubting that which is "not seen" (Heb. 11:1). It is also evident that when we speak of this "absolute" good, this good which is not relative to evil, we are by no means engaging in an idle mythological reconstruction of man's primal state. Our elucidation of original existence is theologically necessary, if for no other reason than to show that the very existence of a state marked by the ethical cleavage between good and evil is itself a sign of the fall.

The evil which is intimated in the divine prohibition in Paradise represents on the other hand a borderline possibility, i.e., a possibility which in the strict sense exists only outside of Paradise, that is, outside the sphere of existence originally given and imposed. This is also why we who live today in fallen history, in the state of cleavage, knowing good and evil, feel this original sphere of existence to be non-historical.

This does not mean, however, that the fall is the laying hold of a "possibility." On the contrary, it is a plunge into "impossibility," into a bondage of the will, into the inability not to sin [*non posse non peccare*]: "For what I want, I do not do" (Rom. 7:15). And here is where the ceaseless struggle begins, the struggle to confine the evil which has broken out and which, as it were, subjectively encompasses me; here begins the unending attempt to change myself. This is the desperate and hopelessly doomed way of "works," which only confirms our embattled state by giving expression to it. "Paradise is locked and barred, and the cherub is behind us; we must make our way through all the world, looking to see whether it may possibly be open again somewhere in the rear." [1]

Thus the final act of man's originally unbroken freedom is that in which freedom destroys itself. It is a delusion on the part of Idealism to think that it is the fall which, by unleashing the very first genuine possibility of decision, makes freedom possible in the first place. What is here made out to be the greatness of human freedom is in reality a sign of the bondage which triumphs over freedom. It indicates that man is no longer able to overcome the guilt-ridden cleavage which has come into being, that he is thus a prisoner of the anti-godly element which he allowed to take over, and which makes this cleavage his destiny.

That which he still describes as his freedom *now*, after the fall, is not the self-assertion of the "intelligible ego" over against "the other," namely, that essentially alien "sensible world." On the contrary, it is in reality the confrontation of the ego with itself. The ego has to admit that that "other" is really its own. The ego has to understand that it is itself standing in the camp of the opposition. Where God's radical command is taken seriously, e.g., where the command to live is made unconditional as in the demands of the Sermon on the Mount, man has not merely to admit that there is present something which is opposed to "him," to his true, intelligible ego. He has also to admit that this "something" is himself, that it is he himself who provides the opposition, i.e., he persists in what is basically a state of opposition because cleavage is the very form of his existence.

Thus man is here forced to make the self-identification which Paul

[1] Heinrich von Kleist, quoted in Erich Rupprecht, *Die Botschaft der Dichter* (Stuttgart, 1947), p. 153.

makes in Romans 7:17 ff. Here the apostle too, looking with "natural eyes," as it were, first speaks of the existence of a hostile factor which has the character of an "it" rather than an "I" (namely, sin). In the end, however, he finds that he must instead speak of that whole complex which is caught up in cleavage and opposition in terms of the first person, as "I."

We have therewith reached the point in our investigation where we must face squarely the question of how evil came into the world. As we turn briefly now to this problem, we would note that it is not and cannot be our task to offer a "solution," but merely to show that the question as such is wrongly put. In theology we constantly come up against problems which cannot be "resolved" at all, but from which we are "released" once we find a way to overcome the *situation* in which the problem arises.

One such question, it may be mentioned in passing, is that of predestination. This is a question which tends to give rise to anxiety about salvation, and to lead to the problem of theodicy. Even Luther's treatment of the problem of predestination does not "solve" it, in the sense of providing smooth answers to such questions as: "Does not foreordination destroy the divine righteousness?" and: "Does faith in Christ make any sense at all in the light of predestination?" Instead Luther argues in such a way that the questions lose their terror. He does this by setting the question of predestination in the shadow of an effected and appropriated justification, so that it is raised concerning the righteous and not the unrighteous. Hence we do not ask: "Granted that I am justified, can I still be saved?" The question is rather: "What has God done that I am thus saved apart from any contribution on my part?" The fact of salvation thus becomes the ground on which I ask the question.[2]

Now the situation is similar in respect of the question how sin came into the world; it is necessary to reach the point or level at which I am liberated from the question. The problem arises through the following line of argument. It is of the nature of temptation to have a point of access in man. Consequently the actualization of evil may be traced back to a disposition or potentiality for it, an inclination thereto

---

[2] See Martin Luther, *Preface to the Epistle of St. Paul to the Romans*, LW 35, 378; also *The Bondage of the Will*, trans. J. I. Packer and O. R. Johnston (Westwood, New Jersey: Revell, 1957), p. 313.

[*Anlage*]. If this is so, however, then evil is inherent in the act of creation; it derives from God the creator. Abstract consistency in the doctrine of *creatio ex nihilo* seems to lead ineluctably to this conclusion.

Now it must be said at the outset that we are here dealing with a problem which is insoluble from the logical standpoint. The reason for this insolubility is to be found, not merely in the fact that our concern is with a problem of origins, a problem whose metalogical structure we have touched upon already in connection with the problem of freedom, but also in the fact that we find ourselves here to be in the realm of the personal. In this realm the possibility of logical consistency breaks down. We may illustrate this by the connection between freedom and necessity, which is also logically insoluble but still highly characteristic of the personal. Freedom and necessity are not to be balanced one against the other, as if freedom could be regarded as a breach of the causal nexus, discernible by objective observation. On the contrary, freedom and necessity arise in two quite different perspectives, perspectives which do not stand in any demonstrable objective relationship to one another. Freedom arises in the non-objective inner world of the ego insofar as it is the active subject in the making of decisions. Necessity arises in the objective external world which I posit for observation. The same process is therefore from the one perspective a function of freedom, and from the other a function of necessity.

Suicide may afford a useful illustration of this. Statistically the annual number of suicides is remarkably constant, and the rhythm of their occurrence throughout the year parallels remarkably the rhythm of the seasons. Here is obviously a case of "macroscopic" necessity where the law of large numbers obtains. At the same time, though, each particular suicide is determined by quite distinctive individual motives. The person who decides to end his life is utilizing his freedom, and can accordingly be addressed with reference to the guilt of such an act. Who would even dream of committing suicide in order to round out the statistics? Such a grotesque idea could only arise through a confusing of the perspectives, whereby the freedom of the individual's own inner world is exchanged for the ordered and statistically bound objective world outside. But this is to forget that a person cannot be treated objectively, as a statistical factor.

It is because this is impossible, because freedom cannot be demonstrated in the objective world but is present only in non-objective subjectivity, that Kant describes freedom as a postulate and not as a category. Freedom is a quality of personal being, and this means that in the personal realm we are confronted by a phenomenon which cannot be explained logically and objectively.

The logical insolubility of these thoughts within the sphere of personal life reaches its climax in Jesus Christ, since he is the Person par excellence and in him the paradoxes of the mystery of personhood accordingly are deepest. It is along these lines that Hebrews 4:15 gives classical expression to this insolubility: "We have not a high priest who is unable to sympathize with our weaknesses, but one who in every respect has been tempted as we are, yet without sin." At this point objective observation comes up against an invincible difficulty, since susceptibility to temptation is itself always sin, the tempter having already gained a foothold in the ego which is tempted. Moreover, to overcome temptation is not necessarily to overcome sin. It is simply to hold it back, to prevent its implementation. Thus the phrase "yet without sin" is inexplicable from a logical standpoint. More precisely, it is inexplicable "from without." It is the mystery of the person of Jesus Christ himself. It is the mystery of his "inner world," of his deity.

In the same sense objective observation comes up against an insoluble problem in respect of Adam's susceptibility to temptation. "From without," the evil which is potentially resident already in the very fact of this susceptibility must be traced back to God, just as, for example, the temptation to suicide can be traced back to a certain law of statistics. Yet if this involves an insolubility in the objective sphere, we can fix logically the basis of the insolubility. It is to be found in the fact that objective observation is simply inappropriate where the personal is concerned.

These observations can suggest the way in which we ought to treat the question of how evil came into the world. I deliberately say "treat" rather than "solve," for to speak in terms of "solution" is to introduce logical categories and thereby to enter the inappropriate sphere of the objective.

We cannot—and this is the important lesson we have learned—treat Adam objectively, i.e., we cannot look at him from without. He must rather be interpreted from within the framework of that reality which is our own inner person. For Adam represents ourselves. He stands for me. Adam—*c'est moi.* On the basis of this starting point, then, we may advance the following propositions.

First, it is "I" myself who have broken loose from God; it is "I" who have set myself against him in thought, word, and deed. I cannot speak merely of an "it" within me (Paul's "sin which dwells within me," Rom. 7:17) that has enabled evil to come upon me. On the contrary, I must confess that I myself am the origin of my wickedness.

Second, in keeping with this, I cannot trace back either myself or the guilt which I bear to a causal nexus of which I am a part—more strictly, of which I am a part if I see myself from without (which would be illegitimately to abandon the perspective from the central point of my world and to surrender the zone of responsibility). This is why Adam cannot attribute his guilt to Eve, nor Eve hers to the serpent. It is of the nature of the personal, and of the inner perspective which it entails, that a person must always regard himself as the first cause. For our personhood is constituted, not by our being secondary to or subsequent to our own genesis or to other determinative causes, but by our being confronted with God (i.e., by responsibility). This is also why the reality of the demonic never appears in the Bible as a *cause,* but only as an expression of self-bondage. The demonic is an expression of self-love in the sense that I am always guilty of it and cannot overcome it, so that it takes on the dimensions of a superpersonal power to which I have subjected myself.

Third, it is in this dimension of the ego, therefore, that I find my will to be in bondage. But it is impossible that I should have received my being from the hands of God in this form. Even the freedom which I still have—that freedom which has its place within the conflict between the intelligible and the sensible ego—even this freedom is for me an expression of cleavage, of the most profound sickness, and hence a sign of the fall from creation rather than of creation itself.

Fourth, this fall from freedom can be expressed only as I am told and shown by way of retrospective prophecy, what was my unbroken relationship to God. But this unbroken relationship involves a situation that lies outside my historical existence in this aeon. For my history is something which always takes place within cleavage. Thus I am compelled to express in historical categories that which lies outside my history. It is for this reason that my thoughts on this subject, when they are projected on the plane of the logical where they are normally competent and at home, take on the appearance of distorted shadows. For they are forced to express in an objective way a non-objective reality.

Fifth, it thus follows from all of this that the question to ask is not how evil came into the world. I have rather to formulate the question in a wholly personal way [*quoad me*]: Why am I evil? But the answer to that question depends on me, for the mystery of person-hood is that the person cannot be traced back to something else.

We see, therefore, that the question about the origin of evil is wrongly put and consequently cannot be solved. Rather than trying to solve it we must let ourselves be set free from that situation in which it necessarily arises in this mistaken form. This situation is characterized, however, not by a wrong way of looking at theoretical problems, but by a wrong "practical" stance which necessarily brings with it the theoretical distortion. This practical stance consists in the fact that the interposing of the objective way of putting the question (asking about the causal origin of evil) carries with it the tendency to do away with one's own personhood, to provide relief by viewing the person as a mere effect.

We are thus confronted by a situation which cannot be corrected epistemologically because in origin the problem it entails is not one of theoretical epistemology at all. What is required is rather that we see the tendency toward self-exculpation which inheres in the situation and then be redeemed from it. In this connection redemption means that man can confess his guilt in its totality only when he knows the gracious God. Thus the radicalized *nomos* of the Sermon on the Mount insists that man is totally guilty, even to the thoughts and intents of the heart, but it does so only after there has first been proclaimed the manifold "blessed"-ness, only when it is certain that the redeemer is present, and the realism of full self-knowledge is therefore possible. For now this knowledge of guilt no longer kills. It leads rather to the confession, "God, be merciful to me a sinner." It becomes the "godly grief" which "produces a repentence that leads to salvation" (II Cor. 7:10).

With these thoughts concerning the person as a "first cause," who cannot be traced back to anything else, we have already advanced, and in part anticipated, essential insights regarding the phenomenon known as "original sin" [*peccatum originale*]. The term is a precise designation so long as we remember that "original" here refers not to priority in time but to that which is in the very "nature" of the case. The concept of "inherited sin" [*Erbsünde*] is misleading in this connection, since it fixes attention exclusively on the first sin, which is then transmitted to others.

It is obvious why something must be said at this juncture about original sin. For original sin implies a particular facet of sin to which our attention has been directed from several angles. I am referring to the logically indissoluble interconnection between the personal and the superpersonal—between our being the first cause of sin and our finding ourselves rather already enmeshed in it at all times from the very outset—and hence also to the question of imputability. At this point, then, where our concern is with the difference between our fallen state and the state of creation, and where there thus arises the question of the origin of it all, i.e., the question of the sin which transcends me and yet is traceable back to me, at this point it is natural that we should speak of original sin.

Original Sin and Responsibility

Original sin means that sin is not something that stands before us as an open possibility which we may lay hold of, but is rather something that is always with us from the very outset. The term thus denotes a state which is "prior" to all concrete acts of sin, and operative in them. To this degree the dogma of original sin expresses the fact that the subject qualifies the deeds, the state qualifies the act. It is for this reason that original sin must be imputed.

When we refer to original sin, then, we do not mean a second form of sin alongside actual sin. We mean rather a specific characteristic of sin, namely, its transsubjectivity. Just as there is basically only one Law—the command that we should love God and our neighbor—so there is, strictly speaking, only one sin—the breaking of this fellowship of love, i.e., unbelief. But this one sin must be seen in all its various aspects. The aspect which specifically concerns us now, as we turn to the problem of original sin itself, is that of transsubjectivity.

The Decalogue—as we shall observe more fully later [3]—being

[3] See pp. 440 ff. below.

couched as it is in predominantly negative formulations, addresses man in terms of his given situation, the state in which at present he always finds himself, and not in terms of a possibility which is still open to him: "You shall not kill, because you are a murderer!" We always find ourselves at the very outset to be constituted thus and so. I am not merely the object or result of my action; I am also the given prior subject. Whenever I approach something objectively (my work, my neighbor, etc.) I am asserting that I am myself the subjective presupposition for both the fact that I step forth out of myself in action and the manner in which I do so. What I am, I do. My nature is always the presupposition and directing force of my will.

This brings us a step closer to the problem posed by the fact that, while I act on my own responsibility, I do so under a compelling necessity. These two things must be simultaneously affirmed.

In the first place, what we are we simply are. We did not fashion for ourselves what we would be: male or female, revolutionary or reactionary, choleric or phlegmatic, ardent or frigid, activist or contemplative and lyrical. These are simply the various ways of being in which we find ourselves. Some say it has been "given" to me to be what I am; others say that is just the way I was "made."

In the second place, however, it is also true that, while I am what I am, I am that only as I take up this being of mine. I have a particular being only insofar as I—as a subject—continually actualize it, and so posit also the character of it. Suppose, for example, that I am prone to drink excessively. This characteristic of my being then is something I have received by way of heredity. I know my "innate disposition," only insofar as I actually implement it! If I were set down on a lonely island where there was no alcohol, my being as a drunkard would never be actualized, and therefore in the strict sense would not exist at all. Thus the state of a man is present only in his act. The possibility comes to exist and to be known only as it is realized. And the particular character of my person—what I am—has reality only as I continually implement and actualize it.

From this it follows that, whenever I do something, I must acknowledge as my own the quality of being expressed in that act. Indeed I am myself present in the act. The act is the form in which my being exists. My being is realized in the performance of the act. This is why Luther can insist again and again, in varied formulations, that the person is not to be evaluated "according to the works, but the works according to the person." [4] This means that the being of the person qualifies the works. It is the person which finds actualization in the works. It is the person which produces them. But since the person for its part is characterized by its attitude to God, and therefore ultimately by faith or unbelief, it is finally faith and unbelief which qualify the works. "As nothing makes righteous but faith alone, so nothing commits sin but unbelief alone." [5]

If we understand faith and unbelief as signs, one or the other of which will always characterize our state, then it becomes evident that the state exists only as it is actualized. For on Luther's view faith—or unbelief—is not just a *habitus* which need not step forth and manifest itself. Faith and unbelief are rather modes of existence which also express themselves, which either affirm or deny. They are a living by the truth or by untruth. In this sense Luther himself can never separate state and act. This may be seen already in the fact that he cannot regard justification and sanctification as two successive stages, as if justification were a given fundamental state and sanctification were the ensuing mode of action. On the contrary, justification and sanctification, as we saw earlier, are simply two sides of the same thing. The state of justification is present only in its actualization. Faith is present only in works. Without works, faith is dead, i.e., it does not exist. Faith cannot be regarded as a mere "possibility," a kind of otiose quality.

When this is taken seriously, light is shed on the relation which here obtains between transsubjectivity and subjectivity. As "I" am always the given prior subject, both of myself and of my action, so it is also true that I am this given factor only as I actualize and implement it; I possess my inherited dispositions only as I actualize them. Here it becomes clear what is meant by the fact that I am responsible for what I am. My being such as it is, is mine. To put the whole mystery in two simple words, it is that wherein "I" actualize "myself."

The mystery and the miracle of my personal being are expressed in this self-imputation which results from the ego's functioning as both

[4] See, for example, Luther's Large Catechism, IV, 12; BC 438.
[5] Luther's Theses of 1520 on the question whether works contribute to justification, No. 1, WA 7, 231.

subject and object. For within the biological and sociological order in which I have my particular position and aptitudes, within that order which runs from my parents right through me to my children and grandchildren, there suddenly arises a phase to which I must say "I." With respect to this phase I cannot take up the objective attitude of a scientist studying a law of nature. On the contrary, this phase is for me non-objective, for it is I myself.

We shall try to clarify these theoretical considerations by means of two illustrations, the one trivial and the other more lofty. We begin with the trivial. Men are inclined to call any of my unreflective, spontaneous reactions a form of activity for which I am not fully responsible but which simply takes place through me as my "nerves" respond to some stimulus. I am quite willing to accept personal responsibility when, for example, I deliberately and intentionally write a public letter or bring a public accusation against someone. There is general agreement that in such a case I am indeed present in my action. But things seem very different in the case of some spontaneous and unreflective action for which my "nerves," my physiological mechanisms of response, seem to be responsible and from which I as an acting and responsible subject am excluded. Is there, for example, any real affinity to sin, to responsible personal being, when someone in a crowd accidentally treads on my corn and I spontaneously punch the offender's nose in return? (Trivialities usually subject theological truths to the hardest test!) Is not this a purely physiological response which might be described as strictly animalistic in nature, a reaction in which "I" am not present and there is expressed only an instantaneous physiological reflex?

Now obviously this can hardly be my own view of the matter if I turn right around and apologize, and make no effort whatever to marshal medical evidence to the effect that it was only my nerves and that I was thus not responsible for the well-aimed movement of my hand. Moreover, it is worth asking why the nerve reflex is specifically directed at the other fellow instead of at myself (at the dentist's office I often dig my nails into my own flesh). Apparently this is due to the fact that I need a scapegoat when I experience rage or pain. I find within myself that primal urge—which is actually connected with what at bottom I really am—to locate outside myself and my own sphere of influence the cause of my acts and sufferings, of what I do to myself and of what others do to me. Small and stupid children are not the only ones who kick and pound the dresser they have stubbed their toe on. The phenomenon of the scapegoat is to be found not only in the case of Nero and other political bigwigs, though one could doubtless write a textbook of human history entitled "Scapegoat"; it can be seen equally well in the case of the bricklayer's apprentice who curses on the job. For the essence of cursing too does not lie in a spontaneous nervous discharge which as such is outside the personal sphere and hence beyond good and evil. As the vocabulary itself shows, the essence of cursing lies in the fact that I make God a scapegoat. I hit my finger with a hammer, and I curse God for causing the hammer to slip. The ladder which gives way beneath me, the man who inadvertently steps on my foot—these all stand in such close proximity to me, indeed I stand in such direct contact with them, that I cannot put the blame on them alone. To attempt that would be to bring myself in for a share of the blame. This is why I curse instead the one who is wholly other, who stands over against the whole apparatus of misfortune and pain, and who as the almighty must bear responsibility for it.

These instances point up the problem of the personal. In both cases (the misstep and the curse) what is involved is a spontaneous act of passion in which "I" am, so to speak, merely the channel through which there streams a particular causal sequence. Both are actions in which conscience, responsibility, or a deliberate plan of action play no role, in which there simply takes place, or seems to take place, on the territory of the ego, what is essentially nothing more than a physiological process of stimulus and response. How can I help it if somebody else does something on my property under cover of darkness? And yet it may be seen that, embedded in this unreflective reaction, a personal process does in fact take place: I take up a very definite, if subconscious, attitude toward the neighbor who gets punched and toward the God who gets cursed. "I" am thus the one who is all wrapped up in this act, "I" together with the entire course of my existence at its deepest level. Even in this phase of my conduct what is involved is a process to which I must say "I."

From this fact we may draw some far-reaching conclusions regarding the ethical evaluation of subconscious processes, for example, regarding the psychoanalysis of events which take place in dreams.

These too express the movement in which the person in his totality is engaged. There is no possibility here of disassociating myself from the "sin which dwells within me" as from something "other" and alien.

Besides this trivial illustration we would mention also the more lofty example of Michelangelo's painting in the Sistine Chapel, in which the breath of life is breathed into Adam. Here Adam is caught up in a process over which he has no control. He still lives in the darkness of unconscious, unreflective life. But his leg is already drawn up. He comes to participate in this process whereby life is inbreathed into him, only as he implements or actualizes the process by a movement of his own, by "standing on his own two feet." Indeed, this is what it means to become a living soul; it means to stand alone, to make oneself independent. The coming alive cannot be abstracted as a mere potentiality from this act in which it is realized. To put it another way, that which is done to Adam cannot be separated from Adam himself as the one who acts. As a living being, Adam is not merely the object or result of a process; he is also the acting subject.

Similarly, I am not merely the bearer of certain biological laws and processes or of certain acts of divine creation and destiny which in their performance move in and through me—I am also "I." These then are the essential points in the relation between subjectivity and transsubjectivity, and we may now sum them up and formulate our findings as follows.

First, transsubjectivity, as we have encountered it in the doctrine of original sin, cannot mean that a power comes over us from without. It means rather that we always find ourselves to be the very people we are. Nor can we separate ourselves from these given selves, for we are continually taking over and actualizing ourselves as we are. The person not only qualifies the works; it has its very existence in these works and only in them, only in the "outward movement" of the person, as Schleiermacher would say. Thus, even in relation to the transsubjective element we ourselves are still responsible.

Now this must first be both clarified and realized. That is, we must grasp and think through that which man unconsciously always knows, namely, that even concerning that which comes upon him transsubjectively he must say "I." Now man repeatedly denies this relationship intellectually; we play over and over again that famous scene from the story of the fall where man passes the buck. The

THEOLOGICAL ETHICS – FOUNDATIONS

reason for this is basically metaintellectual: we have a need to steer clear of any relationship which would threaten our very existence.

When the threat falls, and when conversely man thinks that he can do better by "saying I," he is usually able and ready without reservation to accept identification with the transsubjective element. He then begins to speak in terms of "endowment." For purely etymologically, this term expresses already the transsubjective element; here is something which I have not given myself, but which has been given to me. But by and by linguistic usage makes of that which was once given a quality of the man himself. And if someone says "You are gifted!" he blushes, for he feels flattered. When I feel flattered, though, it is because I think that "I" am being praised, and I assume that "I" am deserving of it. Thus man identifies himself completely with a gift which was only transmitted to him and from which in this instance he ought to maintain his distance, allowing himself to "give thanks" but not to "boast."

Second, transsubjectivity means something more, however. Once I am willing to see myself in it, as one who derives both from God's creation and from the fall, I come to realize that I thereby stand in solidarity with all other men. For here we find ourselves at a point where all distinctions of sex, race, and calling disappear. Beneath the level of all such distinctions there is this oneness in guilt, which is expressed in the fact that the human race is grounded in Adam and at the same time oriented to the second Adam. Mankind is represented by both Adam figures (Rom. 5:14; I Cor. 15:22, 45). We are thus confronted here by the remarkable fact that, while sin isolates me completely insofar as it is mine alone and I am wholly responsible for it, so that I can be represented by none other in respect of it, sin at the same time unites me with all other men in a way that transcends all other conceivable distinctions.

Third, from the two Adams we also learn that the problem of the relation between subjective and transsubjective recurs in the doctrine of redemption. For here the alien righteousness of Christ is imputed to us, and we are permitted to identify ourselves with him as our exemplar: Christ is my sin, my death, and I am Christ's life and righteousness.[6]

[6] For Luther's description of this identification whereby the believer becomes one person with Christ, see WA 40^I, 284-285; cf. Hans Iwand, *Glaubensgerechtigkeit nach Luthers Lehre* (Munich, 1941), pp. 60 ff. and 92.

Fourth, Luther provides the sharpest formulation of this problem of the relation between subjectivity and transsubjectivity, and of the resultant problem of responsibility, when he says in *The Bondage of the Will* that when man is left to himself he is abandoned to his enslaved will. However, this "transsubjectivity" does not imply any exculpating factor of "compulsion," but has rather the character of "necessity." Expressed in Goethean terms, Luther's idea is that it is the "entelechy" of man himself which here necessarily "unfolds," so that for all the compulsion that may be operative I nonetheless have to say "I" to that which does develop; I cannot dissociate myself from it.[7] It is my own will which manifests itself in that necessary opposition to God: "For constraint means rather (as one would say) unwillingness." [8]

We might perhaps regard it as the high point of the idea of imputation, the ultimate in identification, when Luther in this connection describes "man's will" as a "beast" which stands "between" God and Satan and which will be ridden by either the one or the other.[9] And if—when ridden by God—it "may" impute God's righteousness to itself as its own, then—when ridden by the wicked one—it "must" likewise impute his unrighteousness to itself as its own. In neither case is it a question merely of something happening to the ego, in respect of which "I" would be a mere object. In both cases what is involved is rather an appropriation by this ego, in respect of which "I" exercise the functions of subject. I "take over," as it were, the purposes of the evil one and make them my own, by fulfilling them. Terminologically this identification is emphatically underscored by Luther's use at this point of the term "man's will" rather than "I." This usage, which stresses the function of the "I" as subject, is similar to Luther's common use of the term "will" (in contrast to Scholastic usage) for the whole person, whereby testimony is given to the completeness of personal responsibility.[10]

The question of the relation between predestination and ethics, or predestination and responsibility, must likewise be handled along these same lines, in terms of the dialectic of subjective and transsubjective.

7 See *The Bondage of the Will*, *op. cit.*, pp. 102-103.
8 *Ibid.*, p. 103.
9 *Ibid.*, pp. 103-104.
10 See above p. 63, n. 21.

THE PROBLEM OF NORMS: (1) CONSCIENCE

15.

The Evangelical View of Conscience

There is some ground for the complaint of Richard Rothe that linguistic usage in respect of the term "conscience" is so vague and chaotic that we must declare the term to be scientifically useless and advise scholars to avoid it altogether; since its meaning cannot be fixed with precision, the term in fact does not say anything.[1]

For theological reasons, however, it is not possible to stop with the mere assertion of this ambiguity. Nor is it enough to round out the statistical presentation by piously asserting that the fall and the confusion of languages at the Tower of Babel have here, in this matter of conscience, achieved a linguistic masterpiece. While such a pious comment may have some truth in it, it does not get at the basic theological problem underlying the external facts. There is a more profound mystery concealed in this ambiguity concerning conscience.

The differences in the understanding of conscience clearly arise out of differences in the understanding of man. My view of conscience is determined by my understanding of what is the normative factor determining human existence, e.g., practical reason, utility, or the sociological or biological structure. Conscience is always incorporated within the framework of a particular anthropology. Hence it is always indicative of a specific self-understanding of human existence. To this degree the ambiguity of the concept is in fact a sign that the unifying center has been lost, and that there is now a wild and panic-stricken search for substitutes

From the theological standpoint, therefore, what is important is not that we should try to find the "most Christian" among the many extant interpretations and, after certain modifications, give it theological sanction. This is impossible anyway, if the ambiguity really bears witness to a loss of the unifying center. Moreover, it would be as ridiculous to use a comparative or superlative of the adjective "Chris-

---

[1] Richard Rothe, *Theologische Ethik* (2nd ed.; Wittenberg: Koelling, 1869-1871), II, 21.

tian" as it would be of the adjective "married." Our task is rather to seek a common factor in the various interpretations. This must in fact be possible, if it be true that it is the same fallen man who expresses himself in these diverse concepts of conscience. It is this common factor which we must discover and then seek to interpret.

Man in Dialogue

Now there is indeed an element which is constant among all the divergent views. It is the fact that in conscience there is always a dialogue between two authoritative selves. In conscience, man himself comes up against his own being-as-it-is [So-sein] to address it with the claim of an imperative. He may believe that this nature is instinctively evil or that it is fundamentally sound; and he may adjudge the imperative to be the voice of God, the practical logos, the criterion of experience, or the decadence of consciousness. In any case, there is always the same general formal structure involving two partners confronting one another in the ego.

If I am not mistaken, this is the view which underlies the argument of the Apostle Paul in Romans 2:15, where he speaks of the "conscience" of the Gentiles. For it is said of the Gentiles that "the work of the law is written in their hearts," and this is supposed to stop the Jews from "boasting" that they are the only ones to have the Law.

In what precisely does this witness to the "work of the Law" consist? It is described in two expressions which, in the manner of a hendiadys, really mean the same thing. First, the conscience exercises the function of a "witness" whose definitive testimony will be given on the last day; we may accept the exegesis of Hans Lietzmann that on that day conscience will leave the Gentiles too "without excuse" (Rom. 1:20).[2] Second, the function of conscience is discharged in such a way that our "thoughts accuse and excuse one another." Conscience means that witnesses stand up against one another, that there takes place in man a continual fluctuation between accusation and defense, that there is a contest of opposing "parties."

The essential point theologically seems to be that we have here an assertion and a theological clarification of the cleavage which, for

[2] Hans Lietzmann, Einführung in die Textgeschichte der Paulusbriefe an die Römer (3rd ed.; Tübingen: Mohr, 1928), p. 41.

all the variations in which it appears, is everywhere seen to be the formal structure of conscience. The impression of cleavage is further strengthened by the fact that conscience does not have here the role of judge. It is not a unifying factor. On the contrary, it is always one of the contending parties. God alone will discharge the function of "judging" in his own day. There is not the slightest indication that conscience has a proleptic judicial function. (It has this only in a very secondary sense which we shall have to describe later.) Conscience is merely the counsel for one of the conflicting parties, though it is never made clear which one this is.

One thing at least is clear. If conscience is a sign of cleavage, if it is one of the parties which God will judge at the last day, then it cannot simply be equated with the voice of God. It is rather a witness to cleavage, and as such carries with it a reference to the fall in which the unity of creation, the unbroken relationship with God, is destroyed. To this extent conscience is itself a witness to that state of the ego which Paul attests in Romans 7:17 ff.

The function of conscience in the Old Testament is delineated in a way which is wholly in harmony with these statements of Paul, though there is in the Old Testament, of course, no special word for conscience, the reference being instead to the "heart." Very typically, the idea of conscience is present for the most part in passages which refer to some sin that has been committed and to which the conscience is opposed, where again it is a question of the cleavage that has occurred. David's "heart," for example, smites him when he knows that he has done wrong (I Sam. 24:6; II Sam. 24:10; cf. I Kings 2:44; 8:38; Deut. 8:5). This is the point at which the cleavage becomes perceptible—in the heart.

The most profound theological interpretation of these aspects of conscience is to be found in the well-known saying of Luther that conscience is first my defender against the divine accusation, and then—in the Christian state—the accuser before which God himself undertakes my defense (*Deus accusator, cor defensor—cor accusator, Deus defensor*). It is true that we have here a different view of conscience from that of Paul. Prior to faith, conscience for Luther is not the bearer of the imperative. On the contrary, it is the defense counsel which opposes the imperative of the divine Law. It can be that, however, only because the natural conscience is the representative of man, of autarchic man who is grounded in himself and isolated from God. To use Kantian language, conscience adjusts its demands to what man can do ("You can because you ought"), and it defends him against demands which would go beyond that. But then this same conscience, as the conscience of the believer, takes sides with God; it is, as it were, conquered by God, and so joins the attack upon the fortress it had formerly defended.

In this connection it is important to note the way in which Luther brings out the fact that conscience is always one of the contending parties. It is never neutral. It can only change sides. It can never be an impartial judge. Hence conscience itself is one of the most profound expressions of the cleavage within man.

The Unrest of the Unredeemed Conscience

The unredeemed conscience is characterized by the fact that it is restless, uncertain, and insecure. In the Bible this is expressed particularly by the fact that conscience is present as an oppressor and accuser, even where the true and radical nature of the accusation—the fact that it is God's accusation—is not yet perceived.

By way of illustration we may recall not only the Old Testament passages mentioned on page 300, but also Romans 2:15, where the fact that conscience bears witness to the Law of God is, as it were, less important than the way in which it bears this witness. The accusation and defense which take place in conscience are here understood as a prelude to the last judgment. It is at the last judgment that the accusation presently in the foreground, the points and purposes of which are quite false, will be overshadowed by the hidden realities in the background which are then brought to light, namely, the secret but real theme of the whole judicial proceeding. For the moment, conscience knows nothing of these hidden realities. It could be that among these secrets yet to be revealed is the situation of conscience itself, the fact that conscience is in the first place an expression of the cleavage in fallen man, and beyond that, through its defense activity—e.g., by referring to the productive and teleological purpose of the cleavage! (Hegel)—it is not only the prosecutor but also the defender of fallen man.

At any rate, there can be no doubt that conscience is disquieted, and that it is a disturbing force. For on the one hand it anticipates the accusation of the final judgment, and on the other hand it tries at the same time to defend and to excuse itself against this accusation. Either way, therefore, it stands wholly in the shadow of this last judgment, and is consequently a sign of disaster. All this constitutes its restlessness and disquiet. It is, in fact, a wholly ambiguous phenomenon. It points backward to the cleavage introduced at the fall, and it also points forward to the day when the ground of this cleavage—on which, and on the assumption of which, conscience itself operates—will be called in question, and in the most literal meaning of the term brought under "crisis," i.e., judgment.

The Self-Pacification of the Unredeemed Conscience

Since conscience is thus a sign that we are wounded by God, it seeks in passionate acts of autotherapy to heal this wound and to attain to rest. The means readily at hand for this purpose are those of the Law. They consist in good works and sacrifices. Their purpose is to free us from the accusation through attempts to attain moral and cultic righteousness. They do this by seriously trying to "make satisfaction," i.e., by trying to seize the initiative in the relationship between God and man. Now in this concern for satisfaction there is a threefold error.

First, the purpose of works and sacrifices is to make good a deficiency, and hence to restore something. What is overlooked, though, is the fact that we are indebted to God as it is for all that we have and are; hence there are no additional funds on which we could draw to make such supplemental restitution. Anselm has already shown how impossible it is to find anything, beyond what we already owe to God in any case, whereby we might repay and make good what we have stolen from him.[3] "No human being will be justified in his sight by works of the law" (Rom. 3:20).

Second, on the level of the Law, the relationship to God is ordered on the basis of the principle of reciprocity: "As I to you, so you to me." Herein is expressed the typical attempt of the Pharisees (Luke 18:9, 11 f.; cf. also Matt. 19:27; 26:33) to take God by the hand, i.e., to become his partner, and to disguise this attempt to "be as God" by works performed in his service (Matt. 23:25 ff.).

Third, a further crisis of all morality, and of all the righteousness

[3] *Cur deus homo?*, Chapters 20 f.

302

sought through sacrifices, consists in the fact that in these works of repayment or restitution man does not give himself. Instead he lets himself be represented by his sacrifices or good works, that is to say, by his merits. Instead of being reconciled with his brother and so giving "himself" to him, he prefers to omit this self-giving and take a gift to the altar (Matt. 5:23 ff.; Mark 11:25). Instead of realizing in his own life the sacrifice and obedience demanded by the prophets, he prefers to adorn their tombs (Matt. 23:29 ff.). Instead of coming into contact with God by loving his neighbor (which is again self-giving!) he prefers to perform the sacrifice of a long prayer (Matt. 23:14, which is textually uncertain; Matt. 6:7). It is for this reason that Amos attacks liturgical songs, festivals, and sacrifices which men have degraded to the point where they are mere substitutes for the offering of one's own heart and serve merely as representatives for man himself (Amos 5:21 ff.; cf. also Isa. 1:10 ff., and esp. Isa. 15; 59:3). When man does something different from what he is within he becomes a hypocrite (Matt. 23:13, 15, 23, 25, etc.).

Augustine once said, "What thou doest without joy [i.e, legalistically], is only done *through* thee; thou *thyself* doest it not." [4] In other words, you merely cause yourself to be represented by your action as a substitute. Only he who cares for widows and orphans, who loves God with his whole heart and his neighbor as himself (Matt. 22:37; Mark 12:30-33; Luke 10:27), only he really gives himself. But he can give himself only as he knows that he is himself loved by God, and sees in his neighbor the veiled Lord (Matt. 25:35 ff.). He cannot do this if he tries first of all, by means of satisfactions, to draw the attention of the divine love to himself, and to merit it. For to do this would be to treat God as a judge rather than as the one who wishes to love me and grant me his grace. To treat God as judge, or better, to provoke God into judging—which is what actually happens within the sphere of the Law—can only mean that I must then indeed stand in the judgment, precisely where I cannot "stand" at all because I am required to stand there quite alone, without any satisfaction to offer.

In all three respects the restless conscience assumes an attitude towards God which does not do justice to him, and which conse-

---

[4] See Augustine's commentary on Psalm 92:3 (*Enarrationes in psalmos*, XCI, 5), in *Corpus Christianorum, Series Latina* (Turnholti: Brepols, 1953——), XXXIX, 1282.

quently does not do justice to itself, and for this reason cannot bring any true rest. If we try to serve God on our own, and of ourselves to reconcile ourselves to him, without first being served and reconciled by him, we can, so far as our conscience is concerned, know no fulfillment and hence also no peace.

The Epistle to the Hebrews deals explicitly with these connections between the disquieted conscience and justification. With its help the points thus far developed in principle may now be supported from Scripture. Hebrews 9:8 ff. talks about the service of the Law being like the "outer tent" which prevents access to the Holy of Holies. This is said to be "symbolic" of the present situation ("the present age") in which gifts and offerings are brought which are quite incapable of "perfecting the conscience" of the worshiper, i.e., of giving him a good and peaceful conscience.

So long as man tries to force his way to God by means of sacrifices and gifts, he remains imprisoned in the "outer tent" and cannot enter the true sanctuary. To put it non-figuratively, man moves on the level of Law and so falls victim to the curse of that very principle of reciprocity which he was himself trying to use for his own benefit. Within the schema of Law the conscience is constantly engaged in the quest for peace—for Law, cleavage, and restlessness of conscience belong together—but peace is no longer to be found here. It can be found only outside, beyond the Law's field of force.

Hebrews 10 deals with the erroneous sacrificial ministry which is tied to this schema, which constantly tries to escape from it, but which merely chafes against the bonds and thus makes its plight worse. The essential mark of this futile attempt to escape the disquietude is the fact that sacrifice can never cease (Heb. 10:1-2): "For since the law has but a shadow of the good things to come instead of the true form of these realities, it can never, by the same sacrifices which are continually offered year after year, make perfect those who draw near. Otherwise would they not have ceased to offer (those sacrifices)? For if the worshipers had once and for all been cleansed, they would no longer have any consciousness (or conscience) of sins" (cf. also Heb. 10:11).

The daily sacrifice in the forecourt is thus a sign that unrest is not dispelled, that the flaming fire of conscience must be quenched again and again, and that in fact what we really do is pour in oil and cause it to burn the more fiercely. (The way this actually happens is illustrated in Luther's conflicts in the cloister.) This is why the redeeming work of Christ—precisely as regards this unrest of conscience—is contrasted with the daily ministry of the priests in the outer tent, and compared instead with the cultic ministry of the high priest who annually enters the Holy of Holies but "once." The definitive ministry of Christ is accordingly described as "once" [ἅπαξ] and "once for all" [ἐφάπαξ] (Heb. 9:28, 12). God's justifying action in Christ and the consequent pacification of conscience are characterized as being once-and-for-all. By contrast, the human struggle for peace of conscience—in the religious sphere through sacrifices and in the moral sphere through good works—is marked by unceasing repetition and therefore by permanent unrest of conscience.

The Unrest of the Christian Conscience

The same unrest of conscience may also be found of course in the Christian. It does not belong exclusively to the pre-Christian stage of the natural man, man under the Law. Even if there were no New Testament statements or illustrations in support of this truth, we should still be brought up against the problem by pastoral observation both of ourselves and of others.

In the first instance, the alarmed conscience is always the state out of which we come. It is a sign of that state which, in our discussion of the "accusing role" of the law [usus elenchticus legis], we described as that of the "open wound" which is not to be healed.

In the present context, however, we do not refer to this form of the alarmed conscience out of which we thus come and, because of the tension of Law and Gospel, always must come. What we have in view here is rather the pathological form of alarm out of which we not only come but into which we are plunged again and again, that alarm which can take the form of doubt of God's mercy on the ground that our sin is "too great" for it to cope with. This is the alarm which feeds on the tension into which we are betrayed when our regard constantly vacillates, in a spirit of distrustful and despairing comparison and evaluation, between the burden of our guilt and misery on the one hand and God's forgiving mercy on the other, and when the vacillation finally ends with our gaze fixed on the greatness of our own burden. This form of alarm can manifest itself as a disturbance of the assurance of salvation, and consequently as devilish rather than godly grief, as the "melancholy" of Kierkegaard, the inevitable dissatisfaction of a heart that is not pure but divided, that does not will *one* thing—and hence ultimately as the depression of unbelief. This will suffice, then, to fix the concrete and autobiographical reference of this alarm.

Now how does the New Testament answer the question as to the origin of such alarm? It suggests that it is undoubtedly related to a basic disorder, an ultimate cleavage. The decisive passages concerning that disordered state of the Christian which finds expression in an alarmed conscience are to be found in I Corinthians 8:7-13 and 10: 27-29. The point at issue here is that of meat which has been offered in sacrifice to idols. In what follows I shall deal with the verses which are most important for our present theme.

"If one of the unbelievers invites you to dinner and you are disposed to go, eat whatever is set before you without raising any question on the ground of conscience" (I Cor. 10:27). What this means is that your conscience need not be involved here. It can be disinterested and consequently quite at ease since the idols whose sacrificial meat is set before you in truth do not even exist. For belief in idols does not have its root in the existence of false gods; it is merely a projection of the wandering heart which has fallen victim to blasphemy. For this reason idolatry is really belief in demons, since the demonic consists in the decking out of certain spheres of creation or particular created objects (wood, gold, etc.) with the attributes of divine majesty. The idol in itself is nothing. It becomes something only through us. It is, as it were, an objectivized symbol of our allegiance. As such it has only such power as we give to it— but that power it does in fact have!

There are two reasons why idols may be dismissed as nothing. The first is that God alone is Lord, and there is none other beside him. The proclamation of this objective fact, however, is not enough. For, after all, the ungodly are delivered up to the demonic because they make substitute gods out of created things and so become subject to their power (ἐξουσία). Hence there is a second reason, namely, the fact that in faith in the Father of Jesus Christ—in that new field of force into which we have entered—we do not need these substitutes any longer; we can no longer be the victim of any false ἐξουσία.

This is why as Christians we no longer need to take seriously the problem of eating meat sacrificed to idols. For objectively such meat is nothing—it has no magical or demonic power—and subjectively we are delivered from the curse of idolatry inasmuch as we are now related in faith to the one who alone really does exist. This is why, when I return from the mission field, I can place a Buddha or a hideous African idol mask in my study as a memento, without the slightest qualm of conscience.

We quote again: "But if some one says to you, 'This has been offered in sacrifice' then out of consideration for the man who informed you, and for conscience' sake . . . do not eat it" (I Cor. 10: 28-29). That is, if someone comes into my study and points to the Buddha or the African mask and tells me that he believes in the magical symbolical force of these images, that he believes in the materialization of an "idea" as an ἐξουσία just as he believes in icons that "contain" the holy, then I should remove that memento as quickly as possible. For "not all possess this knowledge. But some, because their conscience still cleaves to the idol unto this very hour, eat it as meat really offered to an idol; and their conscience, being weak, is defiled. . . . Take care lest this liberty of yours (namely, to treat idols as nothing, and consequently to eat the meat without religious scruples) somehow become a stumbling-block to the weak. For if any one sees you, a man of knowledge, at table in an idol's temple, might he not be encouraged (literally "edified"), if his conscience is weak, to eat food offered to idols? And so by your knowledge this weak man is destroyed, the brother for whom Christ died" (I Cor. 8:7-11). He is not destroyed because the idol is something after all. He is destroyed because he treats the idol as a real power, and because the demonic element in idolatry consists precisely in this, that that which in itself is nothing is seen to be decked out with the power of God, and so is actually made to be the bearer of this power. The demonic here is not an objective entity; it is an act, namely, an act of investiture.

The advice of Paul may thus be summed up as follows. If someone really regards the meat offered to idols as "something," because he has not yet developed in his Christian faith to the point where he can despise false gods, then he should be protected lest his conscience be unsettled and he be plunged into "melancholy." (The reference here is indeed to the presence not merely of heathen, but also of immature Christians.)

We can now understand rather more precisely that disorder in the Christian life which is accompanied by alarm of conscience. It consists in the belief that idols are nonetheless a reality, and that to eat meat sacrificed to them is thus to renounce God (I Cor. 8:4-5) and to confess demons. The conscience is thus unsettled by something which supposedly stands between God and us, occupying the place which belongs to Christ alone—for it is he alone who gives peace of conscience with his once-and-for-all redemption. Hence unrest and alarm of conscience arise simply from the fact that the comprehensiveness of Christ's lordship is not fully recognized; this is what gives rise to unredeemed forces and spheres of unrest. Hence the unrest, the "melancholy," is once again a sign of cleavage. It indicates that the will does not will *one* thing alone, that alongside "the one" reality others also exist as ἐξουσίαι.

As a result, unrest of conscience arises not through actual sin, in face of which there can never be a "good" conscience, but through failure to fix one's hope wholly and exclusively on Jesus Christ as Lord. Accordingly, the opposite of an unsettled conscience is not a good conscience but a conscience which is comforted and consoled, a conscience which is reconciled to God.

In the light of this we can see that the basis of the unrest, the insecurity, the "melancholy" of conscience in Corinthians is similar to, indeed at bottom the same as, the basis of the unrest of the unredeemed conscience in Hebrews. In both cases the cause of the unrest of conscience, and therefore of its weakness, is that man will not let God be Lord, the one and only sovereign Lord.

In Hebrews, the fact that man will not let God be Lord is seen supremely in man's attempt to placate God by sacrifices and make God like him. To this degree man takes the pacification of conscience, partially at least, into his own hand. The hand of God is not allowed to operate—that hand in which alone this peace may be had inasmuch as Jesus Christ, who symbolizes it, is alone "our peace" (Eph. 2:14). For Hebrews, the time when the age of Christ was foreshadowed was a time in which the grace of God was not yet fully revealed. It was a time in which there was still a place for the liturgical action of man, though only for that which God himself had authorized.

In I Corinthians, the dethronement of God and the consequent unrest of conscience are due to the fact that men see other deities alongside God, and so think that God is not strong enough to take us by the hand and cause his kingdom to come. The gods stand in the way and dispute his power. (This is why superstitious semi-Christians are often more stricken and stirred in conscience than full-blooded neo-pagans. Nothing is more unsettling than "lukewarmness," trying to have it both ways at the same time.) It is somehow up to man himself, we think, to see that these forces do not gain influence over us or over our relationship to God. And man accomplishes this presumably by not serving them, not eating their meat, not accepting public office (for fear of the demon "power"), not becoming entangled with money (for fear of the demon "mammon"), not going to the theater (for fear of the god "world").

But all this avoiding of things, this non-eating, this keeping one's skirts clean, does not bring about the deserved peace and rest anyway. For implicit in these evasions there is a secret recognition of the very powers which we are trying to avoid. So it is that by a negative work, by a withholding of sacrifice, we try to insure that our relationship to God will not be broken, that these powers, these usurping lords, will not come between us and God. Here too, then, the "work" acquires the significance of a *sine qua non*. If despite all this Paul still desires that the consciences of such anxious spirits and "negative" activists should be spared, it is because a demonstration of Christian freedom could not help them. Indeed, it would only confuse them. For they have not yet grasped the basis of this freedom. The fact that God alone is Lord, and that he has dethroned all other powers is still concealed from them. Hence we deny the lordship of God when we secretly reckon with other powers and deities, thinking we can overcome them and attain to God by means of service and sacrifice, or even the withholding of such service and sacrifice.

Thus in both contexts the reference is to the unrest of the unredeemed conscience. In I Corinthians, the conscience is afraid of gods and demons, and hence lives in constant anxiety lest by its service and sacrifice it call them to the scene. In Hebrews, the conscience holds the "outer tent" of the Law still to be valid alongside the once-and-for-all reconciling act of the high priest. It would like to find security in the sacrificial cultus and in moral perfectionism but knows deep down that by this means, i.e., by the simultaneous requisitioning of the "outer tent" along with the Holy of Holies, it cannot please God or drag him over to its side.

These are the two sources of unrest in the conscience, as the pastors of any period could well testify. The conscience is not serene or troubled according to what we have done or not done. Peace of conscience depends solely upon what we *are*, i.e., on whether we believe—and the extent to which we believe—in the boundless unconditioned mercy of God. Pastors would do well to consider the concrete implications of this for their ministry.

It is theologically wrong to try to pacify a conscience-stricken person by talking away his sins. To do so is to try to cure him by means of the "outer tent." But there is no healing here, and cannot be. In fact the heart of his problem is that he is still loitering in this forecourt. The only way we can help is to point him to the ἐφάπαξ, that which took place once-and-for-all for him in the crucifixion of Jesus Christ.

It would appear to be not only permissible but even obligatory to apply this same principle to psychopathic guilt feelings. For in face of this type of sickness it is both medically useless and theologically wrong to "prove" to the patient that his sin is imaginary or illusory. The forecourt cure cannot help anyone. Indeed it is of the very essence of the unredeemed conscience to have lost all sense of proportion, to stumble at imaginary and gigantically inflated difficulties (e.g., the illusory gods), and to be overthrown by blades of straw and the pettiest of sins. Now there is no point in trying to talk out there in the forecourt about this lack of proportion in hopes of clarifying, as it were, the sins committed and reducing accordingly the disquietude of conscience. A proper sense of proportion is a matter of objectivity, and hence of standing off from oneself at a proper distance. But this is possible only in the peace of forgiveness. For only thus can we stand back and look at our lives and judge them in the peace of those who have been snatched from disaster and are now safe in the shelter of forgiveness. In the moment of panic, however, i.e., in the time of the unredeemed man, there is no clear vision, and one is not open to address from without.

We conclude by considering one further fact of great importance.

When we said that the natural conscience—the conscience of the sinner who is not justified—is troubled, this must not be taken to mean that the unrest always assumes a psychical form, that the natural man inevitably gives evidence of visible panic. We know that the reverse is often true. The genuinely agonized conscience is found mostly in the no man's land between Christ and the world, between the Gospel and the Law. It is found especially in that sphere where the call of the Gospel has been indeed perceived but not wholly accepted, where it has for that reason again been turned into Law, where it has become a "Gospel" which is linked with conditions, which is bound to a kind of legal *imitatio,* or which presupposes some kind of asceticism. Wherever the Gospel is robbed of its unconditional and total character, it inevitably and immediately becomes Law, and hence a yoke which is neither easy nor light (Matt. 11:30).

Outside this intermediate zone, however, there in the sphere of the natural man, alarm of conscience is characterized by repression. In the final analysis the ethos of the natural man cannot be understood at all except in terms of this rubric "repression." For the goal of his striving is security, and his means are works and sacrifices. By orienting his ethos in the direction of fulfillment the natural man seeks to justify and vindicate himself so far as his destiny is concerned. To be sure, even he usually has some awareness of the fact that there can never be complete congruence between what he is and what he ought to be, that he will always be at some quantitative remove from the norm. But the resultant remnant of unrest will then be progressively reduced by further works and offerings—at least his efforts will be in that direction—and he will take comfort in the Faustian consolation that the attainment of a goal is not so important as the incessant attempt to reach it. Now it is this very outcome that brings the natural man into sharpest possible antithesis to the ἐφάπαξ of Hebrews, for the point of this ethos is permanent repetition of the process. Works and offerings do not thereby denote the overcoming of the alarmed conscience; they indicate its repression.

One of these acts of repression is to describe as absurd the radical demand of God which reaches down even into the very thoughts of the heart, and to regard oneself as ethically responsible and culpable only within the sphere of overt action. Repression in this sense also includes vindication of the ethical conflict as such by regarding it exclusively in terms of its character as a sign of personhood and of the dignity of personal existence.

In this connection we would at least suggest that, contrary to the way it is usually treated, "repression" actually be made a theme of theological ethics. For from a theological standpoint repression is the characteristic feature of the natural man, not merely as regards sin and anguish of conscience, and not merely as regards the truth of God which is repressed or "suppressed by wickedness" (Rom. 1:18) but also as regards anxiety. Psalm 39:6 says of those who are solicitous about themselves that they are in turmoil "for nought." But this unrest which finds expression in the heaping up of wealth has as its very purpose the concealment and repression of unrest and the bringing of security. We can see this in the case of the rich farmer of Luke 12:16-21, whose anxious filling of his barns was obviously designed to give security (v. 19). He wanted to have solid ground under his feet, but his feet did not know the eternal foundation since he was poor in God (v. 21).

We cannot simply measure what may well be the balanced psychical state of the natural man by some presumed psychical standards of biblical anthropology and then conclude that there are instances of a phlegmatic state of equilibrium rather than the erratic curve of feverish emotion. On the contrary, when we find a state of relative calm, we are to see it as the product of repression. There is nothing in the world more psychically composed than the ordinary Joe sitting at his favorite bar. But what is it that has caused him to take refuge in a glass of beer? Is not his drinking—to use the terminology of Schiller—more often "sentimental" than "naïve"? Is there not frequently in the background, or at least may there not be, a very patent unrest, a need for security, a flight from futility and meaninglessness?

This precisely is the situation of the complacent ungodly who are "in no peril of death but stand strong like a palace," who are "not plagued like other men" and whose "cockiness is supposed to be a precious jewel." [5] Most atheists are very confident; at the intellectual level they are particularly sure of their unbelief. Why is this so? It is because there is in fact no such being as the stormy Prometheus, wildly protesting with his mythical gesture of the upraised fist. The demythologized Prometheus—who today is the average man or the

[5] [Ps. 73:4, 5, 6 according to Luther's translation.—Ed.]

sophisticated intellectual—does not threaten God with his fist; he does not make God the theme of his unrest and melancholy. For that would be to admit that there is a conflict between God and man. Instead he simply reasons God away altogether. Indeed, even to put it in this way may perhaps be misleading, for to speak in terms of reasoning God away is to suggest that God was first of all taken seriously as a hypothesis, and then dismissed. What is actually involved is rather, as exegesis of Romans 1 teaches us, a kind of prophylactic repression, a refusal even to allow any discussion of God at all.

In the terminology of Luther, one might describe these states as follows: Unrest which is repressed leads to "false security" (*securitas*); unrest which is overcome means "peace" (*pax*). "False security" is what results when my nature takes measures to help itself; it leaves me still in the sphere of my inner ego. "Peace," on the other hand, describes an external relationship of the ego to an entity other than itself, namely, to God. This relationship then has a subjective side involving a change "within," namely, the pacified conscience.

Conscience in the Light of Justification and Sanctification

In our discussion we may be brief, since we have already made the essential points from a negative standpoint in our treatment of the unredeemed conscience. The following are the main considerations in respect of the interrelation of conscience and justification.

Not only is conscience in no position to attain to rest by means of the demands it imposes; it is unable even to receive forgiveness, that "peace" which is imparted to it. For the peace of forgiveness is not something which fits smoothly into the structure of conscience; it rather tears it all to pieces. Unrest of conscience cannot simply be done away by the offer of forgiveness. If it is to be pacified, conscience must learn to understand itself in a wholly new way; it must "die," as it were. Like the natural man himself, the unmortified conscience can only stand uncomprehending before the proffered forgiveness. Indeed, it senses in this very offer a threat which runs counter to its own instinct for self-preservation. For conscience, as "moral reason," [6] simply cannot grasp, by way of rational deduction, what takes place in forgiveness, in justification. That God should open his heart to me is absolutely unexpected. It is the miracle par ex-

6 *Ratio moralis;* WA 40I, 410, line 7.

cellence. Rational deduction, for example, in the form of Kant's postulate of God, [7] can think of God only as the author of the Law, and hence also only as the judge who decides whether his law has been fulfilled. If this leads to anything like a sense of God's nearness and friendliness, it can only be in respect of those who fulfill the commandments, not in respect of sinners. Therefore, if conscience is to subject itself to the miracle of the remission of sins, it can do so only by declaring the dialogue between the accusing and excusing "thoughts" (Rom. 2:15) to be unessential, indeed invalid. It can do so only by leaving, closing, and locking this whole courtroom, by forsaking its own fatherland, its own true home, and by going out into a foreign country—in short, by letting go altogether of what it once was.

Justification can thus take place only in face of the accusation of sinfulness which is continually brought against me from the sphere of the conscience. It can take place only in face of conscience itself, only if we "lift up our eyes in spite of sin and call upon God in defiance of conscience." [8] I must believe in spite of the moral assault of conscience, in spite of my bad conscience, even in spite of its postulate of God. For, within the framework of this postulate God would seem to be unrighteous if he is the friend of sinners. The schema of righteousness which conscience espouses is inevitably forced to negate, and hence to describe as unrighteous, anything which transcends or even opposes its own categories. This is why I must not listen to conscience. It will deceive me concerning the miracle which God wills to perform for me.

But what happens to conscience when the miracle of justification actually takes place, when conscience with its legal postulates is, as it were, vanquished by God and we make bold to cling to God's gracious promises and assurances "in defiance of conscience"? If I interpret him aright, Luther describes two aspects of conscience in the state of justification.

First, conscience finds rest in the acceptance of the divine promise, that promise which addresses itself to me and assures me that I personally am the one intended [*quoad me*]. It is precisely this personal reference and application of which conscience could have no assur-

[7] Kant, *Religion within the Limits of Reason Alone,* trans. Theodore M. Greene and Hoyt H. Hudson (Chicago: Open Court, 1934), p. 180: ". . . obedience to all true duties as divine commands."
[8] *WA* 5, 93.

ance within its own schema of righteousness. "Conscience is unable to pacify itself; it has peace only through that grace." [9]

Second, conscience has this peace only as it is constantly spoken to it afresh. "Peace" is present only in act. It never has the form of a given *habitus*. For conscience is still a fountain of unrest in me. It is still a court which continually addresses me in terms of the *usus elenchticus* of the Law. It is still a prosecuting conscience (*cor accusator*). It hits me with the truth that I am in actual fact a sinner [*peccator in re*]. That is why it is conscience, and not just the lower impulses, which for me constitutes a constant source of perplexity and anxiety [*Anfechtungen*]. It sides with God's Law against me. It represents, as it were, a branch office of the divine prosecutor right on the territory of my own ego.

These assaults of conscience now assume, of course, a very different meaning from what they had before. Formerly they portrayed the love of God for sinners, his love for me in particular, as an utter absurdity, indeed as an offense against every imaginable concept of God. Now that I stand within Christ's field of force, however, these assaults necessarily drive me into his arms, just as the Law which accuses me can—in the state of justification—serve only to make me flee from its accusation to the declaration and promise of the Gospel. It is like the gauze which has to keep the wound open, as a constant reminder of the true doctor. The accusing voice of conscience thus serves my own mortification. It startles me and sets me running, so to speak. But it is God who turns the course of my run in the direction of his mercy. In fact, the great change which takes place in the function of conscience is that the accusation of conscience now becomes God's own way of "testing, prodding, and driving a person" to even greater faith in forgiveness. [10]

What does it really mean that conscience impels to mortification of self? It means that God will not let conscience be merely the voice of the devilish accuser and tempter. Instead God uses this voice of conscience to hold before me something which I always need to have before me if faith is truly to look "away from itself" and be "faith in ." Thus Luther says in his 1519 Commentary on Galatians with reference to the question of assurance, whether one can be certain

[9] *WA* 40I, 74.
[10] *LW* 35, 19.

that he is in a state of salvation: "Take care that you never become uncertain but are always perfectly sure that of yourself you are in fact lost!" [11] Thus there would seem to be here an indissoluble connection: I must die to self and despair of self if I am to live in Christ. Uncertainty as to whether I truly live in Christ and have my salvation in him must—paradoxically—be corrected by the negative certainty that in myself at any rate I am lost.

Uncertainty of salvation thus derives from the absence of this negative presupposition, the absence of despair of self. In other words, uncertainty is always the result when I have not yet fully written myself off, when I think that I can still make some small contribution to my own salvation, when I believe that I have a dignity of my own which God must respect. To try to underwrite a positive certainty of salvation on the basis of such arguments about what I am and can do is a mistake; indeed, it always leads to the very opposite result. For even in the form of the most sublime kind of co-operation, trust in myself is always inimical to my salvation, and consequently also to any certainty of salvation that I might have. Uncertainty actually increases to the degree that, in attempts to overcome it, I shift attention to the plane of my own ego as the foundation of certainty. For this plane is slippery ground. Paradoxically, then, it is the certainty that I am lost, it is the abandonment of this slippery foundation, in other words it is the mortification of self which is the presupposition of faith. This presupposition, of course, is not to be understood in the temporal sense of "pre," for in the last resort it is faith itself which first creates this presupposition. Only when I have Jesus Christ, i.e., only when I know what it cost God to help me, only then is the full extent of my lost condition disclosed.

This co-ordination of mortification with faith is nothing other than the co-ordination of the death of Christ with his resurrection. In mortification and faith what is involved is a dying and rising again with Christ (Rom. 6:4, 8; Col. 2:12; I Pet. 3:21). Because I am "in" him, mortification is not to be described existentially as simply failure, despair, anxiety, and bondage to nothingness. It is rather a re-enactment or fulfillment in me of the death of Christ; it is a crucifixion of my own flesh and its hopes, a crucifixion, however, to which the promise of resurrection on the third day also applies. Seen in the light

[11] WA 2, 458.

of Jesus Christ, mortification is no longer "just" a matter of dying. Since what is involved is his dying, an operation of divine grace, it cannot be understood merely as a negative kind of checkmate.

We thus maintain that if there is to be certainty of salvation there must be at the same time the realization that of myself I am lost. For only when I admit that fact do I set my hope exclusively (in the sense of the *particula exclusiva* of the Reformation) on God's redeeming mercy, on the once-and-for-all fact of Jesus Christ. Otherwise I should pander to some quite incongruous optimism or to some other "-ism." Salvation lies "outside myself" [*extra me*]. To find that which is "outside" I must be driven away from that which is within [*in me*]. It is the assigned function of conscience, by bearing witness to my lostness, to perform this task. Taught and overcome by the Law of God, conscience shows me the hopelessness of that perspective which obtains when I insist on seeing myself from the standpoint of my own possibilities, even though that seeing is no longer naïve—involving dreams of unlimited possibilities—but done in the light of that faith which knows that these "possibilities" are really impossibilities. The proposition that in this sense the thing is really impossible for man must always be consciously accepted as a premise of that other proposition which says that with God—and with him alone!— all things are possible (Mark 10:27).

Thus faith can be described only as a movement of flight, flight away from myself and toward the great possibilities of God. Faith has rest only in the object or goal of its flight. It has nothing but unrest in the motivating cause, namely, in the one who does the believing. Hence faith is never a matter of having but of hunting. The moment it becomes a completed quest and I regard myself as a happy possessor, in that moment faith degenerates into a mere psychological event. It becomes peace of mind in the wrong sense, and mere desire for feeling. He who wants to see or feel—in the final analysis they are the same thing—no longer believes. He no longer has his citizenship (Phil. 3:20) "outside himself" [*extra se*]. It is true, of course, that in faith we do feel something, that a forestate of divine bliss fills our hearts. But woe to him who tries to grasp as the main point that which can only have the rank of a by-product! Woe to the man who longs for peace in the heart before any true peace has been concluded between himself and God, the man for whom religious experience is his only goal! For he will not attain it. He will rather fall victim to the worst psychical illusions. For living in the fluctuating psychical states in which faith finds itself rather than in the power of faith's object, he will be blind to the distinction between illusion and reality. This is why during the Third Reich many of those who espoused a theology of experience and emotion succumbed to biological fanaticism. Psychically fluctuating experience has no contours; it allows not only for fishing in troubled waters but also for staggering about in the dark.

For this reason orientation to "experience" is no orientation at all. And it is vital that faith should therefore be driven away from the psychical *in se* and constantly forced to relate itself afresh to its object. In faith we shall of course experience something that is not to be denied, and certainly is not prohibited! But what we experience is something that is merely added to us "as well," after we have "first" sought the kingdom of God (Matt. 6:33). Experience will be ours only as such an increment or by-product. The kind of critical self-appraisal which Luther undertook in his well-known theses concerning the "resignation to hell" [*resignatio ad infernum*], whereby he sought to discover whether his believing and loving were by nature or from the Spirit, must similarly belong to the constant exercise of the life of faith.[12]

Conscience is thus the thorn in "the flesh of my feeling." Conscience sees to it that faith never becomes mere feeling, in the sense of an enjoyment of God or of an optimistically inclined eudaemonism of salvation. To put it rather better and more theologically, conscience is the thorn which sees to it that faith does not become a self-contained, psychically founded *habitus*, but that instead it remains fixed upon its object, and hence in constant motion towards this object. Faith as a flight to the life of God is thus linked with constant dying

---

[12] In commenting on Romans 9:3, Luther amplifies a question raised earlier by Tauler, whether a man should resign himself to hell and let himself be damned. For Luther the test of faith is whether a man will thus submit wholly to the will of God, whatever that will may be, seeking absolutely nothing for himself. Luther concludes that such a believer "cannot possibly remain in hell" for, willing what God wills, he is led by God and so is saved. But such faith does in fact require the complete mortification of the old Adam, eudaemonistic Adam, who even in his piety and faith must first die with Christ if he is, through this death and resignation to hell, to rise with Christ and attain eternal life. *LCC* 15, 262-264; cf. *WA* 40^I, 589 (cited below at p. 330 n. 16).

to self. The open and mortal wound which constantly gives rise to this dying is conscience.

In this connection it must also be pointed out that the conscience of the justified man, which thus bleeds and which must be continually bound up afresh if it is to be a joyful conscience, has in fact become a very different conscience. For, apart from the fact constantly emphasized by Luther, namely, that the natural conscience assents to the Law of God and thus is my *accusator* which "bites and consumes," [13] the conscience also exercises the function of *defensor* against the Law of God by protecting me against its radical claims and by trying to reduce its imperatives to a tolerable measure, i.e., to the radius of my own ability to act. These protective measures could be observed very well in Kant's teaching on conscience.

But one has only to consider the way in which conscience constantly accuses me in the state of justification to realize that it is not a relic of the old Adam, an erratic block from my pre-Christian days which constitutes a disruptive foreign body on the scene of my Christian existence. On the contrary, it is a special, God-given organ of renewal. It has indeed become truly different inasmuch as it now destroys my defenses against God and assents to his accusation in the Law. It is now on God's side—even though he is the author of the Law—and thus shows me that I must overcome God by God. What we said earlier concerning the transformation of the Law in the state of justification is true also of conscience. It can no longer kill me, for now it is not left to its own devices but has been given a pedagogical function and so been transformed.

Conscience has now become in truth the voice of God. I say *become* for it was not that from the outset. And as the tension between Law and Gospel can never cease, without my making the dramatic miracle of God's love into a "Christian world view" and God himself into an idea of divine goodness, so the tension between the hostile, accusing conscience and God's gracious promise can never cease.

To sum up, conscience, even in the justified man, is always the place where the peace and comfort of God must be realized and implemented. The comforted conscience is not a "state" which exists; it is the object and result of a "promise" which is declared. Conscience

---

[13] WA 17II, 106, line 6; cf. LCC 15, 53-54.

is thus an impulse towards, and a form in which there is realized, that mortification of self within which alone the life of God is to be had. Conscience thus becomes something in which *God* acts on *me;* it ceases to be a means of my own entrenchment against God. It thus undergoes a total transformation, and cannot be understood as a constant factor in both "natural" and "spiritual" life.

Since conscience becomes a representative of the author of the Law, and constantly receives anew the promise of forgiveness, it is to be localized theologically within the tension between Law and Gospel. As between Law and Gospel there can be no continuity which may be viewed objectively and formulated theologically, so there is no such continuity between the two forms of conscience. (This fact is significant in relation to the problem of the "point of contact," which can be maintained only on the assumption of continuity.) Since the comforted conscience is identical with faith (I Tim. 1:19) it shares the "punctual" nature of faith, as something which exists only momentarily, at the point in time where horizontal and vertical intersect. As faith is not just a believing state or condition but a "belief in," so a good conscience is not a state or condition but a conscience which cleaves to the Gospel.

Accordingly, comfort of conscience proceeds from that reality of salvation history which is outside of myself [*extra me*]. It is in this sense that we are to understand baptism as the request for a "good conscience" (I Pet. 3:21). For baptism, according to Romans 6:4 and Colossians 2:12, implies that in the name of the resurrection of Jesus Christ I walk in a new life which is won for me by him, and that my old existence is given up to death. Hence I have a good conscience not by arguing away or repressing my sin (by means of works and sacrifices), but by drowning it in the bath of baptism. I thus have it only as I appeal to my baptism.

Thus Luther was expressing a thoroughly New Testament conception when, under the assault of a disquieted conscience at Marburg, he comforted himself by writing in chalk on the table, "I have been baptized" [*baptizatus sum*].

16.

The Impossibility of Conscience as a Point of Contact for the Law of God

We have repeatedly touched in passing on the problem of reason and revelation, and the time has now come to discuss it systematically. The problem runs as follows. Natural conscience, being based on the ideology of disposition, allows for a correspondence between what we *ought* to do and what we are *capable* of [*Sollen und Können*]. In this respect it becomes the defender of a "limited" kind of responsibility over against the radical demand of God. To what extent, then, is this natural conscience still equipped to serve as a "point of contact" [*Anknüpfungspunkt*] for the Law and the Gospel? This is a rhetorical question which, as we have seen already, can be answered only in the negative. For the Law of God does not take up and amplify an existing accusation; on the contrary, it has to destroy an existing defense.

The Barth-Brunner Controversy

Emil Brunner is undoubtedly right, however, not to be satisfied with this cursory answer and to feel that he has not been properly understood by Karl Barth in Barth's dramatic rebuttal, "No!"[1] Although conscience does in fact act as God's enemy, this does not completely set aside the question concerning a point of contact. For there is always the need to take seriously, as Barth does not, the fact that God's Law is not addressed to stones or oxen or asses, but specifically to men. The Law is designed to be heard by man, and by him alone. The Law has in view man as the image of God, even if that image be only in the negative mode.

But this necessarily means that the Law addresses itself to the

[1] [Barth's famous "No! Answer to Emil Brunner," together with the Brunner essay to which it had reference, "Nature and Grace: A Contribution to the Discussion with Karl Barth," both dating from the year 1934, may be seen in *Natural Theology*, trans. Peter Fraenkel (London: Centenary, 1946).–Ed.]

conscience of man. In other words, the Law seeks to overcome the conscience of man. To whom or to what else could it otherwise be speaking?

Now it goes without saying that we must listen to, in fact we have no choice but to accept, the saying of Paul that "no eye has seen, nor ear heard, nor the heart [or conscience!] of man conceived what God has prepared for those who love him" (I Cor. 2:9). But this does not solve the theological question posed by the fact that in the case of those who love him God does in fact address himself to their eyes and ears and hearts (or consciences). To be sure, there is first an opening and changing of these "organs," for "the natural man does not receive the gifts of the Spirit of God" (I Cor. 2:14). But those things which are thus transformed are precisely eyes and ears and hearts. For it is by these, and not by fingers or glands, that the revelation of God is received.

Barth thinks in all seriousness that natural theology is to be ignored: "One can pass by so-called natural theology only as one would pass by an abyss into which it is inadvisable to step if one does not want to fall. All one can do is to turn one's back upon it as upon the great temptation and source of error, by having nothing to do with it and by making it clear to oneself and to others from time to time why one acts that way. . . . If you really reject natural theology, you do not stare at the serpent, with the result that it stares back at you, hypnotizes you, and is ultimately certain to bite you, but you hit it and kill it as soon as you see it! . . . Real rejection of natural theology can . . . only be a complete *lack* of interest in this matter." [2]

But a problem is not removed simply by our declaring a disinterest in it. On the contrary, it becomes thereby all the more serious. For if we do not take cognizance of the power which impels toward a natural knowledge of God, and which uses reason and conscience as raw materials to this end, we run the risk that this power will creep up on us unnoticed and rob us of the dispassionate criteria by which to unmask it.

If appearances do not deceive, Barth himself has fallen victim to this very process. For example, his essay on "Church and State" seems to be exposed in large measure to the danger of passing theological judgments on democracy and totalitarianism which are not derived

[2] Barth, "No!", in *Natural Theology*, pp. 75-76.

from the true center of theological ethics, but are based rather on political and perfectly rational, and sometimes even purely Swiss, considerations which are quite in place but only on the worldly level.[3] Here is a case where natural theology, having been dismissed from view, enjoys a certain measure of triumph.

We do not mention this by way of censure, but simply in order to give a particularly striking example of what we have in mind. From this example one may see how unwise it is to try to ignore completely the secularized attempt at a natural theology rather than to make it a matter of deliberate investigation.

When Barth rejects the whole problem of the "point of contact," when he disclaims theologically all interest in the natural structure of conscience and in what happens to this structure in face of the Law of God, it is hard to see how man can be addressed at all in respect of his natural constitution, and therefore how he can ever be made responsible for it. This is why we have to accept the assurance of Emil Brunner that he is opening up the problem of responsibility not for the sake of propagating natural theology, but rather for the sake of the concepts of responsibility and of judgment. If we have had to tread a different path from Brunner in respect of the *imago Dei,* we believe we must support him at this point in our argument, and not at this "point" alone but also—for all our basic differences— in his general concern for theological ethics as such and in the themes and problems he has advanced.

We hold the following proposition to be decisive: God creates anew the man to whom he speaks and whom he raises to sonship. But if we are to understand the nature of the "new" we have to know the nature of the "old." Now the "old" was a position of defense and self-protection against God. Hence the "old" conscience and the "new" cannot be brought together under the common rubric "conscience," as something which in form remains a constant factor, indifferent in respect of the contents which may fill it from time to time. The reference here is not to two stages in the history of the one conscience but to a total destruction of the old and a wholly fresh structuring of the new. Both take place in the miracle of the Holy Spirit, which is why we can know nothing of "how" this break takes

[3] See Barth, *Community, State, and Church: Three Essays* (Garden City, New York: Doubleday, 1960), pp. 101-148.

place. This is also why we can speak of "contact" only in a derived sense, since the term secretly implies this whole problem of "how" it all takes place. "Point of contact" is a borderline concept, a figure of speech to help our understanding; we use it only in the way a school teacher uses chalk on the blackboard, to make a point and then to rub out immediately what he has written.

Luther on Natural Theology

It is true that in his earlier period Luther could follow medieval and Aristotelian usage and describe conscience (here referred to as *synteresis*)[4] as a matter [*materia*] which awaits a form [*forma*] in virtue of which it can arrive at its goal [telos], or as a spark which, while it glows in the dark, must be fanned into flame by revelation. Here conscience is not regarded as something which must be done away in the sense of *destruere,* but as something which must be perfected in the sense of *perficere.*[5] Yet even here Luther realizes that the *synteresis* may possibly be a mechanism for defending oneself against God, and in the very same sermon—apparently without noticing the happy inconsistency—he rejects these Roman Catholic assumptions. For although, "in consequence of the *synteresis,* will and reason are 'organs' for the apprehension of the invisible and hidden things of God, they prove themselves to be unusable for, indeed obstructive of, that purpose. And this is true of will and reason in their entirety; there are no residual segments within them which would constitute exceptions to the rule. . . . The *synteresis* is a condition for the efficacy of grace. But it can also be an obstacle to the efficacy of grace. . . . We can count on the reality of the *synteresis* only when we do *not* count on it."[6] Nor is Luther thinking only of the misuse of *synteresis,* i.e., as an object of self-confidence. He is thinking of the *synteresis* itself. It is because of the *synteresis* that man desires salvation and fears perdition. But this human desire for eternal bliss can in fact be opposed to the will of God, as we have seen in connection with Luther's discussion of the *resignatio ad infernum.*

[4] [See LCC 15, 24, n. 46–Ed.]

[5] See the sermon *De propria sapientia et voluntate* which Luther preached in the monastery on Dec. 26, 1514; WA 1, 32.

[6] Walther von Loewenich, *Luthers Theologia crucis* (2nd ed.; Munich, 1933), pp. 61 ff.

If, to bring out the parallel to our own train of thought, we go on to say that Luther is here regarding the postulate of eternal bliss as one of the axioms of natural conscience, then it is quite clear that "*synteresis* cannot have for Luther the significance of a divine organ in man." [7] It is "not the bed rock on which our house can be built. What is involved is really a *destrui*, a radical tearing down and a complete re-laying of the foundations." [8] Thus because the *synteresis*, like the "point of contact," seems to persist throughout—and therefore to be above—dying and becoming, the doctrine of *synteresis* can have "no place in the *theologia crucis*." [9]

The theological influence of modern secularism in this whole area becomes apparent once we consider the various collections of quotations from Luther which have been assembled to demonstrate his attitude toward natural theology. In this connection, however, we have to recall not merely his well-known assessment of the "harlot reason," but also his contrary assertions concerning natural law.[10] These two categories of contradictory statements show us that Luther was by no means so aware of the limitations of human knowledge (even in respect of man's ability to know the good and its command) as he was of the limitations of the human will and its works. The contradiction is in fact so great that one can find abundant quotations in Luther to support both the complete rejection of any "natural theology" by Karl Barth and also the national *nomos* teaching of the most rabid of the Nazified "German Christians." The same is largely true also of Calvin, as may be seen from the debate between Emil Brunner and Günther Gloede on the one side, and the brothers Karl and Peter Barth on the other, though obviously Brunner and Gloede are not to be identified in any sense with the camp of the "German Christians." [11]

[7] *Ibid.*, p. 66.

[8] *Ibid.*, p. 67.

[9] *Ibid.*

[10] See esp. Herbert Vossberg, *Luthers Kritik aller Religion* (Leipzig-Erlangen, 1922); Emanuel Hirsch, *Der Offenbarungsglaube* ["Hammer und Nagel," No. 2 (Bordesholm, 1934)], pp. 19 ff.; Ernst Wolf, *Martin Luther* (*Theologische Existenz heute*, No. 6).

[11] See *Natural Theology;* also Emil Brunner, *Man in Revolt*, trans. Olive Wyon (Philadelphia: Westminster, 1947), and *Revelation and Reason*, trans. Olive Wyon (Philadelphia: Westminster, 1946). See also Günther Gloede, *Theologia naturalis bei Calvin* (Stuttgart, 1935) and Peter Barth, *Das Problem der natürlichen Theologie bei Calvin* (Munich, 1935).

It must be recognized that Luther's theological position in this matter cannot be known from a dubious heaping up of quotations intended to disclose "directly" his attitude. What is needed is that his basic principle, the doctrine of justification, should be applied to the questions of natural theology. In this way we may even understand Luther better than he understood himself; we may see how man's incapacity to justify himself by good works is logically to be augmented by, or integrated with, a similar incapacity truly to know the will and commandment of God. Just as there can be no co-operation of the human will in fulfilling the divine commandments, so there can logically be no co-operation of human reason in furnishing knowledge of the true God. The aim of the Reformers—Calvin as well as Luther—undoubtedly consists on its negative side in the rejection of human co-operation, i.e., of man's partnership with God, in any form. If in Luther this does not seem to be so clear in the matter of natural theology as in the doctrine of justification, bondage of the will, Christology, etc., the only sensible way to integrate the problem of the natural knowledge of God into Luther's total theology is to apply the logic of his view of justification to the problem of co-operation in the knowledge of God. In this way we may correct Luther by Luther, by means of his own principles.

Even Karl Barth seems to be adopting this method of interpreting Luther and Calvin when he says that "the Reformers could not clearly perceive the range of the decisive connection which exists in the Roman Catholic system between the problem of justification and the problem of the knowledge of God, between reconciliation and revelation." [12] He says also that although "they saw and attacked the possibility of an intellectual work-righteousness in the basis of theological thought . . . they did not do so as widely, as clearly and as fundamentally as they did with respect to the possibility of a moral work-righteousness in the basis of Christian life. . . . In order to understand that this [i.e., the natural knowledge of God] would really be impossible for Calvin, one has to consider his doctrine of Christ, of the unfree will, of justification . . ." [13]

There can be no doubt that Luther did not see clearly the consequences of his doctrine of justification in the sphere of natural

[12] *Natural Theology*, p. 101.
[13] *Ibid.*, p. 102.

theology. Anyone who is acquainted with his thought in the slightest will have to admit that Luther is very unguarded in the way he expresses himself on these questions. The reason for this is that he was not fighting on the front of natural theology. It was rather moral synergism which posed for him the great challenge, both theologically and in practical life. Now in the final analysis moral synergism and natural theology are very closely related; both can be construed under the rubric "analogy of being" [*analogia entis*]. But this is something that has been made clear to us today only by the orgies of natural theology and activist Faustian synergism through which we have passed, and with which we are still to some extent surrounded.

In any case it is certainly true that a Lutheran theologian can no longer express himself so naïvely and unguardedly as did Luther in problems of natural theology, e.g., as regards the function of conscience and the scope of its action. For we have known in our time the conscience that is emancipated, and therewith also intoxicated and sick. We have seen before our very eyes the titanic paradigm of the secularist experiment. This in itself should have been enough to teach us that here, as in the field of exegesis, it is only *through* the fiery barrier of secularization and natural theology, and not *around* it, that one can return to Luther.

This means, however, that the secularism of our time has not merely meant a setback. To the extent that we are in the church, and observe and evaluate the secularist experiment from this vantage point, it also means an advance. Theology has really been enriched by it, to the extent of a new chapter in the field of anthropology. The consistent self-understanding of man without God, the consistent atheistic ethos of emancipated man actually gives us, in the light of revelation, a profound glimpse into human reality, and opens up biblical insights for us in a way which was hardly possible for previous generations.

From what we have already said about the relation of conscience and justification in Luther, it is evident that he gives particularly clear expression to the truth that conscience does not merely "persist" as it is. My conscience, as we have seen, cannot know God as the one who is gracious to me. It can know God only as judge or else as "the good Lord" whose goodness-in-general is without personal application to me. Even in the regenerate, conscience is therefore always something which must to some extent be overcome; we have to believe in spite of it. As the *cor accusator,* it is a continuing thorn in the flesh, a constant wellspring of doubt. In any case, peace of conscience is never ours as a *habitus;* it is present only in the divine promise.

At this point, then, we would now raise the question as to the implication of these ideas of Luther for the problem of "contact." To begin with, the conscience of the natural man cannot be a "point of contact" for the Law and the Gospel because, to be so, it would have to be able to see of itself that man is lost and in need of forgiveness. But this it cannot do, if in virtue of man's "light of nature" [*lumen naturale*] it understands God as judge. For the criteria of this judging are laid down by conscience itself, and these criteria are determined by the relation between imperative and ability [*Sollen und Können*].

Having a false conception of God, the natural conscience has also a false conception of sin and of guilt. For these are matters which can be understood only in terms of our attitude to God. This is why Luther went beyond merely maintaining that conscience has no insight into the depth of man's plight. He actually gave dialectical precision to this thesis, arguing that man's real plight is that man is in no position whatever even to recognize his plight by the power of his *lumen naturale* alone.[14]

The depth of man's plight does not consist merely in the degree of his actual guilt before God. It consists also in the degree of his blindness to this fact of guilt. If man could see the bottomless depths of his lost estate, the result would be that "godly grief" which produces repentance (II Cor. 7:10). As it is, however, the situation is simply that of hardening, which Luther describes in terms of the *cor defensor.* Man tries to overcome his ignorance by new illusions, and so is involved in a continuous process of trying to extrapolate himself to an ever higher power.

True knowledge of man's plight is possible only on the basis of Christmas and Good Friday. For the profound depth of man's lost estate becomes clear only when we see what it cost God to help us. In other words, it is revealed ultimately only by the Gospel, which in

---

[14] See Luther's commentary on Romans 7:7 ff. in Johannes Ficker (ed.), *Luthers Vorlesung über den Römerbrief 1515-1516* (Leipzig: Dieterich, 1908), I *Die Glosse,* p. 63, lines 14-19; cf. also *LCC* 15, 200-202.

this respect to be sure acts in effect as Law. Thus Luther—and of course others as well—can continually speak of the cross in terms of its significance as Law, as judgment upon our sin. Sin, then, is not just something which I can empirically ascertain and demonstrate. It is something that must be believed. I have to believe in sin as I have to believe in Christ, in whose incarnation and death my sin plays its role and finds its reflection. I must see him who has borne my sin if I am to see what it is that has been borne, and who it is—who I am— from whom the burden has been lifted.[15]

In order to see clearly that the natural conscience cannot be a "point of contact," we must remember that the ignorance of conscience is not just a passive ignorance, like that of a schoolboy. If that were the case, knowledge could always be imparted. The head of man could then be regarded as a vacuum just waiting to be filled. And the head of one who was formerly ignorant but is now learned would still be the same head, even though it was now adorned by intellectual inscriptions and character had replaced vacuity. Conscience's ignorance of its lostness, however, is an ignorance which engages in defense and counterattack. It is an ignorance which produces alibis. In short, it cannot be understood psychologically or epistemologically, but must rather be interpreted as an act of "suppressing in wickedness" (Rom. 1:18).

At this point we can fit into its correct theological place, and understand in context, our earlier thesis that conscience is not just something that is present as such. Conscience cannot be regarded simply as a neutral container for natural *and* divine law. On the contrary, conscience has always to be understood *either* as alarmed and on the defensive *or* as vanquished and comforted.

Thus Luther's theology of conscience inevitably leads us here too to those radical alternatives which must always arise when we think in personalistic rather than ontological terms. Here it indeed holds true that "whatsoever is not of faith (even including the conscience which is not of faith) is sin" (Rom. 14:23; Tit. 1:15). There is no such thing as a conscience that is (quantitatively) more or less good, depending on how close I come to ethically fulfilling the Law (Mark 10:20). There is only that conscience which is *either* blinded by Satan *or* comforted by God. In other words, for Luther conscience is never

---

[15] *LCC* 15, 81 (cf. I Cor. 4:4); cf. *LCC* 15, 294.

alone. It is always qualified from without [*extra se*]. This may be illustrated by two groups of quotations.

In the first place, "Our theology is sure, because it sets us *outside* ourselves. I do not have to rely on my conscience, on the external person, or on works, but on the divine promise, on the truth which cannot deceive." [16] This means that conscience is consoled only when it turns away from itself to the divine promise. When it is alone, or when it thinks it is alone, it is in fact with death, with the wrath of God, with Satan. For it is accused by the Law or plunged by Satan into perplexity and doubt. As Luther puts it, Satan "blows up sin so hard" that it seems as great as heaven and earth; out of one sin he makes a hundred, plunging man into the depths of melancholy or driving him to an equally unhealthy self-confidence [*securitas*].[17] We "lose our red lips and cheeks, and forget how to dance and sing, for the devil is the spirit of melancholy." [18] Thus conscience is always qualified by something outside itself.

Second, this uncompromising either-or necessarily involves rejection of the view that conscience is a middle thing, a link between two stages, an empty and neutral vessel. Whether conscience has faith in God or not, it is "not a middle thing possessing neither God nor the devil." [19] This gives us the decisive answer to our question whether conscience can exercise the function of a "point of contact." For to say that it is "not a middle thing" [*non est medium*] is to say that it is not a connecting point [*non est punctum copulandi*]. Here too it may be seen that we cannot avoid the idea of a point of contact as a borderline concept. We have to postulate something to which the Law of God comes with its demand, and which in so doing it destroys and rebuilds. But we cannot ascribe to this "something" any content, nor can we describe the manner in which it is changed. In this sense Luther can in fact speak of man's will functioning as a *medium* or middle thing (in this connection one might easily substitute "conscience" for "will"). But this is done only in a figure of speech which is dropped the very next moment. We are referring to the well-known passage in *The Bondage of the Will* in which Luther describes man as

16 WA 40^I, 589.
17 WA 27, 96; 29, 572.
18 WA 37, 185.
19 WA 15, 426.

a beast which is ridden either by God or by the devil—which accordingly is not characterized by itself—and thus stands in the "middle" between the two powers.[20]

[20] See *The Bondage of the Will*, trans. J. I. Packer and O. R. Johnston (Westwood, New Jersey: Revell, 1957), pp. 103-104. Here too belongs Luther's famous figure of the two giants duelling one another in the Law, namely, the death of self and the death of Christ; WA 39^I, 427.

17.

Autonomous Conscience and the Relativizing of the Law of God

If it is true that the natural conscience cannot serve as a point of contact for the Law of God, and if it is also true that conscience rests on the ethical maxim "you can, because you ought" whereas the Law of God is linked with the experience "I ought, but I cannot," then the question necessarily arises as to how conscience as the "light of nature" [*lumen naturale*] emerges from its encounter with the Law of God. What is the result of the natural conscience's inability to perceive?

There are two possibilities. On the one hand, when the Law of God is encountered in its radical severity, especially in the eschatological form of the Sermon on the Mount, it may be rejected as absurd, as expressing a negation of life, particularly a negation of human self-assertion. On the other hand, conscience may indeed acknowledge the command of God, but only as it secretly transforms and moralizes this command, invalidating part of it, and assimilating into its own ethical system of co-ordinates what remains and is acknowledged as binding.

Our particular concern is with the second alternative, i.e., with the attempted synthesis. What is involved is an attempted reconciliation between conscience and the command of God, between what is known naturally and what is revealed. It is a question of what concretely transpires when conscience attempts to absorb into itself the Law of God. We have already achieved the important insight that in such an attempt, in such a confrontation, one of the factors—either revelation or conscience—will have to be broken. There can be no analogy between the two. At an earlier point we saw how conscience is broken and must undergo death and renewal. We have now to consider how the authority of revelation, or better, of the revealed Law, is broken when conscience continues to assert itself and refuses to die.

We can best observe this by considering certain essential reinterpretations of the Sermon on the Mount. In consequence of its radicalization of the Mosaic *nomos*, the Sermon on the Mount brings vividly to light the question of the Law of God. It also issues a keen and un-mistakable challenge to conscience with its axiom "you can, because you ought." We may thus conjecture that if conscience wishes to assert itself, and yet continue as a "Christian" conscience which in some way affirms the Sermon on the Mount, it will probably make a particularly strong attempt to overthrow, to reinterpret, and to assim-ilate the Sermon on the Mount. From this standpoint we can bring under a single category many different expositions of the Sermon on the Mount which would otherwise seem to be dominated by the most divergent motives.

The Historicizing of the Sermon on the Mount

Since we are not writing a history of the interpretation of the Sermon on the Mount, we will consider here only certain main types of interpretation which are illustrative of our thesis. We begin with a modern attempt to deprive the Sermon of its severity and to do away with its opposition to autonomous conscience. The attempt is actually an effort to understand the radicalism of the Sermon on the Mount historically, and thus to relativize it. The historical thesis runs as follows. If the radicalism of the Sermon on the Mount is conditioned by time and situation, then it has no power to call in question our conscience. Thus it is argued that the unconditional character of Jesus' demands is to be explained in terms of his attack on the nomistic and cultic theology of the Pharisees, i.e., on Rabbinic teaching.[1] Jesus is said to be dealing a counterblow to the casuistical ossification of piety by demonstrating its hollowness and by showing that its concern is to remove man's guilt vis-à-vis God by empty formalism, so that we may remain the men we are and mean to be. Protest against this formal-istic piety is said to have compelled Jesus to express his demand for personal commitment to God and for a pious disposition in terms of those exaggerated formulations which go beyond even the require-ments of the rabbis. Consequently we would be failing to interpret the Sermon on the Mount in keeping with its own intentions if we were to see in its severe demands absolute and seriously intended theses. For it is basically a matter not of theses but of anti-theses.

[1] Cf. Hans Windisch, *The Meaning of the Sermon on the Mount: A Contribu-tion to the Historical Understanding of the Gospels*, trans. S. MacLean Gilmour (Philadelphia: Westminster, 1951).

But, the debate for us today, it is held, rages on a quite different front from that which engaged Jesus. These antitheses of the Sermon on the Mount are directed against different opponents. Hence its radical demands are not binding on us. To know what Jesus really meant in the absolute sense, and what is thus a binding theology of the Sermon on the Mount for us today, it is presumably necessary to purge out the polemical elements and to engage in a process of assimilation.

The relativization of the Sermon in the extreme proponents of an eschatological understanding, especially Johannes Weiss and Albert Schweitzer, is in principle the same. Here it is understood as a law for a particular situation, namely, that in which the arrival of the eschaton is expected momentarily. A further restriction is that it is meant to apply only to the disciples in the concluding days of this aeon. Since the Parousia has not come, and we must take up our station within this continuing aeon, the eschatological tension of the original situation is no longer present. Therefore the particular radical demands of the Sermon on the Mount are not binding on us.

In face of this historical relativizing of the Sermon on the Mount we would ask but one theological counterquestion. Suppose it be granted that the Sermon on the Mount is to be understood eschatologically in this extreme sense. Suppose it be granted that this eschatological character of the Sermon is due to an erroneous foreshortening of perspective in regard to the end of the world. Suppose it be even granted that—the "error" in perspective having been proved—the Sermon does not apply to us. The question still arises, however, whether this situation of man before the imminent end really constitutes a limitation which is to be understood historically. Is it really an error which we in our "enlightenment" cannot accept, and over against which we must postulate long time periods of mounting evolution? Schweitzer's cultural optimism strongly reminds us of the supposed peace of a time period which is thought to run on indefinitely, and over which there never falls the disturbing shadow of the end. We must say this even though we are naturally reluctant to critize theologically one who in practice has lived so impressively in the seriousness of the "ultimate" end.

Might it not be—this is the question we must ask—might it not be that this "error" of taking the end to be so near actually contained within itself a very productive element which Jesus' believing community ought always to retain? Indeed is it not true that a living church always has a living eschatology and exists in the shadow of the end? Might not this apparently erroneous situation of man before the imminent end give point to the commandment, "Be sober, be watchful" (I Pet. 5:8; I Thess. 5:6; I Cor. 15:34)? Might it not help us more than anything else in the world to take the coming aeon seriously? Might it not cause us to call in question the present aeon as an aeon which is passing away, which goes to pieces under the law of the coming aeon and is thereby shown to be transitory, dependent, and perishing? Are we not in danger of self-deception on all these matters if we do away with the eschatological radicalness of the Sermon by interpreting it purely historically?

To this line of argument, however, the objection might be made that the eschatological radicalness is thus interpreted too pragmatically, rather than in terms of what is really involved theologically. Even though sobriety and readiness for the returning Lord be spiritually fruitful, are they to be bought at the price of an error, namely, that erroneous foreshortening of perspective? Is not this to make truth subservient to secondary ends, indeed to abolish truth altogether? Is it not to close our eyes to the real truth, namely, that we must count on time continuing, and that we must adjust ourselves to the fact that the Parousia will not be a temporal event at all, that it is both "always" and "never," and that from this angle history is to be understood axiologically rather than teleologically?

We could not stand up to this objection, of course, if our purpose were simply to perpetuate error for the sake of some supposed spiritual profit. Indeed, on these terms there could be no such profit at all. Is it not conceivable, however, that in this error so patently exposed by history there is concealed a truth which obviously needs this strange husk, and by virtue of which we must concede to the error a real kinship in substance to certain central truths of the New Testament?

This is what is suggested by Oscar Cullmann.[2] He tells us that the early church's expectation of the end was rooted in the fact of Easter. In the light of Easter the primitive community realized that the decisive victory had already been won in the cosmic conflict between

2 See Oscar Cullmann, *Christ and Time*, trans. Floyd V. Filson (Philadelphia: Westminster, 1950), p. 141.

God and the opposing powers and that, the crucial battle having already been fought, we now live in the period between this great turning point and the final triumph. "V-Day" will not produce anything essentially new; it will simply draw the consequences of that which has already taken place. The more we are convinced that the great turning point is already past—there are many impressive examples of this in the final stages of the Second World War—the nearer we believe ourselves to be to the final triumph; and the less easy it is to grasp the fact that the end has not yet come, that the victory celebration has still to be awaited.

Thus it might well be that the error of the primitive church concerning the end, an error shared by many others up to and including Johannes Bengel, is only a shadow cast by the light of genuine truth, namely, by the light of the truth that the decisive victory over sin and death has already been won in the resurrection and the ascension of Jesus Christ, that the devil has really fallen like lightning from heaven (Luke 10:18), and that we live now in the epilogue period between the provisional and the definitive seizure of power.

But in this case the "ethics" of the Sermon on the Mount would also retain its abiding eschatological significance. It would remain an "interim ethics" for the whole of that intervening period in which the old aeon runs on to its close while the new is already in force. It would constitute today as always—without any historical relativizing!—an abiding corrective for an aeon which posits itself absolutely, which consequently regards itself as ultimate, and which ensconces itself securely in position as if the day of the Lord were not just around the corner, and as if it could find within itself the ultimate criteria and standards. This self-sufficient aeon needs constantly to be "unsettled" by the Sermon on the Mount. It needs constantly to have its provisional character exposed, revealing "our time" to be merely an intervening period located at the point where the two aeons intersect. We plan eventually to take up what this means concretely for existence in this passing aeon, and what it implies for our relations to others, for politics and for culture, though we cannot do so within the compass of this present volume.

The Blunting of the Sermon's Radicalism
along Quantitative Lines

For reasons which are closely linked with the innermost structure of anthropology, and ultimately with the ontological thought pattern, the Aristotelian theology of Roman Catholicism distinguishes different forms and degrees of grace, and also correspondingly different forms and degrees of merits. It distinguishes, for example, such different classes of people as the religious and the secular, or saints and ordinary people. As we have seen, to the extent that grace ceases to be the personal quality of divine favor, and becomes instead the medical infusion of a *habitus,* to that extent there necessarily arises the question as to the degree of this gracious endowment and of the consequent merits.

The difference between this kind of ontological thinking and a personalistically oriented theology naturally emerges with particular clarity in respect of the radical requirements of the Sermon on the Mount. In terms of its presuppositions, Scholastic thinking experiences insuperable difficulties in accepting this radicalism. For the radicalism inevitably implies the sharpest of alternatives. On the basis of the Sermon on the Mount we are forced to the position that *either* I am not angry against my brother and cherish no hostility to him in my heart (which none of us can say!) *or* I am a murderer. There is no other possibility if we truly see ourselves as standing before God [*coram Deo*] and understand the requirement radically. The less radical formulation of the Mosaic *nomos,* on the other hand, at least according to the letter, still allows of distinctions between a potential murderer who does his hating only "inwardly," and an actual murderer who puts his thoughts into action.

There are many different ways of blunting the radicalism along quantitative lines. The four ways most familiar in Catholicism include: first, linking the command eudaemonistically with self-love; second, dividing the commandments into "precepts" and "counsels"; third, postulating various degrees or levels of love for God; and, fourth, differentiating between "intention" and "action." We shall examine each of these briefly in turn.

The Eudaemonistic Link with Self-Love

The blunting of the radicalism is clearly to be seen already in

Augustine. For in him we find a conception of the divine commandments which is limited by eudaemonism, especially in respect of the commandment to love your neighbor as yourself (Matt. 19:19; 22:39; Mark 12:31, 33; Lev. 19:18). Augustine interprets this commandment in such a way that love of neighbor is bound up with, and in some sense conditioned by, love of self: ". . . you ought also love yourself thus [i.e., as you—in keeping with the love of God—love your neighbor] that there be no error in the rule: 'Love your neighbor as yourself.'"[3] The commandment that we love our neighbor is thus taken to imply also the demand for love of self, whereas correct exegesis surely suggests that love of self is taken to be the dominant natural impulse of man, and that as the strongest impulse, e.g., in the instinct of self-preservation or in the search for happiness, it ought to be the standard for our love of neighbor. It is as if the unconditional character of our love of neighbor could not be more sharply brought out than by relating it to this elemental impulse in man.

Though self-love is impressively spiritualized by Augustine, and though it is characterized by love for God, one cannot say that it thereby passes over into, or is absorbed by, this love for God. For we must press the question whether what is here called love for God is not infected by that which, even in its most sublime form, is still self-love. We are forced to conclude that it is, since in love for God as Augustine understands it there is a very palpable element of desire for happiness. In this connection Karl Holl can even say that the desire for happiness is the "ultimate motivation on which the whole of religion rests for Augustine. . . . Conversion is basically only a change in taste; the desire for earthly goods is replaced by the sweeter desire for heavenly goods."[4] It is only when we are aware of this sublime eudaemonistic content of man's love for God that we can see why Augustine is so interested in the matter of self-love, and why self-love can never be completely absorbed by love for God. How can self-love pass over into something which is itself essentially

[3] Augustine, *Enarrationes in psalmos*, CXL, 2, in *Corpus Christianorum, Series Latina* (Turnholti: Brepols, 1953—), XL, 2026. Cf. Augustine *The City of God* xix. 14; *A Select library of the Nicene and Post-Nicene Fathers of the Christian Church* (First Series), ed. Philip Schaff (14 vols.; Buffalo and New York: Christian Literature Co., 1886-1889), II, 410. (Hereinafter cited as *NPF*.)

[4] Karl Holl, *Gesammelte Aufsätze zur Kirchengeschichte*, III (Tübingen: Mohr, 1928), 85.

determined by self-love? What is really involved is an expressly dialectical relation between the two forms of love. Self-love is defined by love for God (not vice versa), but only after self-love has first exerted a sublime eudaemonistic influence on love for God.

Nevertheless, to avoid too crass a schematization, we must insist that in intention at least Augustine is speaking of a spiritual or theological love of self which serves as a presupposition for the love of God. We do justice to the dialectical relation between the two forms of love only if we also say that in some sense love of self can only follow love for God. Yet it still remains that love of self is plainly envisaged by those who love. Augustine thus introduces into the sphere of love a theme which necessarily works itself out with some degree of autonomy, and which for this reason finds no explication whatever in the New Testament. And if Augustine is right in saying that man enjoys no greater satisfaction than when he forgets himself,[5] then it is also true that love of self would be more congruously integrated into the total nexus of love if it would forget itself instead of being made an independent—however highly spiritualized—theme of its own. *Either* it is, as spiritual self-love, implicit in love for God, and we are thus forbidden to make it the object of an imperative; *or* it must itself be commanded, instead of constituting merely an illustration of that which is indeed commanded, namely, love for God. In the latter case, a measure of rivalry between the two strands of love is unavoidable, and self-love acquires at least the suspicion of having a normative significance which is restrictive of love for God.

There is thus introduced into the command of love a conditioning element which undercuts its radicalism and makes it dependent on a eudaemonistically emphasized love of self. A further consequence is that the command to love can have only partial validity, that is, to the degree that it may be united with self-love. This gradation finds further expression in the extent of the reward which is earned in any given instance. Indeed, the very idea of reward contains within itself by nature the idea of gradations from greater to lesser, whereby the greatest in degree can also be the most sublimated in form, as may be seen from the thought of "enjoying God" [*frui Deo*] in Augustine.

The logic of this kind of "quantitative thinking," however, leads

---

[5] Augustine *On Free Will* 3. 76; *LCC* VI, 216.

to even further consequences. In connection with the command to love, Scholasticism carries the idea of gradation to the extreme of maintaining that there are differences in rank—and therefore in degrees of obligation—among our various "neighbors," so that from this angle too the unconditional character of the command to love one's neighbor is threatened. Thus Thomas says that "those who are nearer to us are to be loved more."[6] It belongs presumably to the "order of nature" that we should distinguish degrees of proximity, such as family, relations, friends, etc. This brings us to the sharpest deradicalization of the command to love and of the Sermon on the Mount, whereby the command is caught in a wide net of conditions and its original thrust is altered.

Precepts and Counsels

The differentiation between specific acts of obedience is carried to its extreme when a distinction is made between "precepts" [praecepta] which relate directly to duties imposed by command, and "counsels" [consilia] which are oriented to true perfection. This is an ancient distinction, found as early as Origen. For Origen, what is done on the basis of counsels involves a higher degree of perfection than what is done on the basis of the minimal demands of precepts. The commandments require of us only that which is our duty anyway [debita]; they cannot therefore be the basis of merit. The counsels, on the contrary, are concerned with that which goes beyond what we already owe; this alone is meritorious in the true sense. (As a rule virginity is what he has primarily in view.)[7] The quantitative scheme which underlies this whole distinction finds particularly precise and vivid formulation in Augustine (again in respect of virginity). He says concerning I Corinthians 7:10-11 that "this is a command, not to obey which is sin: not a counsel, which if you shall be unwilling to use, you will obtain less good, not do any ill."[8]

It is worth considering this statement of Augustine more closely,

---

[6] Thomas, *Summa Theologica*, part 2^{II}, ques. 44, art. 8, reply obj. 3; *The "Summa Theologica" of St. Thomas Aquinas*, trans. Fathers of the English Dominican Province (New York: Benziger, 1917). Cf. Holl, *op. cit.*, I (Tübingen, 1923), 165, n. 3.

[7] See Origen, *Commentaria in Epistolam ad Romanos*, X, 14, where he refers also to Luke 17:10.

[8] Augustine *Of Holy Virginity* xv, trans. C. I. Cornish, in *NPF* III, 421.

for it poses a real alternative only in respect of "precepts." Only here is it a question of *either* sin *or* righteousness. Only here is the matter conceived qualitatively. But over and above this foundation of the precepts, conceived as minimal demands, there then arises the super-structure of the counsels, which have to do with a "higher" right-eousness. At this point the thinking is unequivocally quantitative. The alternative is not that of good or evil; it is that of a greater good or a lesser good. But even when—within the system of counsels—the "goodness" is reduced to zero, we still do not arrive at "evil" in the sense of the precepts. On the contrary, we have done what is our duty in any case, and so have satisfied the claims of righteousness. The consequences of this approach for our understanding of the Sermon on the Mount are immediately apparent: Whatever lies beyond the hard core of precepts necessarily has the character of counsels and is thus binding only on the "perfect," and even in their case is subject to gradations. Radical obedience is required at best only in respect of the foundation constituted by the true precepts.[9]

Once we accept the distinction between precepts and counsels, we no longer see man in terms of the fellowship with God, which he either has or does not have. We see him instead in terms of a process of growth. We see him, as it were, *coram se ipso*, and can thus speak of approximations to the norm.

The Three Levels of Love for God

Augustine had already pointed out that perfect love for God is not attainable in this life. This is because we can love perfectly only what we know perfectly, i.e., only what we see face to face.[10] Consequently our love is the greater—that is, it is less imperfect—the more we deny this life; the degrees of perfection usually have a negative standard and are attained in a progressive exercise of self-abnegation.

What is hardest to achieve—and thus demands the greatest abnega-tion—carries us furthest in perfection. The implied deradicalization

---

[9] Luther objected as early as his 1518 *Explanations of the Ninety-Five Theses* to interpreting the Sermon on the Mount in terms of "counsels": "If this were only a counsel [*consilium*] (as even many theologians erroneously seem to think) . . . " *WA* 1, 619; cf. *LW* 31, 235.

[10] Augustine *On the Spirit and the Letter* lxiv, trans. Benjamin B. Warfield, in *NPF* V, 112. On the distinction between perfection in this life and in the life to come, see *ibid* lxv, in *NPF* V, 113; also Augustine *On Man's Perfection in Righteousness* vii-viii, in *NPF* V, 163-165.

of the command to be perfect is apparent here, not merely in the fact that levels of perfection are introduced, but also in the fact that the higher levels of perfection cease to have universal validity. In some cases they have no force in "this life" at all, or they can be attained only by specific groups, e.g., the monks, and are pointless for all others. For the higher levels are covered only by the "counsels." But these belong to the non-obligatory zone which rises above the sphere of the true "precepts." This becomes clear when we consider the three main levels of perfection. In this connection we shall be following for the most part the classical exposition of Thomas Aquinas.[11]

The highest level is when the "whole heart" of man is continually and "actually" oriented to God. This uninterrupted actual participation of the whole heart in love for God is not possible, however, in this life, since the "infirmity" of human life prevents it from producing of itself the "stability" and "constancy" of an unchanging and total movement of love. Thus is done away the unconditional character of the command to love with one's whole heart. The fact of the commandment's unattainability is not noted in connection with a confession that "I ought, but cannot." It is not regarded as a confession of sin. On the contrary, it is simply "explained," in terms of the given ontic structure of human life, a structure which is characterized by "infirmity."

We see here at the same time the consequences of Roman Catholic anthropology, which we have already discussed in some detail in connection with the doctrine of the *imago Dei*. If the human person is not understood in relationship, but instead is given ontic autonomy as the bearer of demonstrable components of nature and of ontic substances of grace, then loving with one's whole heart can no longer be understood personally in terms of existence in fellowship with God. It has to be understood instead as the ontically demonstrable content of the heart, as a psychodynamic state of being filled with loving impulses. The ontology of man leads at once to psychology, and indeed to a theological-speculative kind of psychology. Once we enter upon this path, we cannot avoid the conclusion that in terms of psychical structure man's "infirmity"—i.e., the fact that the soul is

---

[11] See Thomas, *Summa Theologica*, part 2[II], ques. 24, art. 8, "Whether charity can be perfect in this life?"

necessarily filled with shifting contents, with varied impressions and desires as well as with tasks and duties which claim our attention and devotion—man's "infirmity" is completely incompatible with his being totally filled with love for God. Love for God as a psychical "content" then enters at once into competition with other potential contents of the soul. In any case love's claim to be total is wholly illusory—until these competing contents of the psyche are eliminated in the uninterrupted and undeflected vision of God in the life to come. Thus the command to love loses its unconditional character the moment we enter the ontological-psychological plane of thought. For on that level it is no longer possible to grasp the personal category of love.

It is precisely at this point that we can find help in the formulation of Luther whereby the subject of faith—in this context we may with equal propriety say: the subject of love—is a "mathematical point" [*punctum mathematicum*].[12] Luther uses this extreme formulation to combat the fatal psychologizing of faith and of love. He refuses to regard the subject of faith and of love as merely an extended psychical tract filled with diverse forces and aspirations. There is thus no place for such questions as whether we are supposed to believe or to love at every single moment of our lives, or how such permanence is to be understood psychologically. Faith and love are characterized by their object, not by the psychical accomplishment of believing and loving. I confess him who loved me, and I believe in him who has given me his promises.

Thus the life of piety has as its goal not that all our time should be filled up with "conscious" faith and love, i.e., with conscious acts of believing and loving, so that everything else is dismissed from our minds as we move to at least approximate that total filling of the psyche. On the contrary, Luther recommends periods of prayer and devotion merely as "signs" that my time is under God, not as devices for filling up my time—at least partially—with thoughts of love and faith. The object of my faith, God's love and righteousness, is still effectively present even when I "feel" nothing, even when I am "empty" or terrorized by doubt.

Hence love for God, if we may put it epigrammatically, is not a "content," but a "sign" which stands over all contents. It does not consist in being psychically filled with love for God, with pious

12 See above, p. 182, n. 2.

thoughts, with an enjoyment of God [*frui Deo*]. Instead it places all our acts—our love of neighbor, our work, joy, sleeping and waking, eating and drinking—everything under the commandment of love. It makes them all functions of love. It groups them into a single category above which stands the caption "love." "Whatever your task, work heartily, as serving the Lord and not men" (Col. 3:23). Here the unconditional character of the command to love remains intact in its twofold significance: first as the Law which accuses, and which cannot be relativized by any psychical structure, and second as the Gospel, which addresses to me that love of God which alone can unleash and activate my love (I John 4:19). We thus see once again that even to enter upon the ontological plane, no matter how we may proceed once we are there, is to weaken the unconditional character of the commandments.

The second level of perfection consists in the attempt to be free and open to God and the things of God by dropping everything else except what is required as a bare minimum for the continuation of our present life. This level thus consists in a relative emptying (*vacare*) of the soul of all other contents.

Here again we cannot understand this kind of thinking at all unless we start with the concept of ontic "contents" which dominates it. If, in competition with other potential contents of the soul, love for God cannot establish itself in the sense of itself alone totally filling the soul, then it becomes important, at least as far as possible, to banish these other rivals, and in the form of abnegation to renounce concern for worldly matters. The highest degree of abnegation attainable in this life consists in ascetically restricting oneself to what is required simply to exist. That the thought of competition amongst the various possible contents of the soul is explicitly in view may be seen already from the fact that the term *vacare* is used in such cases,[13] and that there is thus a reference to the freeing or emptying of the soul for love.

Now Thomas does not suggest that this degree of a relative filling of the soul with love for God is attainable by all of us. For it presupposes a negative exercise which only monks can perform, since others, simply by virtue of their calling, are necessarily concerned with

[13] See Thomas, *Summa Theologica*, part 2^{II}, ques. 24, art. 8, where he speaks of "giving his time to [or freeing himself for (*vacandum*)] God and divine things." Cf. Holl, *op. cit.*, I, 164, n. 2.

worldly matters. The relativization of the command to love is accordingly considered from the following angles. First, since love is regarded as a "content" of the soul, and since space has to be made available for this content, the exercise of love takes primarily the negative form of removing obstacles. Second, this removal can proceed only up to a certain point, ultimately to that point which is fixed by what is required simply to exist. Third, this point is to be imposed not on all, by way of precepts, but only on the "perfect," on ascetics, in the form of counsels. The command to love can thus apply in the strict and absolute sense to no one "in this life"; and in the relatively highest degree to which it does apply in this life it is restricted to an esoteric circle of the elect. Finally, since a predominantly negative form of fulfillment is envisaged, the command to love is deprived of its true object, namely, a fellowship with God in love. Perfection of the ego is made the object instead, though this does of course include that fellowship in love as a goal. At all points we see here the same trend towards introversion of perspective which we noted earlier in connection with the Roman Catholic doctrine of grace, where what was involved was not the gracious God and the *extra me* but the "graces" which become properties or qualities of our own and so contribute to the work of sanctification in us.

The third and lowest level of perfection, to which all persons may attain, is that "a man gives his whole heart to God habitually." [14] When we reflect that this is a *habitus* infused *ex opere operato* by the sacraments, and that it may exist in a state of latency whereby the possessor is not even conscious of its existence, we quickly see the strange alteration which has here taken place in the radicalness of the New Testament command to love with one's whole heart. This alteration may be seen even more clearly when we consider the forms in which this *habitus* is actualized. If for monks the acts of love consist in restricting all other contents of the soul to the minimum in order to make room for love of God, then what is involved at this lowest level of love is simply that we "neither think nor desire anything contrary to the love of God." [15] We are thus brought to the most extreme form of negation to which the weakening of the radical character of the commandment can attain.

[14] *Ibid.*
[15] *Ibid.*

Intention and Action

Augustine had pointed out already that a work can have value and be meritorious before God only if it proceeds from the motive of love. What I do legalistically, i.e., under compulsion, is done without joy; it is as if I myself were not present and involved in the doing of it. "What thou doest without joy, is only done *through* thee; thou *thyself* doest it not." [16] An act can be meritorious only if I am myself really the acting subject, doing it in free spontaneity, out of love. For only in love is compulsion by the Law excluded. Only when we "praise God in love" do we act out of free will. "When thou lovest [in this case, feelest thyself drawn to] that which thou praisest; compulsion is not involved, because you are only doing what you want to do." [17] In all other cases I am not the subject but simply an object, namely, the object of that which is awakened in me by fear and hope.

This thought has a Reformation ring—we have only to substitute "faith" for "love" to find many parallel statements among the Reformers. Love is here a total movement of the ego, and hence the only true fulfillment of the Law. Nevertheless, this personalistic understanding of love is inevitably modified by the Scholastic scheme of thought. On this point Karl Holl has collected the essential materials,[18] and in what follows we shall simply attempt a dogmatic evaluation.

Even at a later period love is still recognized to be a motive, or formal cause. It is very largely depersonalized, however, for it denotes a "power of supernatural grace infused in the sacrament." [19] As such, love is a *habitus* which is to be characterized ontologically, not personalistically. Hence for true fulfillment of the commandments it is not necessary that love should be present as a conscious, personal motive; love can also be the unconscious sacramental character of a person, of which the person himself is by no means expressly aware. This sphere of the *habitus* thus constitutes a zone of immunity for the ego, in which the ego cannot be called in question by God's unconditional demand. This *habitus* thus represents a blunting of the radical character of the command to love.

[16] See above, p. 303, n. 4.
[17] See Augustine's commentary on Psalm 135:6 (*Enarrationes in psalmos*, CXXXIV, 11), *NPF* XL, 1945.
[18] See Holl, *op. cit.*, I, 165 ff.
[19] *Ibid.*, I, 141.

In this way the *de facto* opposition between my conduct and God's commandments (e.g., those of Colossians 3:17 and I Corinthians 10:31) may be so crass—there may be so little evidence of my being personally involved and of love being actually operative as a motive—that the question naturally arises whether and how far a certain actualization and awareness of the love motive ought to be required, whether and how far one must emerge from the mere *habitus* if he is to do justice to the true significance of the command to love.

The Scholastics' answer to this question is typical of their ontological mode of thinking. They seek refuge in the concept of what is called "virtual intention" [*virtualis intentio*]. This denotes an original purpose or intention once determined upon, but a purpose which—this is signified by the term "virtual"—does not have to be continually produced anew during the course of the ensuing actions. Bonaventura used the well known illustration of the man who wanted to donate a hundred gulden for a certain charitable purpose. It would suffice if his intention was serious at the moment he took out his purse. It was not necessary for him to have his charitable intention consciously in mind each time he laid another coin on the table. In fact, it would be more sensible for him to forget his "ultimate purpose" and concentrate from moment to moment on the "immediate goal" of actually getting that next coin safely onto the table. This virtual intention might be compared with the concept of "religion" in Schleiermacher, which is not adapted to furnish the motive for moral action (though for reasons other than in the case of Scholasticism's virtual "intention") but which simply "accompanies" all action like a kind of music.

Since the underlying purpose of so restricting the commandment of love to the sphere of intention is to avert any challenge to man in his totality, there is an inevitable tendency towards increasingly refined casuistical considerations and distinctions. Thus Thomas deals with the question whether it suffices to refer to God just "once a year" all that we expect to do that year, and thus to act out of the "intention" of love.[20] His answer to this question is for our purposes irrelevant. The question itself is enough to show us to what extremes of casuistry one is driven once one begins to distinguish between *habitus* or "in-

[20] Thomas, *Commentary on the Sentences*, II, dist. 40, ques. 1, art. 5, 6; cited in Holl, *op. cit.*, I, 168, n. 3.

tention" on the one side and "act" on the other.

This question brings us back again to the distinction between precepts and counsels. "Precepts" have reference in the first place to that ultimate intention which has been given us as a *habitus* by the infusion of grace, and which is actualized by virtual intention, i.e., by the act of conscious and deliberate love which takes place from time to time. And precepts also have reference to the fact that acts which contradict this intention, and which are thus "opposed" to the commandment, are forbidden. This relativized requirement, so drastically divested of its original absoluteness, is then supplemented in a relative way by the "counsels," to which the monks, the *homines religiosi,* subject themselves by way of supplementary meritorious acts. Here too, then, what is involved is that particular kind of thinking which is done in terms of degrees and levels, which finds expression in the system of various degrees of merit, and which even exercises control over the "other world." For here too, in the world to come, there is another gradated structure of punishments and blessings, i.e., in purgatory.

We may thus formulate the following law: Deradicalizing the commandments by differentiating them into precepts and counsels leads to gradated forms of fulfillment. This type of thinking in terms of degrees is itself conditioned and made possible by the general ontological thought pattern as such. For in this ontological scheme grace is not just the favor of God [*favor Dei*]. It is also at the same time a material quality, a "medicine," which must needs be dispensed in various doses and according to varying degrees of effectiveness. This kind of theological reflection always has its ultimate root in the ontological way of thinking which is its starting point.

It is thus no real help if certain monastic orders, like the Franciscans, happen to tackle the question of intention with greater precision and responsibility than others. For here too we can at best only arrive at comparative levels or degrees within the accepted relativity.

The Legalistic and Literal Absolutizing of the Sermon

Diametrically opposed to these Roman attempts at relativizing the Sermon are the attempts of the fanatics to fulfill literally, already in this life, the unconditional command of the Sermon on the Mount, and thereby to demonstrate its applicability to this aeon. Here there is no trace of any awareness that the Sermon on the Mount is actually giving expression to the tension between this aeon and the coming aeon. There is no realization that the Sermon on the Mount is, as it were, a guard standing sentry at the border where the two aeons meet, and that it contains a reminder of the provisional nature of "this aeon." The fanatics will not admit that the laws enunciated for this aeon stand under the divine patience and forbearance, i.e., that they are given because of "the hardness of our hearts." Instead, they see in the Sermon on the Mount a Law which is divorced from its context in salvation history and which is thus invested with a false absolutism, a timeless validity, and is to be fulfilled to the letter.

As it is a mistake on the one hand to limit in advance the demands of the Sermon on the Mount and to refuse to apply them to ourselves in all their rigor, to take the teeth right out of them as it were, so it is no less a mistake on the other hand to stop asking why these demands, even when they are taken seriously, are never fulfilled in this aeon. We have seen that this question has in fact been raised in Scholasticism, and that the answer is that because of its "instability" man's life cannot be totally and at all times referred to God. Thus Scholasticism sees very precisely the limits of attainability, and for this reason makes significant attempts to find theological sanction for a relativizing of those absolute demands. But we have seen also why this explanation is inadequate. Scholasticism thinks it is the constitution of the human psyche, i.e., its instability, which brings to grief the absoluteness of the demands; it thereby psychologizes the Sermon on the Mount itself.

In fact, however, the question why "this aeon" cannot of itself provide any total fulfillment may be answered only in terms of salvation history. The laws proclaimed in the Sermon on the Mount are those of the coming world, and these laws continually draw the attention of this aeon to a gaping wound which it bears. This aeon is the world of the Noachic covenant; it is the provisional, passing world. If this provisional character of the world is not to be lost sight of, then the radicalness of the Sermon on the Mount must be maintained unimpaired. If the Sermon's radical claim is restricted in any way, the world may easily come to think of itself as intact, a "self-contained infinity."

In this connection I would call attention once again to our earlier proposition that the Sermon on the Mount addresses us as though we were still in our primal state, and as though the new aeon had already definitively dawned. Only when the Sermon on the Mount is understood in this way—only when we do not weaken its radical character—can we perceive that moment in salvation history at which we and the world now find ourselves. This is the moment in which we realize that we are not only fallen but also summoned home, so that our old existence stands under the patience of God and we are already, even in this existence, accepted in grace and set in the new aeon. That the Sermon on the Mount does not hereby represent merely the "accusing function" [usus elenchticus] of the Law, that it does not merely bring us under judgment but also gives us concrete directives and orders, will have to be shown in our development of practical ethics, and especially in the portion dealing with politics. There we shall have to face seriously the question asked by Bismarck, whether one can actually rule the world with the Sermon on the Mount.

But we are getting ahead of ourselves. Our task at this point is to recognize that there is a certain parallel between the Scholastic and the fanatic points of view. Neither pays attention to the situation in salvation history to which the Sermon on the Mount applies. Neither has in view the sphere where the two aeons intersect. As a result Scholasticism, on the one hand, limits the demands of the Sermon with a view to the ontic structure of man, trying to show that these demands cannot in this life be harmonized with this structure. The fanatics, on the other hand, adopt for the same reasons an equally undiscriminating and legalistically absolute conception of the radicalness of the Sermon on the Mount.

It is of a piece with this understanding that one may see at work among the fanatics that law of nomistic casuistry which is operative in all sectarian movements. In monomaniacal fashion they pander to certain pet preferences, for which they believe they have found authorization in the Sermon on the Mount. Thus for one thing, the prohibition of oaths is singled out and overemphasized. Above all, however, it is held that the Sermon on the Mount demands pure passivity in relation to the world, that we should "not resist evil." This rule of passivity is specifically applied—even to our own day—in respect of the attitude one adopts towards public life, which is thought to show an evident affinity to the "ruler of this world." Thus one is not to hold public office, or be a soldier, or acquiesce in the usurious aspects of a capitalist economy, or employ other people as servants. In conflict situations there must be no fighting, only patient suffering in the sense of "bearing one's cross."

It is precisely their legalistic understanding of the Sermon on the Mount which causes them to miss the decidedly positive thrust which is the Sermon's constant and overriding concern, namely, the unconditional command to love, a command which embraces even our enemies. The many negative prohibitions, such as those just mentioned, cannot by any means be isolated and understood as independent laws. On the contrary, they exist solely to serve the positive commandment, and their meaning and spirit are to be interpreted accordingly. If this is not perceived, then in our utilitarian fulfillment we will miss the spirit of the Sermon on the Mount, and in our casuistical radicalism we will miss the unconditional nature of its spiritual claim. Thus, when the fanatics ask who is my neighbor, they give an answer which leaves only a narrow circle of men, namely, those who belong to their own group. "They will hardly speak to people who do not belong to their own sect," [21] let alone extend their love to an enemy.

We find much the same features in Tolstoi, particularly the characteristic coupling of negatives with the postulate that we are not to resist evil. Among these negatives particular stress is laid on the fact that we are not to defend life or property against injustice or attack by force, and that consequently policemen and laws are to be regarded as contrary to the commandment. No less radical in its effect is the application to our world's disordered state of the command to love our enemies, for war is forbidden, and thereby indirectly also all group egoism—such as patriotic nationalism—which serves as a fertile soil for warlike tensions.

In essence the radicalism of Tolstoi too has nothing to do with the Sermon on the Mount. It does not derive in the first instance from love, nor does it live eschatologically by the law of the coming aeon, that aeon which is signified and provisionally reflected in the Sermon on the Mount and in the miracles of Jesus. Moreover, the basic impulse of Tolstoi is not connected with the person of Jesus, who is already here and now the representative of the coming aeon. These several constitutive features of the Sermon on the Mount are almost totally

---

[21] Sebastian Franck, *Chronika*, 444, cited in Holl, *op. cit.*, I, 459, n. 2.

lacking in Tolstoi. Instead, he is impelled to his convictions by the supposedly rational nature of this radicalism. He lays particular stress on this in his diary. Christianity is for him a "metaphysics of moral economy." [22] Active struggle for the good is not all the direct why to achieving it; it is indeed a bypath. For where there is struggle and conflict, contrary forces are unleashed, and the whole field of conflict is atmospherically polluted. The only attitude which promises success, even in the practical sense, is passive endurance. It is to allow the enemy's attack to exhaust itself deep inside one's own territory—a conception which may be partially understood, at least psychologically, in terms of the infinitude of the Russian soul. This glorification of passivity almost reminds us of the cynical comment of Oscar Wilde in *The Picture of Dorian Gray*, that the best way to have done with temptation is to yield to it. In any case what is involved here is a metaphysically grounded technique of living and working which corresponds to the mysticism of pure immanence which is its background.

Now it is wholly in keeping with this "economic" principle that love is understood almost Stoically as an attitude which conserves energy. For it would be a dissipation of energy, and therefore uneconomical, to bind myself with real love to another person, thereby subjecting myself to that incessant intercourse which is made inevitable the moment I give myself to solidarity with him. Tolstoi is surely thinking of the pertinent fact that the greatest and most exhausting exertions and efforts in life are not occasioned by hate but by love, i.e., by the fact that we bear the less fortunate person as a burden on our own soul. If love is to correspond to that principle of economy, it can be nothing more than a general disposition which does not commit me, an attitude of mere friendliness which does not bind me to someone else in a relationship of personal solidarity. In his essay on Tolstoi, Karl Holl quotes some typical statements from *Plato Karataiev*: "He lived in love with men—not with any particular man, but with the men whom he encountered—yet Pierre felt that, for all his warm friendliness towards him, Karataiev would not have experienced the slightest pang at their parting." [23] If I may adapt an expression of Boris Sapir,[24] love

<hr />

[22] Holl, *op cit.*, II (Tübingen: Siebeck, 1928), 446.

[23] *Ibid.*

[24] See Boris Sapir, *Dostojewski und Tolstoi über Probleme des Rechtes* (Tübingen, 1932), pp. 78 ff.

as an economic principle thus involves the same "schematization of man" which Tolstoi sees to be dominant in the orders of society against which he revolts. Any economic view of man and his functions is bound to result in this kind of impersonality, since man is here seen primarily from the standpoint of his being the bearer and object of a process which ends with all living happily ever after. The point of the whole teaching is that all men are brothers, and that if they will but show love and pity for one another all will be well. The economic goal of "the greatest possible well-being of all" is thus clearly taken to be the intention of the command to love!

How far removed this is from the Sermon on the Mount may be seen from the fact that the basic principle of Jesus, namely, that love for my neighbor rests on the fact that he too is a child of God, is here replaced by a Utopian social reconstruction which is at root impersonal. There is in Tolstoi no distinctively personal relationship to God. This is because God himself is not regarded as personal. He is simply a symbol of such reason and spirit as are present in all men. Hence God is not a personal "Thou" to whom man is related. He is an impersonal "It" immanent in man himself. This is the underlying basis of the impersonal character of man's love for man. The other person is not seen in his uniqueness, as a person, but merely as an instance of the species which possesses reason and spirit.

This depersonalization is perhaps displayed most crassly when Tolstoi says that, while it is within our capacity to abstain from harming our enemy, it is not within our capacity to love him, yet Christ cannot have been demanding an impossibility. This "abstain from harming" is again a kind of economic concept. It means that we do not exhaust ourselves in struggling against someone, but follow instead the more rational path of putting up with him. That the command to love our enemies means something very different on the lips of Jesus, that it relates not to world amelioration but wholly to the Thou of the other person seen as a child of God and a brother of Jesus Christ—and therefore apart from the hostile system which controls him—all this is either forgotten or not perceived by Tolstoi. For him, love in the Sermon on the Mount is divorced from the Person who preached it; it is made instead into a principle. And this principle is always timeless, whether the reference be to a valid law, an economic procedure, or an ideal construction, for example, of society. It is for this reason that Tolstoi, in spite of his very different concept of God, must be grouped with the fanatics of the Reformation period.

We thus maintain that, if the Sermon on the Mount is understood in terms of literalistic casuistry, if it is taken legalistically, then it is divorced from its situation in salvation history. And to do this is to miss the whole point of the Sermon, namely, that it is inseparably bound up with Jesus Christ as the representative of the coming aeon, and that it proclaims a love which has meaning only with reference to him. That this love has meaning only with reference to him implies that the love required of us is only a response to or re-enactment of that love which has come to us in Jesus Christ. It implies too that the love which is called for is directed toward a neighbor, or even an enemy, who stands in the light of that same love of Jesus Christ. Thus one cannot make a timeless principle, or a mere disposition, out of that which the Sermon on the Mount understands by "love." This brings us to yet a further misunderstanding of the Sermon on the Mount.

The Restricting of the Sermon to the Area of Disposition

Johannes Müller is perhaps the most typical modern representative of this particular tendency; at least his interpretation has exerted a considerable influence on recent discussion.[25] Müller finds in the Sermon on the Mount not a law which is to be kept literally, but an appeal for that "disposition" out of which a Christian is supposed to act, namely, the disposition of love. The concrete form in which such a disposition will be realized can vary greatly with varying circumstances. It can be very different from those specific and extreme forms which Jesus advanced by way of illustration. Under certain circumstances the disposition of love may require that I not turn the other cheek to the one who is striking me, but that I give him instead a powerful blow in return. For it could be that in patiently offering the other cheek I would be doing him a disservice. I might thereby be merely enhancing his rudeness and contemptuousness. A counterblow, on the other hand, could conceivably teach him manners. The field of teaching seems to provide many impressive examples of the fact that love can go hand in hand, not merely with passive endurance, but

[25] See Johannes Müller, *Die Bergpredigt verdeutscht und vergegenwärtigt* (Munich, 1906).

also with active intervention, with discipline, and with an uncompromising insistence on authority. The main thing of course would be that such counteraction should truly have as its motive love, and not the law of revenge, "an eye for an eye and a tooth for a tooth" (Matt. 5:38; Lev. 24:19 f.). This is what takes the venom out of it whenever force has to be used.

Now we must be impressed by the way in which this interpretation tries to go by the spirit of the Sermon on the Mount rather than the letter. We must also acknowledge a measure of validity in the directives it gives for concretely applying the Sermon on the Mount in the present aeon. Nevertheless, this line of understanding seems to overlook certain essential features which are integral precisely to the "spirit" of the Sermon on the Mount.

In the first place, the radicalization of the Mosaic *nomos*, so far as the Sermon on the Mount is concerned, takes place in the *radix* or root of "thought" as well as "disposition." Jesus extends the Mosaic requirement to the total person by emphasizing the sovereignty of God and the responsibility of man even in respect of that stratum of primal spontaneity which is man's heart. What is called in question is not merely murder nor yet a murderous disposition, but anger as a "thought," and hence the most profound depths of the human heart (Matt. 5:22). This is something my will cannot control, because it is something that precedes all will. The same is true of the lustful thoughts which lie well below the surface of actual adultery, and even of an adulterous disposition (Matt. 5:28).

There is thus a far-reaching difference between "thoughts" and "disposition." For "disposition" is already the product of an action. It is something over which I have a measure of control. Though it is inward, it belongs to the sphere which is within my competence and at my disposal. It has to do with the way a man shapes his life inwardly. To this degree it belongs to the objective sphere of my ego.

In the case of "thoughts" on the other hand the situation is quite different. Thoughts are not objective to me. They stand at the very root of my ego. They are, as it were, that which is always part and parcel of me, the starting point from which I proceed. Now it is precisely here, in this innermost part over which I have no control, that I am called in question. This can only mean that what is innermost is imputed to me; I am responsible for it and must acknowledge it as "mine." But this decisive point is not brought out if the demands of the Sermon on the Mount are limited only to the area of "disposition." When this is done, they are surreptitiously made into a moral law such as Kant would also affirm. What is involved is simply a deeper morality which is not limited to acts but which reaches down to the level of motives, a morality which is concerned more for the maxims of action than for the action itself. What this morality fails to see, however, is that far beneath and behind the level of motives in man there is always something more; there is yet another sphere of the ego for which man is equally responsible *coram Deo*. When mere "disposition" is understood as being the real central core of the person, then the Sermon on the Mount becomes a disposition-shaping principle. But this is to ignore the fact that in the Sermon man is viewed *coram Deo*, as one who has his existence from the hand of God and is thus totally indebted to God, as one who stands in the dilemma between the original state and the coming kingdom. And this is also to ignore the situation in salvation history, and to proclaim instead a timeless principle of "disposition."

In the second place, to restrict the Sermon on the Mount to disposition is to rob it of the function which we have described as that of "challenging" or calling in question both our own individual persons and the whole of this present aeon. After all, we have to ask the question how it happens that Jesus should define appropriate performance in terms of such strange and disconcerting examples as "turning the other cheek" or "plucking out the eye." Can we really render these extreme demands innocuous by saying that what we have here are merely rhetorical statements which utilize the techniques of exaggeration and epigrammatic formulations? Surely this difficulty cannot be solved in so banal a fashion! But that being true, then the problem is posed in quite different terms: Do we not have to ask why it is that we find it necessary to display our love in such a very different way from that which Jesus seems to insist upon? Why is it that we cannot simply duplicate his explicit examples in our daily life?

The heart of my objection to Müller—who does not even raise this question—is not to be found in the belief that we have to follow the examples of Jesus literally and without the slightest deviation, and that the love which is commanded cannot also be realized in such forms as punishment, resistance, or even the pedagogic use of force.

The fanatics' misunderstanding of the Sermon on the Mount ought to have been warning enough for us not to espouse a legalistic and literal interpretation. What I fail to see in Müller is rather any consideration of the question how it comes about that we find it necessary to fulfill the Sermon on the Mount in such very different forms from those indicated by Jesus himself in the examples he cited.

Theologically, however, this is the decisive question. For it directs our attention to the character of this aeon as a whole, which in principle is of itself incapable of producing the radical results demanded, and which is thus exposed by the Sermon on the Mount, both in its totality and especially as regards its particular "orders" [Ordnungen] as a fallen aeon. For this aeon lives in contradiction and is ruled by the law of conflict. This is why there can indeed be such a thing as self-assertion, punishment, or even self-defense that proceeds from love. When the commandment of God enters the medium of this aeon, it undergoes, as it were, a kind of refraction, similar to that of a stick in water. If it were to insist on its unconditional character, it would blow the world to bits or bring it to a standstill. That it does not do so is due to the patience of God, who out of regard for the structure of the fallen world graciously refuses to let it fail.

But the fact that the commandment does not insist on its unconditional character does not mean that we are excused from perceiving and acknowledging that unconditional character. It is one thing to have to actualize the unconditional character, and another to have to perceive and acknowledge it as such. The fact that on account of the "hardness of men's hearts" God can permit divorce does not exempt us from seeing but actually requires that we must see behind it God's real will, that which he commands unconditionally. We must see that from the creation "it was not so," and that in making his concession, in refusing to insist on the unconditional requirement, God is simply having regard for the weakness of our nature and the difficulties of the fallen aeon. (We shall be pursuing this question more closely in a moment as we discuss Luther's doctrine of the "two kingdoms.")

To say all this is not to relativize the Sermon on the Mount. On the contrary, we are taking the absoluteness of its requirements with utter seriousness when we see how they compel us to understand this aeon as the zone of relativities, and for this very reason as the fallen aeon.

When it comes to the question of concretely fulfilling the demands of the Sermon on the Mount we shall always have to enter into a consideration of various degrees, and for this reason the actual fulfillment will, so far as form is concerned, strongly resemble the fulfillments recommended by Scholasticism. On the level of deeds, as in all other respects as well, things will be much the same, even to the point of interchangeability. Nevertheless, it should have become sufficiently clear by now that this type of thinking in terms of degrees has a very different theological basis from that of Scholasticism. What is involved is not the resultant in a parallelogram of forces, a compromise between the thrust of the commandment and the laws of "this life." What we have here is rather a sign of the fallen aeon, which stands before the law of both the original and the coming world in its inadequacy and guilt, confessing that "I ought, but cannot."

This view of the world, this perspective in terms of "aeons," is completely missing in Müller. He refers the problem of the Sermon on the Mount solely to the sphere of "inwardness," and within this sphere to the single point of disposition. What he fails completely to bring out is the fact that man is radically claimed in his totality.

18.

Luther's Doctrine of the "Two Kingdoms"

After all these restrictive interpretations of the Sermon on the Mount, it is with great expectation that we turn now to Luther's position in respect to this problem. To treat the question properly, it would really be necessary to include a history of the interpretation of Luther's doctrine of the "two kingdoms." For if, as regards Luther, the decisive issue in the dogmatic sphere is the *sola fide*, then in the sphere of ethics the decisive issue is surely his teaching on the "two kingdoms." The din of battle is accordingly all the more confused with reference to this teaching. And anyone who would seek to sketch it even in its barest outlines is immediately plunged into the thick of the polemics. So even as we try to clear up certain misunderstandings of the doctrine, we shall also be posing some critical questions of our own, questions however which are directed not so much to Luther's basic principle as to the forms in which he expounded it.[1]

God's Will in the Orders and in the Radical Commandment of the Sermon on the Mount

We best start with the discrepancy Luther seems to find between the radical demands of the Sermon on the Mount and the structure of this world's orders [*Ordnungen*] which seem to stand in opposition to them. Can the commandment not to resist evil, for example, really apply in the matter of rearing children? Is a father really supposed to spare the rod in the name of this commandment? Luther says, "It

[1] The more important works on this issue include Ernst Troeltsch, *The Social Teaching of the Christian Churches*, trans. Olive Wyon (New York: Macmillan, 1931), II, 494 ff.; Georg Wünsch, *Die Bergpredigt bei Luther: Eine Studie zum Verhältnis von Christentum und Welt* (Tübingen: Mohr, 1920), and *Evangelische Ethik des Politischen* (Tübingen, 1936), esp. pp. 65 ff.; Franz Lau, *"Äusserliche Ordnung" und "weltlich Ding" in Luthers Theologie* (Göttingen, 1933); Harald Diem, *Luthers Lehre von den zwei Reichen* (Munich, 1938); Herman Beyer, *Der Christ und die Bergpredigt nach Luthers Deutung* (Munich, 1935); Gustaf Törnvall, *Geistliches und weltliches Regiment bei Luther* (Munich, 1947); Herman Diem, *Karl Barths Kritik am deutschen Luthertum* (Zollikon-Zürich, 1947); and Otto Dibelius, *Die Grenze des Staates* (Berlin, 1949).

would be most cruel, indeed gruesomely homicidal, if a father were to let his child go unpunished; it would be like choking it with his own hands." [2] To fulfill it literally would thus be not to keep the commandment of love but to break it and pervert it into its very opposite. The tolerant act of "love" would actually be murder. And "what mother would be so foolish as to stand idly by while a wolf or a dog attacked her child, and then explain her inaction by saying, 'A Christian is not supposed to defend himself'?" [3]

The same problem arises in respect of an aggressive action undertaken against us by an enemy. If the Turks attack us, should we not dispatch to them our learned truce negotiators (the theologians of Louvain) who will show them how unjustifiable their war is and threaten them with an anathema? Should we really let them ravage unopposed, and in so doing feel that "inwardly" it is we ourselves who are the victors? [4] Obviously, it would be consonant with the command to love also those who are threatened that we not permit this but take active steps to ward off the threat.

Luther thus sees that the moment we enter the sphere of the "orders" certain factual requirements obtain which contain directives for the shape which our concrete fulfillment of the law of love is to take. Now the first thing of importance to notice here is that this question of the mode or shape of our obedience is in fact raised. That this should happen is not to be taken for granted. It is not something implicit in the commandment itself. It is rather something that is determined in part by the sphere in which the commandment is to be kept. If we do not take into account this sphere, and its influence on the mode of fulfillment, then the commandment may be perverted into its very opposite. The situation is similar to that of a plant which is good in itself, divinely created, but may become poisonous in its effect when assimilated into the human organism because it is not meant to be taken thus, directly, even though in some cases it may be made into a useful medication by means of certain pharmaceutical processes. Now this is apparently the situation with respect to the demands of the Sermon on the Mount. They are valid for the new world of God, but if they are applied without sense or reason

2 WA 41, 325.
3 WA 32, 391.
4 See LW 32, 149.

to our world they act as destructive poison. In the matter of rearing children, for example, they bring murder and chaos, as we have seen.

For the moment we will skip over the fact that the failure of God's radical commandment in our present world is really a failure of our present world in respect of God's commandment, that the commandment decisively challenges our world and calls it into question. This aspect of the matter unfortunately is very secondary in Luther. Indeed his lack of regard for it has contributed significantly to the impression that Luther actually ascribed autonomy to the kingdom of this world and was thus the forbear of modern secularism. What we are concerned with at the moment is the reason why Luther could not believe that God wills by means of these radical commandments to bring chaos.

What God wills is in fact the preservation of the world in face of the destructive onslaughts of the devil. He wills to bring his world through to the last day. One of his means of preserving it is the "orders," and especially political authority [*Obrigkeit*], which for Luther is "a sign of the divine grace, of the mercy of God, who has no pleasure in murdering, killing, and strangling," [5] i.e., in destruction. "Temporal authority was not instituted by God to break the peace and to initiate war, but to keep the peace and repress the fighters. As Paul says in Romans 13 [:4], the office of the sword is to protect and to punish, to protect the good in peace and to punish the wicked with war." [6]

In other words, what I encounter in the temporal order is God's will to sustain his world and not allow it to perish. Hence it cannot possibly be the will of God at the same time to bring in chaos and to disrupt this world by his radical commandments. The will of God expressed in these commandments cannot possibly contradict the will of God expressed in his establishing of governmental authority and of parenthood with its attendant responsibilities in the rearing of children. Hence, in this sphere of the offices and orders of preservation what is involved can be nothing more than a mode or form of obedience which carries over the "absolute" commandment in a mean-

---

[5] *The Table Talk of Martin Luther*, trans. William Hazlitt (Philadelphia: United Lutheran Publication House, n.d.), p. 371; WA, TR 1, 77.

[6] *PE* 5, 56. See also *LW* 13, 45, and WA 31^{II}, 590 where in line 11 the political order is said to be a "gift" of God, a "miracle" sanctified by his Word.

ingful way into the sphere where God's will is expressed through the orders.

Of course, alongside these mediate expressions of God's will in the orders there is also a sphere of immediacy in the relationship of I and thou, of person and person. In this latter sphere such a transposition of the radical commandment is neither necessary nor indeed possible: "Apart from that office [of rebuking and denouncing], let everyone stick to this teaching: You shall not denounce or curse anyone, but you shall wish him everything good and show this in your actions, even though he may act badly; thus you will disclaim all right to mete out punishment, and you will assign it to those whose office it is." [7]

Luther thus distinguishes here between conduct in the "official" sphere and conduct in the "personal" sphere. What is good in the former, e.g., offering resistance, may be disobedience in the latter. Thus far all interpreters of Luther, from Troeltsch to Törnvall, are in agreement. However, Törnvall correctly argues against Troeltsch that this is not yet the decisive point in Luther's teaching, and that Troeltsch and Wünsch are mistaken in regarding this distinction as the essence of Luther's view and on that basis establishing a double standard of "official" morality and "personal" morality. In point of fact, the real problems first begin to arise when we ask *how* then this distinction is actually to be drawn between conduct in the official sphere on the one hand and conduct in the personal sphere on the other.

In the first place it must be admitted that Luther often expresses himself rather unguardedly, so that certain passages may perhaps be found which seem to support directly the thesis of Troeltsch. Thus Luther sometimes differentiates so sharply between the temporal kingdom and the kingdom of God that "secular men" and "Christians" seem to be two distinct classes of men belonging either to the one kingdom or to the other. Such statements often seem to suggest that the Christian—who must also participate in the temporal kingdom as a father or mother, a citizen or soldier—surrenders, as it were, the identifying marks of his Christianity the moment he enters the worldly sphere.[8] Thus, one would have to agree that there is something

[7] LW 21, 124.
[8] See Luther's *Temporal Authority: To What Extent it Should be Obeyed* (1523), LW 45, 88 ff.

suspect about the statement that "a Christian should not resist any evil; but within the limits of his office, a secular person should oppose *every* evil." [9]

How are we to understand this? Is not the Christian too, apart from the personal relationships in which he stands, a "secular person," an office bearer? And are we not thus forced to the interpretation that in his "quality" as a secular person he must resist every evil, whereas in his "quality" as a Christian (in the narrower sense) he must not? And does not this imply that one's Christianity is simply a "quality" which we have to set aside or ignore in certain situations? Or, conversely, has it no bearing whatever on the duty performed by a secular person that the person who thus acts is also a Christian? Is it really of no consequence that the person here discharging the office of statesman, judge, or father is a Christian? Or do these functions performed by the secular person—or by the Christian in his quality as secular person—have no affinity whatsoever to the man's Christianity, and hence also, for example, to the Sermon on the Mount? Do they simply "follow their own laws"?

When Luther makes bold to assert that Matthew 5 properly applies only to "Christians" and not to those who are "un-Christian," [10] one is inclined to see here a differentiation not merely between two classes of men (Christians and secular persons) but also between two spheres of existence which have nothing to do with one another, spheres which are subject to very different laws and which divide the Christian person—through whom the dividing frontier passes—into two completely separate and isolated segments. At any rate such unguarded statements as these have contributed to the notorious misunderstandings of Luther's doctrine of the two kingdoms.

Now if this were in fact Luther's view, we should indeed have to ask in all seriousness whether he too has not relativized the Sermon on the Mount, and beyond that opened the way to certain developments which are essentially more dangerous even than the relativization which we found in the Scholastic doctrine of "virtual intention." For according to such a view, could there be more than merely virtual influence of the Sermon on the Mount and of Christianity on secular affairs? Would there not develop a privileged secular zone of im-

[9] *LW* 21, 113.
[10] *LW* 45, 92.

munity whose thoroughgoing secularization would have theological sanction? Did not the Lutherans of Germany all too often teach this view during Hitler's Third Reich, tacitly giving free rein to the state in what was supposed to be its sphere of operation? The very fact that the question of the extent of this "secular" sphere, and of the limit of the state, was seldom raised shows how dominant was this line of thinking. For the state has no limit whatever (for the church to keep a "watchful" eye on!) unless the commands of God apply also on its territory—and so can precipitate a conflict when that territory is overextended! In the light of these experiences in Nazi Germany, Lutheranism has every reason to examine its basic position. Whether the criteria are theologically sound by which most of these examinations are being conducted, including those of Barth and his disciples, is a very different question, to which we will now address ourselves.

Possible Dangers in Luther's Doctrine

From what has been said, it is clear that we have to consider three potential dangers attendant upon Luther's doctrine of the "two kingdoms": bifurcation, secularization, and harmonization. At the same time we must ask what possibilities exist within the doctrine itself for meeting these dangers.

The Danger of a Double Morality: Personal versus "Official"

The first danger is that of a "double morality" such as that expressed by Troeltsch in his well-known distinction between the official morality of the world and the private Christianity of disposition.[11] This interpretation of Luther, which purports to set forth not a "danger" in Luther but Luther's avowed thesis, finds its bluntest expression in Georg Wünsch's very dubious work on the Sermon on the Mount in Luther.[12] The main value of Wünsch's work is that it shows us how Luther is interpreted from the standpoint of certain deviations in modern Lutheranism. In this sense it poses some very serious questions for certain theological tendencies. Wünsch sees in the distinction between the two kingdoms "a permanent influence of his [Luther's] early mysticism, which in The Freedom of the Christian

[11] See Troeltsch, op. cit., II, 506 ff.
[12] See Wünsch, Die Bergpredigt.

distinguishes between the outward and the inner man, and which carries over this distinction into the relation of the Christian to worldly events." [13] Wünsch then follows Troeltsch in his description of this distinction: "As inner man the Christian acts within the kingdom of God wholly intent upon fulfilling the morality of the divine goodness, but as secular man he follows in his office the autonomy of the world in pursuing a morality of force and of power." [14] The cleavage into which the Christian is thereby plunged is said to be something he accepts only reluctantly, attempting at the same time to mitigate it by a relative aloofness from the autonomous world.

For Wünsch's Luther, the relation of the Christian to the world is thus marked by two main features. On the one hand, there is a pronounced quietistic element which indeed tolerates the autonomy of the world but does not allow of "joyful, willing, and active participation in and transformation of it." On the other hand, there is a positivistic belief in the creator which just "takes things as they come." It does not call things in question on the basis of the commandments of God, nor try to order everything according to the commandments. It simply accepts the things that are as morally neutral. [15]

Try as we will, we cannot see here a true picture of Luther's theology; but we do see with startling clarity certain features of modern Lutheranism in Germany. The theological foundations of a document like the Ansbach Ratschlag, for example, could not be more strikingly revealed. On June 11, 1934, the Ratschlag was addressed "primarily to our brethren in the National Socialist Evangelical Union of Pastors." It was intended to express the authentic voice, not of the "German Christians," but of genuine Lutheranism. If it were merely the claim of the "German Christians" that was here raised, we should have to regard it as beneath our dignity to quote the document—posthumously at that—in a theological work. But here is the self-proclaimed "voice of genuine Lutheranism" with the typical thesis that the basic constitution of the world has been sanctioned by creation, and that it is thus sanctified in even the most questionable of its detailed manifestations.

We are told in this document that the Law "binds all men to the

[13] *Ibid.*, p. 222; the reference in Luther is to *LW* 31, 344 ff.
[14] Wünsch, *loc. cit.*
[15] *Ibid.*

estate into which they are called by God, and obligates us to the natural orders to which we are subject, such as family, nation, and race." Even in this general form the thesis is highly dubious. However, the Ratschlag goes on to work it out in specific detail: "Since the will of God always comes to us here and now, it binds us also to a particular moment in the history of that family, nation, and race, i.e., to a particular moment of history." But how can we dare to speak of the providential character of "a particular moment of history" without at the same time saying that within this moment there also is at work another factor which is not the providential will of God, namely, man with his sin and rebellion against God's will, and that this sin and rebellion are expressed structurally in the historical nexus and institutionally in specific perversions of family, nation, etc.?

At this point we see already where weakness in the doctrine of Law and creation leads. It is thus not surprising that in what follows the idea of "secondary causes" [*causa secunda*] is introduced, and we are told that "the natural orders . . . are . . . also the means whereby God creates and sustains our earthly life." Now it is true that God works through the orders. But does not the sin of man also work through them? If in spite of this we still wish to assert that God is supreme even in human action, that he "rides the lame horse and carves the rotten wood" (Luther), then we ought to do so in terms of the really "genuine Lutheran" concept of the "alien work" [*opus alienum*] of God. In any case the Ratschlag is at least consistent in finally concluding with a statement which everybody today would reject, but the theological root of which is still fed and watered with watchful care in many circles: "In this knowledge[!] we as believing Christians thank the Lord God that in its hour of need he has given our people the *Führer* as a 'good and faithful sovereign,' and that in the National Socialist state He is endeavoring to provide us with disciplined and honorable 'good government.'"

If Troeltsch's interpretation of Luther were correct, it would mean the establishment of a secure and unimpeachable base in the world sphere comparable to that zone in the individual sphere which we have already described in terms of "disposition," a zone which cannot be called in question. The laws of judicial retribution, the principle of force in politics, such matters as punishment, resistance, and conflict, would all then belong so to speak to the "disposition" of the world. And ethics would become relevant only at the point where we ask what then we are to make of these "dispositions" and how we are to act on the basis of them.

The Danger of Secularization: Autonomy for the Various Spheres of Life

The second potential danger attendant upon Luther's doctrine of the "two kingdoms" is that of making the "world" a nexus of "autonomies" [*Eigengesetzlichkeiten*] which a person cannot penetrate and influence but, as Wünsch says, only take cognizance of in a positivistic way, acknowledging it as a fact. But this would be to hand over the economic sphere wholly to the political economists—or to some ruthless captains of industry or even to some Department of Commerce in a totalitarian regime. It would be to make Machiavelli the ideological pope of politics. It would be to abandon education to a pragmatism or humanism of who knows what hue. In short, it would be to farm out every sphere of life ideologically, and to withdraw into a tortuously rationalized quietism on the grounds of one's own personal incompetence. It would be to declare even the commandments of God incompetent, since what would be involved would be spheres which, as a modern Lutheran once said of war, "cannot be moralized." Secularization would then have become a movement endorsed by the church—until such time as that secularization should presume to claim as its own, as "world," and therefore to smoke out, even the sheltering catacombs and esoteric precincts of the church itself.

The thesis that it was Luther who betrayed the world and sold it out to its supposed autonomies—or finally delivered it from its theocratic and Scholastic bondage and thereby helped it to attain its legitimate autonomy—has in the modern period been asserted by two rather opposed schools of thought. The first is that of Karl Barth. Barth too blames Luther's doctrine of the "two kingdoms" for this error, and he finds the root of this fatal doctrine in what he regards as a false separation of Law and Gospel. We have already shown the peculiar way in which Barth himself would distinguish but not separate the two. He, and his followers to an even greater extent, fear or even assume that Luther's radical distinction implies two Gods, and that the cleavage in man as a member of two kingdoms is linked with a dualism in God himself. What is the connection between the hidden God [*Deus absconditus*] and the revealed God [*Deus revelatus*], between the *Deus extra Christum* and the *Deus in Christo?* Do these different dimensions admit at all of being brought together in one person? [16]

It does appear that the main affirmations of Lutheran theology are more or less threatened by the implicit dualism with which they are all weighted down. As we have said, Barth makes Luther's definition of the Law-Gospel relationship responsible for the fatal ethical dualism. In a letter to France in 1939 he writes, "The German people suffer from the legacy of the greatest of all Christian Germans, from the mistake of Martin Luther regarding the relation of Law and Gospel, of temporal and spiritual order and power, by which the Germans' natural paganism has been ideologically clarified, confirmed. and strengthened rather than being limited and contained." [17]

Barth expresses the same view even more clearly in a letter addressed to friends in Holland in 1940; in this letter he asserts a direct connection between this Lutheran doctrine and the nihilism of the National Socialist revolution and the rise of "German paganism." [18] On the other hand, the fact that similar developments have not taken place in other Lutheran countries—as Barth himself admits—should be indication enough that these frightful aberrations are more closely linked with a particular manifestation of Lutheranism than with Luther himself. Yet this is obviously a cheap assertion if we are not prepared to investigate the matter further, and to consider whether these particular forms of Lutheranism are to be interpreted simply as a departure from Luther, or whether there may not be in Luther himself at least certain suggestions or impulses which could have fostered such a development. We can, in fact, hardly avoid some criticism of Luther, though ours will necessarily be at a different point from that of Barth.

But first let us hear Barth again on the connection between the nihilist revolution and Lutheranism: "Lutheranism has to some degree paved the way from German paganism, allotting it a sacral sphere by

[16] On this question see the valuable study by the Finnish theologian Lennart Pinomaa, *Der Zorn Gottes in der Theologie Luthers* (Helsinki, 1938), esp. pp. 202 ff.

[17] *Eine Schweizer Stimme, 1938-1945* (Zollikon-Zürich, 1945), p. 113.

[18] *Ibid.*, p. 118.

its separation of creation and the Law from the Gospel. The German pagan can use the Lutheran doctrine of the authority of the state as a Christian justification of National Socialism, and by the same doctrine the Christian in Germany can feel himself summoned to recognize National Socialism. Both these things have actually happened." [19]

We have seen already that Barth tries to correct this error by connecting Law and Gospel as closely as possible: the Law is the "form" of the Gospel, and the Gospel is the "content" of the Law. On this basis Barth hopes to be able to relate reality to the Gospel in a very different way, and to bring it under the norm of the Second Article of the Creed.[20] It is the Gospel, i.e., the good news of Christ, not the Lutheran *usus politicus* of the Law, which informs us concerning the basic principles of the state. It does so by virtue of the state's existence as an analogue to the kingdom of God.[21] Political principles of every variety are thus deduced "christologically," from the Gospel. But the fact that these christological deductions are so forced and artificial [22] indicates that along these lines a concrete relating of the world to the Gospel and its norms is hardly possible. If we try to correct Luther in terms of such an approach we avoid the Charybdis of an autonomous world only to fall victim to the Scylla of a wholly obscure Christology. The older theocratic conceptions at least had the advantage of confessing openly the analogy of being [*analogia entis*] and, within this erroneous construction, of speaking clearly. (We have already drawn attention to Barth's failure to appreciate the concern of Luther for a separation of Law and Gospel, and to the implications of this failure.)

In any case, if in his separation of the two kingdoms Luther was pursuing a theological intention which can never be surrendered, then we shall have to provide a very different explanation for the observation that with this doctrine Luther makes it dangerously easy for the world to be dissociated from the Gospel. We shall try to suggest such an explanation very shortly when we inquire concerning the objective

[19] *Ibid.*, p. 122.

[20] See "The Christian Community and the Civil Community" in Karl Barth, *Community, State, and Church: Three Essays* (Garden City, New York: Doubleday, 1960), pp. 149 ff.

[21] Cf. *ibid.*, p. 176.

[22] See our discussion of certain examples on p. 275 above.

and subjective linking of the two kingdoms, and the validity of such interrelation.

The second and, as it were, opposite corner from which the thesis of a world independent of the Gospel is advanced—and is thought to be championed in Luther—is that of Arno Deutelmoser.[23] According to Deutelmoser, the main principle of Luther's theology, as evidenced especially in *The Bondage of the Will*, is that the God of the Gospel and the Bible is secondary to the figure of the omnipotent God who alone is the cause of all things. This omnipotent God expresses himself in all of concrete reality, which is his mask. Within this reality the "orders" declare his creative will, and therefore do not come under the specific norm of his Word. For example, the state, which is the foremost order of historical reality, is "divine by itself and in its own power. It needs no priestly justification or spiritual blessing. It bears its own right within itself," since God is "essentially one with his masks."[24] For this reason Luther expressly did not apply to "this divine state the validity of the Christian commandments [i.e., of the Word of God], and thereby limited the application of Christian morality to the private life of man."[25] Deutelmoser thus presses to the limit the main point of Troeltsch's interpretation of Luther: "The state is the work of power. The church is the work of inwardness."[26]

To be sure, on Deutelmoser's view the validity of the commandments can hardly be maintained even within the private sphere. For if conscience is regarded as belonging to this private sphere, then conscience too is a mask of God, a manifestation of his omnicausality, and hence one of those worldly forms which pass themselves off as "autonomous," as "divine sparks in the soul."[27] Obviously Deutelmoser himself does not see that his interpretation of Luther involves conscience itself in a conflict between its "private" commitment to the divine commandments and its quality as an agent of that will of God which is expressed in the orders. It can hardly be contested, however, that, as Deutelmoser sees it, the emphasis in Luther's theology clearly passes from a commitment to the Word to a commitment to reality.

[23] See Arno Deutelmoser, *Luther, Staat und Glaube* (Jena, 1937).
[24] *Ibid.*, pp. 30, 28.
[25] *Ibid.*, p. 99.
[26] *Ibid.*, p. 201.
[27] *Ibid.*, pp. 60 ff.

As a result, even that most private of all spheres, the Word, that sphere which at the beginning was tended by conscience, also disappears. Even this last remaining chapel of Philemon and Baucis must be removed to make way for the expansion of the secular kingdom.[28]

It may be noted finally that the Roman Catholic interpretation of Luther is basically much the same. According to Luther's doctrine of the two kingdoms as Schilling understands it, "nature and the natural man evince no positive relation to the saving will of God. External circumstances and relationships are purely natural, and social acts are therefore never religious acts . . . the destruction of man's moral powers leads inevitably to complete passivity, while on the other hand the work he does in the world becomes an independent end in itself." [29]

In all these interpretations, that which in Luther's teaching constitutes a real danger is treated as if it were an accomplished fact. The assigning of autonomy to the world is thus described as a "theological disaster" rather than as a "theological possibility" in Luther.

The Danger of Harmonization: Static and Timeless Coexistence

The third element of danger in Luther's teaching undoubtedly consists in the fact that for him time and again the two kingdoms seem to stand side by side in mutual harmony. It almost seems as if Luther accepted the fact that these two dimensions of existence must necessarily stand side by side, and that different laws must obtain in each —or that the same law must be expounded and implemented differently in the one dimension from what it would be in the other.

Törnvall has rightly pointed out, however, that in Luther the kingdoms do not actually stand side by side, nor yet the one above the other. Indeed, Luther speaks of two "governments" [*Regimenten*] rather than two "kingdoms" [*Reichen*] in order to show that we are not dealing with two spheres but with two modes of the divine rule. Writes Törnvall, "For Luther there is only one religious and ethical principle, namely, that which is expressed in the spiritual government. His purpose in distinguishing in principle between the two kinds of government is not to denote an ethical difference but simply to express

---

[28] For a critical appraisal of Deutelmoser see Harald Diem, *op. cit.*, pp. 7 ff.
[29] O. Schilling, *Katholische Sozialethik* (1929), p. 68.

THEOLOGICAL ETHICS – FOUNDATIONS

the fact that man must occupy two distinct positions in his relationship to God: Man must rule in the created orders, and at the same time be the servant of God's supreme rule." [30] Törnvall thus regards it as a mistake to define the two spheres in terms of an ethical distinction, as though the spiritual government were to be called "the higher and specifically religious sphere" while the temporal is a subordinate moral sphere representing merely a "general foundation for the religious." [31] For this reason Törnvall rejects any distinction between the two kingdoms in terms of locus ("public" versus "private") or of a quantitative degree (limited "ethical" versus radical "religious").

Though Törnvall is right in this, he still does not seem to touch the most sensitive spot. He has simply replaced the fatal differentiation of the kingdoms in terms of locus by the contention that what we have here are simultaneous modes of divine rule and human obedience. The "local" relation thus gives way to a "temporal" relation. But is this temporal relation correct?

So far as I can see, Törnvall's interpretation of Luther is correct at this point. But assuming that it is, the question remains whether Luther himself is right in emphasizing the simultaneity of the two modes of divine government. Is not the temporal relation between the two governments seen very differently in the New Testament? According to the New Testament it is not a question of two spheres of existence lying spatially side by side, nor yet of two simultaneous modes of divine "government," but rather of two "aeons" which temporally follow one upon the other.

Only when we abide by the New Testament's definition of the time relationship can we avoid reconciling the two governments in terms of a static and timeless coexistence. Only then is our aeon constantly called in question by the coming aeon. Only then does the challenge to all its orders and to its fallen nature remain. Only then can we be dissatisfied with the view that, because power reigns in politics and retribution dominates law, these are supposedly the will of God. Only then are these things seen to be provisional and open to question. Only then does the coming world constantly break in upon the present world, to disturb it, to keep it from absolutizing itself and fashioning for itself a "good conscience," a conscience which

[30] See Törnvall, op. cit., p. 81.
[31] Ibid., p. 80.

no longer lives in—expectation. Only then is there any continuing awareness that in the orders of our world we are confronted not merely by the commandment of creation, but also by God's patient and gracious toleration of our hardness of heart, by the emergency measures [*Notverordnungen*] God has taken on behalf of our stricken world.

Since we live in the area where the two aeons intersect it is true that there is such a thing as a coincidence in time between the two spheres of command. But this coincidence can hardly be interpreted in terms of a simultaneity which would be of continuing character and suggest a lasting state. This is precisely what is not involved. It is rather a case of our being caught between two lines of fire and therefore having to press on incessantly without a moment's rest. This is why it is necessary to insist that the two aeons succeed one another, in order to prevent not merely that misunderstanding in terms of locus, but also any understanding in terms of some logical principle of timeless simultaneity.

Built-in Safeguards in Luther's Doctrine

The question we must address to Luther therefore is whether and to what degree his doctrine of the two kingdoms includes the necessary built-in safeguards to assure that the world will be neither "locally" bifurcated nor "modally" conjoined in terms of mere simultaneity, and so in a measure put at ease. In other words, we have to ask whether and to what degree Luther really maintains the radicalism of the Sermon on the Mount. In this respect it is to be noted first that Luther has two ways of holding the two kingdoms together in order to prevent their being separated in terms of locus. He provides for an objective link on the one hand and a subjective link on the other.

God Works and Rules in Every Sphere

On the objective side the link is provided by the fact that the state, indeed every order of this aeon, is regarded as being in a very specific way the work of God. We have already cited numerous quotations from Luther to the effect that the state is ordained in order to repel the forces of chaos which have been radically threatening the very existence of the world ever since the fall.[32] The state is thus a nor-

32 See also *LW* 1, 104.

mative instrument in God's hand to protect humanity against the de-structive consequences of sin.

Now this purpose of God does not embrace Christians alone; it applies to all men. For all must be accorded the *kairos*, the space in which to repent. It is in keeping with this understanding that Luther refuses to derive either the state or its norm from the Gospel, or from the Sermon on the Mount. For the state is an order also in the non-Christian or pagan sphere. This is why Luther believes that the Gospel as such tells us nothing concerning the way in which the state is to be maintained and governed. This too is why pagans are said to be capable of making valid statements concerning the state. Indeed, Luther thinks that pagans are often more successful at running things than Christians.[33] His statement to that effect must be taken to imply that pagans who do not know the Gospel are not exposed to the temptation of confusing Law and Gospel, and of entertaining theocratic illusions. They accordingly go about political matters in a more realistic and objective way, if we may put it thus with tongue in cheek.

We obviously misinterpret the concern of Luther in this matter if we conclude that for him the state has nothing whatever to do with the Word of God. Neither is the state for Luther to be understood—as Deutelmoser thinks—in terms of created reality in general, or of some abstract concept of divine omni-causality, and hence independently of that Word. For the factual requirements of politics have meaning not in themselves but in relation to the divine purpose toward which they are directed, namely, to preserve man by restraining evil and to give man the physical opportunity to attain to his goal in salvation history. This purpose is that the peace established by the state [34] should make possible the preaching of the Gospel,[35] and thus help physically to preserve the *kairos*.

Thus the state does not exist simply as a matter of course. It cannot be explained in terms of immanence. For Luther it is rather a "gift" and "miracle" of God's sustaining grace.[36] Without it the world would sink into the abyss, and the external presuppositions of the age of grace would be missing.

---

33 See above p. 141, n. 17 and p. 142, n. 18 WA 40III, 204; and Karl Holl, *Gesammelte Aufsätze zur Kirchengeschichte*, I (Tübingen, 1923), 265.

34 *PE* 5, 56.

35 *WA* 16, 339. See Lau, *op. cit.*, pp. 132 ff.

36 See above p. 361, n. 6, and p. 373, n. 32.

To be sure, none but the Christian is aware of this "final cause" [*causa finalis*] of the state. Only the Christian can see, on the basis of the Word, that the state has this function in salvation history, this place within the divine economy of salvation. The Christian thus understands the state better even than it understands itself. For the mystery of the state is not illumined in terms of the economy of cosmic order, but in terms of the economy of salvation history, within which the cosmic order itself also has its place. Consequently the Christian has a relation to the state which differs not ontically or ethically but noetically from that of the non-Christian. This does not rule out but rather implies the fact that the "final cause" of the state is something that is operative quite apart from any knowledge of it, i.e., even when it is hidden from the perception of the non-Christian.

Accordingly, it cannot be a matter of indifference either for the believing community in its relation to the state, or for the Christian in government, that there is in fact this relationship of the state to God and to his sustaining grace. For when they are aware of this relationship, this very awareness gives them a criterion for assessing the limits of the state. They can distinguish the areas of its actual competence and legitimate authority from all forms of idolatrous absolutization. They can estimate the autonomy of the political, e.g., the friend-foe relation, in terms of necessity rather than virtue, and this will constantly give them sufficient drafts of cooling water to pour into the foaming wine of political fanaticism.

From one angle Luther may well be right when he says that pagans are not exposed to the temptation of confusing Law and Gospel, and that they can thus the more easily achieve political realism. Yet this is only half the story. The danger of becoming unrealistic is always present where idolatry and false absolutizations hold sway, and to this danger the pagan state too is exposed. (We have only to think of the conflict between Creon and Antigone, or the trial of Socrates, or perhaps even the trial of Jesus himself.)

True realism is possible only when the state is aware of its assigned limits. But these limits can be seen only from without. They come into view only when the state is aware of its significance as an instrument serving a particular purpose within salvation history. That is to say, they come into view only under the Word of God. Realism is thus in the last analysis a theological concept. Even such restrictions of the state as derive from humanism, e.g., in modern democracies of the western world, owe more to Christian traditions than is realized.

In thus ascribing to the state a function within salvation history, Luther obviously provides a safeguard, in principle at least, against political autonomy, against an understanding of the state in terms of itself. The question which must be asked, however, is whether the safeguard thus provided is adequate. Our answer to this question will finally depend on an exact analysis of what Luther means by the state as an "order" of God, and of the manner in which for him the will of God is manifested in it. We plan to deal directly with these questions, though not in the present volume.

In any case, the division of God's government into that on the "left hand" and that on the "right hand" [37] does not as such establish a zone which is beyond the reach of the divine commandment. This distinction in no way implies the existence of spheres of life which are to be treated "realistically" (which really means pseudorealistically!), in accordance with their own immanent economic, political, or aesthetic autonomy, and which in their ethical neutrality would be comparable to what we found in connection with Kant's concept of "disposition."

The Command to Love is Fulfilled Also in the Orders

In addition to the objective link which Luther sees between the two kingdoms, there is also a subjective link, which is developed in two ways. First, whoever fulfills an office must realize that he is not engaged in his own business but in God's. When the maid swings her broom, the mother brings up her child, the politician runs the government, and the judge pronounces sentence, their work is a kind of service rendered to God. It is this sense of performing one's official tasks at the behest of God which enhances the joy and spontaneity of the action. Luther is thinking especially of the kind of action, e.g., political or judicial, which for Christians is always subject to certain reservations inasmuch as it seems to bring them into conflict with at least the letter of the unconditional commandment: "This is said for the comfort and encouragement of magistrates. They should know that they are not carrying out their own business but God's. They should be certain that they are the servants of God and the executors of a divine work, and that they are in a blessed estate [i.e., their

[37] WA 52, 26.

politics or jurisprudence does not automatically exclude them from salvation]. Therefore they should have all the more courage to judge justly and with integrity [*recte et simpliciter*]." [38]

Second, this referring of one's action in the conduct of his office to the commandment of God is even plainer in the many passages in which Luther states that such action too must be rooted in and determined by the motive of love. In this way Luther attempts to avoid the kind of twofold morality which with respect to personal life proclaims the radicalness of the commandment and its relevance to the whole person, but which with respect to actions performed in one's official capacity prescribes duties gauged to what is realistic and more or less independent of the commandment of God. For Luther the law of love is *the* determinative principle of all the orders.[39] In this respect Luther can even appeal to Matthew 7:12,[40] thus avoiding the view that over and above what we do in the orders there is a "particular" and radical commandment, the "true" command to love, which applies to personal life. No, love is exercised also *in* the orders.[41] "Love in the strictest sense is the telos of all laws." [42] Hence for Luther the concept of "calling" (*vocatio*) has as its criterion service to our fellow men, the ministry of love. This is why it requires of a person total dedication to his office, a giving of self to the point of an almost ascetic "renunciation of the easy life." [43] To bring out the relationship between the objective and the subjective link, one might almost say that love as a subjective attitude of those engaged in fulfilling their offices corresponds to the objectivity of that love which God exercises and implements when he ordains these offices and orders as a means of preservation.

We shall not pursue our examination of the material further, since what we have said is enough for our purpose. The details are already available in numerous monographs. Our concern has simply been to note the safeguards Luther himself provides against, first, a "double morality," and, second, the establishment of a temporal sphere in

[38] *LW* 9, 19-20; see also *LW* 13, 44.
[39] *LW* 45, 98; see also Lau, *op. cit.*, p. 116.
[40] *WA* 51, 393.
[41] In the treatise *Whether Soldiers, Too, Can Be Saved* (*PE* 5, 32-74) there are some astonishing and almost grotesque statements in this respect.
[42] *WA* 15, 691, line 19.
[43] See *LW* 21, 357.

which the radical commandments of the Sermon on the Mount do not seem to apply, a sphere which consequently cannot be called in question. We have seen that Luther, contrary to the interpretation of Troeltsch and others, does not restrict to the personal sphere the radicalness of the commandment of love and its demand for selfless dedication. On the contrary, he holds that the command applies also to the sphere on the "left hand," the temporal kingdom and the orders. This is how Luther seeks to preclude all self-assurance and undercut any attempt to secure immunity from the divine challenge, even though the precise manner in which the commandment is to be fulfilled will always depend on the relationship to the right hand of God or to the left.

The Eschatological Corrective

At this point the question becomes urgent whether Luther does not lose sight of an essential element in the Sermon on the Mount, namely, the fact that the Sermon actually calls in question this whole aeon. We have seen that Luther does in fact apply the commandment of love to the sphere on the left hand. But the suspicion still remains that in this sphere a variety of ways are prescribed for fulfilling the commandment, and that here the commandment as such is no longer able radically to call in question. Luther clearly overlooks the fact that the significance of the commandment is not limited to its meaning within the orders, but goes beyond that to show how questionable the orders themselves really are, failing as they do to measure up to the radicalness of the divine requirement. The commandment in effect characterizes the orders as "emergency" or "interim" solutions. Luther obviously does not perceive the problem posed by the fact that the commandment of love is modified by "the form of this world," that this is a symptom of sickness, and that when God allows for any kind of "fulfillment" within the compass of this reality called "world" this is a sign of his patience and forbearance with our hardness of heart. This ability of the commandment to call radically in question is certainly present in the Sermon on the Mount and must necessarily be brought out. But Luther seems to miss this significance of the commandment. In his view it has in the main only a positive significance, affording the believer a "good conscience" and reassuring him that in the conduct of his office it is possible for him to fulfill the command to love and so remain in a state of grace. What is missing is any trace of that moment when a man must come to see that he is a murderer and adulterer. But this is something a man is bound to discover if he takes the Sermon on the Mount and its radical commandment at all seriously. He will have to perceive that any love which is to be lived out "in this world" is obviously going to be all tied up with "hacking" and "strangling," with "judging," "punishing," and "repaying," and that in all these things the powers, motives, and laws at work are very different from those which God "originally" instituted in the world. Where in Luther's understanding does that moment appear in which man fearfully admits that "from the beginning [i.e., from creation] it was not so" (Matt. 19:8)?

In any case we cannot but be surprised that as regards the orders the command to love has for Luther only one significance, that of positive motivation. It does not call in question. It functions only to loose but not to bind. This seems to be a fault in the structure of Luther's doctrine of the orders, to which we shall return later.

As we pursue the problem to which we have now come we will set forth in a brief and systematic way some criticisms of Luther's understanding of the Sermon on the Mount which have already been intimated in what has preceded. The conclusion seems inescapable that despite the objective and subjective links to which we have referred there is still the possibility that in Luther the temporal kingdom is understood to be too isolated, too insulated, vis-à-vis the Law of God. We thus have here—contrary to Luther's true and original intention—the seed of a fateful movement of emancipation on the part of the temporal kingdom whereby a follower of Luther can appeal all too easily and hastily to the "good conscience" he enjoys in his secular position. We shall develop our criticism here in four points, leaving to ensuing chapters a corresponding statement of our own constructive view of the matter.

First, Luther's doctrine of the "two kingdoms" lacks the eschatological element which plays such a formative role in the Sermon on the Mount—at any rate this is not, in Luther's scheme, a dominant factor. As we noted earlier, the situation of theological ethics is determined by the delay, or postponement, of the imminent Parousia. This means that the believing community must take up a position in the world and find a *modus vivendi* within its orders. This is why in the matter of practical conduct there is always developed some kind of a doctrine of the two kingdoms, or better, of the overlapping of two aeons. In some way ethics must express the fact that when the commandment of God enters the element of this aeon it suffers a certain "refraction," assuming the character of a necessary emergency measure. In this connection, however, it is vital that the "necessity" should really be understood as a necessity and not as a virtue, and that a distinction should thus be made between the "proper" and the "improper" will of God. All this is possible only when eschatology is taken with utter seriousness.

In the form in which Luther bears witness to the refraction, this eschatological element is missing. In the New Testament the temporal kingdom and the kingdom of God are sequential in point of time, and the *kairos*, the "acceptable time," is the plane where the two aeons intersect. In Luther, however, the two kingdoms stand side by side, sometimes with emphasis on the "local" aspect and sometimes with emphasis on the "modal" aspect. Luther is concerned primarily not with two time continuums but with two spheres of reality. And this is due to his unmistakable de-eschatologizing of the Sermon on the Mount.

Second, action in the temporal sphere hereby assumes something of the character of a "compromise," of what we called earlier the "resultant in a parallelogram of forces." The two norms which are "compromised" are on the one hand the command to love, and on the other the independent law inherent in the orders, which allows that command to find fulfillment only in a prescribed way, e.g., in the action proper to a particular political or judicial office. Now we have already come up against this matter of alteration or "refraction" of the divine commandments in our consideration of the quantitative problem in ethics. We have seen that in matters of actual conduct the way of counterbalance, the way of the lesser evil, is the one which is followed most generally, i.e., in the sphere where presumably one's Christian confession [*status confessionis*] is not at stake and everything is ethically neutral or adiaphorous. Meanwhile we have also seen clearly the error in sectarian absolutisms. Nevertheless, we have deliberately avoided using the term "compromise" for this way, even though, for all the divergences in theological undergirding, the concrete forms of action always seem to follow a path astonishingly similar to that of "compromise." (Indeed this is why the concrete forms of action are so equivocal and interchangeable—in practice the moral theology of Roman Catholicism, of Evangelical ethics, and even of the moral metaphysics of secularism often give surprisingly similar advice.) We have avoided the term "compromise" because it suggests an ultimate norm which is merely the average of two other norms more extreme than itself, an average which a man can accept and so put himself at ease. When the two kingdoms are regarded as succeeding one another, however, and the eschatological tension remains, then there is none of this putting oneself at ease, none of this geometrically calculated finality and sense of inevitability. As we have seen, the coming aeon "breaks in upon" this present aeon like a "disturbing fire." So what is involved is not a resultant of two forces, but a zone of tension within the intersecting aeons.

Third, all peaceful coexistence between the two aeons is thereby ruled out. To be sure, I must de facto pay tribute to the world and its order, and I can do so quite willingly and gladly, knowing of the divine patience and the saving purpose of these orders. But this does not mean that I can therefore acknowledge the various spheres and orders as "laws unto themselves" [Eigengesetzlichkeiten]. For I am constantly reminded that the laws which are native and proper [Eigengesetze] in this world, e.g., the laws of jurisprudence and politics, are really strange and alien [Fremdgesetze] so far as the kingdom of God is concerned.

Fourth, this leads us to a specific concept of order which is different from that of Luther. The emergency orders [Notverordnungen] of this aeon themselves belong to the fallen world. They are therefore ambiguous. They are not only directed against but are also expressive of the fallen nature of the world and its destructive forces. Inasmuch as they comprise the structure of the fallen world, they are also representative of that world. In this sense all ethics is an emergency discipline following upon the fall—i.e., following upon the loss of our original state, our original fellowship with God—and yet taking place also within the promise and the dawning power of the new aeon. Thus all the propositions of theological ethics are enclosed within brackets behind and before: behind they are determined by the fact that the original fellowship no longer exists; and before they are determined by the fact that the promised dominion of God, with its new and total fellowship between God and his children, does not yet exist but is only dawning. Behind us stands the cherub with the flaming sword which smites in all directions, and before us stands the Last Day, beyond which alone lies unbeclouded vision, and hence immediacy in judgment and in grace.

D.

THE PROBLEM OF NORMS: (2) NATURAL LAW AND THE DOCTRINE OF THE ORDERS

19.

The Problem of Natural Law

In what has preceded we have spoken at length of man's ability to know by nature the Law of God, and we have assembled from many different contexts the arguments which prove that God's Law cannot be known in this way. Natural man knows the imperative, and one may regard this knowledge as a recollection of the Law of God. But we have seen how depraved this naturally known imperative is, how alien to the Law of God. We may summarize the decisive points in the following way.

The relation of imperative to ability ("You can, because you ought") gives a certain defensive character to the natural imperative in face of the radicalness of the Law of God. Moreover, the *usus politicus legis* can be grasped in a material sense, but this very fact, as we discovered, is an indication that the actual commandment of God is not taken into account by those who know only the "political use." For in the first place the author of the Law remains unknown to them, and in the second place they accordingly fail to see the efficient and final cause—the fact that the Law is determined by salvation history— and so miss the very point of the Law altogether. Finally, in keeping with this, there is the fact that the natural imperative cannot be regarded directly as a given "point of contact" for the claim of the divine commandment. The conscience, which is the bearer of this imperative, shows us clearly how these natural presuppositions are broken when confronted by the divine commandment.

The whole question posed by the concept of "conscience" leads us necessarily to another aspect of the problem, namely, the question whether in the objective structure of the world itself there are not given and discernible certain indications of the eternal orders, and therefore of the will of God. In other words, we must go beyond what we have done in examining the problem of natural theology from the subjective standpoint of conscience, and extend our investigation also to the objective side. So far as ethics is concerned, the objective side of natural theology has to do with natural law.

What is called "natural law" always arises out of the question of the "order" [*ordo*] of the cosmos, a question which may be stated as follows: Are there not concretely demonstrable constituents of the structure of the cosmos (e.g., the law *suum cuique*, "to each his own," or the correspondence between the freedom of the individual and his solidarity with society, etc.) which have about them intrinsically an imperative character, and which enable us to discern an ultimate, divinely willed order in the world—an order whose existence is demonstrated primarily in the fact that, when it is disregarded, it is replaced by the direct opposite of order, namely, chaos? We have good reasons for weighing this question carefully, not least of which is that there has always been this kind of natural law. Beyond that, there is certainly a desperate longing for it—and not only in Germany—in the wake of the devastation of World War II. Even such a thinker as Emil Brunner is now concerned to establish an Evangelical natural law. Finally, the Roman Catholic church continues to offer a most impressive form of natural law. In the modern world there is added to lasting theoretical interest in the problem one further basic motivation, namely, the question whether we may not here find the decisive forces for healing a deranged world.

The Urgency of the Problem Today

The urgency of the problem of natural law emerged with particular clarity at the first of the Nuremberg trials, which was the only "war crimes trial" deserving of really serious consideration. Here the problem of natural law arose in two respects.

First, whereas the accused appealed to laws and commands which were in force at the time of the alleged offense, and by which they felt themselves to be covered, the prosecution argued that these laws and commands (e.g., the extermination of the mentally ill, the massacre of the Jews, the shooting of hostages, etc.) are immoral, and hence that the acts themselves are also immoral. There is consequently such a thing as ethical and legal "illegitimacy" even where the corresponding acts are technically "legal." The governing theory here is that written laws and edicts stand under the norm and control of unwritten and eternal laws. We are reminded of the well-known issue in Sophocles' *Antigone*, in which there is a conflict between Creon's prohibition of burial and the divine law of piety. Here the matter under discussion is nothing less than this tension between "legality" and "legitimacy."

A second aspect of the problem of natural law cropped up in the arguments of those who criticized the Nuremberg trial. They pointed to the basic legal principle *nulla poena sine lege*, "no punishment where there is no law." On this ground they disputed essential points in the prosecution, for some of the laws on which both the prosecution and the verdict rested were only laid down or accepted after the fact —in part not before the time of the trial itself—and so were not in force at the time the alleged offenses were committed. Against this the prosecutor again argued that the laws were nonetheless in force at the time the offenses were committed because they had *always* been in force, even if "only" as unwritten law. Appeal was thus made to basic laws of the cosmos, and especially to certain axioms of humanity and to their negative counterpart, the "crimes against humanity." The problem of natural law was thereby posed most sharply.

This question was raised with particular urgency in the indictment read by the Chief of Counsel for the United States, Robert H. Jackson, who argued that "when the law evolves by the case method, as did the common law and as international law must do if it is to advance at all, it advances at the expense of those who wrongly guessed the law and learned too late their error." [1] His statement almost has the force and grandeur of antiquity. It reminds us of some of the mysteries of Greek tragedy: Why must Oedipus incur guilt through patricide and incest? Why must there be such a thing as tragic guilt at all? Apparently because the avenging wrath of the world order is provoked by offenses against it such as patricide and incest—whether they be committed wittingly or unwittingly—and because what is otherwise a seemingly invisible and latent order is thereby compelled to manifest itself. The orders thus reveal themselves as they react to crimes committed against them. This is why there is in such crime that meta-

[1] See the Opening Statement for the United States of America on the subject of "Crimes in the Conduct of War" made at Nuremberg on November 21, 1945, by Justice Robert H. Jackson, in the official English text of the proceedings: *Trial of the Major War Criminals before the International Military Tribunal* (42 volumes; Nuremberg, 1947-1949), II, 147.

ethical element of tragic guilt which modern ears, accustomed to moralizing, are so slow to hear. In tragedy there is thus an implicit exculpation of guilt. It lies concealed in the teleological purpose of guilt, whereby such guilt compels being to yield its true secrets, and helps men to respect and reverence these secrets.

As Jackson sees it, the order of the world is in some sense not manifest in advance. Otherwise it would not have to be discovered and spelled out in connection with concrete crimes which are its negative counterpart. All the same, that order—though still latent— does in fact exist. Otherwise it could not be violated.

It would seem as though the same must be said of natural law. On the one hand, it has not yet been laid down, in the sense that every detail has been formulated and every crime which will bring us into conflict with it (such as genocide or the liquidation of the "unfit") has been codified. On the other hand, it undoubtedly exists; ultimately all positive laws derive their legitimacy from it alone. And even those intermediate spheres which are not yet legislatively demarcated are potentially subject to it. Or so, at least, its proponents believe.

It is thus understandable why the problem of natural law, the question as to the order of the cosmos and the norm of law, should become so urgent in times of utter chaos. For then men are bound to inquire after the tie which "inwardly holds the world together." Here the "provocation" offered by wrongdoing becomes acute. Here there is genuine helplessness and anxiety in face of sinister retributions, and consequently an understandable urge to return to the ultimate and healing foundations of being. Men begin to search, as it were, for the plane from which we have fallen.

This is the sense in which the problem of natural law was posed by the Chief Prosecutor for the French Republic, M. Francois de Menthon: "There can be no well-balanced and enduring nation without a common consent in the essential rules of social living, without a general standard of behavior before the claims of conscience, without the adherence of all citizens to identical concepts of good and evil." [2] Menthon apparently assumes the existence of some kind of collective morality which ultimately binds men together despite their divergent

[2] See the statement of M. Francois de Menthon made at Nuremberg on January 17, 1946, in *Trial of the Major War Criminals*, V, 369.

religions and world views. Accordingly there must necessarily be certain fundamental orders of life whose decisive characteristic is that they are obligatory for all, for if they were not thus obligatory the world would be delivered over to chaos. In content, the marks of this world order are respect for human life—indeed for every single human being—and the resultant limitation of state sovereignty by the inviolable basic rights of the individual.

Belief that such a world order may be discovered and put into practice implies both a strengthening of the position of conscience, and also a challenge to it. It implies a strengthening of conscience to the degree that conscience plays a vital part in discovering this order. For the order derives not from a theoretical analysis of the structure of the world, but from a perception of ultimate ethical obligations, those which strike us with the force of an imperative the moment we encounter them. Scholasticism, and especially Thomas Aquinas, worked out very precisely this connection between imperative and being. Their point was not that the imperative posits a certain being, as if there were originally no being and no reality until they were raised up and actualized by the imperative. On the contrary, the order of being is already present, even if only in a hidden, obstructed, and potential form; and because it exists it must needs be actualized.

In the present context our primary interest is in the fact that the existence of the world order is not the object of purely theoretical analysis. On the contrary, it meets us not as a thesis but as an imperative or claim. In this sense human worth, for example, is not the object of a theoretical proposition; it is the content of a demand, a demand that we respect it. It is in this way that the position of conscience is strengthened when we accept the existence of such an order. For conscience is the responsible instrument of this imperative claim. We know of the order, not with our intelligence, but with our conscience. Conscience is the source of all subsequent propositions regarding it.

To accept the existence of a world order, however, is not only to strengthen the position of conscience but also to challenge it. For to the degree that we regard such an order as given, we necessarily become skeptical in respect of direct and unmediated guidance by conscience. Such guidance would in effect be merely a kind of improvisation on the basis of not too well established findings. In itself, conscience is quite without guidance. It cannot rely merely on its own voice. Instead it must constantly refer to the order, and its commands can be issued only within the framework of the order. The order alone gives some guarantee against the continual fallibility of conscience, against its inconsistency, its fickleness, its dependence on fear and hope. It is for this reason that conscience too, according to Roman Catholic teaching, must be bound to the one legitimate interpretation of natural law, i.e., to the authority of the church. The order which in natural law is the object of reflection and which is established by the church, constitutes this safeguard not only because it is an objective and fixed entity not subject to the immediacy of acute improvisation, but also because it is, or *ought* to be universally recognized. The fact that there is this universal validity offers some protection against individual and subjective caprice. This is why the concept of the "universal" also plays so vital a role in the categorical imperative of Kant.

Already, then, we see two of the essential features of natural law. The first is that it implies a perceptible order of being which underlies every imperative. The second is that this order can and must be recognized by all. Both these features are implicit in the concept "nature," which we shall examine for a moment.

The Meaning of "Nature" in Natural Law

What is "nature"? A variety of definitions are given in the various systems of natural law. For Aristotle—perhaps for everybody else as well, but we will come to that question in due time—for Aristotle "nature" in an artificial abstraction which results when certain conditions of communal life are regarded as universally binding, i.e., when they are regarded not as "accidental" (θέσει) but as "essential" or "natural" (φύσει). It is in this sense that slavery, for example, is regarded as something "natural." From the time of the Enlightenment "natural" has come to mean much the same as "rational." But here again the center of interest is the fact that the law exalted by reason is not accidental, but of universal validity.

We shall return later to the various definitions of "nature." For the moment, considerations of methodology demand that we concentrate on the Roman Catholic or Scholastic understanding of the term. In the course of our earlier discussion of the *imago* doctrine we have already come upon this particular concept of nature. In what follows here, we shall take up again the theses already advanced and try to develop them in relation to the understanding of "natural" law.

In Roman Catholic dogmatics the term "nature" is used primarily to mark off one particular sphere of being from two other dimensions, on the one hand that of "super"-nature, and on the other that of what may be called "sub"-nature, i.e., the sphere of *tyche,* the contingent or accidental, that which is simply "given." Nature thus embraces everything this-wordly, including all the orders of this world, and that means the orders that are bound to matter.[3]

In this sense nature has the following characteristics. First, since *natura* and *nasci* relate etymologically to conception and birth, and since the origin of a creature is closely linked with its non-contingent essence, the law of nature has reference to that fundamental law which is linked with the essence, origin, or nature of man. In this connection it makes no difference that many changes may have been made in this essential law as a result of the contingent circumstances of positive law. Sometimes that basic law may even have been silenced, as in the case of the laws against humanity formulated during the Third Reich. Thus the concept "nature" ought not to cause us mistakenly to think only of the biological side of man. To the essence and origin of man belongs the personal aspect as well as the physical or biological. For this reason the law of nature is frequently called "the law which is born with us." The phrase serves to remind us that this law has to do with origin and essence. It is to be affirmed as an order which is permanently valid, in contradistinction to the fluctuations of *tyche.*

Second, to the nature of generation belongs the similarity between that which does the generating and that which is generated. What is generated carries within it a declaration of its origin and a manifestation of the original essence. The biological content of this likeness is in itself a reminder that that which does the generating represents the final cause as well as the efficient cause. Our origin is not merely the physical cause which makes possible our coming into being—here we may think of the process of procreation in which our parents engaged—it also posits, together with the conception, the

3 See M. J. Scheeben, "Natur und Gnade," *Gesammelte Schriften* (Freiburg, 1941), I, 15 ff.

goal towards which we develop in the sense of entelechy. It is with reference to this goal of similarity as regards character traits or physical traits that we say, "That boy is his father all over again."

As the concept of nature causes us to think of the origin and goal of man, so in the theological sphere we must substitute the term "creation" for that of "procreation," considering what it means that as regards both his origin and his goal man is referred to God. In this sense whenever we speak about natural law we are reminded that that which is created is like the One doing the creating; he is patterned after Him. Thus we can speak of "natural" law—using the term nature in the sense in which we have just developed it—only if we take as our starting point the likeness between creator and creature.

This statement may also be formulated negatively in terms of the understanding of the *imago Dei* arrived at earlier in our debate with Roman Catholicism. We can speak of "natural" law—using the term nature in the sense in which we have just developed it—only if we assume that sin involves merely a limited disturbance, but in no sense a total corruption or loss, of this likeness. We are thus confronted at the very outset by that decisive concept which determines the framework within which alone there can be any "Christian" natural law. This is the concept of the analogy of being [*analogia entis*].[4]

Natural law thus relates to the essence of man, and to the law which he is thereby given. It does so because man in his very essence, in origin and goal, is characterized by likeness to the creator. It does so because man, to the degree that he is aware of this likeness, also bears within himself a knowledge of this natural law. Because natural law is thus grounded in the essence of man as such, and because the essence of man includes this conscious reference back to his origin and goal, natural law is also something that can be known by all men. It is thus independent of such contingent facts as whether or not I have been confronted by the declaration of God's will in his commandments, whether or not I have received Christian instruction, etc.

[4] Important modern discussions of the *analogia entis* are to be found in Erich Przywara, S. J., "Religionsphilosophie katholischer Theologie," in *Handbuch der Philosophie* (Munich and Berlin, 1927), esp. pp. 22 ff.; Jacques Maritain, *Von der christlichen Philosophie*, trans. Balduin Schwarz (Salzburg-Leipzig, 1935), pp. 55 ff.; Johannes Hessen, *Religionsphilosophie* (Essen and Freiburg, 1948), II, esp. 231 ff.

Third, the same thought occurs in the Roman Catholic doctrine of natural law from yet another angle. Since the species generates only within its limits, and since the individual member—in this case man—is determined by the species, the implicit reference of natural law to the species implies also an explicit suprapersonal factor. What is involved here is a legal relationship which applies to "all" who bear the countenance of man and who thus belong to the human species. Hence the claim of natural law to universal validity is grounded in the fact, not merely that it is independent of such a contingency as encounter with the Word of God, but also that it is independent of such contingencies as personal quality or particular situation.

Fourth, whereas in the case of these first three characteristics we have been differentiating the "natural"—as Roman Catholic theology understands it—more particularly from such "lower" spheres as that of "contingency," we must now differentiate it also from the "higher" sphere of the "supernatural." "A good can be called natural *either* insofar as it is originally granted to a being by birth—or else insofar as this being can and must attain it by its own natural powers—*or* insofar as this being is capable of receiving it, and by actually receiving it does attain to its highest perfection, not just that perfection which it can achieve through its own powers but in general such perfection as is at all possible by any means whatever." [5]

The demarcation from the supernatural thus means that man receives the law which is essential to him, and he is enabled to know it, without the necessity for any external help such as that of a supernatural revelation. (This is basically the same point we just made in the preceding paragraph, but it is important to state it from this angle as well.) Only to the degree that man's nature is incapacitated—as it is for Scholasticism by sin, by loss of the order of rectitude—is there need of the supernatural correction afforded by grace. The essential point is that the first task of grace is to heal nature. It is not a case of nature, after the fall, after the disruption of its functional capability, now being dependent on supernatural grace for the accomplishment of what it had formerly done itself before the fall. It is not that natural law, following the fall, must now be "revealed" or "brought in from outside" whereas prior to the fall nature was capable of getting at it by itself. No, it is a case of grace having

[5] Scheeben, *op. cit.*, p. 18.

as its first task to rectify nature, and thus to restore the original scope of its activity.[6]

While nature in its present condition thus needs to be supernaturally corrected if it is to be restored to itself, the fact is that by this supernatural correction nature is indeed restored to *itself*. Only to the degree that disrupted nature needs the supernatural help of grace to be itself, only to this degree then is there also need of supernatural help in the unleashing of natural law. But the fact that such unleashing is necessary cannot mean, according to the Roman Catholic view, that natural law must be "imparted" supernaturally—e.g., by concrete, revealed commandments of God—that it is consequently tied to and limited by the contingent encounter of man with this supernature, in other words that it loses its universal validity for all men. No, it is truly "natural" law, no more and no less. The only point is that a situation may arise—as has happened after the fall—in which nature has to be restored to itself, and in which, within nature's own sphere, certitude of knowledge has to be restored.

We may thus sum up as follows our provisional analysis of natural law as conceived in the Aristotelian and Thomistic system. Natural law is the law which belongs to man by nature, i.e., not by virtue of his bios but by virtue of what he essentially is. For this reason it is binding on all. It is independent of all outside influences, including those of existent positive laws. Indeed, so far from being influenced by these laws it is itself normative for them. It is grounded in the relatively undisturbed likeness between creator and creature. Finally, it can be known by all, and thus constitutes a norm to which all are accountable "by nature."

To put it in more Scholastic terms, the *lex naturae*—in keeping with the system of analogy—derives from the *lex aeterna*, i.e., from that

[6] Cf. Chapter 2 of the Dogmatic Constitution concerning the Catholic Faith adopted by Session III (April 24, 1870) of The First Vatican Council of 1869-70 on the necessity of revelation: "It is owing to this divine revelation, assuredly, that even in the present condition of the human race, those religious truths which are by their nature accessible to human reason can easily be known by all men with solid certitude and with no trace of error. Nevertheless, it must not be argued that revelation is, for that reason, absolutely necessary. It is necessary only because God, out of his infinite goodness, destined man to a supernatural end. . . ." *The Church Teaches: Documents of the Church in English Translation*, trans. John F. Clarkson, John H. Edwards, William J. Kelly, and John J. Welch (St. Louis: Herder, 1955), No. 59.

order which God envisioned and provisionally established in the world, and which may rightly be called "law" because of its stable and unchanging nature. (We are perhaps reminded at this point of the scientific concept "laws of nature," a term which can indeed be used here though with tongue in cheek.) This law can be called "natural law" because it is immanent in the created cosmos, the world of nature, and that in a twofold sense.

In the first place, natural law embraces the unconscious force of order operative in all creaturely life, e.g., in the laws of astronomy, in the laws of the evolution of plants and animals, or in the laws of physics. Concrete illustration of these unconscious "natural" laws may be found in the structure of an anthill or in the regularity of the annual bird migrations.

In the second place, natural law embraces theoretical and practical reason, and thus becomes freedom's final court of appeal. As natural moral law in this sense it is different in character and function from unconscious natural law in the narrower sense. For it frees man from the compulsion of natural necessity—"must" is replaced by "ought to"—and man is admonished through reason and conscience to integrate himself into the order of creation.

Here already there arises the question which is of determinative significance in theological discussion of the problem of natural law, and we may briefly allude to it in passing: Where and how can I recognize this order of creation? From what authoritative source is it to be deduced? Is it to be read off from the world as it was originally, before the fall—for it was undoubtedly present there, present and intact—or is it to be read off from "this aeon," in which, to put it mildly, it has undergone some disruption? Is it really so plain and accessible in this aeon? And if it is, can we from this general order casuistically deduce directives for specific situations here and now? For example, even if in the political sphere this order were to suggest that government ought to take the form of a constitutional or democratic monarchy (as seems indeed to be implied in many papal encyclicals), would we not still have to take into account the concrete historical circumstances in any given situation? And would not the mere fact that we do this involve a modification, and therewith a warping, of the original order?

So far as our knowing and applying it is concerned, that order is thus called in question from two directions. It is challenged first by the relation between the original state and the fallen world. (Though they may use other terms, even secular doctrines of natural law recognize this problem. They all have in view an original model of what ought to be, and of what once was but is now destroyed.) An order of being which can serve as a model for concrete historical configurations is obviously possible only if the difference between the original state and the fallen world is regarded not as total and qualitative but as quantitative and limited in scope, so that some possibility of comparison still obtains. The very notion of a model or pattern depends essentially on there being some similarity, and on the difference being purely quantitative. An example provides incentive only because one can make comparisons with it. Standing basically on the same level with it, one can see there the superiority which he is called upon to emulate. The likeness between the original state and the fallen world depends, however, on our understanding of the fall, or, for the secularist view of natural law, on the extent to which history has gone wrong. Thus we see from the very outset that the understanding of sin plays a normative role in any discussion of natural law.

The second challenge to this order of creation, and to our apprehension and application of it, is posed by the problem of "the concrete." When we try to bring a specific historical situation, with all its complexity, under the norms of such an order, are we not faced by insuperable difficulties?

Nowhere are these decisive theological problems of natural law more clearly displayed than in the Roman Catholic system. It is peculiarly fitted to reveal with classical precision all the essential questions and problems which theology has to raise in respect of any idea of natural law. By the same token, however, it is not affected by the host of subsidiary problems which arise in connection with the increasing perversion of natural law in secularism. This is why we have started with the Roman Catholic doctrine of natural law. For the most part it takes up the basic problems for itself and tries to wrestle with them, especially in its distinction between so-called "absolute" and "relative" natural law. We shall return to this whole matter later.

The Historical Presuppositions of Natural Law

The very idea of natural law might lead us to suppose that it first arose in mankind's initial state of "natural" innocence. Wherever men make reference to it, however, natural law is clearly something that is deeply rooted in reflection. Hence, even though it may not be exclusively the "product" of man's self-awareness, it has at least traversed this stage. Moreover, it presupposes that provocation of the world orders by the transgressions committed against them. For these reasons we would do well to heed the warning of Heinrich von Kleist, and not date it back too far.[7]

The fact is that we can inquire concerning natural law only on the basis of two presuppositions. The first is that of the possibility of and capacity for historical observation and reflection. Only with observation of the interplay of law and custom does the question of constant and valid norms arise. "No one in the pre-Sophistic period thought of examining the law and asking in what its claim to valid authority consists."[8] Before this question could be raised there had to be productive skepticism and a certain threat of relativism in consequence of the rapid changes in laws and constitutions. Thus Greek ethics, or more precisely the question of natural law, began with a problem which corresponded exactly with the starting point of ancient physics. The philosophers of the first period in physics undersood by "nature" ($\phi \acute{\upsilon} \sigma \iota \varsigma$) the eternally constant essence of things which outlasts all change; its logical counterpart accordingly was to be found in the transitory and contingent. Similarly, in the first stage of ethics it is asked whether that eternally constant nature ($\phi \acute{\upsilon} \sigma \iota \varsigma$) does not also constitute within the sphere of history an unchanging law which is above everything contingent. Is there not a "natural" law which is to be distinguished from the finite, contingent, and hence only partially valid prescriptions of all merely human statutes ($\theta \acute{\epsilon} \sigma \epsilon \iota \varsigma$)?

The second presupposition basic to any reflection on natural law is that of a geographical widening of the field of vision beyond the horizon of one's own state or nation. As comparison is thereby made possible, laws and norms are no longer accepted as a matter of

---

[7] See Heinrich von Kleist, *Betrachtungen über den Weltlauf* (Leipzig, 1926), III, 354 f.

[8] Wilhelm Windelband, *A History of Philosophy*, trans. James H. Tufts (New York: Macmillan, 1923), p. 72.

course, and the assaults of relativism open wide the quest for that which is enduring. A gulf opens up in what was formerly thought to be an integrated and self-contained world, and there is a corresponding cleavage in consciousness.

Two questions are hereby suggested which the theological observer of these origins is forced to ask, and which may help us to attain to a critical starting point for future discussion. In the first place, it would appear that the question of eternal law and its eternal constancy arises only when there is awareness of the incongruity between the actual state of law and the eternal norms, i.e., between relative law and absolute law. The very fact that it is the poet Hesiod who probably has to be regarded as the true instigator of natural law in antiquity suggests that an awareness of this cleavage is necessary if the question of natural law is to be raised. For in this "rhapsode of Ascra" [9] the harmony of the Homeric world is disrupted. Hesiod had a profound sense of this cleavage, based on personal experiences recounted in the didactic pastoral *Works and Days*. The external occasion for this poem is the lawsuit with his lazy, quarrelsome, and covetous brother Perses, who after wasting part of their father's inheritance was continually coming forward with new demands, and who in the process did not scruple to bribe the judges. Hesiod sees himself to be involved in a struggle between right and brutal might. This struggle is for him the symbol of the general state of society in which he finds himself. It becomes the basis for his attack on the "gift-eating nobles" whom he charges with avarice and with harsh misuse of their powerful position in the community. He is pessimist enough to see in history a process which continually widens the gulf between right and might. For Hesiod, history unfolds in five ages, the first being the "golden" age under the reign of Kronos, and the last being the present age devoid of justice and happiness: "Aidos and Nemesis have veiled their faces, and quitted the earth to rejoin the other gods on Olympus, leaving nothing to mankind but suffering and endless strife." [10]

Hesiod thus knows *diké* from the perspective of its perverted form. But right is falsified and "relativized" when might enters the picture,

[9] Hesiod is thus referred to by Werner Jaeger in *Paideia: the Ideals of Greek Culture,* trans. Gilbert Highet (2nd ed.; New York: Oxford University Press, 1945), I, 58.
[10] *Ibid.,* p. 59.

for might uses law, and tries by means of law to conceal and legalize its own brutality. In mythological terms, Themis triumphs over Dike.[11] We may thus venture already a statement whose significance we shall have to develop later, namely, that the question of right arises out of the longing for liberation from wrong. This is why the question of *diké* does not yet arise in the unbroken world ruled by the blessed Olympian deities. For the person who inquires about righteousness and law, the person who yearns for them, is not the aristocratic lord who lives like the gods in joyful opulence and knows nothing of social jeopardy or the petty concern for daily bread. It is the Boeotian, the small peasant, who tills his field in the sweat of his brow—or a man like Hesiod, whose rights are trampled under foot in a lawsuit. In the happy Homeric world, therefore, it is still Themis alone who rules, Themis who embodies might more than right even though she is also the mother of Dike. It is Hesiod who sets the daughter against the mother, right against might, and sees in them two antagonistic powers.

If it is true, however, that the question of the eternity of right, of eternal law, first arises because of the oppression wrought by wrong—and right that has been warped by might is indeed wrong! —then we have to consider seriously whether the question of natural law does not have to be seen against the bleak background of a fallen world, a world in which eternal law is no longer in force. To use the terminology of Hesiod, this means that the question of eternal law arises only in the "iron" age. In the "golden" age it is not yet even a matter for discussion. We would in addition have to put the broader, the theologically critical, question whether the supposed eternal law, which men think they may explicitly delineate and codify, is not perhaps a negative abstraction. Does it not arise by the *via negationis,* i.e., by negating obvious elements of wrong and perversions of power? There is room for serious question, indeed for doubt, whether the criticism of falsified actual norms really presupposes a positive knowledge of the true absolute norm, whether such criticism proceeds on the

---

[11] See F. G. Jünger, *Griechische Mythen* (Frankfurt, 1947), pp. 51 ff. [In its fundamental meaning *diké* is much the same as "due share." Following Hesiod's complaints against corrupt princes who turned justice askew the common people made *diké* the battle cry of their class-conflict, in which they demanded codified law and equal justice. Themis, which etymologically means "institution," had reference to the authority imposed on the people from above, the justice dispensed by the early kings and nobles in virtue of their established position. See Jaeger, *op. cit.,* pp. 102-103.—Ed.]

basis of such prior knowledge. Might it not be otherwise? Might it not be that, while we know perfectly well that certain things ought not to be, we cannot say with the same assurance what ought positively to be? In other words, we may indeed know what is wrong, but still not have the same clarity when it comes to knowing what is right in terms of the order of being. Are we really justified in assuming that because we know what is wrong we can also reconstruct the eternal order of being with the help of the geometric loci derived from our various negations?

We may already disclose at this point our own conviction that it is not possible to work back in this way from the negative; such an approach is quite out of the question—if we accept the fallen nature of "this aeon." In this whole problem of natural law, and specifically in the question of the possibility or impossibility of a recognizable order of being, I see a basic difference between Roman Catholic and Reformation theology. For the possibility or impossibility of working back to the eternal order depends upon the understanding of sin, upon the degree to which we think the being of our world is altered and impaired by the rent of the fall. If sin involves only a kind of subtraction from original perfection, as in Thomistic thinking, and if there is thus a kind of continuing analogy between the two states, then that working back is quite possible; it is not difficult to reconstruct the order of natural law.

This then poses for us very clearly the first of the two questions we said we are forced to ask: Does not the question of eternal law arise out of the fallen world of wrong, and would it not be transgressing the limits of theological discourse to go beyond this merely negative assertion? We shall consider later whether the predominantly negative form of the Decalogue does not have an important bearing on this problem.

The second question we are forced to ask is as follows: Is there not implicit in the relation of the Greeks to eternal law a factor similar to that which is said to obtain in their relation to the timeless beauty of form in art? For the theory of Goethe and Johann Joachim Winckelmann, that in his art the Greek is expressing himself "naïvely" (in Schiller's meaning of the term) and objectifying his inner harmony, has been largely exploded by Nietzsche and J. Burckhardt. According to Nietzsche's interpretation, the Greek created the

Apollinian dreamworld of beautiful form because he was aware of the Dionysian abysses of being and could not have endured them without that transfiguration, that restraint found in form. For Nietzsche, the Hellenic is the only possible form of life, because it banishes what is horrible by masking it as beautiful. Greek art is said to have taught us that there is no truly beautiful surface without horrible depths. The closest approximation to our understanding of the Greek view of natural law is perhaps to be found in Nietzsche's thesis that any Greek statue can tell us that the beautiful is mere negation. Perhaps in this sense it is no accident that the most profound utterances on the cosmos of the good and the beautiful in Plato's *Symposium* are made by Socrates, who regards Eros not as possessing the beautiful and the good but as lacking them, and out of this negativity longing for them.

Hence one might describe as follows the secret of the Greeks, those first great teachers of natural law. Like the aesthetic order of art, the order of law is a kind of exorcism of chaos, of the menacing abyss. The world is surrounded by Ocean, by one of the deposed Titans; he is himself a threatening, primitive force, but it is behind him that the real threat stands, namely, the terror of the illimitable. Hence the cosmic and aesthetic order of the Greeks is in fact a kind of exorcism formula. It is posited as order because the world is at root menacing, Dionysian. It is thus a postulate based on negation of the terrible, and cannot be regarded as a "positive" order known from the very first.

20.

Natural Law in Scholasticism

It is with a certain measure of skepticism then that we come now to deal with the concept of an order of being which is susceptible of positive reconstruction, and with the natural law to which it gives rise. Our concern will be primarily with Scholasticism's scheme of Christian natural law. We begin by recalling what we said earlier about the Scholastic concept of "nature" in relation to the terms φύειν and *nasci*, namely, that a specific goal is implanted in that which is generated. In every thing and in every person there is at work an essential law or law of being, and εἶδος, which takes shape and which helps the thing or person to achieve its intended "form." In Thomas this law is call the "formal cause" [*causa formalis*]; genealogically it goes back to the "pure form" [εἶδος] of Aristotle. Now this form is a constituent part of the very order of being itself. For this reason we must distinguish the Aristotelian and Thomistic concept of entelechy from that of Goethe, for whom the entelechy is the immanent character of an individual which simply develops from within.

The Abiding Validity of the Order of Nature

For Goethe, the individual is a self-enclosed cosmos which is autarchic and which obeys only its own law of development. (We can say this even though the use of astrological symbolism by Goethe does at least betray overtones of the cosmic reference of individuality.) In Thomas, however, the formal cause is not regarded from the standpoint of the mere immanence of individuality. It relates rather to the telos, to the position which the individual occupies in the totality of the cosmos so far as the hierarchy of meaning and purpose is concerned. The two, however, are not contradictory. For the ego can come into its own only by fulfilling itself as a member of the order. The self, or, as one might say today, the particularity of the ego, cannot be determined by its own immanent teleology.

It can be determined only by the teleology of the cosmos, within which it has its place and function. Thus, for example, the nature of conscience cannot be discovered by an analysis of its immanent structure. It can be discovered only by virtue of the fact that conscience has its place in the world order as a whole, where it functions to make this order "audible." If it once loses this reference to the order, and thus is subject to no authority but its own teleology, it does not cease to be conscience, but it is a misguided conscience which mistakes both the order and therefore also itself.

Epigrammatically one might say that the true nature of a thing, the true nature of individuality, resides in its relatedness to the order. Hence the "formal cause" [*causa formalis*] is also the "final cause" [*causa finalis*], i.e., the goal to be actualized. But the goal, or, in the personal world, the "purpose," always transcends that which is referred to it. The ultimate "form" is supernature, into which the whole order of nature is itself integrated.

Now this order of the sphere of nature and creation persists even in the fallen world. We have already brought out earlier the essential points in this connection. We need only recall that, while the fall deprives us of the supernatural and praeternatural gifts, it leaves nature itself intact. Hence the ordered arrangement, which after all is still present "virtually," simply has to be reconstituted out of these intact elements. Each of these jumbled elements has, as it were, a number which indicates its place in the hierarchy of the order's individual strata. (For example, within the phenomenon "natural man" reason is number one.) Even the element of "concupiscence," as we saw, is neutral in relation to sin. It contains sin only as a possibility, or, more precisely, as a possibility which presses for realization (the idea of *fomes*). This possibility is suppressed when the elements of reason and sensuality are again integrated into the order, specifically by means of the restored *rectitudo*. Since this order persists with permanent validity, and since its scattered elements are reintegrated into it by the saving action of the church, there must necessarily be norms of natural law on the basis of which it may be actualized.

Primary and Secondary Natural Law

What we have said concerning "individual" nature applies also to the social nature of law, economics, society, the state, and the various cultural spheres. In these too there is an evident order of nature. That which has been required from the very beginning by the will of the creator is clear and can be formulated in terms of natural law.

Doubts might arise on this score when one considers that in the state of the fallen world these social constructs, e.g., the state or law, have been very radically changed. The church fathers were aware of this change. They felt, for example, that the fallen nature of man was responsible for the fact that freedom no longer exists as it once did in man's primal state, and that there are now unavoidable restrictions of freedom. It was for the same reason, they thought, that there no longer obtains the equipoise of the original hierarchy; instead state and law have to be granted the use of force in order that, if necessary, they may meet with superior power the insubordination which has now broken out. Thus the situation is no longer what it was at the beginning; it is not that of creation and of natural law in the true sense. Victor von Cathrein, the Roman Catholic moral theologian, can even say that Schopenhauer was right when he spoke of man as a wild beast that must be locked up in the cage of law.[1]

Now this would appear to mean that the original order is shattered. On the Roman Catholic view, however, this shattering is indeed only apparent. For, even though sin has brought into this altered order the use of force, something that could not have existed in the hierarchical order of the original state, the principles of this original order nonetheless remain, such as "one should live in peace" [*pacifice est vivendum*] or "to each his own" [*suum cuique*]. Thus the order which as a matter of fact has been changed by sin may still be established on the plainly recognizable principles of the original order. Formally, there is here a certain parallel to Rousseau: The absolute "good" of the original order no longer exists, hence we must strive to achieve what is relatively "best" in righteousness.

The problem in this connection consists in the mode of transformation. In what respect can principles from the sphere of the "absolute" be carried over into that of the "relative"? From what has been said earlier about the Roman Catholic concept of sin and the way its meaning is restricted in comparison with the Reformation view of sin, we may suspect already that for Scholastic thinking there are no insuperable obstacles to this transformation. For, after all, the two

[1] Victor von Cathrein, *Moralphilosophie* (Leipzig, 1924), II, 477.

spheres involved in the transfer are not totally distinct; they stand in a relation of quantitative gradation. All the same, the problem of transformation is not without its difficulties.

Before we take up this question, however, we may sum up in a single sentence the most important result of our attempt to discover the basis of the Roman Catholic doctrine of natural law: An order which remains intact in idea stands in juxtaposition to a reason which remains intact in idea and is capable of being reincorporated into the *rectitudo*. To put it even more simply: there is an order of being, and this order may be known.

The very fact that we had to speak of the alteration of this order after the fall, itself gave intimation of that important distinction to which the problem of transformation must inevitably lead, namely, the distinction between the original order and the actual state of the orders in the fallen world. How are the two related? In any case the relation is not such that there could be one natural law which would apply both to the original state and also to the fallen world. On the contrary, one is compelled here to make an important distinction between "primary" and "secondary" natural law.

"Primary" natural law refers to that sphere of natural law which applies to the unenfeebled nature of man; it is thus natural law in its true and unimpaired form. "Secondary" natural law, on the other hand, refers to that sphere of natural law which is "tailored" or "accommodated" (Lessing) to the enfeebled nature of man, and which thus takes into account such phenomena of our fallenness as servitude, and at times slavery, also the element of force in the state and in law.

The nature of fallen man, which is wounded but not destroyed, presumably can still recognize the axioms of the original order. But the form of the reality corresponding to these axioms is not so evident. The assumed congruity between idea and reality, principle and actualization, has been broken. The soul of the original order no longer has a directly commensurate body. Hence the necessity of what we have called "transformation."

This transformation of principles from the level of the original state to the level of the fallen world entails for natural law a loss of perspicuity. It is no longer evident at first glance what is demanded by natural law, e.g., what the principle of "to each his own" means concretely, here and now, in the situation after the fall. For instance, what does *suum cuique* imply in a land where there are millions of refugees? When one thinks of the older resident, or again of the newly arrived refugee, what is "his own"? Nor need we think only of such exceptional circumstances. Even in "normal" circumstances we live in a world of inequality. Ought all men, for example, to receive equal pay? Are not their various jobs of different worth? But, on the other hand, has not all work the same dignity?

We have only to ask a comparatively simple question like this and natural law sets off a veritable chain reaction of further questions. For to decide the question of equal pay we have to consider the criteria by which work should be evaluated. Is it to be in terms of the value of this work to society, its social and economic utility? But what would then be the place of a lyric poem or of Mozart's *Kleine Nachtmusik?* Or does the value of work lie simply in the doing of it, on the principle that a thing must be done for its own sake? But then the question of value would not even arise; the work of Sisyphus—and his reward—would be exactly equal to that of the farmer who sows and reaps. Or is work to be evaluated, quite apart from any "objective" worth, purely as a personal function, as a fulfillment of duty, and according to the measure of what is put into it?

In all probability what is involved here is none of these neat alternatives. On the contrary, the value of work is probably determined by a complex of factors involving what is actually accomplished in a variety of dimensions. But this means that we will have to assess the relative importance of these various dimensions. Is it even possible to draw up a point-system here? And what good would it do? On what basis would we decide how the points are to to be allotted? Thus we see how one gets bogged down in an infinity of uncertain concepts and unspecified relationships the moment he tries to translate the *suum cuique* principle out of the simplicity of the original state into the multiplicity of the world after the fall.

Similar problems arise with the respect to the concept of "freedom" in natural law. Can all men be granted the same measure of freedom in respect of self-realization, from the field hand on a plantation to the great artist dedicated to his genius at the very pinnacle of culture? Can we simply ignore these accrued distinctions in the name of an equality which no longer exists?

The regulation of the fallen world by natural law can thus be achieved only by relative approximation to the original principles. Complicated transpositions are necessary to make concrete approximations possible. To try to make a "direct" transfer from the primal state into our own world is to be as ridiculous as the man who tries to drive his automobile down the lines of longitude from the North Pole to the South. In the concrete world in which we live longitudes are an artificial abstraction. If we wish to journey from the North Pole to the South, it is advisable, given our world as it is, to translate the abstract way of the straight line into a complicated system of roads adapted to the actual territories which must be traversed.

It may be well to pause here for a moment to survey the several problem strata which the Roman Catholic doctrine of natural law, by dint of its massive conceptual effort, has actually penetrated, and of which we have here given a general exposition. As the motion picture film involves a projection of individual frames in rapid time sequence, so we may project the following theses in consecutive thought sequence as characteristic of the questions involved in the Roman Catholic doctrine of natural law. First, the world is seen from the standpoint of an ultimate order which is directly displayed in the original state. Second, since in the fallen world this order has been impaired, the connection between the original state and the fallen world is jeopardized; this means too that the possibility of receiving ultimate ethical directives and obligations directly from the original order as a matter of course no longer exists. Third, it is consequently necessary, with the help of the concept of nature, to establish a corridor leading from the original state to the fallen world, in order that the connection between the two may be maintained; for nature is that which remains as a relic of the original state. Fourth, at the same time, however, it is seen that we are up against the problem of transferring what is valid in the original sphere over into the sphere of the fallen world (the problem of transformation), and this is not easy because the differences, in spite of the constancy of nature, i.e., in spite of that corridor, are very great. Fifth, recourse is thus had to the distinction between "primary" and "secondary" natural law, the former having to do with pure nature as originally created and being therefore perspicuous, the latter having to do with impaired nature and being necessarily derivative. Sixth, a key has thus to be found by means of which the original forms may be transposed into the fallen world.

This whole movement of thought has displayed unparalleled energy and intensity, as over the centuries efforts have been made to solve this cardinal ethical problem of "man and reality." The hurdles and barriers which constantly arise have been met undauntedly. We are reluctantly forced to allow that Reformation ethics in contrast has not shown anything like the same energy, but has been suspiciously willing to stop with Luther's doctrine of the "two kingdoms." Only in our day, when excessive secularization, e.g., in totalitarian regimes, has brought the problems of this doctrine unmistakably to light, does it seem that a new movement of thought is beginning.

But let us return once more to the Roman Catholic doctrine of natural law. Transition from primary natural law to secondary natural law is undoubtedly complicated. The deductions are intricate. There is a constant groping in obscurity and ambiguity. A host of alternatives repeatedly present themselves whenever it is asked in a given case what might here correspond to natural law. In face of the danger of complete dissolution it is felt that there must be a teaching office of the church which can hand down authoritative decisions. The seriousness of all these difficulties emerges plainly in two classical concepts which are central in the Roman Catholic teaching on natural law, that of "conclusion" on the one hand and that of "circumstances" on the other.

The Concept of "Conclusion"

The concept of *conclusio* refers to the attempt by means of deductive reasoning to apply to concrete cases the fundamental norm of natural law, namely, to do good and not to do evil. This attempt takes two forms: first, to subsume the specific case under the general law, and second, to effect the transfer from the level of the primal state (in which the mode of actualizing the norm is evident) to the level of the fallen world (in which the mode of actualizing the norm is problematic).

How this "conclusion" works out in detail, we shall see later. For the moment our purpose is simply to note that it has to do with the question of certainty. What is desired in the *conclusio* is to transfer to the individual case the evident certainty of the basic norm of natural law, and to do so with the least possible loss, that is, by means of a conclusion which will be as exact as possible, maintaining as far as it can the degree of axiomatic certainty. It is obvious, though, that in the course of the transfer process there must inevitably be some diminution of certainty. An axiom is always more certain than the conclusion deduced from it. Hence there arises the differentiation between particular grades and ranks of certainty, in a manner consistent with the over-all character of Roman Catholicism's gradated thought concerning order. There is, as it were, a hierarchy of degrees of certainty. How deep are the roots of this hierarchical thinking in the basic soil of Roman Catholic theology may be seen from the fact that there is here a strict parallelism, as it seems to me, between the dogmatic sphere on the one hand and the sphere of ethics and natural law on the other.

We shall have to make mention at least briefly of the various degrees of certainty in the sphere of dogmatics, since we shall be following these very closely as we attempt to systematize the various certainties involved in the *conclusio* of natural law. In the dogmatic sphere the teaching office grades the various dogmatic statements according to the degree of certainty which presumably attaches to them. The various grades are listed as follows in Roman Catholic dogmatics.

The highest degree of certainty is to be found in the so-called "statements of faith" [*propositio de fide*]. These are the indisputably revealed dogmas, irrespective of whether they have been expressly defined by the infallible teaching office of the church as binding truth for all believers.

The next highest degree of certainty is that of the so-called "proximate statements of faith" [*propositio fidei proxima*]. These are propositions which in the opinion of the overwhelming majority of theologians declare a truth the substance of which is contained in the sources of revelation, even though the teaching office of the church may not yet have expressly defined those propositions as binding statements of faith.

The third degree is that of "statements which are theologically certain" [*propositio theologice certa*]. These are propositions which are closely linked with dogma. Finally, the lowest degree of certainty is that of the "common notions" [*sententia communis*], which are generally held by theologians, on dogmatic grounds.

With the help of these four categories we shall now try to define the axiomatic degrees of certainty which in the sphere of natural law may be deduced by means of "conclusion."

To the highest degree, that of the "statements of faith," there correspond in the ethical sphere the "axioms of natural law." Basically these consist of two norms: first, the principle that we are to do what is right or good and not do what is evil, and second, the principle of *suum cuique.* These foremost rules of practical reason are self-evident. They are immediately perspicuous, and do not need to be proved. For to prove is to deduce, and to deduce is to derive from superior norms. But there are no norms superior to these self-evident truths. Hence they cannot be demonstrated, any more than the main principles of theoretical reason can be demonstrated, e.g., the validity of causality as a concept of understanding, or the Cartesian *cogito, ergo sum.*

As a parallel to the second degree of dogmatic certainty, we might propose what could be called "proximate statements of natural law." It must of course be pointed out that we are here introducing a terminology which is not actually taken from Roman Catholic teaching, but which we believe to be important for a proper understanding of *conclusio.* What we have in mind is as follows. If the doing of what is "good" or "right" is said to be the highest norm, then it becomes necessary to spell out what this implies specifically, first in general in the various spheres of the human I-Thou relationship, and then concretely in particular cases. We can do this only with the help of a "conclusion."

But it is of the nature of conclusions that they always involve a continuing chain in which b follows from a, and c follows from b, etc. Now in all spheres except mathematics, the longer the chain of conclusions, the less certain they become; the less they are, as it were, firsthand or self-evident truths. In penal law circumstantial evidence, building as it does on conclusions, has a lower degree of certainty than when the wrongdoer has been caught red-handed. Hence Thomas says of conclusions that they are right "in the majority of cases [!] but fail in a few." [2] So there is after all such a thing as a fallacious conclusion!

[2] Thomas, *Summa Theologica,* part 1^{II}, ques. 94, art. 4, reply obj. 2.

On the other hand, it may be said that at least the first, the direct and immediate conclusions, those which are not yet concerned with application to individual and specific cases, do possess a very high degree of certainty approximating as closely as possible the self-evident certainty of the original basic norms. This is particularly true when the sources of revelation themselves draw these direct conclusions, thus adding the weight of their own authority to the self-evident certainty implicit in immediacy. Among these direct conclusions from principles Thomas lists the Decalogue, which describes love for God and neighbor and which, accepting the fact that man is a free, rational, and social being, defines what this means for the preservation of life (not kill), for marriage (not commit adultery), for property (not steal), and for honor (not bear false witness).

The reference here is to the general implications of the ultimate, self-evident norms for the social structure. But the question of their significance and application refuses to stop here. It goes on, as we have said, in a continuing chain, requiring ever more detailed conclusions until an answer is provided to the question of the concrete implications of the original norms in particular instances. And it is at this point that we come upon what we called the third and fourth of the levels or degrees of certainty in the dogmatic sphere. Since we did not feel required to make any precise distinction between them there, we shall treat them here in a single category. This is the category of "notions which are both common and certain," although—in distinction from dogmatics!—they often seem in the sphere of natural law to be very uncertain and individual. Even Thomas admits that natural law can undergo quite a change when it comes to particular instances. The individual case cannot always be subsumed under an appropriate and convincing conclusion.[3]

Why is this? It is because the individual case is bound to the concrete and contingent circumstances within which it arises, and these have to be taken into special account so far as any directives issuing from the norm of natural law are concerned. This brings us then to the second important concept, that of "circumstances," without which the problems involved in *conclusio* cannot be understood.

[3] "The natural law is altogether unchangeable in its first principles: but in its secondary principles . . . it may be changed in some particular cases of rare occurrence, through some special causes hindering the observance of such precepts, as stated above." *Ibid.*, ques. 94, art. 5.

The Concept of "Circumstances"

The concept of *circumstantiae* suggests the existence of two possibilities, only one of which can actually obtain. Think, for example, of the situation in war-devastated Germany: The food shortage is so severe that I have to decide whether, to keep my family from starving, I will engage in illegal "black market" operations—and in so doing further augment the already inequitable distribution of food. The question then is whether or not this concrete case can be "subsumed" under the main self-evident or proximate propositions in the form of a conclusion. If the answer to this question is affirmative, then logically deducible, casuistical rules are indeed possible, just as in the field of physics, where such a primary norm as the law of gravitation allows us to calculate such individual differences as the speeds at which a bullet and a feather will fall. That would be the first possibility.

The second possibility would be that the individual case cannot be subsumed in this way under the original norm. In that event an alteration of the original norm itself would be necessary, because the individual case lies on a plane which is radically different from that of the original state, the state to which the original norm genuinely applies. It is for this reason that the original basic norm must be trimmed to suit the altered circumstances.

In any case it is in these relatively later stages of the conclusion chain, when application is ultimately made to concrete cases, that we thus come up against the critical problem in the whole doctrine of natural law, the problem to which we have already alluded and which demands further discussion. This is the problem of whether the fallen world really can be understood at all in terms of natural law. Can the fallen world, even with respect to the borderline cases we have in view, be that simply and directly referred to an order of being? Such referral would be possible only if the fall is accorded a merely relative significance. Conversely, to the degree that we insist that no such order can be seen, the radicalism of the Reformation understanding of sin becomes important. Whatever else may be said about it theologically, we should have to take seriously the concept of the world which finds expression in this radical view of sin.

It is for this reason that we must put the decisive question: Can the individual case be subsumed? We put the question in the form in which Roman Catholicism asks it of itself, and we will listen to how Roman Catholicism deals with the question on its own level of thought. What is at issue is again this matter of "circumstances," and perhaps our best mode of procedure is to study some of the border-line cases in which the difficulties come to head.

Our first example is as follows. In keeping with the axiom *suum cuique,* property which is borrowed or held in trust is supposed to be restored to its owner. This duty is clear in principle, but it can give rise to problems "under special circumstances." For example, the property should not be returned but retained if the rightful owner intends to use it for purposes of treason.[4] Similarly, the weapon taken from one who is running amok should be kept even though it is his property.[5] There are other cases in which restitution to the owner is unlawful because "the giving itself was illicit" in the first place. This is true, for example, "when a man gives a thing simoniacally [*simoniace:* simony, bribery]. Such a man deserves to lose what he gave, wherefore restitution should not be made to him." [6]

On this basis we may state the matter thus. Whereas conclusions diminish in certainty the further removed they are from the original basic norms, the loss in certainty immediately takes on gigantic proportions and becomes all but complete the moment individual cases are subsumed. This is because the transition from conclusion to subsumption involves a complete change in the manner of cognition. Conclusions arise by a process of deduction, whereas subsumption proceeds from that which is inductively apprehended. Conclusions embody, as it were, a priori statements; subsumption on the other hand proceeds a posteriori on the basis of experience. Consequently, subsumption lies in the sphere where deduction and empiricism intersect.

Empiricism for its part, however, has to "interpret" facts as well as to "observe" them. There is thus involved a broad complex of cognitive processes, of which a comprehensive overview is very difficult to obtain. Our previous illustrations, however, indicate plainly enough what is at issue. I must first note, for example, simply in the factual manner of a secretary's minutes, the purpose for which the borrowed

4 *Ibid.,* art. 4.
5 *Ibid.,* part 2^II, ques. 62, art. 5, reply obj. 1.
6 *Ibid.,* reply obj. 2.

property, if returned to its owner, would be used. I must then evaluate that purpose, considering, for example, whether in substance it is actually treasonable. And in order to make this judgment I would also have to know what treason is, whether or not I go into the further and very complicated question of whether the owner's ideas on this subject differ from my own, so that what to me is treason is to him a moral act.

We might enumerate still other factors of observation and interpretation which would also be involved, for there would be such a vast number of them at work complicating the cognitive process as a whole that the final subsumption would ultimately appear to be altogether arbitrary, a venture or act of decision. At any rate, the processes of cognition in this sphere where deduction and empiricism intersect by no means share any longer the unalterability of common principles or the self-evident character of axioms.

This is perhaps even more obvious with respect to the demand of natural law that life be held inviolate ("You shall not kill"). In concrete application, this commandment too undergoes a crisis through the addition of empirical components. The state, for example, can execute a murderer, and in self-defense an individual can kill an attacker. If with respect to capital punishment and self-defense the ambivalence of the situation is still somewhat restricted—though it is still there and often makes necessary a complicated web of observations, inferences, psychological interpretations, and legal deliberations —it often reaches the point of utter obscurity when it comes to the question of a "just war."

For example, it must be established that the enemy to be killed is indeed an enemy, since the slaughter of innocents is forbidden. Prisoners of war, for example, are not to be slain, because they have ceased to be enemies. But legally to identify someone as an "enemy" is possible only if the war is "just," i.e., if it is a defensive war. For only in a defensive war does my opponent from the very outset come at me as an enemy. In an offensive war it is I who *make* him my enemy; strictly speaking he is not my enemy; I am his enemy. Proof that the one who confronts me is in the true sense an enemy whom I may kill thus depends entirely on the justness of the war in question.

Thus, to apply concretely to the matter of war the axiom of natural law concerning the sanctity of life and so to determine the manner in which the commandment not to kill is to be fulfilled, it is necessary first to reach a judgment on whether or not the war in question is "just." In our present context we would simply point to the almost impossibly complicated nature of the processes involved in arriving at such a judgment. In order to make that judgment I must take the following steps.

In the first place I must have accurate information on all the essential "circumstances" leading to the outbreak of war. How difficult this is to secure may be seen from the fact that even after several decades politicians and historians are not agreed as to the guilt for the First World War. How difficult it is then at the moment the war breaks out—which is when the individual is compelled to make his decision to participate or to abstain—is quite obvious when we remember that the occasion which precipitates the conflict, and the responsibility for it, are usually concealed by means of the political, diplomatic, and military rules of the game. In an age when whole populations war against one another it is almost impossible for the average man to have access even to the bare facts, since only garbled information is placed at his disposal. There is also the further complicating factor that behind the occasion for war stand deeper causes which can be appreciated only against the background of a well-founded historical view of the matter, and this historical view is itself, in turn, embedded in philosophical foundations, where what is involved is basically a venture on the part of the men making the decision. But even in respect of the "occasion" alone we are confronted by almost insuperable difficulties in trying to ascertain the facts. One has only to recall that an "occasion," e.g., hostile acts in the country slated for attack, or the pretended oppression of minorities, etc., can often be artificially manufactured in order to create an excuse for going to war and at the same time avoid the appearance of aggression. In other words, the facts themselves can be falsified as well as the accounts given of them.

Assuming though that I have accurate information on all the circumstances, I still have to interpret that information. But my interpretation may go in any of a variety of directions, depending on the criteria with which I approach the problem, criteria which in their turn can also be highly diverse. For example, a particular attitude towards the problem of "preventive" war may determine my judgment as to the justness of a particular war. And this question of "preventive" war is one for which there is in turn no final answer. In the days of Thomas people may have thought that they had the problem solved: an aggressive war can never be a just war. But in the age of the atom bomb, when both the outbreak and the outcome of a war are determined in a matter of minutes, it is no longer possible to decide the question in these terms. What is required is rather an assessment of the enemy's intentions, and an attempt to anticipate his resultant actions. In this sense a preventive war may very well be an anticipatory action of defensive character, a defensive war conducted in advance.

These considerations will perhaps suffice to show that we are here involved in a sphere clouded with uncertainties, and that the above-mentioned dictum of Thomas that conclusions are right "in the majority of cases but fail in a few" must be regarded as highly optimistic. Thomas of course admits that in the application of the general principle to concrete circumstances errors may be committed because of erroneous conclusions, passions, or evil habits.[7] The question is, however, whether his mentioning of these subjective sources of error really touches the decisive problem. For when we consider more closely the uncertainties to which we have alluded, we shall find that they are grounded in the objective structure of the world itself.

War is obscure in its objective nature. It is not obscure merely because those who are to observe and judge it are clouded in intellect, blinded by passions, and radically deceived so far as their information is concerned. It belongs rather to the very nature of war that there should be an endless changing of opponents, that there should be concealment and deceit, and that there should be a deep involvement in a historical guilt which recedes further and further into the distance the more we pursue it. The events of past decades and centuries and even of most primitive times continue to play their role in precipitating every new war. What is manifested here is nothing less than the world-after-the-fall, whose injustices cannot be clearly demonstrated because there are in this world no zones of unequivocal righteousness with which to contrast them. There is no such thing as a wholly just war, and my decision to endorse a given war and participate in it can be made only from the standpoint that I see, or think I see,

[7] *Ibid.*, part 1^{II}, ques. 100, art. 1.

greater wrong on the one side than the other, and that when I plunge into this confused tangle of unrighteousness I do so in the confident assurance of God's forgiveness.

It would thus be a mistake to ascribe wholly to the subjective imperfection of our cognitive acts, as they take place within that zone where deduction and empiricism intersect, all the uncertainty involved in making concrete applications of natural law. No, we must see that this uncertainty is grounded rather in the very structure of the world itself, an objective structure which is so permeated by guilt that there are in fact no "pure" cases at all, such as those which the Thomistic view of order and nature is compelled to assume. We shall have to bring out this interconnection between sin and world structure in our later discussion of the orders. But in saying what we have, we touch already on the question as to the final reason why concrete cases cannot be subsumed. We thus draw gradually nearer to a concept which is essential both here and in the doctrine of the orders, namely, the concept of "this aeon," αἰὼν οὗτος.

In this connection we are obviously dealing with a very different level from that on which the highest axiomatic norms are valid. These basic norms are valid in the "original state," whereas the sphere in which they are applied is the disordered world of the fall, the world of injustice. We must now consider more closely this distinction between the two levels, or rather between the two systems of co-ordinates within which values and the related forms of reality arise.

We may illustrate what is meant by a "system of co-ordinates" from the world of baseball. A well-coached baserunner properly touches all bases with his right foot rather than his left on an extra base hit, but only because the underlying pattern of the game calls for a counterclockwise circling of the bases. If the basic pattern were just the opposite, clockwise, it would require a corresponding adjustment to a left-foot pivot on the part of the skilled baserunner. This rather banal example is a useful illustration, for it suggests that appropriate application on the level of the individual case is wholly determined on the more basic level of the system as a whole. What is right and proper, what is good and just, changes according to the system of co-ordinates. In ordinary matters it is relatively easy to find the formula of transformation by means of which what is laid down in one system may be translated into another. Drivers who sit left and drive right in New York, for example, ordinarily manage to adjust with relative ease to the sit right and drive left requirements of certain other countries when they travel abroad. But no clear formula of transformation is perceptible when it comes to the relation of the original state and the fallen world. The two situations are so totally different.

Only in this light can we understand why the divergence between the *proton*—or eschaton—on the one hand and "this aeon" on the other has been the subject of rather blasphemous humor, such as that of Nietzsche's reference to the monotony of hearing the eternal Alleluias of the white-robed angels, or that of the comic cartoons in their treatment of grass-eating lions living at peace with lambs. Similar illustrations drawn from the area of man's primal state are practically beaten to death in modern films and novels. But here too the humor itself, so far as aesthetics is concerned, arises precisely from the exploitation of the contrast between the two states. The very fact that the law of opposites, more strictly the law of contrasting systems of coordinates, can be used at all to such comic effect, is naturally from a theological standpoint very revealing and full of implications.

To test this business of contrasting systems one might ask: Can any direct norm be deduced from the original state, for example, in respect of politics? But even if one could postulate the existence of a political order in the days of innocence, is not the concrete political configuration of our own day something completely different? If we were to speak at all of a "state" existing in the days of innocence, our reference could only be to the sociological expansion of freely undertaken relationships of subordination and superordination. For when man lived at peace with God he maintained a free inclination to obedience, he paid strict attention to God, and the primitive "state" may thus be regarded as nothing more than an order of voluntary and enjoyable "play." By contrast, the concrete political order of this aeon, even as ideally conceived, is always marked by the possession of force, the might by virtue of which it can maintain itself against evil and the powers of chaos.

In face of this qualitative difference between the two, is it possible to transfer the norm from one level to the other? Can it be said, for example, that by natural law, i.e., in the name of the original order, it is required that there be but one specific form of political order, namely, the "democratic" state, since this form more than any other fosters the free obedience of individuals and comes nearest to the original voluntary spontaneity? Is this really so? Might it not be that in an age of disintegration and centrifugal tendencies there can be no alternative to some form of dictatorship and planned economy? I am simply raising the question. To be sure, in such a case we would be bound to realize that ultimately such a dictatorial form of government is simply not right, that it is not—to put it in secular terms—"ideal." But could we then say with equal certainty what this "ideal" order actually is, in positive terms? We have already spoken of this before: What *ought* to be cannot be known simply by postulating its opposite, what ought *not* to be. And is not our inability to declare or postulate the positive side an indication that sin is a "total" phenomenon? By "total" we mean not that everything is sin, in the sense of Flacius, but that sin permeates all things so that we cannot even conceive of that which ought unequivocally and unreservedly to be.

The Reformers understood this full well so far as the individual is concerned; we demonstrated that rather amply in our discussion of the personalism of their thinking. We would supplement what has so far been said only at one point, by noting that for Luther even a man's piety, indeed his faith as a psychic act, is construed in closest proximity with his existence as a sinner. In this connection one need think only of Luther's discussion of the "resignation to hell" in his Commentary on Romans.[8]

Now at this point there thus arises the crucial and pressing question whether this total permeation with sin, which is so complete as to necessitate the *resignatio ad infernum,* is not present also in the "orders," in the sense that every political organization too, for example, must reckon with egoism, and is itself egoistic. The state not only *reckons* with egoism when it uses force to meet the centrifugally oriented egoisms of individuals and groups. It is egoistic; its own egoism lies in what has usually been called from the days of Machiavelli a "sacred egoism."

What is meant by this is perhaps seen most plainly in the matter of the ethical status of treaties between states. According to ancient political wisdom, treaties are meaningful and lasting only to the

---

8 See p. 318, n. 12 above; also Erich Seeberg, *Grundzüge der Theologie Luthers* (Stuttgart, 1940), pp. 31 ff.

degree that they control and protect mutual interests, and it is thus advisable to conclude treaties only on the basis of such mutual interests. It is a fact that treaties which only serve the interests of one side—perhaps because, being concluded after a war, they were actually dictated by the victor—are not usually kept. The same is true of treaties whose underlying presuppositions change substantially during the period of the treaties' validity. Ordinarily treaties are not kept merely because of the sanctity of the pledged word or the validity of a promise. For their function is to protect national interests or egoisms, and they are therefore immediately affected or modified by any change in those interests. It is for this reason that statesmen and ambassadors utilize both official and unofficial occasions to ask their counterparts on the other side whether they still stand where they stood when the treaty was first concluded, and to assure them that they themselves still stand on the same ground. Oaths and promises are often renewed in politics, precisely because they do not have eternal validity but depend on shifting interests. This illustration from the world of diplomacy shows us that, while the state provides a safeguard against diverse egoisms within its own bailiwick, it must also be regarded as the bearer of an egoism of its own.

We must now ask why this is so. And for the moment it must suffice to say that the state very obviously bears certain egoistic traits which belong to its nature, or, more precisely, to its nature as a phenomenon subsequent to the fall. For this reason these traits cannot be described simply as a departure from its true nature. On the other hand, one cannot take the "total" permeation of the state by egoism to mean an equal and uniform denigration of all forms of statecraft without distinction. It goes without saying, for example, that there is a difference between an imperialistic drive for expansion on the one hand, and on the other a readiness to act in concert with other states. The expansionist state must certainly be regarded as a perversion. But the more responsible behavior of the non-expansionist state ought not to mislead us and make us overlook the egoistic permeation, and thereby the essential link with the fall, which is present in it too. It ought not to make us forget or deny how impossible it is for us even to imagine, let alone realize, an ideal situation in which the state is only accidentally but not fundamentally linked with this egoism.

This brings us squarely to the theological question concerning the

418

orders as such, which we shall deal with later in a more basic and concrete way. It will be our purpose to show that the orders, e.g., that of the state, do not confront man as a representation of the pure will of God the creator. They represent also an objectification of fallen humanity. Thus the state bears not merely the characteristics of God's gracious will to preserve, but also the features of egoistic, self-exalting man. For this reason there is no such thing as an ideal state, discharging its function as a kind of altruistic ministry. Such a state is not found even in abstract speculation, let alone in reality. We would be literally at a loss for ideas and for words if we were to try to say what such a state would be like. We simply cannot describe such a pure and primal state, any more than we can describe or even conceive of man before the fall. At least, we cannot do so if we understand sin in qualitative and personal terms, without misinterpreting it in the ontological sense as merely a subtractive factor.

If the Reformation view of sin as qualitative and personal is right, then this implies a fundamental criticism of the doctrine of natural law and its concept of order. For there is then an infinite qualitative distinction between man's original state and the fall such that no formula of transformation is possible. The cited difficulties which confront the doctrine of natural law when it attempts such a transfer would seem to be a significant indication of this fact.

21.

A Critique of the Roman Catholic
View of Natural Law

David Friedrich Strauss once said that the history of dogma is the
judgment of dogma. Similarly, and probably with even greater
truth, we may say that the history of natural law is its most impressive
criticism. The staggering variety in the conceptions of natural law
is itself a cause for doubt as to whether there really is any such thing
as a constancy of ultimate norms. If, then, we are indeed not capable
of perceiving any such constant axioms, we are now able, on the
basis of our historical survey, to enumerate in summary fashion the
reasons why this is so.

Inconstancy in the View of Man

In the first place there is, on the level of natural cognition, i.e.,
apart from the revelation of the one man Jesus Christ (*ecce homo*), no
way of knowing the nature of man. The reason for this is that man
is to be defined as a being in relation.[1] What is generally supposed
—apart from this relation—to be his nature is determined by the requirements and prescriptions of (indeed the θέσεις or statutes posited
by) the situation at any given time, by its contingency and mutability.

This is evident already in Aristotle's view of natural law. Here
the idea of man is identical with the idea of the citizen, the *civis*.
This means that the idea of humanity arises only within a very
definite "political" relation, which is itself conditioned by the contingency of the social order as it existed at that time. Accordingly,
ethics itself is politically determined. The political relation replaces
that of God and man, and constitutes the nerve of the value system
of natural law. The result is that slavery, for example, is necessarily
sanctioned by natural law, for he who does not stand with respect
to the state in the relation of a citizen is not a true man and may

[1] See Chapters 9-12 above.

thus be "used" as a means to an end, which is to say inhumanly, as a slave.

Similarly, tribes who eat human flesh also possess a kind of natural law, within which such eating is a virtue, or at least does not come under the reproachful judgment which we civilized Westerners express in the word "cannibalism." For these tribes each man has a *mana* which can be appropriated by killing and eating him. It is the *mana* anthropology which leads directly to a corresponding natural law, or something analogous to it.

It may be that future space flights will give rise to interplanetary conflict as hostile earth-dwellers try to establish outposts on other planets. Then the question may conceivably arise as to how we are to treat Martians and other "men." Do they come under the protection of the earthly concept of humanity, of earthly "citizenship"? The whole situation might give rise to further complications in natural law. For Martians might have snouts instead of noses. They might lay eggs instead of reproducing in the manner of earthly mammals. This would certainly give rise to different schools of natural law, some arguing that the Martians are men and insisting that they be covered by all international conventions on human rights, others being prepared to grant them at best only the protection afforded by statutes designed to prevent cruelty to animals.

We thus maintain that natural law in any given instance depends on the view of man which underlies it. But this view of man is something which is quite inconstant, apart from the relation between God and man. And even the relation between God and man cannot be perceived apart from faith. Hence there can never be such a thing as a natural law which is unchanging and of universal validity.

Inconstancy in the Understanding of Justice

This variability in the concept of man, and in the corresponding concept of natural law, is readily seen in the case of one of the main axioms of natural law, namely, the principle of *suum cuique*, "to each his own." [2] This principle seems indeed to be constant, since it is found as a purely formal principle in all ages and places. But

2 On the *suum cuique* see Dietrich Bonhoeffer, *Ethics*, ed. Eberhard Bethge, trans. Neville Horton Smith (New York: Macmillan, 1955), pp. 108 ff.

of what avail is the constancy of form if the content is always different, varying as it does in accordance with the unending variations in the concept of humanity? Does not the *suum* continually vary according as man is seen as either citizen or slave, either beast of prey or human brother?

Now it might be objected that, while the content of the *suum* admittedly depends on something which is inconstant, namely, the view of man, and that while the fact of this inconstancy is itself grounded in the natural unknowability of man, nevertheless, once the view of man is firmly fixed, as it is in the Roman Catholic doctrine of natural law, a clear definition of the *suum* can then be given. This argument is impressive only at first glance. For even in the Roman Catholic doctrine of Christian natural law there can never be anything more than a conditional constancy of the *suum*, as has been shown especially by Emil Brunner.[3]

In every human society there is both equality and inequality. All men are equal before God [*coram Deo*], for all are sinners together and all are righteous together, i.e., without differentiation, without respect to individual merits. Moreover, all are equal before the law, i.e., in the legal sphere. And in the most general and formal sense all are equal as regards their quality as moral beings. Yet all are not equal among men [*coram hominibus*]. The inequality exists in many different respects. Thus men are unequal as a result of historical contingencies ($\theta\acute{\epsilon}\sigma\epsilon\iota\varsigma$), e.g., the contingency that one is the son of a king and the other of a worker, that one is Brahman and the other a pariah in the Indian caste system. There are also natural inequalities ($\phi\acute{\upsilon}\sigma\epsilon\iota$); in this respect one has only to think of the differences of sex, age, gifts, and psychical temperaments.

There is obviously, then, no supreme authority by which to decide the content of the *suum* in any given instance. Sometimes it will be decided *in actu*. Thus there can be no certainty that a better position will automatically go to the more gifted, and a worse position, e.g., that of an unskilled laborer, to the less gifted. On the contrary, any position has to be won. And if a man gains it, so that it is not something that has been handed to him but something he has at-

---

[3] See esp. Emil Brunner, *Justice and the Social Order*, trans. Mary Hottinger (New York: Harper, 1945), pp. 29 f., 42 ff. On this whole problem see also Walter M. Horton, "Natural Law and International Order," in *Christendom*, IX (Winter, 1944), No. 1, 2-21.

tained, then this "something" represents as a rule the *suum* which is appropriately due him. It is shown to be "his own" by the very fact that he has created it for himself. Communist doctrine, of course, would dispute this whole line of thought on the ground that it proceeds from the individual, at the expense of the group, which illustrates very well how controversial the *suum* concept is in its very roots as between the divergent world views.

We thus maintain that there is no universal or constant *suum*, not even where the Christian view of man may have attained a certain constancy, as in the Roman Catholic world. At the very most, the content of the *suum* can be determined to some extent only when we are dealing with spheres in which men are equal, and it must be added at once that the extent of this equality is understood very differently in different systems. On the other hand, the content of the *suum* cannot be determined in that other sphere in which men are not equal, since in the main we do not know what the distinctions are. The distinctions in endowment, for example, only come to light as men actually create that which belongs to them. Whether a particular man will be a minister or a beadle cannot be decided in advance, i.e., before the divergent endowments disclose themselves through the movement actually made towards these offices.

We must not make the mistake of thinking that, because these distinctions are largely unknown, no firm distinctions at all can be made. The fact that there are such firm distinctions comes out very clearly in such a thing as war-time rationing, where it becomes obvious that in terms of nourishment, for example, an infant has a different *suum* from that of a common laborer. This is something that cannot be denied, even by the frequently arbitrary scales of ration allotments which take no account of individual differences.

Yet even in respect of such a clear distinction as that between the sexes, the appropriate *suum* is not always clear. Certainly—or should we say "probably," or even "possibly"?—it is not the *suum* of the man to darn stockings or to wear his hair long. But the question can be very complicated, often insoluble, when for example we ask what vocations are suitable or unsuitable for women. This is particularly true in relation to such public positions as those of statesman, jurist, or minister. Women can presumably represent a healthy corrective in such professions, even when their services are in no sense restricted to members of their own sex. But even if that be admitted it is still doubtful whether we should ignore sex distinctions altogether and grant women full and unrestricted equality in the choice of a vocation, as Communism believes. At least this is a question that can well be argued, the answer to which is by no means a foregone conclusion. To state the matter even more precisely, there are three main arguments against the *suum* in this particular regard.

First, there is a "sociological" argument. The question might be raised whether it is not in greater accord with the organism of human society that the vocations of motherhood and nursing should be reserved for women, and that women should be largely restricted to these vocations. It is a fact, however, that the structure of a particular society may be such as to require a departure from this principle; this becomes even clearer in the argument posed in our next paragraph. Hence there can be no doctrinaire insistence on this natural *suum* of women in a situation where the structure of society no longer corresponds to the supposed natural stage, when increased industrialization, for example, makes inevitable an increase in the percentage of lower-paid women in the working force.

Second, there is a related "historical" argument which favors the view that women should be allowed to occupy the public positions usually held by men, and therefore disputes any allocation to them of a so-called "natural" *suum*. We have in mind, for example, ·the surplus of women over men in the population which is generally evident after a war. In such times, which are often times of increased poverty as well, there are simply too few openings in the occupations "naturally" open to women for this surplus to be absorbed. The *suum* which is "natural" in ordinary circumstances has thus to be distinguished from the *suum* which arises in emergency situations, the *suum* character of which, however, is nonetheless incontestable. At any rate it is clear enough from this example that the *suum* is always subject to alteration according to specific, temporally conditioned "circumstances."

Third, there is the matter of "personality traits." Among both men and women there is such a broad variety of personality types as almost to invalidate the traditional distinctions drawn between the sexes. Among women, for example, we have the motherly type who shuns publicity and wants no part of the tasks normally reserved for men. But on the other hand we have the bluestocking who despises kitchen work and looks down on the man whose love is influenced by a woman's skill with pots and pans and who, if he has to choose, prefers a wife who can cook to one who has a better knowledge of the poets.

Whatever example we choose, the variable character of the *suum* is always evident, at least in the sphere *coram hominibus*. Undoubtedly there is an appropriate *suum* for each individual. Of that there is no question. The only problem is that we men are unable to know what it is, and that consequently we are not in a position to lay down general rules on the basis of which the *suum* can be ascertained and granted in any given instance. In this whole matter of the *suum* what is involved is not an ontic, but a noetic problem.

The Limitations of Man's Knowledge

In summary we may list three reasons why access to the *suum* is always barred to us. First, it is impossible for us to get a comprehensive picture of all the distinctions, and of all the various arrangements which would be appropriate in the light of them. The picture we do get is simply not clear enough. Second, pedagogical considerations complicate the matter further. We can never be absolutely sure that what seems on the face of it—by nature—to be best for a person will really be helpful in any given case. By nature artistic work might seem to be most appropriate for a tender person devoted to the Muses. It could be, though, that his true and deeper powers may be released in all their fullness only when he is exposed to a bitter struggle for existence, the kind which only the hardy can survive. Which of the two forms of life then is really appropriate to him? Which of the two represents his *suum?* Which of the two brings out this *suum*, this true being of his? These pedagogical considerations come into play even in the question of sickness and health. Intrinsically one might say that health is the *suum* of every man. Yet many have come to true maturity, to their real selves, in the course of suffering—quite apart from the more specifically theological question as to the bearing health and sickness may have on our relation to eternity. Third, we simply do not know the *ordo* into which all these distinctions are integrated. And not knowing that order we cannot know the key to their distribution, the key which could help us determine what is appropriate to the individual in relation to the whole. These things are all known only to God.

The statement that the *suum* is known only to God and cannot be deduced by man from an order has important implications in two areas, eschatology and prayer. As regards prayer, any petition can be made only with the reservation that we leave the manner of its fulfillment to the will of God. To be sure, we do ask for a particular *suum*, which in prayer we seek to establish and specify. We ask for food and drink, e.g., for shoes and clothing, for daily bread and the coming of God's kingdom. Now in so doing, we are actually presupposing a definite *suum*, even when we realize full well that it is ours not by our own merits but by the grace of God. The mother who prays for the safe return of her son reported missing in action is assuming that it belongs to his *suum* to be at home with his family instead of languishing in an enemy prison camp.

But everything that we say about our *suum* in prayer can be said only if we are at the same time willing ourselves to be told, "Your Father knows what you need" (Matt. 6:8). Hence our prayer is spoken with the proviso, "Not my will, but thine, be done" (Luke 22:42). What this means is simply that in the final analysis the *suum* is known only to God. It may even consist in the very opposite of that for which I pray, that which my prayer takes it to be. The *suum* may involve for that missing airman the fact that he must remain a prisoner of war, that only in this way can he be tempered to fulfill his true destiny, and that later he may even be grateful for this period in captivity.

Consequently the statement that the *suum* is known only to God is a constant theological reservation in respect of the ethical concept of the *suum*. This reservation dare not keep us from concrete prayer or action of course. But the unguarded claim that natural law can know and specify the *suum* in any given instance is sheer blasphemy if it is made without this theological reservation.

The statement that the *suum* is known only to God has implications also in the area of eschatology. Only the last judgment—in which God, unlike man, sees the heart rather than merely the outward appearance (I Sam. 16:7; Acts 10:34; Rom. 2:11; etc.)—will bring to light the true *suum* proper to a man. Seeing as we do only the outward appearance, men do not have the necessary criteria. The true relation of reward and punishment, and hence that which is truly appropriate, is only disclosed eschatologically.

This aeon does not yet know the testing fire whose flames will reveal what is "gold, silver, precious stones, wood, hay, stubble" (I Cor. 3:12-15). World history is emphatically not world judgment. It is this impossibility of perceiving the *suum*, and of doing justice to it, which repeatedly raises the problem of theodicy, a problem which only eschatology can solve. To the degree that we think we know the many distinctions, and the key to their distribution, eschatology recedes into the background. This is why the comparative lack of emphasis on eschatology in Roman Catholic thought may be related —not exclusively but surely in part—to its emphasis on natural law and its presumed insight into the order of the world. A truncation of eschatology seems to go hand in hand with thinking in terms of natural law. One ought to be very careful about asserting that such formal elements as the principle of *suum cuique* remain constant. For the assertion of constancy in form means very little if the content is as variable as we have seen it to be, and if the mode of variation is unknown.

The same objection may be made against the concept of "freedom" in natural law. Here again there is a very complicated variety as regards the actual content of the concept. Here again it is relatively easy to determine that content in respect of those spheres in which man is equal, especially in the dimension *coram Deo*. For in this dimension one will doubtless arrive at the postulate—understood of course in a purely formal sense—of freedom of belief and conscience. Yet even here there are problems in detail: Will Roman Catholicism, for example, when it has the power, concede freedom to atheism or even to anti-Catholicism?

In the dimension *coram hominibus*, however, i.e., in the sphere of distinctions, it seems to be completely impossible to determine and concretely to grant the freedom which each person should have. One has only to think of the reciprocal claims of the individual and of society. Thus in times of emergency it may be necessary to curtail the liberty of the individual, e.g., by billeting or rationing, so that it is hardly possible to make any general and abiding pronouncements on freedom. In particular, the measure of freedom which may be allotted to the individual is subject to the same uncertainties and reservations as we have noted in respect of the *suum cuique.*

This does not mean, of course, that the principle of the *suum cuique* is flatly contested. It goes without saying that the principle does in fact have a certain heuristic significance. It helps us to discover what is relatively just in a given situation, or better, it keeps open and never allows to be silenced the question of what is equitably proper. What we are contesting is simply the view that this *suum* may be concretely ascertained and defined by just reading it off from the scale of a given order. In other words, we do not believe that the individual case, the distinctions within which embrace the whole plenitude of uncertain factors, can, by means of certain formulae of transposition, be subsumed under general norms. We have already seen to what difficult and insoluble problems this attempt leads, even in respect of such relatively general matters as the restoration of some-one else's property and the justification of war. With what unknowable data these attempts at subsumption and transposition must work when they try to deal with individual cases!

Our main objection to Roman Catholic natural law is that it ascribes a false rank to the *suum cuique,* regarding it as an imperative, the expression of a given and knowable order. The *suum cuique* is thus accorded more than merely heuristic significance. It becomes a symbol of that order of being which we on our part believe, for dogmatic reasons, to be not given and knowable.

What is proper to man in general may be established only in the venture of a subordinated order (or, better, a subordinated emergency *ordo*) and with the help of rules of procedure which are to be fixed according to the measure of man's restricted knowledge. We shall have to speak of this later in connection with our discussion of law. What is proper to man individually can be determined only in the venture of love, because love affords the greatest possibility of understanding. We shall have to speak of this later in connection with our discussion of the direct I-Thou relationship.

We may thus sum up our polemic against natural law's axiom of the *suum cuique* in the following propositions:

First, the formula cannot be described as constant if the contents are always variable, necessarily variable because of the sphere of distinctions. In this respect, we must distinguish between the formulae of natural law and those of mathematics or physics. The latter may be constant even though they refer to changing contents, because the possibility always obtains of subsuming the concrete contents under the relevant laws. But this is exactly what is not possible in respect of natural law.

Second, what is specifically proper to any given individual is unknown to the degree that the distinctions are unknown and we therefore lack the ultimate prerequisites for ascertaining the *suum*.

Third, the whole matter of what is proper to the individual is hence restricted exclusively to the righteousness of God. It is left to His will. It is His to distribute, and will be revealed only at the last judgment. The *suum cuique* is thus seen to be an eschatological postulate.

Fourth, the *suum cuique* has, of course, a heuristic significance, but not in the sense of natural law, whereby it is said to give expression to a given and knowable order into which the individual case may validly be integrated. Our protest, then, is not against attempts to establish the *suum cuique* and the concrete commandment of freedom. It is against the authoritative rank accorded to the arguments used in these attempts, and in the attempts to settle the question of how concrete applications are to be made.

The Validity of the Secular Quest for Natural Law

In spite of the impossibility of perceiving an order of being, and in spite of the resultant crisis of natural law, it is nonetheless vital that the question of an order of being be kept alive, and that some kind of an impulse towards natural law be maintained. It would be madness to link a rejection of this impulse with our rejection of natural law. For in this impulse, which continually drives man to inquire after that which is right "by nature" or "from creation," are expressed two essential concerns which cannot be surrendered.

There is first the feeling that human will and human caprice can never posit their own legal or ethical norms. A dictator, for example, cannot establish what is right and wrong simply by autocratic decree. He must rather give heed to the norms which themselves determine his will, the norms to which he must be obedient and before which he must validate his own authority. There is thus posed the question of ultimate validities which are secure against all caprice. This is a question which must never be silenced. As a question, it is an antidote to the secret poison of pride. When genuinely put, the question leads man to repentance. To be sure, it can be genuinely put only when man recognizes the distance between the will of God and the orders of this aeon, i.e., only when man believes. Here as elsewhere, repentance is a function of faith. Nevertheless, even when asked apart from faith, the question of ultimate norms is an antidote to the worst consequences of pride because, in intention at least, its quest always reaches out beyond man. The question continues to have a wholesome significance even when it is falsely resolved, e.g., along the lines of natural law.

The second concern expressed in the impulse toward natural law is a kind of involuntary confession that, living as we do in a world of relativities, the good is something which has to be sought. For in the fallen world it is impossible not merely to do the good, but even to know what is good. The good is beyond our knowing not merely because our cognitive function has been reduced, but primarily because this good is not objectively at hand in "this aeon." This aeon, being incompatible with the will of God, can no longer produce the good from within itself. This is why, for example, a certificate of divorce has to be given in this aeon, because of "hardness of heart" (Matt. 19:8; Mark 10:5).

Karl Barth is surely right when he maintains that secularized politicians—poor wretches!—have no option in respect of human dignity, the *suum cuique*, and freedom but to seek something like a natural law. He writes, "The civil community as such—the civil community which is not yet or is no longer illuminated from its centre—undoubtedly has no other choice but to think, speak, and act on the basis of this allegedly natural law, or rather of a particular conception of this court of appeal which is passed off as *the* natural law. The civil community is reduced to guessing or to accepting some powerful assertion of this or that interpretation of natural law. All it can do is to grope around and experiment with the convictions which it derives from 'natural law,' never certain whether it may not in the end be an illusion to rely on it as the final authority and therefore always making vigorous use, openly or secretly, of a more or less refined positivism." [4]

[4] Karl Barth, *Community, State, and Church: Three Essays* (Garden City, New York: Doubleday, 1960), p. 164.

When men know nothing of a revealed will of God, what other possibility do they have of expressing both the limitations of their own caprice and also their respect for superior norms? Even as Christians, as church, we should respect these references to natural law in the secular sphere. We should respect them as a kind of symbol in which are expressed both "recollection" and, if not fear of God, at least something similar, namely, reverence. For here is declared that which transcends man in his autonomy.

For this reason, it would be wrong for Evangelical churches to protest against the overtones of natural law found in the preambles to so many state constitutions. For in the secular sphere what can we put in place of natural law? A concern for humanity in terms of natural law and human rights would seem to be both more honest and more modest in such preambles than would a direct appeal to the name of God, that God in whose name a specific people is ostensibly setting forth its constitution. It is better not to mention at all this "God" who means different things to different individuals, and nothing at all to some. It is better to restrict oneself instead to such concepts of natural law as human dignity, freedom, and justice. For such references have symbolical significance along the two lines just indicated.

The church's task is not to destroy this secular appeal to natural law. The church's task is to preach. Of course, this means that it must address man also in terms of his political life and not merely in terms of some private, inner, non-political sphere. It cannot limit itself to a message about "personal salvation." In other words the church's task is to proclaim—though on the basis of very different presuppositions—that which natural law has in view but in respect of which it entertains illusions. The church has to preach the patience of God with the fallen world, as this patience is expressed in his covenants with Israel and in the new covenant with the Christian church and the whole world, as well as in the commandments which have been given in connection with these covenants. Hence the Christian must act on very different grounds from those of natural law. He acts in obedience to the commandments of God. He acts in the knowledge that even those actions which conform to the ultimate norms perceptible in this aeon must stand under forgiveness, and that these perceptible norms are not identical with the divine commandments, or, more precisely, with the "true" will (the *voluntas propria*) of God.

The external, "practical" form of action will of course not always be fundamentally altered by this new motivation and purpose. Sometimes it will not be altered at all. In general, the closer the basic attitudes, the fewer differences there will be on the level of deeds. This is why in practice action can often be undertaken in common by Protestants and Roman Catholics, e.g., in the political and economic spheres. Even between Christians and "secular men," or between the church and an atheistic outlook, there may often be parallels in practical programs and practical action. We have only to think of the many proposed solutions to various social problems, and of what is said about them in papal encyclicals.[5] It is for this reason that in the secular sphere Christian statesmen and economists can be associated with the "sons of this world" in the same systems.

The differences rest essentially in the basic presuppositions, though they can often erupt violently into the sphere of action, especially in borderline cases, e.g., the question of whether someone is "unfit" to live. It is true, of course, that if the salt of the Christian message were taken from this secularized earth, if the light on the hill no longer shone over the darkness of earth, then the secularistic presuppositions of action would necessarily come to the fore with incomparably greater strength both in the action itself and in the whole structure of the world. We have seen what is the final end of "natural law" when it is abandoned to its own devices, when it is completely secularized and no longer salted. It leads to the dictatorship of autonomous man, who ultimately describes as "nature" that which pleases him, that in which he can work out his own will to power. In this case, to be sure, the moment can come very quickly when the Christian can no longer co-operate in the same political system with those who represent this perversion of natural law; at least he cannot go along in those spheres which involve fundamental responsibilities relating to the image of man. (In relatively neutral and objective spheres, especially those of a secondary character, these decisions do not present themselves with the same acuteness; a junior officer, for example, does not confront the decisive question in the

---

[5] See esp. the encyclical *Quadragesimo anno* ("On Reconstructing the Social Order," 1931).

same way as a high-ranking general, nor a petty economic official in the same way as the minister of economics.)

These questions have taken on tragic actuality under modern dictatorships. As we shall have to show more fully when we deal with the doctrine of "orders," and especially with political ethics, it is here that the impossibility of Barth's conception becomes evident, according to which, in his *The Christian Community and the Civil Community*, the two spheres are said to be related like two concentric circles. No: in the secularized state there is expressed, not merely natural law with its recollection, but also the power of the fallen angel, Antichrist.[6] In state and church we do not merely have two divine orders with different purposes. To ascribe this significance to the state is to see only one aspect of the total problem. In state and church we have also the confrontation of two different forces. On the one hand there is the political representation of the world of death, epitomized by Pilate. On the other there is the spiritual representation of the kingdom of God in the church, epitomized in that same scene by the king of truth.

[6] This has been demonstrated by Günther Dehn in his excellent essay on "Angel and Authority," in *Theologische Aufsätze. Karl Barth zum 50. Geburtstag* (1936), pp. 90 ff.

22.

"This World" and Its Orders[1]

The Bible and the Reformation do not conceive the world to be a cosmos of hierarchical values, as do those who engage in speculations involving natural law. Instead of being the bearer of a created order the world, in biblical and Reformation thought, is understood as the aeon between the fall and the judgment. This aeon is characterized "totally" by the fall, in the sense in which we have already defined that term when we set it over against Flacius' view of the matter. In every part of its structure and in every expression of its life, the world is "world-after-the-fall." As Christ is the prototype of the new aeon, so Adam is the typical or representative figure of "this aeon." The world is never understood as a sphere whose materiality sets it apart from man, as having a significance of its own. On the contrary, the world is the macrocosmically extended sphere of man, which in all its aspects bears the features of its human original.

One might put it as follows: The world is not regarded as a sphere which simply stands over against man; it is rather the very form of man's being. To use the terminology of Heidegger, man is not merely characterized by his "being-in-the-world," as if he were merely "thrust" into the world as into an element quite alien to him. On the contrary, being-in-the-world is the sphere proper to man, the sphere in which man is "at home." It is not merely his "environment" [Umwelt]; it is his "world" [Welt]. He takes his stance within this alien sphere and makes it his "home." To "be-in-the-world" is to possess worldliness and be qualified by it.

The Orders as the Structural Form of Man's Fallen Existence

Worldliness is thus a quality of man. As such it bears all the marks of man. The world, and hence also the *order* of the world, is wholly

---

[1] Most of this chapter is based on the corresponding chapter in my *Kirche und Öffentlichkeit* (Tübingen, 1948). Sometimes there are direct quotations, though these are not expressly marked as such.

human. Just as man is both creature and sinner at one and the same time, and no ontic division of the two is possible, so the orders are "at once" both creaturely and sinful. In short, they are human.

The New Testament expresses this humanity of the orders in two ways. First, "the world, heaven, earth, creation, and humanity are all interfused" [2] in the New Testament. The unity of man and cosmos is regarded as so far-reaching that there can be no thought of innocent man being led into evil by the badness of the world. That would be the view of Gnosticism according to which man, belonging to a divine and transcendent sphere, stands over against the fallen and anti-godly world. On this view the very same word, *pneuma* or *mana* for example, can be used on the one hand for the divine substance and on the other for the essential core of man. In contrast, the anthropology of the New Testament is so structured that, when the terms "body," "soul," "flesh," and "reason" are used of man, they do not imply that man is divided into different spheres in the sense of strata of the ego. On the contrary, each of these terms denotes the whole man but from a different angle or standpoint. For all the difference in standpoints, they all have one point at least in common: they all denote man as he is permeated by sin. In any case we must not try to see in these various concepts differing affinities to sin. In principle, there cannot be in New Testament anthropology spheres of the ego which are either nearer to sin or more distant from it. Matter is not from the very outset evil in the Platonic and Gnostic sense; it is not the basis of man's guilty corruption.

Neither can we follow Rousseau and modern Romantic sociology in ascribing the perversion of intrinsically good man to the corrupting influence of human society. Rudolf Bultmann writes: "Man does not stand over *against* the world; he *is* world. That is, the world is not a given, self-existent factor which man can confront in theoretical contemplation . . . the world becomes what men make of it . . . Since the being of the world is constituted by those who are in it and who are responsible for it, it is not a state or contingency which can be viewed as a fate that men have to endure, as in Gnostic mythology." [3] Hence, men

---

[2] Richard Löwe, *Kosmos und Aion. Ein Beitrag zur Dialektik des urchristlichen Weltverständnisses* (1935).

[3] Rudolf Bultmann, "Die Eschatologie des Johannes-Evangeliums" (1928), in *Glauben und Verstehen* (Tübingen, 1933), pp. 135-138.

do not endure the world as something distinct from themselves. In confronting the world as something apparently distinct they actually endure themselves, for they themselves "are" the world. And it is necessary to undergird theologically this notion of identifying with "this aeon," this idea of its worldliness being the form of our own existence, if we ever hope to quit holding ourselves aloof from the world as from something "outside."

Now when we assert that the Christian community needs to "identify" itself with the world, this statement needs qualification. From what we have previously said about the Pauline and Reformation doctrine of justification, the reference is obviously a kind of "indirect" identity. For a member of Christ's believing community may now say of himself: "I am the righteousness of Christ" instead of "I am the world." As one who belongs to Christ, the Christian is called out of this world, and this can only mean that the identity between him and the world is broken.

This removal of identity must be more narrowly defined. Identity with the world is broken inasmuch as I can no longer be nailed to it in the divine judgment. Because I am justified, I am identified instead with the righteousness of Jesus Christ. The relation between these two identities is to be described in terms of the well-known *simul* of Luther, which we have already defined. On the basis of our previous deliberations we may now say that our identity with the world is "indirect" insofar as it relates only to the *de facto* or *res* form of our existence (*peccator in re*). In "essence," however, we are determined by the righteousness of Christ.

Yet the Law of God forces us to keep in view our worldliness, the *res* form of our existence. It compels us to realize that, so long as we are here below, we are implicated in innumerable, suprapersonal webs of guilt, that we live in the half-light of the orders of this world, that we are actors in a thousand plays which we individually have not staged, which we might wish would never be enacted, but in which we have to appear and play our parts. The Law of God forces us constantly to keep in mind whence we come, who it is from whose clutches we have been snatched, and what it is that can no longer do us harm. The Law of God teaches us this in order thereby to announce the wonder of the Gospel, namely, that God sees us differently, that for him we are no longer representatives of these orders but brothers of

Jesus Christ. We could not appreciate the significance of being brothers of Jesus Christ, however, if we did not realize, and were not continually told, that to the very roots of our being we belong *de facto* and from the very outset to the half-light of this world and its orders.

Here too it is again a matter of perspective. I have to see myself both from the standpoint of the Law and also from the standpoint of the Gospel. The two belong together, and it is only in this interrelationship that I can appreciate the miraculous character of the new righteousness and the wonder of the glad tidings. What might seem externally to be a very pessimistic view, namely, that I must identify myself with the sin of my time, that I must bluntly acknowledge the spirit of the age to be my spirit, and recognize the "Hitler in me" (M. Piccard), is in fact the only way in which I can be glad and free, and laugh and sing in this world of guilt and death. Indeed I may even wonder *"that* I am so merry" (Hans Thoma). For the more guilt I heap upon myself, the more forgiveness I am privileged to receive.

How else should the royal pardon become effective in this world except as there are people who come with their guilt and allow themselves to be pardoned? How else can the power of remission be summoned into this world except as I let it be addressed to me? To whom else does it apply if not to me? To the "world" or the nation as such? Sin cannot be forgiven the "world" as such because the "world" as such can neither know sin nor confess it.[4] It is as I confess my solidarity with the guilty world of my time that I summon the forces of remission into this world. This is done, of course, not in the sense that through my representative penitence and reception of mercy forgiveness is summarily ascribed to the "world" as such, but in the sense that I bear witness to this forgiveness as something which has come to me and my worldliness, and is offered in the same way to this whole world. Only in the shadow of this forgiveness is there freedom and joy.

In all other cases there is merely the convulsion of the tragic. For in the sphere of tragedy I transfer the guilt from myself to suprapersonal forces, especially to the law of fate, and hence to the zone of

[4] In the period after 1945 it was one of the grotesque consequences of the grotesque debate about "collective guilt" that theologians who must otherwise be taken seriously began to speak about collective absolution. They seem to have forgotten the distinction between guilt and confession, world and church, judgment and grace.

that which is beyond good and evil. Why is it that the faces in Greek sculpture seem always to be wrapped in a kind of ultimate sadness? And why is it that the face of Stephen, when he was confronted by accusations which could mean his death, shone like "the face of an angel" (Acts 6:15)? Why is it that at his execution he saw the heavens opened (7:55-56)? There is nothing comparable in the whole world of tragedy, that world which holds aloof from guilt. Connected, then, with this identification of man and this aeon is the fact that the world bears the features of the human original, and consequently cannot be neutral. As the New Testament sees it, the whole historical sphere, the milieu of man's immediate existence, is loaded with demonic possibilities. The Johannine Apocalypse in particular bears witness to the way in which these potentialities can be realized.

The sphere of man's outward extension, however, is not limited to history alone but includes the whole expanse of the creaturely world in general. In this connection we may refer to the "groaning in travail" of the whole creation (Rom. 8:22), and also to the concept of the στοιχεία (especially in Col. 2:8 but cf. also Gal. 4:3; II Pet. 3:10, 12), i.e., the "elements" and "elemental spirits," by which are meant atmospheric forces. The powers of chaos and destruction, which threaten the human sphere (in the narrower sense) as this is ravaged by sin, affect also the spheres of non-human nature. The stars, the sea, fire, and all the elements have a part in the cosmic catastrophes which, from the flood to the events of eschatology, break over the Adamic world in judgment and determine its end.

In the present context it is not our task to consider in detail the significance of the fact that Adamic man discharges a kind of cosmic representation through his fallen existence. In particular, it is not our task here to ask what might be the existential meaning of this inclusion of non-human cosmic spheres, or whether and how far we can attribute to it only a speculative and mythological significance. What is important for us at the moment is to maintain that the supra-individual spheres of history do come within the magnetic field of human guilt in a particular way, and that they are thus to be interpreted as a zone to which this guilt is transmitted.

The theological insight to which this leads us is that the world cannot be separated from man. It is man's "sphere." It is permeated by him. The world is the worldliness of man; and man is his own worldliness. That this assertion of the humanity of the orders is wholly biblical is plain enough from the various materials which we have already assembled and presented in this volume. It will suffice, therefore, if we simply list the most important arguments in the form of three theses.

First, the truly "radical" understanding of the requirements of the Sermon on the Mount which we reached made it impossible for us to separate ourselves from the historical basis of our existence, or to assume that the claim laid upon us applies only within the "mobile sphere of the ego." It is impossible for us to appeal to some "disposition" of our world which compels us to a certain course of action fatalistically, thus absolving us from all responsibility. The very fact that the Sermon on the Mount claims us totally makes it clear that we must impute the structure of history to ourselves, regarding it as a dimension of our own ego and as the macrocosmic objectification of our "Babylonian heart" (Francis Thompson).

The laws of life and of movement in the history of the aeon are profoundly linked with the law of oppression, force, and egoism, so that a complete *de facto* fulfillment of the law of love would be identical with the cessation of history altogether. With this is linked the fact that both the eschatological and the protological images are, as we have seen, incompatible with nature and history. With this is also linked the fact that there is no biblical foundation whatever for the concept of historical progress. By nature, i.e., in its innermost structure, history is implicated in the fall. This is why it cannot sever itself from its own roots and approximate more and more to the kingdom of God. On the contrary, its demonic potentiality is never more fully disclosed than at the very end.

Second, it is for this reason that the orders of history, e.g., the state, law, the economic order, etc., cannot be presented as "orders of creation" [*Schöpfungsordnungen*]. They are rather orders of the divine patience, given because of our "hardness of heart" (Matt. 19:8). In them God so elaborates and adapts the laws of the fallen aeon—e.g., that fear and terror should rule in it, and the principle of force prevail —that in virtue of his miraculous preservation and blessing these laws cannot lead to the destruction of the world but must instead serve its preservation. As von Rad says in his exposition of the "Noachic covenant," [5] they are orders which "obtain in the relationship of the creatures to one another, a relationship that has been perverted by acts of violence." They are "emergency orders" [*Notordnungen*] whereby killing is opposed with killing.

Third, the orders of this aeon therefore cannot be described as having been created such as to be good by nature, or morally neutral in the sense of their belonging to a sphere of pure causal necessity as factual requirements beyond good and evil. These orders are rather the structural form of fallen existence. This explains the ambiguity which envelops them. In the first place they bear the marks of "this aeon," being a kind of objectification of man's creaturely and fallen nature. But at the same time they are also a representation of God's gracious and patient will. These two aspects are interrelated. The miracle of these orders of the "Noachic covenant" is that God uses the elements of the fallen world for his own purposes. He uses the egoism of the state to restrain the egoism of individuals and of groups. He uses killing to prevent killing. He does not allow the world to destroy itself. For over the world, i.e., beyond the wars and rumors of wars, beyond the killing and being killed, beyond the anxiety and terror, there is always the overarching bow of reconciliation and of peace. Under this bow the world will assuredly reach the longed-for Last Day.

The Negativity of the Commandments and Natural "Lawlessness"

With our assertion of the "humanness" of the orders and the "worldliness" of man we have arrived at the direct antithesis of the conception of order in terms of natural law. The question arises, however, how we are to act in keeping with this antithesis, indeed, whether on this basis we can even act at all.

To proceed at this point, we must refer once again to something to which we have already drawn attention in many different contexts, namely, the predominantly negative structure of the Decalogue. Our task is to see how this structure expresses a certain remoteness from all conceptions of natural law, and how the Decalogue is to be understood in spite of this remoteness—or perhaps because of it—as a directive for conduct.

It is hardly adequate to adduce in explanation of this negativity the

[5] Gerhard von Rad, *Das erste Buch Mose, 1-12* ("Das Alte Testament Deutsch" [Göttingen, 1949]), pp. 108 ff.

cultic and ethical exclusiveness of Israel: "You shall not kill—as others do." Such an explanation affords no real answer. Instead it simply raises a fresh question, a question which is formally not so much a specifically historical question as a question of substance raised in a historical context: Why must there be such persistent and vehement emphasis on this exclusiveness? Is it because the threats from without (idolatry, the vitalism of orgiastic Baal religion, etc.) do not really come "from without" but are already rooted within the people themselves? Is it because that which comes from without finds within Israel itself a prepared and natural foothold to which it may resort and which it can immediately claim? But if so, how can there be any immunization against heathenism by way of exclusivism? More precisely, how can we explain the necessity of such immunization except on the ground that there are potential inward threats which, if they are brought into contact with the lurking dangers without, will catch fire and lead to disaster?

To explain the predominantly negative structure of the Decalogue then as an expression of the impulse towards exclusivism is not to explain it at all. It is merely to present the problem in a rather more developed form of the same question.

The negative character of the Decalogue is to be understood only in terms of the First Commandment, which, after the self-presentation of God, immediately states the prohibition, thereby reflecting once again in this negative form that positive self-presentation, "You shall have no other gods before me." God's self-presentation necessarily involves protest against false human presentations. God confesses himself in face of an anti-godly front which is already present. This model of confession-protest is copied later in the confessions of the church. For the confessions of the church are themselves provoked by the protest of the world. This may be seen in their very structure, whereby each particular confessional affirmation is usually accompanied by a corresponding anathematizing of opposing positions. The confession is not primarily a statement against these opponents, essentially negative in form. Nevertheless, it is a declaration in face of the devil, and as such it is compelled to draw with reference to its opponents the negative consequences of its positive affirmation.

Now it is along similar lines that we must interpret the negative character of the Decalogue, as indeed we have already done. There is within this negativity a protest against man as he actually is. "You shall not kill—because you *are* a murderer"; "You shall not commit adultery—because you *are* an adulterer." The negative character is thus to be understood in terms of the specific background of the Decalogue.

In the first place, the Decalogue presupposes the fall of man. It thus occupies a definite point in salvation history. It cannot be regarded as non-historical natural law, or as a set of timeless ethical maxims. The man here addressed is fallen man, man in revolt.

In the second place, the Decalogue is expressly set down within the context of a dialogue. Now in any dialogue there must obviously be at least two parties, in this case God and man. Rebellious man is here called to order. In thus calling him, God at the same time presents himself to man as his ("your") God. In addition to this statement there is also the human speech which has already preceded—the speech of our first parents with the serpent and with God. And human speech must also follow: either there will be acceptance of God's commandments, man countersigning the signature of God in his covenant whereby he has promised to be gracious to his people, or there will be man's refusal and rejection. The dialogical character of the divine commandments is just as significant as its predominantly negative structure. Indeed, there is a connection between the two.

This is apparent at once when we compare the structure of the Decalogue with that of a secularized ethics like Kant's. The first point to strike us in such a comparison is that the two-faceted dialogue of the Bible is replaced by the solitary ego of man. The intelligible ego is for Kant the bearer of the practical logos. It is the practical logos that thus finds expression in this monologue of the ego. The reference to human society, and above all the reference to God, is only introduced later. In Kant's *Critique of Practical Reason* the idea of God is first introduced as a postulate in connection with the doctrine of the *summum bonum*. In Kant's *Religion within the Limits of Reason Alone* God is merely brought in as the author of those laws which, as the epitome of my "duties," I already know on the basis of the practical logos alone. If, however, the ego is alone, deriving its norm solely from within itself, then the norm can and indeed must have a positive form. The ego itself is the model to which ethical self-realization must attain.

This is even plainer in the popular ethics of secularism, which is characterized by monologue even to the pathological form of solipsism. It is further characteristic of secularized ethics that the concept of a "model" appears as a decisive ethical impulse. This means that man himself in various distinguished forms exemplifies the sum of the qualities which are to be attained. As this model or guiding image, he thus provides the possibility of bringing these qualities into a proportioned relationship. In other words, it is "the ideal of man," or idealized man, which constitutes the model.

In regard to this presupposition of the monological ethics of secularism, two points stand out in the light of what we have just said. First, its imperatives have to be defined positively: "You shall be . . . upright, honorable, brave, noble, etc.," for what is intended is that positive model of man; whenever commandments are structured in such a positive form we may thus trace them back to an ultimate philosophical background. Second, idealized man can be a model, however, only when there is in the background a picture of man which is not affected by the fall, a picture in which the *imago Dei* is intact and which thus has the possibility of furnishing on its own a true pattern for ethical self-realization. When the fall is not taken seriously in the Pauline and Reformation sense, when it is not regarded as a reality which characterizes human existence in its totality, then the imperfection of this existence can be construed, not as qualitative revolt against God, but as a quantitative failure to reach the norm. With the help of idealizing fancy, the imperfection can then be transcended, and there is then no reason why a congruence between idea and reality should not at least be conceivable, a possibility which the Reformation doctrine of the fall would in principle altogether exclude.

Implicit in this ideal image of man is also an ideal image of human society, as a kind of ideal fellowship of these ideal individuals. Hence the positive formulation of ethical imperatives always has within it an ontological pathos (inasmuch as it refers finally to a demonstrable order of being). However, it does not have the personalistic reference which we found in the Decalogue.

It is because of this personalistic reference—the fact that man is addressed after the fall and by the divine Thou—that the Decalogue cannot be understood in terms of natural law. This must be strongly asserted in face of every erroneous interpretation. For the order of being presupposed in all conceptions of natural law can be assumed only on the presupposition that the fall has only a comparatively accidental but not an essential significance. It would thus be gross misunderstanding to argue that because of certain similarities in content, the Decalogue is simply a codex of natural law, distinct not in content but solely in the formal sense of being couched mainly in prohibitions. Natural law and the Decalogue in fact belong to completely different worlds.[6]

We shall now try to express this insight in terms of a highly pointed formulation which will prove useful again when we come to deal theologically with the matter of economics: The Decalogue contains no statements concerning natural law [*Natur-Recht*]; on the contrary—if we now apply logically the fact of its negative character—it speaks rather of "natural law-lessness" [*Natur-Unrecht*], not of what is natural and right but of what is natural and wrong. This lawlessness consists in killing, stealing, committing adultery, etc. It may thus be rightly said that the commandments of God do not furnish us with a casuistical system of behavior patterns which may be deduced from self-evident first principles. They rather give us a criterion on the basis of which the proffered rules and solutions, including those which refer to public life, may be critically appraised. Thus the commandments maintain very strictly their position vis-à-vis reality without being integrated into it, as if they were the rules or legal structure of reality. The commandments stand over against reality and are not a part of it. Their negative form is an expression of their sovereignty.[7]

In a rather different context, A.C.F. Beales seems to have something similar in view when he says, "Christianity cannot of itself solve our political problems. . . . But the Christian spirit can at every turn point us along one path or close to us another; if we bring every thought into

[6] One should not try to prove the identity of natural law and Decalogue in Luther by merely adducing a few unguarded statements from his writings. Cf. Emanuel Hirsch, *Der Offenbarungsglaube* (Bordesholm, 1934), pp. 19 ff., and Franz Xavier Arnold, *Zur Frage des Naturrechts bei Luther* (Munich, 1937). Luther's position in relation to natural law and natural theology must be developed from the standpoint of his "theology of the cross" [*theologia crucis*].

[7] This is why I do not rule out the possibility that the Socratic *daimonion*, with its negative function of warning (as distinct from a positive function of directing), had the sovereign power, by virtue of this particular structural affinity to the Decalogue, to maintain its position vis-à-vis a total, self-enclosed, moral and religious world, never allowing itself to be integrated into this world as its ethical impulse and hence to speak from within it.

captivity to Christ, we are not thereby given a criterion by which to judge all that are offered." [8]

The objection might be raised against us at this point that even in coining the term "natural lawlessness" we are implying a positive concept of natural law. One can hardly deny the logic of this objection, yet it must be emphasized that everything depends on the rank accorded this positive concept, and on the content embraced by it.

As concerns the rank accorded such a positive concept of natural law, it obviously makes a great deal of difference whether a concept is the starting point from which certain deductions are made (as in the case of the Thomistic concept of natural law with its casuistic deductions), or whether it occurs as the logical consequence of a line of thought, and thus forms, as it were, the horizon which bounds our view (as in our own case).

We may clarify this matter of rank by reference to certain theological and philosophical parallels. For theological thought, man's "original state" [*Urstand*] is a borderline concept in the sense that it cannot be developed positively and hence can never assume the role of a starting point. It is impossible to state what is meant explicitly by the original state because our terminology, oriented as it is to the legal structure of this aeon, here reaches its limit and is reduced to silence. How, for example, can we even conceive much less describe a life without antitheses, or conflict, or the tension between good and evil? How can the fall be explained logically, i.e., with the modes of thought of "this aeon," as a transition from the primal state to the fallen world? The manifest impossibility of all such attempts must indicate that we have chosen a false starting point for our deliberations. Hence, such a borderline concept as the original state can never have the significance of a first principle from which various other things may be deduced. It constitutes instead only the horizon of that thinking which has its inception in *our* concrete situation. It constitutes this horizon, i.e., this extreme limit of our field of vision, because the fallen nature of our world, the concrete original sin of the individual, and the suprapersonal nexus of universal guilt cannot be set forth except against the foil of this primal state as a background. The statement that we are not now as we came from the hands of God is not logically possible without the further statement that we came thus

---

[8] A. C. F. Beales, "The Sword of the Spirit," *The Month* (October, 1940), p. 7.

from the hands of God. In this connection, however, what is important is not simply the sequence of the statements but also the divergence between them in terms of specificity of content. The first statement, which describes our present existence, can be filled out theologically. The second, however, is essentially an assertion which says little about what came from the hand of God, only that it was from his hand we came. It represents an affirmation arrived at mainly through negation of what we presently are, but one that is itself incapable of positive explication.

We are brought to exactly the same conclusion when we look at the philosophical concept of the "beautiful soul" as Schiller develops it.[9] The harmony between duty and inclination, reason and sensuality, cannot be logically described in terms of its nature, because Kant, who worked out this thought of Schiller to its conclusion, rightly describes the imperative as a polemical concept directed against the mere inclination of the sensible world. The imperative necessarily contains a protest against another impulse which is posited "naturally." Furthermore, the imperative has in view not merely individual deeds which it wrings from that impulse as ethical acts, but also the forming of the total person. Hence the harmony of imperative and inclination necessarily becomes the content of a postulate. One might express this postulate paradoxically be saying that the purpose of ethics is to make itself superfluous, i.e., to replace the stage of preliminary fulfillment—which is always represented by an open imperative—by that in which the imperative is fulfilled and hence superfluous. It is characteristic of the rank assigned this ethical concept of the "beautiful soul," however, that the concept can never be the starting point for the development of ethical theories. It can only be a deduction, a postulate, a horizon at the limits of our power of ethical conception.

Similarly, we may say of "positive natural law" that it is a concept which cannot be avoided, even in an ethics which, pursuing the intentions of the Reformation, is consistently personalistic. Nevertheless, the fact that it is regarded as a borderline concept, a horizon which bounds our view, means that it is understood and evaluated quite differently from the way the concept is treated in a positive and systematic doctrine of natural law in which it is actually ranked as a principal starting point.

[9] See p. 33, n. 10 above.

Not only do we rank the concept differently, but we also construe its content to be radically different from what the Thomistic doctrine of natural law conceives it to be. For we have seen that in terms of the Reformation doctrine of sin the order of being is so chaotically disrupted that even in abstract theory it can of itself provide no intact order. There are three important consequences of this for the concept of natural law.

In the first place, natural law can have the significance only of an "emergency order" with the result that the concept of natural law is in the strict sense dissolved; at any rate it can no longer be described as felicitous or meaningful. In no case can natural law have the rank of an "order of creation" in the sense of the analogy of being [*analogia entis*]. In the second place, natural law cannot be an "order of salvation" but only an "order of protection," restraining evil rather than promoting good, unless here too we would push to the point of utter absurdity the argument that the idea of a positive and describable good is implicit in the concept of evil. Finally, natural law does not apply to persons in the sense of ethically restoring them or even of being able to suggest the form of such restoration. On the contrary, it simply contains a veto of certain structures of society and its orders which are to be described as objectified forms of law-lessness and wrong.

We thus see what theological significance is to be ascribed to the negative character of the Decalogue. It shows that an evangelical ethics must begin with the doctrine of natural law-lessness understood in personalistic terms.

The Implications for Economics and Politics

When we take these insights into the theologically important negativity of the Decalogue and apply them to the problem of how far this negativity may have normative significance for the economic order —this example is particularly fruitful from the standpoint of method— we are led to the following results. First, and negatively, the Decalogue yields no specific economic program in the sense of natural law. On the contrary, when we interpret the Decalogue in terms of its application to economics, we see that it undoubtedly addresses man in terms of his natural law-lessness. In other words, what the Decalogue discloses are the forms of natural *wrong* to which the economic order is repeatedly prone.

As I see it there are two main forms of lawlessness which find expression in specific economic systems. At least the threat of them is constantly present. The one comes out in capitalist economies to the degree that their system of unrestricted expansion promotes the interests of particular individuals and groups, and is powerfully abetted by the technology of the industrial age. The trend towards a monopolizing of economic power, by means of privileged and autocratic trusts, for example, leads on the one hand to a hybrid "will to power" and a permeation of all spheres of life by economic concerns—in this connection the relation between causes of war and economic movements is instructive—and on the other hand to a corresponding degradation of labor to the status of a "means of production."

But the opposite kind of economy, which we may broadly call socialist,[10] finds its own particular form of lawlessness in collectivism and depersonalization. Man is in danger of becoming a mere economic functionary, and the unethical element in such a course is that it makes of man no more than the means to an end (such an end as the work itself or the good of the group). The end result of this form of lawlessness is much the same as that of the capitalistic variety, namely, the dehumanization of man.

The way in which both parties appeal to the slogan of a "new humanism" shows that in both camps there is some sense of this wrong to which we refer. It also shows, of course, how little can be achieved with the help of this humanitarian dream, which at best usually succeeds only in advancing man a notch within the hierarchy of means, skilled workers, for example, being prized even more highly than the best machines. But this does not mean that the process of dehumanization to which man is subjected is in any way arrested. It simply means that man is gradually given a different position within this process, or within the hierarchy of economic means. It does not involve any new discovery of the "infinite value of the human soul." Man is simply advanced within the framework of his economic usefulness.

Now in referring to these two kinds of lawlessness we are not expressing a judgment concerning the two economic systems as such. It

[10] Capitalism and socialism have long since lost their pure forms. They exist only in various admixtures. Hence the terms have now taken on the unwelcome character of slogans. This need not prevent us, however, from using them to bring into focus the problem of natural lawlessness.

would not be the duty of the church anyway to make such a judgment. We are simply drawing attention to certain fundamental possibilities which are present in these as in all economic systems; and in doing so we have cited these two extremes as examples—no more!

These statements about the economic order have to do primarily with the church. They represent a theology that is truly relevant and appropriate, for the simple reason that they regard the economy as the product of man's working, while at the same time regarding man as the object of the divine Word dispensed by the church. Consequently, our primary concern here is not with "economics" but with "anthropology." The problem is a pastoral one. Beyond that, whatever the church may have to say will always be said indirectly.

The church will always be interested in having political economists —hopefully Christian political economists—work out an order that will fall somewhere between these two forms of lawlessness. A broad field of operation is thereby opened up for the constructive efforts of those who are technically competent in such matters. There can be no question of trying to sanction one particular order as specifically "Christian." The commandment of God has no abiding affinity either to a specific economic order or to a specific political order.

When we speak of the political order we immediately come to the problem of a Christian political party, one that has its own "Christian" economic, political, and cultural programs. To be sure, this involves many other suggestions which we cannot now pursue. Nevertheless, it may be pointed out that, should evangelical Christians desire to engage in political activity by way of an organized "Christian" party, there is one permanent and insoluble problem which can always lead to complications, particularly at the point of their making common cause with Roman Catholics who build on a completely different foundation in this respect. These practical complications of co-operation, of course, only mirror the more basic dubiousness of the whole procedure. A "Christian" party is at best only an "emergency solution," and that in a twofold sense.

In the first place, there are many spheres of life which must necessarily be taken into account in the program of a political party for which the predicate "Christian" cannot possibly provide any kind of normative direction. (We have only to think of certain practical or even fundamental questions of political action, such as compliance

449

or passive resistance, etc.) Here the predicate "Christian" repeatedly runs the risk of being used inappropriately and of being tacked on like a label that does not apply.

In the second place, while the representation of Christian concerns within the other parties and through their representatives in the legislature—in principle a sound form of Christian political action— has repeatedly proved itself in practice to be insufficient, this insufficiency appears to derive not merely from the inadequacy of those party members who are Christians but from the strongly ideological character of the parties themselves. Since the representation of Christian concerns within all parties, however, remains a basic requirement which can never be surrendered, even a "Christian" party can never claim a privileged position within the Christian community, claiming to speak for or include all Christian believers. The church must repeatedly emphasize this, even more than the party itself does, and then draw the necessary conclusion that there cannot be too close an identification between Christianity and any given party.

Hence from the standpoint of theological ethics, or the church's office as "watchman," the main thing is that specific anthropological limits be set to economic, political, and other configurations, while at the same time preserving the possibility of a broad variety of solutions in any given field. To put it even more pointedly, there is no such thing as "Christian" economics or politics. But there is something more than merely a "Christian" economist or politician. There are in fact specifically non-Christian forms of economics and politics. The theological task therefore can consist only in the marking out of a course or path which is only indirectly demonstrable in terms of the pitfalls and abysses which flank it on the right hand and on the left.

For all these reasons the statements made by Protestant churches in such areas as politics or economics cannot lay claim to the authoritarian rank of papal encyclicals—for one thing they have a very different theological basis—but they do nonetheless bear the character of semi-official documents. On the one hand, in the name of the freedom which exists vis-à-vis the orders to the degree that these are not "orders of creation" and do not represent a divine will which may be known directly, these statements avoid the theocratic tendency inherent in the doctrine and development of the orders. On the other hand, they refrain from regarding this freedom as a dispensation by virtue of which the church permits itself to evade these questions. Their task is rather to implement this freedom vis-à-vis the orders in a positive way by commending such concrete structure or order "as may befit the time and occasion." [11] Thus they allay any suspicion that in basic intention they are nothing more than a mere Platonic commonplace without any interest in, or capacity for, realization.

The "illustrative" character of the structures thus commended seems to me in this connection to be particularly important, since it establishes the connection with the ethical situation of man generally. For all man's ethical action (interpreted from the standpoint of theological ethics!), has the character of demonstration, of analogical indication, rather than of de facto fulfillment or accomplished righteousness in re. What does apply in re is rather the fact that we always are and remain sinners.

In other words, when I seek to fulfill the commandments of God, this does not mean that in my de facto action I am really "fearing and loving God" in the sense in which Luther explained the commandments. It means rather that on the level of deeds what I do is only an imperfect likeness or copy of complete fulfillment. My concrete fulfillment has symbolical significance to the degree that it is always improper, and does not efface the actual incongruity which always obtains between what is demanded and what is achieved. In the fundamental conception of an economic or political order this same character must also be maintained: the res form is only illustrative, symbolically demonstrative.

[11] *Secundum occasiones et tempora*, WA 44, 461, line 32.

Part Three

MAN'S RELATION TO THE WORLD

23.

Theme and Method in Ethics

If as a result of our dealing with basic principles in Part Two the reader should now expect to find in Part Three a firm, casuistically mapped out line of march for what is customarily called "Christian conduct," he will, I suspect, be disappointed. The reason for this disappointment certainly cannot be that all further development of ethics is to remain at the level of theoretical principle and never become practical. The perspicacious reader will hardly fail to come upon many concrete problems concretely discussed in the course of what follows now in Part Three. Disappointment—if such there be—will more likely result from the fact that our earnest analysis of concrete cases does not lead to the desired casuistical directives, and that as a consequence of our study ethical decision is not made easier—as one might have expected—but more difficult.

That this is so is not due to some kind of irresolution on the part of the author, nor to an unwillingness on his part to offer solutions for concrete problems or to give the most precise possible exposition of the divine commandments in respect of the many difficult ethical situations of our time. The reason that theological ethics cannot prescribe exact lines of action, nor obviate the need for personal decision but only insist on the necessity for it, is inherent in the very nature and structure of evangelical ethics.

Christian Freedom versus Casuistry

The structure of evangelical ethics is essentially determined by the fact that its sole task is to spell out and expound what Luther calls the "freedom of a Christian man." This freedom implies that we are free from the Law and from tutelage, that we are the children of God and hence are of age, mature [mündig]. This means above all that we are always the subjects of our action. Those who are under the Law are not the subjects of their own action but merely objects of an alien will; they are "functionaries."

What Paul and Luther call the "obedience of a servant" or the "service of the letter," in contrast to the love of a son or the freedom of the Spirit, may be described in modern, sociological terms as the functional reliability of a functionary. When the functionary operates it is not he himself who is acting. He is simply the implement of whoever or whatever exercises its will "through him." The legalist fulfills the will of God formally, but not in the sense of its being his own will. The powerful seventh chapter of Romans shows us that my own will, the "law in my members," resists the Law which is alien to me, and lives in permanent conflict with it. For this reason alone the Law can never lead to total obedience, to a true application of my whole being to the will of God. Because my own will never coincides perfectly with the demanding will of God, there always remains an element of opposition. Hence the ego in its totality is never drawn into the act of obedience. Once we realize this it becomes clear why the New Testament talks about love alone being "the fulfilling of the Law." In other words, love alone does justice to the true intention of the Law, which is to bring about a conformity with the will of God. He who loves finds that he is in agreement with the one who is the object of his love. Hence an action done in love is an action performed out of conformity. But this action cannot be casuistically predetermined by laws. Because it is the objectification and expression of the love which motivates it, it has the freedom to take any of the most varied forms. This is what Augustine meant with his famous dictum, "Love, and do what you want" [*Dilige et fac quod vis*]. The commandment of love is thus the end of casuistry. It is what empowers to freedom.

Ethics does not teach us what we are obliged to do; strictly speaking, it teaches us what we are permitted to do. It surveys the sphere of freedom. It reflects on illustrative situations in order to understand what is involved in our living in freedom and forgiveness here and now, within the engulfing autonomies, conflicts of value, collisions of interest, and domineering factualities of this aeon. It shows us how the prodigal son lives once he has left the bondage of the far country and outgrown the legalistic virtue of the elder brother.

The word "freedom," to be sure, is not altogether a pleasing one for modern ears. To modern man it seems to be asking too much that he should be required to make a decision for himself, that he should be denied the possibility of having this decision made for him and hence of being merely the functionary who executes a law. In moments of tragic ecstasy, in blind transports to some kind of love of moral destiny, he can even speak with Sartre of being "condemned" to freedom. He secretly believes that it is desirable to be a functionary, a mere executor. How blessed to be an object!

Evangelical Christians who today look with envy at the casuistry of Roman Catholic moral theology, and deplore the fact that the Reformation has denied them any such solid undergirding, should consider well whether these longings do not derive from the darkest cellars of the modern world and have utterly nothing whatever to do with a genuine readiness for ultimate sacrifice. The flight into libertinism is not the only kind of escapism. There is also the flight into obedience, into the quest for authority, into the security of the functionary. The Son of God who "makes us free indeed" did not die in order to lead us to that kind of sacrifice which draws its strength from fear of freedom. This evil and self-deluding desire for authority really represents both a misunderstanding and a missing of the freedom of the children of God, and in all fairness it should be said that the concern for law in Roman Catholic moral theology is much too sublime, and—thank God!—much too permeated by an awareness of the mystery of grace, to be truly receptive to this desire on the part of a fallen and degenerating world.

If the theme of ethics is thus the freedom of a Christian man, it is important that we come first of all to a new understanding of what this freedom is. For us today to understand it, however, can only mean that we learn to see it not as something to which we are "condemned" but as a gift, the possibility of a new existence. For the fear of freedom from which we suffer derives from a diminution of existence. If we no longer stand on eternal foundations, if instead of existing we merely continue to vegetate, then any appeal to our freedom is necessarily experienced as a painful and discrediting exposure of our faults. For the demand to be free is first and foremost the demand to "be oneself," in the sense of being a "subject." [1] If I

---

[1] By now the reader will recognize that what we intend here is not the concept of autonomy advanced on a Kantian understanding, and that we are not unaware of the many theological questions which have been raised concerning the subject-object relation.

am no longer myself, if I am simply the anonymous instrument of the one who works through me, or the executor of an alien will, or one who is driven by anxieties and longings, then the appeal to my freedom forces me to the embarrassing confession that there is nothing left here to be free, and that it is because of the default of my own true being that I thus simply vegetate.

This is why we must respect the honesty of Sartre's saying that we are "condemned" to freedom. What is gift and fulfillment for those who exist, is torture and shame for those who do not. Only for those who exist is freedom the radiant sign of victory over the Law and of freedom to "be oneself." For those who do not exist any longer, freedom itself is the Law which judges and condemns, and their striving against it is an understandable and logical reaction.

The appeal to freedom becomes good news only when our existence has been given a new foundation and so is made capable of freedom. But it is a law of God's dealing with us that the gift [Gabe] precedes the task [Aufgabe]. The indicative positing of the new creature is completed before the imperative appeal reaches us.

Concretely this means that before we are commanded to love—before we receive in this love the freedom of movement, of doing what we want—God makes himself an "object that can be loved," an *objectum amabile* to use the neat term of the Apology.[2] Through his disclosing of his heart, whereby he becomes an "object that can be loved," *our* love is kindled, and with it the new existence which can now—but only now and not before—be addressed in terms of its freedom. For those who do not exist—which is to say, for those who are not yet freed for love—freedom is indeed too big a demand, and hence a ground for flight. For those who exist, however, freedom is a gift which enables them to be themselves, a gift which projects them towards their destiny.

Thus we know very well what we are about when we refuse to be casuistical. For to engage in casuistry at this point would be to see in ethics an enterprise intended to rob man of his freedom instead of leading him to be free. In what follows here in Part Three we will further develop the basic principles dealt with in Part Two and illustrate them in terms of the most varied situations. At this point I

[2] See the Apology of the Augsburg Confession, IV, 129; *BC* 125.

have simply been concerned to correct any possible false expectations on the part of the reader.

A Theological Approach to Methodology

Our method will be to work out the problems of ethics in the light of specific cases, of typical concrete situations. This approach to methodology is governed by the following considerations.

First, the problems of ethical conduct do not arise merely out of our confrontation with norms and the resultant tension between Spirit and flesh, morality and impulse, the intelligible world and the sensible world. They arise also out of our confrontation with reality. For it is a fact that in the various spheres of life there is—to put it cautiously and provisionally—a certain "trend," and in some circumstances even a certain necessity, imposed upon our conduct. The political and economic worlds, the worlds of work and of art, are pervaded by certain factual requirements which to a large extent help to determine our conduct. The scope of ethics is thus illegitimately reduced if it is limited only to the sphere of disposition and does not also embrace the problem of that transsubjective reality into which man is thrust and by which the free play of his action is restricted. It is thus of supreme importance that we find a method which allows for a theological analysis of reality.

Second, the theological character of such an analysis will consist in the fact that the reality which is to be described in any given instance will be shown to be a situation typical of "this aeon," i.e., of the world between the fall and the last judgment. As this is done we shall see that the secret of the world is disclosed in an illustrative way precisely in the most unusual situations and extreme possibilities which this aeon offers. Hence we cannot begin with normal situations, but must take as our starting point the borderline situations [Grenzsituationen]. For example, we shall come to regard the totalitarian state and its ideological tyranny as the most highly compressed form of the problem of how the Christian, as one who is justified and "called out" of the world, can at the same time live in the midst of the rebellious world and still remain true to his salvation. It is precisely this borderline case in the area of politics which protects us against that genteel and respectable religion whose champions believe they can live with a supposedly "good conscience" in what they take to be "divine orders." Such a pseudo-Lutheran armistice between God and the world, which is often based on a badly mangled doctrine of the "two kingdoms," is mere fantasy to those for whom the world enslaved by the totalitarian state has become the occasion for discovering the extreme possibilities of the world between the fall and the last judgment, and for asking—in that light!—concerning the justification of the sinner and his possibilities of action. They alone learn to catch the full meaning of the high-priestly words of Jesus when he prays not that the Father should take his own "out of the world," but that he should "keep them from the evil one."

Only against this background of the borderline cases do phenomena come to light which are generally concealed by the more normal cases—and therefore by the picture given in most ethics—such phenomena as "conflict situations" and "compromise." The fact that our action in the concrete, *in re,* is sinful, that it is, as it were, by its very constitution opposed to the divine will, is to be seen not only from the standpoint of the rebellious disposition but also from that of the reality of the world which lives in constant tension with God. We have thus to indicate the ontological structure of concrete sin. Compromise is one of the normative features of this ontological structure. It seems to me that the almost complete absence of this concept, this basic ethical phenomenon, in theological ethics generally is an eloquent sign that in such ethics—we shall speak later of the very few exceptions—the reality of the world is excluded from ethical consideration.

By thus extending much further than usual the scope of ethical inquiry, we are compelled at many points to do our work in wholly new territory, and so to take corresponding risks. In this connection there has weighed heavily upon me not merely the magnitude of the task but also the necessity of laying myself open to public scrutiny. Such exposure would be unbearable were it not for the fact that the task imposed upon us makes such thinking necessary, and we have the consolation of knowing that inherent in our very assuming of the task there is an element of help in getting on with it. For quite apart from the compulsion involved in the very nature of the case, in the ethical problem itself, the questions raised by the borderline situation are a bitter and supremely relevant fact for those Christians who do presently live under ideological tyranny. It is my hope that, by not restricting myself to problems of the West, I may render to brethren behind the iron curtain a vicarious service of theological reflection, the service of thinking *of* them as well as *with* them and *for* them, and that by this extension of subject matter the ethics itself will be materially enriched, indeed actually made a viable possibility. This existence of Christians in the borderline situation of the totalitarian state imposes tasks not merely upon our hands, to serve, but also upon our heads, to think. To renege on this task of thought is not simply to do damage to the soul but also to lose our grasp on a matter of really vital concern for the entire ethical enterprise. We can speak of the "freedom of the Christian man" only if our field of vision is specifically not restricted to the regrettably saturated Western world and its self-complacent Christianity (though we recognize that this very saturation, no less than tyranny and the perversion of the orders, can become a borderline situation in which that very freedom is threatened and has to be protected).

Third, our "typical situations" have been chosen according to whether or not they have paradigmatic significance for our development of the fundamental ethical problems. That is, they have been selected on the basis of theological considerations and not with reference to the phenomenology of life in general. This is why, at least from the standpoint of such a phenomenology, they thus embrace the most diverse and apparently disparate themes. But the constant factor in all this diversity is always the theological approach; it alone determined the selection of the various concrete cases here discussed. It is precisely because this approach is rooted in a strict theological systematics that it gives us the possibility and the freedom to take our various cross-sections from every area of human life. If the reader bears this in mind—and the work itself will constantly draw his attention to it—then the guiding thread will not slip from his hand even though the phenomena change in rapid succession.

In contrast to this theological point of departure—which we are concerned clearly to indicate—many theological ethicists allow the development of their work to be controlled, not by theological enquiry, but by the law of that phenomenology of life to which we have just referred. The essential phenomena of man's individual and collective existence are assembled, and then, either with or occasionally even without the assistance of the corresponding sciences,

461

they are first illumined phenomenologically and only later, as in a second operation, made the subject of theological appraisal. Thus the human body, society, the neighbor, the callings, work and play, indulgence and abstinence, nation and state, economics and social questions are all treated in this way. This mode of procedure seems to be open to question on two counts, as regards both the matter treated and the method of treating it.

In the first place, with respect to the matter itself, it is a question whether the task of theology is simply that of commenting on phenomena which have been singled out and set forth by other authorities. And in this respect it makes no difference whether these other authorities are certain writers in science and philosophy or simply one's own two eyes with their powers of observation, their "visual act" (as in the case of Adolf Schlatter, who by the way displays an exemplary openness and nearness to life).[3]

As a matter of fact, however, the theological meaning and relevance of these phenomena does not first begin with our commentary on them, or with our critical evaluation of the commentaries of others. It begins already with the initial observation and selection of the phenomena. Even the most naïve visual act—in the sense of Schlatter —even what appears·to be the most objective and natural human understanding of that which is observed, is not really without its prior assumptions. As R. H. Lotze has made clear once and for all, it contains in fact an act of evaluation which precedes the observation, an act which itself presupposes a scale of values and therefore a particular view of things, in short, a world view. There is thus resident in the very mode of observation itself a critical and evaluating function which, precisely through the naïve form in which it is always present in advance of the visual act, disrupts the naïve objectivity of that act.

This is why there has to be something like a theological critique of cognition or of pure reason which deals critically with the presuppositions underlying perception. Ultimately it would have to be a theological critique of those world views which underlie all natural knowledge, including the apparently non-metaphysical enterprise of Kantian epistemology. We mention this problem in the present context only to show that theological meaning and relevance is not first

---

[3] See Adolf Schlatter, *Die christliche Ethik* (Stuttgart, 1929).

attained by an ex post facto appraisal of selected phenomena but is already present in the initial observation and selection of the phenomena themselves. It is at this point that the emphatically theological character of the act of observation must be recognized, where the visual act becomes one that is consciously purified—or at least seeks to be purified—from the encumbrance of any world view.

In the second place, with respect to methodology, the phenomenological approach is open to question because it never lets us get beyond the orbit of these pre-interpreted phenomena. We repeatedly get stuck, for example, within the confines of certain social teachings or certain economic theories. We have first to hear the voice of the "experts," among whom theologians will often play the role of a dilettante, agreeing a little here and amending a little there. Answers are thus given to questions which, though supposedly directed to Christianity, were never actually put, and in turn there is a failure to address to the world those questions which ought to be put and which would probably gain a hearing.

We for our part shall try to take the very opposite path. We shall put at the very center of the discussion certain decisive theological problems which we have uncovered in our consideration of foundational principles, and we shall then try to illustrate these concerns in terms of various concrete spheres of life. Thus, for example, we will treat of the "borderline situation" [Grenzsituation] of this aeon, namely, its extreme disorder, in order thereby to bring to light in its extreme form the tension between time and eternity, between God and man, between righteousness and sin, a tension which is always present and cannot be resolved by any "analogy of being." By means of this borderline situation we shall then focus attention on the cardinal ethical problem of how in a fallen world there can be any obedience whatsoever to the commandments of God, indeed how any congruity between demand and fulfillment can even be theoretically conceived, and what right we thus have to speak of the freedom of a Christian man.

Our primary question, then, will not be with respect to the perverted order of the present world, nor shall we attempt to give *ad hoc* casuistical advice on how one is to act as a Christian in this or that confused and complicated situation; to do so would again be to come under the law of phenomenology. We shall instead take the diametrically opposed course of bringing to bear upon this perversion the specific questions of a theological diagnosis. This will be our procedure in all the remaining sections of this ethics.

24.

The Three Zones of Ethical Validity

We would begin our elaboration and development of the foundational principles of ethics at this point with a short consideration of the question in what spheres of being something like "ethics" is even possible. Before ethical relations can arise at all there are three conditions which must exist: man's relation to himself, to his neighbor, and to his world.

Personhood as Man's Relation to Himself

In the first place, the agent of an ethically relevant act can only be a human person. Whatever may be the basis of a particular ethos, be it Christian, Platonic, or Shintoist, its concern will always be with a decision between what ought to be and what ought not to be, and hence with a subject who can be addressed in terms of that decision which is his to make. Two conditions are thereby posited which must be fulfilled in such a subject.

First, he must stand in some kind of ultimately binding relationship from which he receives his imperatives, and therewith his criteria for distinguishing between good and evil. He must be conscious of some commandment or some source of law, and in either case of its having been confirmed by his own conscience.

Second, this subject must have the capacity—and as he is addressed in terms of the decision to be made he must also come to know that he has the capacity—to stand over against himself and over against the law which binds him. That is to say, he is aware of having been called out of the bondage to impulses and instincts—and out of the law of that society in which he finds himself, the slough of conformity—and of having been set down instead in an ultimate responsibility which demands that he be free. Ethical freedom as a primal moral phenomenon in this sense is very basic. It stands at an even more basic level than that level of differentiation on which freedom is divided up into different complexes of meaning and accorded correspondingly different interpretations, being regarded as the essence of Fichtean "action," for example, or perhaps relegated to the sphere of Luther's "bondage of the will." What is meant by this "primal phenomenon" is simply that man as a person is excised from the whole nexus of the reality around him. He may no longer understand himself as a mere transitional link in the chain of causal necessity. Instead he is forced to tear himself free from the nexus of determination wherein he is himself determined by the events which take place around him and through him, and to stand over against them—and over against himself—and so to recognize his own distinctiveness. Ethos never arises except in such a dissociation from nature. The being that ought to be always stands in opposition to that being in which we presently find ourselves, that being which seeks to determine and control us.

Relationship is consequently one of the essential features of personal being. The human person [*Dasein*] can and must stand in relationship to himself. To use Heidegger's terminology, *Dasein*—in distinction from all other mere entities [*Seienden*]—is always something that understands itself and is thereby related to itself,[1] i.e., it either lays hold of or misses its destiny (as the being that ought to be). Only by laying hold of my *Dasein* as truly my own [*je Meiniges*][2] do I elevate it above all the entities round about me and break through that nexus of determination by the chain of causal necessity with which the processes around me and the physical and psychical laws of my own nature seek to hold me fast. We may thus say that personal being is achieved only where there is a relation to myself.

From these two conditions it may be deduced that the ethical is inoperative except within the circle of human existence. There is no "ethos" in the animal kingdom. The training and tricks of animals have nothing to do with ethics. Training does not imply an overcoming of animal instinct or of the animal mode of life through the instrumentality of improved insight, a sense of duty, or loyalty to the trainer. Training is simply a matter of utilizing a variety of instincts, actually employing the instinct of fear (fear of punishment), and of

[1] See Martin Heidegger, *Being and Time,* trans. John Macquarrie and Edward Robinson (New York: Harper, 1962), p. 32.
[2] *Ibid.,* p. 67.

certain animal needs (eating, stroking, etc.), and finally of the law of least resistance, habit.

The animal may form a personal attachment to his trainer which in some ways reminds us of personal relations among people, the more so as the human partner in the relation involuntarily sees in the attachment of the animal a reflection of his own personally tinged love, and thus ascribes to such attachment the human quality of his own attitude. In fact, however, the animal is comporting itself on a quite different level of relationship. This divergence appears most clearly wherever the analogies between animal and human conduct are most fully pressed, as, for example, in certain poems and stories which attribute to animals such conflicts in duty as are commonly felt by men.

The I-Thou Relation

There is a second condition which must be met before ethical situations can arise, namely, the existence of an I-Thou relation. This condition is very closely related to that of personhood just discussed. For the personal being of man is something that is realized only in encounter with other persons. One might even say it is realized only insofar as it exists "historically," i.e., is referred back to a togetherness in time.

Kant expresses this by saying that personal worth rests on the fact that the person represents humanity. In this concept of humanity there is a twofold emphasis. On the one hand, it denotes humanity in the sense of "the human." On the other hand, it denotes humanity as a corporate image, as the nexus of all men. "The human" could not be understood as such if it were not abstracted from the type, and if consequently it did not comprehend the totality of all human individuals.

It is in the light of this that we can understand the fact that ethical acts are possible only when there is posited a "Thou" who is to be respected as an end in himself. If this Thou is not present, then we are either in the sphere of the autonomous life of the individual entelechy, where no limit is set to my own development by the claim and dignity of the other, or we move in the circle of mere pragmatism, i.e., of purposive action in which we set our own goals and use everything else as means to their attainment. The fact that

the "other" should be an end in himself, a fact which for me constitutes a limitation, is guaranteed, however, only by the fact that he
is a "man."

To see this is all important at this point in our thinking. We shall
have occasion later to note that the term "man" or "humanity" can
mean very different things. It can derive its significance, for example,
from the *imago Dei,* or from the fact that Christ died for me. Or
again, it may derive its meaning from humanity (in the sense of
Kant) or through what is called "existence." For the moment, however, the main thing is to see that there has to be this counterpart,
this "other" who is an end in himself, if the ethical situation is not
to be replaced by the mere development of one's own entelechy—
a neutral, natural process—or by a non-ethical pursuit of ends, a pursuit which may, of course, take place on the human level and even
make use of pseudoethical categories, as in the political theory of
Locke and Hobbes.

From what has been said, it is clear that the complete solipsist,
who as such is, of course, an artificial abstraction, is incapable of
ethical action. Robinson Crusoe would not be a good example for
the ethical situation. For the consistent solipsist—let us say, someone
who has lived in lonely isolation from early childhood and grown
up without any real human communication—would simply not exist
as an ethical person, or ever know himself to be such a person. I
learn who I am only in encounter with the Thou. Only here does
it become apparent that I love and hate, act and am acted upon,
fulfill my obligations or fail in my responsibilities. Apart from this
encounter there obtains only the law of entelechy and pragmatism,
as we have said and as we have here tried to illustrate.

The I-World Relation

The third condition which must be met if ethical existence is to
be at all possible, and which is again contained in the first two but
must now be formulated in its most general form, consists in the
existence of something outside the ego, a vis-à-vis or counterpart to
the ego. Very provisionally, we best state this by saying that there
must be a "world" in which the "I" is present and over against which
it stands.

The I always finds itself to be *in* something, in which and in rela-

468

tion to which it exists. To say this is simply to describe, again very provisionally, what Heidegger calls "being-in-the-world." This implies first that I am in encounter with the Thou, and with human groupings of all kinds. But it also implies that I have dealings with the world of things: I live in a certain house in a certain country, I use certain tools and instruments, I am in touch with nature, and I find myself in specific historical situations to which I must relate and with which I must come to terms. The last of these relations implies that my being-in-the-world is set within the horizon of temporality, i.e., that I have my "history" in the world. This dimension of history opens up to me a fullness of new modes of encounter with the world.

First, I am always coming up against my own past, and that of "my world." The latter, e.g., the past of my nation, has created the historical situation in which I now find myself, into which I am, as it were, "thrust." I have no choice on this score. I cannot, for example, argue that the period of the Renaissance or the Enlightenment would suit me better, and so make it "my own time." The fact that I cannot do this but must accept my given historical moment as it is, that I must come to terms with the present age of the mass society, even though I may be Wertherian by nature, sets me at once in tension with my situation and imposes a multitude of "tasks." In the same sense I am in tension with my own personal past. In my present and future I am very largely dependent upon, and even determined by, that which is behind me, by my education, by the vocational choices I have made, by the guilt and obligations with which I am burdened.

Second, I encounter at each moment my own present in the sense that I have to make decisions. I have to do some things and not do others.

Third, I encounter my own future and the future of "my world" in the sense that I am called to assume responsibility for shaping it, and also in the sense that I look upward and outward towards the succession of events which transpire independently of me, and toward that future which is always coming upon me without my doing. This forward look always involves certain ethically relevant attitudes, whether of hope or anxiety or the apathy of laissez faire. Since these attitudes always carry with them the question whether I accept or reject them, whether I must overcome them or can regard them as ethically legitimate, they have their roots in responsible decisions which are constantly being demanded of me.

Thus the temporal aspects of my being-in-the-world also point in the same direction. They too indicate that in its fundamental constitution this being is a being-in-relation, and therewith ethically relevant.

The Theological Basis of the Zones of Ethical Validity

Now that we have presented the various forms of encounter in which the ego finds itself, vis-à-vis itself, the Thou, and the world, and now that we have affirmed these three relations as being the necessary prerequisites for the ethical as such, the question arises: How do we know this? Have we merely been pursuing a type of phenomenology? Have we simply brought forth facts and observations which anyone can immediately corroborate for himself? Have our reflections simply been those which are inevitable for anyone who thinks about the matter at all? Or is all this that we have been saying distinctly a matter of theology?

The very fact that this whole matter of encounter, of the vis-à-vis, is minimized in the monistic tendencies of both materialism and idealism, philosophies which for all their antitheses are linked in common opposition to Christianity, would seem to suggest, and indeed makes it very likely, that it is the Christian message itself which plays no small part in establishing this threefold confrontation, and which leads us accordingly to a theological understanding of it. At any rate, the Christian faith in its very root is linked with the conviction that man cannot be understood immanently, in terms of the world, but only in terms of God,[3] and that man is thereby set over against the world. This conviction is actualized every moment of his life, for the claim of God, as it comes to him incessantly in Law and Gospel, directs him continually into certain relationships.

One might say that the vis-à-vis in which man finds himself with respect to God, the encounter which continually requires of him new decisions (e.g., faith or doubt, obedience or resistance, hearing or hardening), finds constant concretion in

[3] John 17:14, 16; 15:19; Rom. 12:2; I Cor. 1:27; Eph. 2:2; Col. 2:8; I John 2:15; 5:4; Jas. 1:27.

1) the way in which he confronts himself (e.g., his body in sexuality); [4]
2) the way in which he encounters his neighbor,[5] whether as friend,[6] enemy,[7] fellow citizen,[8] or spouse; [9] and
3) the way in which he deals with things, with the entities and orders of this world, such as money,[10] worldly wisdom,[11] or the powers that be.[12]

Obedience to God cannot be rendered abstractly. It can be rendered only in these relationships. Only when we realize that there is, as it were, no absolute subject of faith—that faith is always the faith of the concrete man who precisely as believing subject lives in the world of relationships—only then will the necessary inner connection between faith and works become clear to us (and that without impugning in the least the *sola fide*). Perhaps in this light too the difference between Paul and James in defining this connection (see Jas. 2:14-17) will appear much less violent, and can be regarded as a mere difference in perspective.

The point is that faith is always actualized—wholly actualized—within the concrete relationships of the world. This may be illustrated in terms of the concept "world" as it is understood theologically.

Man's Understanding of Himself in Terms of the World

All these relationships of the ego, these various forms of confrontation in which we stand, may be summarily expressed in the fact that we have a relation to "the world." The way in which this particular relation is defined determines whether within these several relationships we believe or disbelieve, obey or disobey. The relation in question can be defined in such a way that we understand ourselves either in terms of the world or in terms of God.

If, on the one hand, we understand ourselves in terms of the world, then we exist, as it were, in "conformity" to the world. That is, we

[4] I Cor. 6:13, 18.
[5] Exod. 20:16; Lev. 19:18; Matt. 5:34; 19:19; 22:39; Luke 10:27.
[6] Ps. 38:11; Prov. 14:20; Mic. 7:5; John 15:13.
[7] Exod. 23:4; I Kings 3:11; Prov. 25:21; Matt. 5:44; Rom. 12:20.
[8] Luke 14:12.
[9] I Cor. 7:3.
[10] Matt. 6:24; Luke 16:9, 11; Mark 12:42.
[11] I Cor. 1:19-25; 2:4.
[12] Matt. 22:21; John 19:12; Rom. 13:1; Titus 3:1; I Pet. 2:13.

receive the norms which determine our lives from the "autonomies" [*Eigengesetzlichkeiten*] of the several relationships. This means that we work out our relation to ourselves on the basis of this very "self," e.g., on the basis of an artificial conception of the isolated ego such as that which underlies the Cartesian *cogito, ergo sum,* or the Fichtean "ego," or "the individual" of Max Stirner. It means also that we regard our relation to the Thou as a purely instinctive relationship (in the sense of sympathy or erotic attachment), or as a purely rational relationship (involving mutuality of interests and with a common concern to keep life under control by means of collaboration). Finally, it means in respect of our relation to the world that certain autonomous norms are taken from the political, social, and cultural orders of the world, not simply for purposes of explaining these orders in terms of themselves but also—once the norms have been extended in the form of a world view—to determine altogether the structural nexus of the world. Hence it is that there arise political ideologies which make the friend-foe relation a cosmic law, or economic ideologies which regard the world as a mere economic process, or aesthetic ideologies which proclaim that life is lived out not between what is good and what is evil but between what is amusing and what is boring (Oscar Wilde), or that art is autonomous in terms of the principle of "art for art's sake."

If, on the other hand, we understand ourselves in terms of God, we still exist only within worldly relationships, and it is precisely within this worldly existence that we are to believe and to obey. Hence we are not called out of the world (John 17:15) to a life of asceticism. Nor are we to be saved from the world by death (Phil. 1:23 f.; II Cor. 5:2, 6, 9). Nor are we to regard ourselves as abstract subjects of faith. We are rather set into the world in a new and different way. We are in the world as those who do not receive their norms from the world and who are not products of the world itself, but who stand over against the world, the world being the place or theater of our obedience, where faith is tried and proved.

Although we stand at every moment within some worldly relationship, in all such relationships we nonetheless stand also under the same overriding command: "Do not be conformed to this world" (Rom. 12:2). For the determining ground of your being is not the form of this world, but the will of God (Rom. 12:2). Because from beginning to end, however, we do live in conformity to the world (Eph. 2:2), the beginning of the Christian life involves a transforming of mind (Rom. 12:2) whereby we take our orientation, not from the law of the world but from the will of God. Whenever I conform to God's will, the world becomes aware of its having been invaded by a kind of "extra-planetary material," my Christianity, and the world attempts to expel the foreign body. In other words, the world's reaction to my Christianity is one of hostility (Jas. 4:4).

This "hostility" of the world—and it is frequently described in the New Testament (cf. Matt. 10:16 ff.)—would make no sense, however, unless my Christianity had points of contact and consequently points of actual friction with the world, simply by virtue of the fact of its existence, its being there in the world. Such hostility indeed becomes quite unintelligible if we regard the believing community as some-thing which is largely removed from the world and has renounced all worldly concerns.

Faith and Worldliness

We thus note already here that the relation between my Chris-tianity and the world is wholly and utterly dialectical. My Christianity is removed from all conformity to the world. It truly stands over against the world because its determining ground is the will of God. Nevertheless, it is in the world. And it is so essentially, not as a mere addendum. We cannot say, for example, that despite its stance over against the world my Christianity is "still" in the world too. To speak of it thus is to relativize the relationship. It is the world which after all makes available to me in the first place the various forms of con-frontation or encounter whereby personhood exists. And inasmuch as the "I" which has its very being in such encounter is also the believing subject of faith, faith must have—essentially!—a worldly reference. Es-sentially faith can have its place only in the world. It must have the world—essentially—as its atmosphere. Death which takes me from the world, and the Last Day which shatters the form of this world there-fore imply the end of faith; they replace faith by sight. Thus faith and world belong together. Release from the world and the cessation of faith are similarly related.

Inasmuch as it has within it a law which claims us, inasmuch as it seeks to bring us into conformity with itself, the world is actually that *against* which we must believe. Only as we actually believe *against* it, and against the assaults of doubt to which it gives rise, is the world set in its true form and its proper place. For only thus does it become the stuff whereby and the place wherein our faith is tested and proved. Only thus does it become the state of pilgrimage and the sphere of salvation history. That *against* which I believe belongs as essentially to faith as that *in* which I believe, and to that degree the world is also a constituent part of faith.

This is why we must not relativize it in the name of an eschatological understanding of its being "still" with us. Worldliness must endure as long as faith endures, i.e., to the Last Day.[13] The eschaton which is the end of worldliness is consequently the end of faith as well.

Only in this sense is faith relative; and in this sense—but only in this sense—worldliness is also relative. Until the Last Day breaks in upon us, our Christianity is worldly or it is nothing; for without the world, and even in our relative attempts to dispense with the world, our Christianity departs from that sphere in which alone faith is at all possible.

This is why any attempt to forsake worldly pursuits or to despise politics and culture in the name of faith leads necessarily to arbitrary forms of worship, and to servitude under certain laws of sanctification which supplant justification by "faith" with an anticipated form of "sight," namely, the sight of a realizable and realized holiness, an anticipated form of the kingdom of God. In place of the Gospel which we are privileged to believe there is then set the Law which we are obliged to obey.

The World as Cosmos and as Aeon

Our immediate task is to define certain basic, essential, and universal Christian attitudes with respect to the world. The relation to "the world" is particularly appropriate because, as we have seen, it includes as well all other relations on the horizontal level. It thus serves as a model in which we may discern outstanding attitudes which apply in other relations as well.

Before we address ourselves to this task, however, we may briefly summarize our findings on the nature of the world, and fill out the

---

[13] The only modern theological ethicist who seems to have perceived this clearly and weighed it thoroughly is Dietrich Bonhoeffer.

picture with further details. It should be noted in the first place that in the New Testament the concept "world" takes the two forms κόσμος and αἰὼν οὗτος.

In terms of the second of these nuances ("aeon") the world is primarily the time of the world, a time which is passing away as it runs its limited course from creation to the day of judgment. What is involved here is not just a certain quantity of time, but a particular quality of time: the world "passes away." It is the sphere of transitoriness, of corruption (φθορά, II Pet. 1:4).[14] It is ruled by the rulers of this age (I Cor. 2:6, 8) [15] who themselves perish and are overcome, just like the intellectual "superstructure" of this aeon, namely, the wisdom of this age (I Cor. 2:6).[16]

This conception of time in terms of its quality, a concept which speaks of the "form" (σχῆνα) which is earmarked for transitoriness, judgment, and the end, is not linked exclusively to the term aeon, but may also be combined with the term cosmos. In Paul the two words are used synonymously in this sense. In John the word cosmos carries this whole freight of meaning which is elsewhere expressed by the term aeon.

From the standpoint of "cosmos," the world is in addition the sphere in which encounter takes place. In Greek usage, cosmos expresses generally the order through which all individual things are mutually related, the whole in the sense of a world system.[17] In particular the term denotes the order which obtains among men, especially that order of life constituted by the state. The term cosmos thus expresses both an order of the world and also, by derivation as it were, the particular orders in which the totality of the meaningful and purposive structure of the world is manifested.

To the degree that what is intended is the world as sphere of human encounter, the main reference is thus to relations of the suprapersonal variety (the state, economy, society) within which the various encounters take place. In this sense the cosmos comprises, even in New Testament thinking, the various places or settings in which men and groups of men live, the forms and patterns in which they exist. It

14 Cf. I John 2:17; I Cor. 7:31.
15 Cf. John 12:31; 14:30; 16:11.
16 Cf. I Cor. 1:20; 3:19.
17 Gerhard Kittel (ed.), *Theological Dictionary of the New Testament*, edited and translated by Geoffrey W. Bromiley (Grand Rapids, Michigan: Eerdmans, 1964——), III, 870.

seems to be characteristic that in this connection the world is never understood as the sphere in which a number of individuals are simply thrown together indiscriminately. Instead it is always the sphere in which their several individual structures of order are jointly comprehended.

For this reason the cosmos includes the kingdoms of this world (Luke 4:5) and the nations of earth (Luke 12:30; Matt. 6:32). This is already indicative of the fact that the world represents a power potential, and is thereby delivered up to the ambiguity which attaches to power. Those who wish to "gain the whole world" (Mark 8:36) and accept "all these" kingdoms and dominions which the devil offers them (Matt. 4:9) may indeed come to possess all that man can handle. They may well come to harness and control the power potential invested in the world. But in so doing they will also fall a prey and sacrifice to this very power, because by falling away from the true Lord of the world they slip automatically into subjection to false lords or usurpers.

In the development of the concept "world" there can therefore be no thought whatever of a purely formal and neutral scheme of formal and neutral orders and spheres of power. Incipient neutrality is always discarded and qualified at once. The power built into the world, together with its forms of organization, is always a power which is poised for action; it is a power which draws me to itself, and thus involves me in decision between God and Satan. The world which is in some sense meaningfully ordered can in a moment turn into a force to which I fall victim, thereby destroying the integrating purpose of my life and so making life meaningless. The world's power potential is a demonic potential. It has a secret lord who affects me through the orders as surely as does the true Lord who established the orders for my preservation.

Since the world—and my relation to it—stands, even as I do, between God and Satan, and since the world is nothing other than my own worldliness (to put it epigrammatically), we may say that the world is the macrocosmic form of the human heart, and that the heart is the microcosmic form of the world. The world is not distinct from me, it is not the isolated representation of evil which Gnosticism takes it to be.[18]

---

[18] See p. 435, n. 3 above.

We ourselves are the world. As we observe the world and speak of it, we must say, "This is I." Our relation to the world is thus a relation to ourselves. To know the world is to know oneself. Since every relation, whether it be to neighbor, enemy, nature, or things, is a relation to self and is expressive of the particular character of our being, our relation to the world is a particularly significant paradigm for this relation to self. It is significant because "the world" includes all that is outside us, everything to which we could relate. Moreover, we ourselves are the world, and therefore our relation to self is most directly expressed in our relation to the world. Finally, from all this it is clear that we always stand at every moment in relation to ourselves.

Nor is the world illustrative only of our relation on the horizontal level. It also represents our relation on the vertical plane, our relation to God. In Christ God loved the world (John 3:16); it is the world that has been visited by him. Hence my relation to God is revealed in my relation to the world.

There is, for example, a love for the world which is modelled on the attitude of Jesus towards the world. This is a love which prizes the world's divine destiny, the goal of its visitation. It is a love which involves pity for the world's lostness, plus a recognition that all created things are through Christ and for Christ, that all things hold together only in him (Col. 1:16-17), that in him the world is reconciled (II Cor. 5:19), and that all things, i.e., all things which exist in the world, are united under one head (Eph. 1:10).[19] That the possibility of the world's being loved in this way should even exist is expressive of the fact that "he who believes in Christ, and is baptized, lives under Christ both in the church and in the world." [20]

There is, however, another kind of love for the world which springs out of enmity against God. In this case the world is loved for its own sake (Jas. 4:4) with a love that does not derive from God's love for the world. In this case the world recognizes in me something akin to itself. It acknowledges me as a fellow rebel, thereby subjecting me to the dominion of alien and anti-godly powers.

Thus by the very fact of my being-in-the-world there is constantly put to me the question: with which love do I love the world? My relation to the world continually implies a decision with respect to God.

[19] Dietrich Bonhoeffer, *Ethics,* ed. Eberhard Bethge, trans. Neville Horton Smith (New York: Macmillan, 1955), pp. 287 ff.
[20] Alfred de Quervain, *Ethik,* II[1] (Zollikon-Zürich, 1945), p. 24.

It carries within it not merely a relation to myself but also a relation to God. Indeed the world is the summation of all conceivable relations. This significance of the world for the vertical dimension of history is biblically expressed in the fact that in the Bible the world is understood as the sphere and theater of salvation history. It is never a matter merely of an isolated secular history, but of the event of revelation—of salvation history—which is always wrapped up within the secular history.

The world is, finally, the sphere of darkness in which Christ appears as the light (John 8:12; 9:5; 12:46). It is that sphere which is forced by confrontation with this light to recognize that it is itself in fact darkness (John 1:5; 3:19). Similarly, the disciples are described as those who, being related to the light, also shine in the darkness of the world (Matt. 5:14).

If the world can be called the "field" on which the good seed and the weeds both grow (Matt. 13:38), we certainly cannot take this metaphor to mean that the world is itself a neutral ground "on" which the mutually hostile plants are to be found. This field—and here the metaphor breaks down—is itself always one of the contending parties. For here too it is the lost world which has to be saved (I John 4:14). The field is the extended zone of the fall of man, comprehending both the groaning sphere of creation (Rom. 8:22) and the summit of human history, indeed even the cosmic powers of darkness (Col. 1:13). It is the world after the fall, the world which is visited by the redeemer Christ.

Hence the world is not some kind of neutral operational sphere in which the event of redemption takes place, nor is it a neutral "point of contact" for the liberating work of Christ. On the contrary, it stands in an infinite qualitative opposition to Christ, so that all that takes place in salvation history involves a calling and snatching out of the world. The spirit of the world thus stands in a mutually exclusive antithesis to the Spirit of God (I Cor. 2:12). The wisdom of God is unintelligible to the wise of this world (I Cor. 2:6 ff., 14), and the wisdom of this world is mere folly to God (I Cor. 1:20 f.; 3:19). God's criteria for evaluating men are fundamentally different from those of men themselves (I Cor. 1:26 ff.).

This is why, when we consider Christ's coming down into this world, we must always maintain that his origin is not in this world. In distinction from men, who are "from below" and "of this world" (John 8:23), Christ and his kingship are "from above" (John 8:23; 18:36).

One indication of the fact that the world cannot be regarded as a neutral platform for natural and spiritual existence is afforded in the emphatic statement concerning believers in Christ that they are not "of the world" (ἐκ τοῦ κόσμου, John 15:19; 17:14, 16), but that Christ has chosen them "out of the world" (ἐκ τοῦ κόσμου, John 15:19), i.e., that they have been given to him by the Father as those who have been thus taken "out of the world" (John 17:6). [21] When it is said further that those who believe in Christ are still "in the world" (ἐν τῷ κόσμῳ, John 17:11; 13:1; I John 4:17), here too we are not to understand this being-in-the-world as a neutral and colorless indication of location. The connecting of the "in" with the "not of" suggests rather that the relation of believers to the world is completely altered, that within a world which is hostile to God the believers are set apart, that with respect to its downward plunge their fall has been arrested, that the pull of this aeon—the attraction of its standards and spiritual values and of the power and status it promises to deliver—no longer has any dominion over them. The lordship of the world yields before the lordship of Christ to which the believer subjects himself in faith.[22]

Thus the Christian is "extra-territorial," even though he lives on the territory of this world. Nevertheless, as we have seen and as we shall have to show more fully later, this does not mean that he lives in isolation. On the contrary, he is one who shapes the world, is active within it, and in Christ loves the world. The fact that this is possible, that what is intended is not something like a Christian fortress in an alien land—though in exceptional circumstances this may be a necessity— but that the Christian can live in the freedom of which Paul spoke ("all things are yours," I Cor. 3:21-22), rests essentially on two main grounds.

In the first place, this possibility of openness to the world rests on

[21] The same phrase, ἐκ τοῦ κόσμου, can thus be used in the one case to signify living by the world and in intimate dependence upon it, and in the other case to signify being taken out of the world. In what follows we are adopting the latter sense except where there is specific indication to the contrary.

[22] See the excellent treatment of this distinction between the "in" and the "not of" in terms of the specifically American church situation—identification with the world vs. escape from the world—in Reinhold Niebuhr, *Does Civilization Need Religion?* (New York: Macmillan, 1927), pp. 165-167; and cf. Hans Hofmann, *The Theology of Reinhold Niebuhr*, trans. Louise Pettibone Smith (New York: Scribner's, 1956), pp. 27-29.

the fact that the "all things are yours" is accompanied by a safeguard: "and you are Christ's." The Head does not leave his members. We have the freedom of those who are secure. This is why there can be a kind of Christian abandon. Those who transcend the world can be more free and creative in respect of it than those who are subject to the world, whose struggle to live consists either in the defensiveness of anxiety or in the despairing offensive of intoxication with the world, of lusting for its power and glory, of fanatical and frantic activism.

Alone, however, this safeguard cannot suffice as the ground for Christian freedom. For even at best it could only mean that the Christian will stride erect in unencumbered sovereignty through the hail of enemy bullets, confident that they can no longer harm him. It would not include the possibility that he will be warmly and enthusiastically active on the "enemy front" in free and creative self-offering. There must be yet another ground for this possibility of Christian openness to the world, one that is admittedly linked with the first, and it is this: that for those who belong to Christ the world itself has been changed. It can no longer be called only an enemy front, for it is at the same time also the object of God's gracious visitation, and of the concrete manifestations of his love. Among these manifestations we may number, for example—though only as an "example"—the gift of political order, by means of which in a world of sin and impending chaos God creates a protective shield which will preserve mankind until the day of the Lord.

For the Christian the world is integrated not merely into the order of redemption, but also into the order of creation, not in the sense that he can perceive and soberly mark out within it pure spheres and intact orders of creation, but in the sense that he knows God's plan for creation and therewith the true goal of its forms. The Christian knows why there is in creation the possibility of culture, work, and play. He knows the meaning of bios and logos. And even though sin has made rents and gaps and distortions in all of this, the Christian can see even in the distorted image the great intentions of God. To take the example of art, the Christian can see God's intentions no less in a perverted and debased art than in the sublime error of tragedy, the artless innocence of Mozart, or the lofty chorale of praise—just as he can see man's orientation to God even in such perversions of it as the most abstruse magic or the most blatant idolatry.

He cannot, of course, conclude from this discovery of the divine intentions that there is a possibility of excavating from the ruins unspotted values of creation, in the sense of a nature which has remained intact and can therefore be established ontologically.[23] Thus there can be no question, for example, of man's distinguishing and manifesting the core of a true and lasting philosophy [*philosophia perennis*] in the midst of hybrid worldly wisdom, or of true art [*ars perennis*] in the midst of "fallen art," or of a durable religion [*religio perennis*] in the midst of idolatry. The fall of man permeates totally everything that derives from man. It mixes creation and sin to the point where they are no longer distinguishable from one another.[24]

Consequently we cannot in the name of creation give a pure and clear representation of what the world is supposed to be. On the other hand, the ambiguity of worldly phenomena does not prevent us from showing analogically and indirectly, on the basis of such phenomena, what God's purpose for the world is, namely, that man should rule it, that he should shape it to the glory of God, and that from its forms and shapes there should sound forth that praise which is still rendered in the liturgy of mountain, wind, and fire (cf. Ps. 18:10 f.; 104, 8).

23 See above pp. 199 ff. and 388 ff.

24 Cf. our critique of Flacius on p. 214. We would hope that our statements about the "intentions" of God being discernable even in the perverted forms of creation have been sufficiently safeguarded as to preclude anything like a religious interpretation of existence or an interpretation in terms of the analogical form of the world. They are meant rather in the sense of an analogy of faith, not an analogy of being. The creator is not known on the basis of the creation. On the contrary, the creation is disclosed in terms of the creator. With respect to creation too the saying holds true that it is the person who makes the works. And only he who knows the person understands His work, the creation.

A.

THE CONFLICT SITUATION AS TYPICAL
OF MAN'S ETHICAL RELATION TO THE WORLD

25.

Compromise as an Ethical Problem

"All our works are in vain, even in the best life." In this statement
Luther denotes the heart of the Reformation doctrine of justification
on its negative side. The moralism of the Enlightenment would corrupt
this acknowledgment of the *sola gratia* into the banal statement that
"all our works are *impure,* even in the best life." It does not require a
great deal of thought to ascertain—all polemics aside—what is here
asserted in this dubious paraphrase. There is in fact a partial truth
expressed, one which is not relevant, however, in the present context.
What is expressed is in effect the principle which is so frequently repeated in Kantian ethics, namely, that our ethical practice always falls
far short of what the imperative intends, and further that ethical motivation is highly complex, involving always a eudaemonistic admixture which is alien to ethics.[1]

It can, of course, be said—indeed it has to be admitted—that this im-
purity of conduct is not due only to inner conditions in the acting
subject, but is determined also by the objective structure of reality
within which the acting must be done. Our action hardly ever takes
place exclusively in the area of a man's inner intention. As regards
both motivation and performance it is also determined at least in part
by environmental factors.

The statement that even "the best man cannot live in peace if it does
not please his evil neighbor" shows analogically the change to which
our conduct is subject once it leaves the abstract sphere of mere disposition and steps out into the world. And since this abstract sphere of
disposition has to take into account the environmental destiny of action, since it has to be relevant and realistic, the question arises whether in the long run it can itself be protected against this very change.
Does not our disposition too have to reckon from the very outset with

[1] Cf. Immanuel Kant, *The Fundamental Principles of the Metaphysics of
Ethics* (New York, Appleton-Century, 1938).

the existence of this "evil neighbor"? Does it not have to reckon similarly with the necessity of some measure of self-assertion, and consequently with a peace which is not "pure" peace but at best only an armed truce?

And what about the "good" neighbors? Do they offer me a sphere in which pure motives can find pure actualization? Even as regards the best of neighbors, do I not come constantly into conflicts of interest with them? Is not my sphere of operation continually restricted by theirs?

In any case, we live in a world in which we can never arrive directly at any goal, but must follow the various paths which lead to the various goals. Thus in certain circumstances political encirclement or the use of force to stop a troublemaker may be the way which leads to the ethical goal of international peace. So, too, may be the enhancement of the power of one's own nation, if that nation is to become the dominant factor in a system of alliances and use its strength to guarantee the peace. The goal of becoming this dominant factor may in turn require the pursuit of a variety of political, economic, and sometimes military means toward that end, in the selection of which one cannot of course draw the line too narrowly. Shady deals will have to be made, and some of the people who get hurt by them will point out—and rightly so!—that "from the moral standpoint" this is a pretty foul way to begin.

For even if we can assume that back of this far-reaching political strategy there is the noble motive of peace, this does not mean that that motive will be apparent in every action at every preliminary stage along the way. Quite understandably, the preliminary action will instead be judged with respect to the means chosen and the particular goals enunciated, which will, in the early stages, perhaps consist only in the securing and building up of extensive power establishments. What would be more obvious and natural than to view the ostensible goal of "peace" as nothing more than an ideological smoke screen, the whitewashing of what is basically an out and out power grab, a kind of deceitful propaganda? Furthermore, on the level where the particular action takes place, among the various smaller nations themselves, for example, the ethical imperatives involved may assume a very different character. Here a very different imperative might well arise because of the very different political interests.

In any case there are exceedingly complicated, extensive, and difficult obstacles to be surmounted or overcome before a statesman can
reach a goal which for him has ethical rank, for example, the goal of
peace. If as a leader in war he sets peace as his goal, then he may
even have to ask himself—there are, for example, statements in von
Ludendorff to this effect—whether the most vigorous prosecution of
the war, including use of the most destructive and terrible weapons,
may not be the most merciful thing he could do, since such terror
would put the quickest possible end to the otherwise endless terror,
and in so doing perhaps make possible a new and positive beginning.

Wherever we act with reference to a goal which is subject to ethical
intentions we thus find that we must make use of devious and indirect
ways. The ends which I seek to attain are always separated from me
by a system of means. A particular end can usually be attained only
by a very specific means. I can perhaps attain the ethical end of peace,
for example, only by the above-mentioned means of most "unethically"
playing off the various interests against one another, by self-assertion,
cunning, force, and tricks or tactics which naturally are not at all
compatible with Kant's categorical imperative.

At least there is one way in which I certainly cannot reach the intended goal, and that is by trying to convince the others involved that
peace is a high ethical goal and that all who ethically desire it must
resolve to attain it by giving up their legitimate interests, sacrificing
their accrued advantages, and pulling back from their established positions of power. The very fact that such a notion appears so strange
and fanciful serves to underscore the truth that ends can be attained
only by a specific system of means, and that no ethical goal, however
noble, can enable me to skip over this preliminary sphere of means.

Now this introduces at once a new ethical problem which may be
formulated in the following questions. First, can the means which I
am compelled to use be brought into harmony with the ethical imperative which motivates me? Can I choose both ultimate good and
penultimate evil? Is it only my goal—but not the way to it—which
stands under the control of the imperative? Second, can the means
which I am compelled to use be brought into harmony with the goal
which is ethically demanded of me? Does the end justify the means?
Even before we engage in any fuller analysis, we can see at once from
these two sets of questions themselves that irreconcilable contradic-

tions arise in this area. We also remember from our discussion of the Sermon on the Mount [2] and of the problem of transposition in natural law [3] that these difficulties exist for Christians too, and that there is no easy solution to the problem of how means relate to ends. Hence there arises the eminently theological question whether this refraction of all ethical action can be justified in terms of Law and Gospel, and how this refraction is to be interpreted theologically.

The Provisional Character of Life in this World

The philosophical, or better, the "natural" ethicist is inclined to contest the possibility of pure ethical fulfillments, and consequently to come up with a doctrine of "compromise" which teaches that the purity of ethical volition must make concessions to things as they really are. The imperative, it is held, is limited by the autonomy of the available means, and any realistic ethics is thus an ethics of compromise.

The Christian is confronted by the same problem. But he cannot simply surrender the unconditional character of the command in the name of a so-called realism. The "natural" ethicist can always make appeal—and this appeal is one of the ethical axioms of Kant—to the fact that volition and capability [Wollen und Können] must necessarily correspond: "You can, because you ought!" The moment our capability comes up against a demonstrable and fundamental limit, such as the wall imposed by the autonomous world of means, the imperative must necessarily be reduced to whatever sphere of action is thereby staked out. For in addition to the imperative itself, the objective world within which the imperative is to be realized is also a normative component of the ethical. Compromise is thus the final conclusion of ethical wisdom.

In the face of this situation the Christian is partly in the same position as the so-called "natural man," and partly in a very different position. On the one hand, he is in the same difficulty as anybody else who is confronted by a world in which certain goals are attainable only by specific and often very dubious means. On the other hand, he is not in a position to water down the imperative accordingly, and thus to make it dependent on what seems to be fundamentally "possible" in this world. For the Christian, the duty of obedience is

[2] See Chapter 17 above.
[3] See Chapter 20 above.

not a composite of the imperative on the one hand and that which is posited by the stuff of the world on the other. Hence the Christian is not in a position to regard the resultant compromise as the tragic product of pure goals and questionable means, of noble volition and the "force of circumstances." For him there is no axiomatic connection between imperative and ability along the lines of the thesis, "You can, because you ought." The radical requirement of the Sermon on the Mount compels him rather to confess, "I ought, but I cannot." Such a confession is in principle impossible in Kantian or any other secular ethics.

That Christian ethics does not acknowledge this tragic compromise, that on the contrary the commands of the Sermon on the Mount are quite unconditional and make no allowance for the conditioning of all goals by the sphere of means, that they demand the ultimate and ignore the penultimate, all this would be completely misunderstood if we were to explain it in terms of a formal radicalism or even in terms of remoteness from the world. Actually, the Sermon on the Mount with its unconditional demands is explicitly passing judgment on the world, judgment to the effect that the world in its concrete state is not the world which came from the hands of God at creation but is a perverted form of the creation. Consequently, the unconditional imperative—which in this instance means the will of God—is not to be inferred from this world's conditions, autonomies, and spheres of means, nor is it to be modified or watered down on the ground of these encumbrances of a fallen world. Man cannot get off the hook of God's unconditional requirement by pleading that the world in which he is set is an evil world. For man is himself this world. The world is his work. It is the objectification of his own ego.[4] He is thus robbed of any alibi in terms of the tragic ineluctability of fate. The Sermon on the Mount does not overlook the reality of the world; it protests against it.

Hence the Christian is prevented from any simple affirmation of compromise. For the commandments of God are uncompromising in their demands. We search them in vain for any concession to the effect that "You shall . . . insofar as is humanly possible," or "You shall . . . insofar as existent circumstances allow."

When it comes to ascertaining what is required of us, we are no

[4] See pp. 434-440 above.

more permitted a side glance at the existent circumstances—at the available means of fulfillment—than at our own psychical structure, which is also obviously incapable of accomplishing what is demanded. For although we can come to some sort of compromise with our lusts, e.g., by contesting and quantitatively restraining them, we cannot extirpate them. Here too we cannot be completely radical, for radicalness would be identical with a total surrendering of the self. Yet once again the commandment is not, "Reduce your lusts, combat them as far as possible." With piercing simplicity—and simplicity always means no compromise—we are simply told, "You shall not lust" (Matt. 5:28).

This puts squarely before us the real problem facing theological ethics. As Christians, we live in the same world as all other men. We live in a world in which goals can be reached only by way of specific means. And the further we move out from the constricted sphere of personal living to act in the orders, the more deeply implicated we become in the contradiction between means and ends. This is especially true in the area of public life. The politician and the business man, for example, feel they are tied to certain rules of the game which they simply cannot break.

In this respect the Christian is no exception.[5] He is an exception only to the degree that he is not allowed to say, "Since this world is a forest full of wolves, let us howl with the wolves and respect the laws of the jungle." God's majestic command does not admit of integration into this world's system of co-ordinates. In dealing with God's command we are dealing with "extraplanetary material."

Even though this command is in force, however, the Christian too acts always in the form of compromises. Nor are these compromises linked merely with a weakness which in theory at least can be overcome. They derive rather with demonstrable and objective regularity from the structure of the world, and the Christian is compelled to orient himself on the basis of this fact. He must know what it means *coram Deo* that in principle he never gets in this aeon beyond the stage of compromise. He must also know what it means that he can-

---

[5] Reinhold Niebuhr is right, however, when he points out that the Christian easily evades a realistic recognition of his non-exceptional character by making "devotion to the ideal a substitute for its realization" and by becoming "oblivious to the inevitable compromise between [his] ideal and the brute facts of life." *Does Civilization need Religion?* (New York: Macmillan, 1927), pp. 63-64.

THEOLOGICAL ETHICS – FOUNDATIONS

not adopt the tragic interpretation of this ineluctability, i.e., that he is not allowed to make a virtue of this necessity and assume that God will be satisfied with a watered down imperative. He must know what it means that God's radical demand, even with reference to the admittedly unsuitable material of this world, nonetheless still stands, so that, even though this material is incapable of the obedience of uncompromising action, he—the Christian—is still required to act. He has also to realize that any attempt to break free from this dichotomy—whether in the form of fanatical radicalism or of a tragic interpretation of the world—is an inadmissible illusion, a coming to terms with the world on which God has declared war and judgment, that world in which God's Son has died. He has to realize that the enduring of this dichotomy alone is ethical realism, the only realism there is, and that every attempt to take into account the world and its structure is in fact a flight into illegitimate ideas which are truly alien to the world. He must know that this truth is devastating, and that it corresponds exactly to the death to which the Law of God when it is taken seriously condemns man, by taking from him every excuse, every rationalization of his situation, and every possible self-vindication.

He must know finally how it is that the sting is taken from this death, that he can live joyfully under the condemning majesty of God, that this dichotomy and the *de facto* compromise of his life need not harm him and can in fact no longer separate him from the love of God. He must, no, he is permitted to know that only this preservation under the fatherly grace of God—only the certainty that his alienation cannot kill him—can give him the power to be truly realistic and to take his orientation from a truth which immunizes him against spurious consolations, against illusions, and against the meretricious peace of this world.

We live in a perishing world, a world in which God nevertheless has tasks for us to do. And we pray, "Thy kingdom come!" The whole secret is here—to wait and to leave behind, to know the world of death and yet be empowered for life, to leave the requirement of God intact and to live provisionally under forgiveness. In order to make as concrete as possible what we mean theologically by the provisional character of life in this world we shall first address ourselves to a phenomenology of compromise.

The Phenomenology of Compromise

To proceed with our dissection of the problem of compromise, we must first make mention of the fact that life consists of one huge mass of compromises which are ethically neutral. These can best be illustrated in terms of various technological processes.

Every motor vehicle, for example, represents a compromise between the need for speed and the need for reduced weight, between the greatest possible horsepower and maximum maneuverability, between top performance and maximum gasoline economy. Every business firm with any sense of social responsibility works out compromises between profits on the one hand and expenditures for employee benefits and community welfare on the other. Every scholarly book is a compromise between the exact and exhaustive treatment desired by the author on the one hand and such factors as time, the publisher's question of marketability, and in some instances the actual needs of the reader on the other. A work of art is a compromise between what is desired and what is possible, between the visionary conception and what material, time, space, and circumstances allow. Law is a compromise between the freedom of the individual and the necessity for regulating a group of such free individuals by means of rules which will protect that freedom even as they limit it.[6]

The list of such examples could be extended indefinitely. This indefiniteness is the infinity of life itself. For, whether in nature or history, all life involves movement and resistance, opposition and the working out of the same, the conflict between matter and energy. What is attained in any particular instance is only one possibility within the sphere of the given. Hence at every stage and in every form there is compromise between the various participants in the particular interplay of forces. When Bismarck describes politics as "the art of the possible," what he has in view are these factors in the interplay of forces. He means there is no such thing as an action which is pure and absolute, devoid of all presuppositions. In effect, his expressive description is a good definition of the nature of compromise. What else is the ability to live but the art of the possible? Is not politics simply a paradigm—particularly blatant, pointed, and striking to be sure but nonetheless merely a paradigm—of life itself, in

[6] Cf. Kant, *op. cit.*, p. 35.

which the sublime and often divergent laws of existence are disclosed to an extraordinary degree?

It would be a complete mistake, however, to conclude that because compromise is a law of life it is also ethically neutral, and hence to sanction it in the name of life itself. This would be equivalent to saying that death or sexuality is a biological phenomenon which occurs everywhere in nature, and that it is thus to be described as ethically neutral. Those who contend for such neutrality might well ask how anything can be ethically relevant—and hence the subject of personal decision—which is to be found also outside the sphere of human life, and which is even a law of life for such metaethical creatures as insects and reptiles.

Nevertheless, this rhetorical question is an unwitting reminder that even such natural phenomena as sex and death undergo a change the moment they enter the sphere of human existence. We have only to consider what are the implications for ethical decision of our "being unto death" in the sense of Kierkegaard, i.e., the limitation of our span of existence, and our awareness of this limitation. We have only to consider further what death itself—that biological event!—implies as a limitation of illimitable man, grasping from the very first at equality with God.[7] We have only to consider what a basic and ultimate decision with respect to God is reflected in that fateful judgment which is our death. Finally, we have only to consider that in the human sphere sexuality is divested of its purely biological character and becomes, as it were, merely a medium within which the I may encounter the Thou, a medium within which persons either love or exploit one another.

The same is true in respect of the "natural law" of compromise. It undergoes refraction and is essentially changed when it ceases to be a mere process of natural arrangement and becomes instead an element in human decision. The frontier where this change takes place fluctuates a great deal in the human sphere. For example, although it is men who actually work them out, the compromises in the sphere of technology to which we made reference a moment ago are still purely objective matters which have not yet reached the frontier of which we are speaking. In many cases we cannot locate

[7] On the connection between sin and death in the Bible see my *Tod und Leben* (2nd ed.; Tübingen, 1949).

this frontier with any precision at all, for it is fluid. This is particularly true, for example, in regard to our second illustration, the business firm caught between the demands of profit and of social responsibility. For here is a tension that cannot be resolved in an objectively attained equilibrium. In addition to simple calculation it requires the making of decisions and the taking of risks, and sometimes even the courage to do something which cannot be defended on purely financial grounds. We have only to think of the not uncommon situation when the market is flooded, sales decline, and a business can maintain its profits only by laying off more employees. In such instances the calculable compromise between profits and a concern for community and employee welfare (which would demand the longest possible postponement of such a layoff) will often not be the only criterion on which those responsible for the firm must base their decision. The point is that management then is simply not in a position to base a decision on mere calculation—and so deprive it of its character as real decision!

Nevertheless, as a matter of course business men repeatedly try to do exactly that. Now this is quite understandable. For, in addition to the evident duty of basing economic decisions on sober calculation, and so safeguarding ethics from getting lost in the clouds of theoretical abstraction, there is also that fundamental human urge to avoid the venture of decision and evade the responsibility for it by seeking refuge in the sphere of the supposedly calculable. This is why calculation has so great an attraction. For what I can calculate is "necessary," and hence, as it were, automatic. It demands no decision. By means of it I make my escape into a sphere which is, as it were, beyond good and evil.

Yet we must not make the mistake of thinking that in the distinctively human sphere compromise really is the object of calculation, and can thus be executed with "assurance." To think this is to deceive ourselves. The fact is that compromise, as an action taken in human history, is always based on decision. This may be seen from the following consideration: The material and personal values between which compromise has to mediate may be so heterogeneous that objective criteria are inadequate and only a venture of decision is of any avail.

To switch to a very different example of this conflict of values, we may ask: By what objective standard is a pastor to effect a compromise between the demands placed upon his time for pastoral visitation and the demands of solid sermon preparation? Is this really a case of values which can be calculated objectively, which have only to be subtracted from one another in order to arrive at a result which will clearly indicate the proper line of action? If this were so, the pastor might easily banish his tormenting sense of inadequacy and guilt in the name of such objective calculation. The claims, however, and the harassing accusations of his conscience tell him that what is really involved here are decisions which entail guilt. The simple division whereby the different demands, being necessarily subsumed within a definite and very limited quantity of time and energy, have to be curtailed in the form of compromise, just does not come out even.

When autonomous, compromise leads either to despair [*desperatio*] or to false security [*securitas*]. It leads to despair if I try to hold my own against the unconditional demand that, as a preacher of the Word of God, I give the utmost in hard work, gathering my material, thinking it through, and reverently trying to formulate it, and that at the same time I burn myself out in service of the sick and needy and spiritually weak. On the other hand, it leads to false security if I think that compromise is justified as the only possible rational solution, and if I thus draw a calculated line between the different demands.

Before we proceed further, we may simply formulate in this connection two questions which we shall have to take up again after we have concluded our discussion of the phenomenology of compromise. First, what is the significance of the fact that here in connection with the problem of compromise we meet the same two concepts—despair and false security—which we encountered in the doctrine of the Law as the basic attitudes to which we are driven by the Law? There would seem to be an indication here that what is brought to light in compromise is in fact a fundamental aspect of existence in the world, existence between the fall and the last judgment. This is the only thing that could account for the fact that compromise affords such a classic demonstration of the condemning and destroying power of the Law.

The second question is this: Why must the calculability of compromise be ruled out as an illusion? Why is *securitas* an impossible solution? From a purely "natural" standpoint this question is meaningless, for so far as the world is concerned—seen in purely immanent terms—there have to be in the human, historical sphere the same balancings and counterbalancings as are found in the sphere of nature. If the question is posed notwithstanding, this must be because another factor is now taken into account, a factor which cannot be conceived in immanent terms but rather calls the world in question from a standpoint outside itself, a factor which makes it plain that the false security into which the world naturally lulls itself is a mortal peril indicative of the fact that the world has failed in respect of that for which it was intended.

But we must leave these two questions for the moment, simply bearing them in mind as we pursue our phenomenological investigation. Our attempt at the moment will be to develop further, by means of typical illustrations drawn from various representative spheres of life, the situation of compromise. We shall take these examples from the distinctively human "historical" sphere in order to show that beneath every compromise there is always a conflict which reaches down to the very foundations of human existence and which is basically insoluble.

26.

Forms of Compromise in the
Religious Sphere

It is natural, when in our pursuit of the phenomenology of compromise we come to an analysis of the forms of compromise in the religious sphere, to begin with Roman Catholic moral theology, because one suspects that here the problem of compromise plays a considerable role. This suspicion arises on two counts. First, the Roman Catholic doctrine of the analogy of being [*analogia entis*] both permits and demands that supernatural requirements be referred in large measure to the factualities of human nature, and therewith to the "conditions" which affect a given situation, i.e., to the concrete, given form of man's nature; the lack of any "infinite qualitative distinction between time and eternity" (Kierkegaard) means that here there are less radical antitheses between the divine and the human, and that balance and "compromise" are thus more likely possibilities. Second, there is the related factor that the ontological scheme of thought which is predominant in Roman Catholic theology [1] favors the relatively possible at the expense of absolute decisions.

Roman Catholic Casuistry:
The Problem of the "Grand Inquisitor"

The problem of compromise in the sense indicated is classically expressed in Dostoevski's story of "the Grand Inquisitor" in *The Brothers Karamazov.*[2] The Grand Inquisitor is the representative of a church which in its preaching has effected a compromise with the fallen nature of man in order that by such adjustment and accommodation it may enjoy the greatest possible success in its handling of that nature. The advocate of this compromise is the devil. He

---

[1] See Chapter 12.

[2] See Fyodor Dostoevski, *The Brothers Karamazov*, trans. Constance Garnett (New York: Macmillan, 1948), pp. 253-272.

argues his case for it during his confrontation with Christ in the three great temptations (Matt. 4:1-11).

The tempter's first appeal is that Christ should change the stones into bread, which means in effect, "Feed men, and then ask of them virtues." [3] Christ is supposed to make a pact—a pact is compromise! —with human nature, that human nature which always acts on the principle that "he who pays the piper calls the tune," i.e., whoever does the most to meet material needs will make the most headway with his ideological demands.

The second suggestion, that Christ should work a spectacular miracle by leaping from a pinnacle of the temple, points in much the same direction. Basically "man seeks not so much God as the miraculous." [4] He does not want a personal conscience. He is determined not to make the personal decision of conscience to which an encounter with God would drive him. His search is simply for "miracle, mystery, and authority." [5] He wants nothing more than to succumb to the overwhelming power of suggestion. He does *not* want to be touched in the central core of his person.

The third temptation is that the Christian message should entrench itself in earthly power, that it should grasp "the sword of Caesar," whose "world" and "purple" [6] Christ himself spurns in order to tread the impotent way of the cross-bearer. If the church is ready to seize power, it will enchant the hearts of men and make them its own. None can withstand this power, for it rests on two primal needs of man. In the first place, this supreme power will be bold, not merely to take from men the punishment of their sin, but also, in answer to the deepest longings of us all, to take the sin too upon itself, thus setting aside in its usurped kingdom the profoundest laws of guilt and expiation, and taking the law into its own hands. In the second place, men are tormented by the desire to unite the whole world in a "universal unity" [7] in which the antithesis between God and world, between inner and outer, is resolved, and in which the whole cosmos is grasped, as it were, in a single powerful hand. By seizing power the church can meet both these needs.

[3] *Ibid.*, p. 260.
[4] *Ibid.*, p. 263.
[5] *Ibid.*, pp. 262, 264.
[6] *Ibid.*, p. 265.
[7] *Ibid.*

It is during the Spanish Inquisition—according to the story—that Christ moves amongst the people "with a gentle smile of infinite compassion." [8] And as he did during his earthly ministry, he raises a maiden from the dead. But this act of love arouses the wrathful opposition of the Grand Inquisitor, "an old man, almost ninety, tall and erect, with a withered face and sunken eyes," [9] who in face of this gracious miracle does not know what else to ask but, "Why art Thou come to hinder us?" [10] Jesus Christ is a serious hindrance to that calculating conception in which the message of the church is adjusted to the weaknesses, miseries, longings, desires, and anxieties of humanity. In Christ eternity is manifested not as the elevation of man's nature but as the contradiction of it, and in this contradiction as victorious, sacrificial, enduring—but not compromising!—love.

Although the Grand Inquisitor is obviously characterized here as being in a pact with the devil, we do him great injustice if we regard him merely as a diabolical intriguer or as one possessed by the devil. We have to take into account the goal which the Grand Inquisitor has in view. His concern is to find ways and means to bring men to God. Deep down he knows that the immediacy of God must be fatal in its effects—like the hindrance caused by Christ—or that the Word of God will never get beyond the stage of impotent and unsuccessful wooing.

The "end" of union with God is thus to be attained only by use of the "means" available. But these means must take account of human nature, drawing out of it what is already there. This implies, however, that the driving forces of man's nature must be mobilized, that even its lower instincts—which may well be its strongest impulses!—must be "utilized," in the interests of a good cause. Man's hunger, his need of authority, his longing to find someone with power enough to bear the burden of his guilt, must all be exploited.

What is thus needed is something like an act of higher homeopathy, in which healing is effected in accordance with the principle of "like to like." Man can be cured only by powers which resemble him, powers which are akin to the organism of his soul. Are not these means, then, these all too human means, justified by the end which

8 *Ibid.*, p. 255.
9 *Ibid.*, p. 256.
10 *Ibid.*

496

is to be attained through them? Of what avail would it be to come to man with something which is not like unto himself, but only like unto the Word of God? Would not man's inner organism resist this alien body? Then even with the absolute truth absolutely nothing would be gained. To be sure, the ultimate goal may be a little corrupted when an adjustment is made to human nature, and to the means it has placed at our disposal. The goal may be a little too deeply immersed in the mire of earth. But then there is always the chance that what remains of the goal will at least have an effect, and this is better than nothing—which is all that comes of the attempt to bring to man God in his immediacy. What that leads to can be seen from the bankruptcy of the cross on which the *un*compromising Son of God died!

Thus this adjustment to human nature is undertaken "to the greater glory of God" [*ad majorem gloriam Dei*].[11] But is it not true that this attempt involves what might be called a "revolt of means"? Do not the means obscure and alter the real end? Does not every attempt to take human nature into account in preaching, and so to integrate conversion into a calculable strategy, imply an inadmissible "co-operation" on the part of man? Is not this co-operation forbidden precisely because it calculates instead of believing, precisely because it fails to trust God to find the ways and means for vanquishing the human heart? Is it not man's purpose thereby to do the work himself instead of letting God do it? And does not man thereby secretly put himself in the place of God? Is this not—as Luther interpreted this kind of work-righteousness, this co-operation—a case of idolatry? Only when we realize that this desire to reckon with and adjust to human nature leads to the decline of faith itself, and finally to idolatry, only then can we see why this process is harshly described as a pact with the devil.

In any case what we have here is the most dubious feature of the doctrine of *analogia entis*. When grace links up with nature it is committed to such means as are available to nature, and hence leads to this business of reckoning and of utilizing these means. This is, as it were, the central danger in the Roman Catholic approach to dogmatics and moral theology. This is also the root of the perennial calumny—and it is a calumny!—that according to Roman Catholic

11 *Ibid.*, p. 255.

teaching "the end justifies the means." [12] Roman Catholicism does not actually go this far. But we can understand how the suspicion that it does constantly arises.

It is not by chance that in these last paragraphs we have been setting forth our argument in the form of questions. For we are here confronted by problems which in many respects are more complicated than might appear in this presentation. It is obvious that even a theology which tries to take the *sola fide* and *sola gratia* seriously, and which is thus opposed in principle to any defense of co-operation, must still take human nature into account in its preaching. The big difference of course is in the way the resultant adjustment—the compromise —is understood.

That the compromise and adjustment which actually takes place in Reformation churches is also deliberate may be seen quite simply in the fact that we do not preach to adults as we do to children, and that it is one of the axioms of catechetical instruction to adapt oneself to the imaginative and volitional capacities of the children. The "point of contact" [*Anknüpfungspunkt*], which is such a hotly debated point in the theoretical definition of nature and grace,[13] is no problem at all when it comes to the concrete teaching and preaching. In every presentation of the message the person addressed simply has to be taken into account.

At this point there arises at once the question as to wherein this adjustment and accommodation consists. Does it concern only the formal presentation and actualization of certain contents of the message? (For example, in the case of a senile old miser I would emphasize the commandment "You shall not steal" more than the commandment "You shall not commit adultery"; whereas in the case of a child I would usually emphasize aspects of the message which are likely to be—though they are not always—more relevant for a child than for an adult.) Or does the adjustment involve rather an accommodation of the very substance of the message to the existent needs and desires, whereby the message itself is altered or watered down? Those who take the *sola fide* seriously will regard only the first of these two al-

[12] See the note of Victor von Cathrein showing that literary prizes offered to anyone who could demonstrate a Jesuit use of this principle have gone uncollected. *Moralphilosophie* (Leipzig, 1924), I, 327.

[13] See Chapter 16.

ternatives as legitimate, whereas those who subscribe to the principle of co-operation may in some circumstances view with relative unconcern even an assimilation of content.

Perhaps the difference in respect of accommodation may be expressed as follows. For the Reformation churches man is confronted with the clearcut decision whether he will accept in faith the God revealed in his Word, or reject him and become his enemy. Between these two choices there are no bridges, for example, in the form of works which are more or less good, for "whatsoever is not of faith is sin" (Rom. 14:23). As Luther put it, "If adultery could take place in faith, it would be no sin. If you try to worship God in unbelief, you will perform an act of idolatry." [14] For Roman Catholicism, on the other hand, education is more important than decision. For education makes use of specific means in order to exercise an influence upon the one who is under instruction, at times even bringing him to the point where he is one day led across the threshold of decision unconsciously and without realizing it.

In this distinction concerning accommodation there may be discerned the radically different understanding of sin. Reformation ethics perceives that not merely individual acts, but even the laws and orders of this aeon, within which they are performed, are infected by sin. This is true for the simple reason that this aeon is man's world; it is, as it were, the objectification of man. The form of this world is no more able to produce absolute righteousness than is our human heart. We see this not least of all from the fact that the Sermon on the Mount, precisely because it cannot be fulfilled within the laws of this aeon, is one long protest against the world of fallen man. [15]

The consequence is that fundamentally there is no ontically righteous form of conduct. Conduct is *de facto* a compromise between the divine requirement and what is permitted by the form of this world, by the autonomy of its orders, and by the manifold conflicts of duty. Luther's doctrine of the "two kingdoms" is a sign that the divine commandments are altered in accordance with the different spheres of existence in which they are to be fulfilled. [16]

---

[14] See Luther's Theses of 1520 on whether works contribute to justification, Nos. 10 and 11 (WA 7, 231).

[15] See pp. 378-382 above.

[16] See Chapter 18 above and cf. Gustaf Wingren, *Luther on Vocation*, trans. Carl C. Rasmussen (Philadelphia: Muhlenberg, 1957).

It is for this reason that my conduct can never claim to stand justified before God in its *de facto* form. Even the doctrine of the two kingdoms is not to be understood as though the alteration in the divine commandments effected on account of the world were "justified" by the notion of vocation and calling within the orders, as though my conscience could be pacified, for example, by the fact that within the divinely sanctioned estate of the military it is necessary to kill. It is not merely by reason of our constantly defiant heart, but also by reason of the form of this world, that we can live only under forgiveness, by "grace alone" and without any co-operation on our part. Compromise does not mean that we have an excuse. It means that we participate in the suprapersonal guilt of this aeon.

Conscience, then, can never be put at ease through compromise. It cannot but remain disquieted, even though the divine patience sanctions, as it were, the defective and often very questionable orders in which we must act. The fact that we would see this unrest of conscience perpetuated, however, leads at once to a new problem. Worldly affairs, such as those of the politician, demand that we enter into them wholeheartedly. They will not allow us to be hampered or, as it were, reduced to "half-throttle" by the dubious character of the compromises which we must make and by the accompanying disquiet of conscience. Yet the miracle of justification in the very midst of such disquietude of conscience is that we are liberated precisely for this questionable world. We perceive—and our analyses will inevitably take us further into this area—that it is God's will to bring this world, in all its dubiety and even with the help of that dubiety, to the eschaton of its destiny. God solidly affirms even the questionable "means" of this world, those means (such as the miraculous but still questionable factor of political order) whereby he sustains the world; he affirms them in his "patience" ($\dot{\alpha}\nu o\chi\acute{\eta}$) and on the condition that they represent his "alien will." It is thus that we are freed to enter creatively and wholeheartedly into the world and its affairs.

Yet even as we do so there must always remain at least some slight realization of the fact that this world is not in order. This realization must be the salt in all our intense activity. It was present in all Bismarck's realpolitik to such an extent that it acted as an accompanying brake and corrective. Hitler did not have it, and thus venerated this world's "laws of the jungle" as if they were ultimate norms, believing that the conformity of his supposed realism would necessarily make the most successful impact on the world. In his realism of consistent immanentism he failed to discern the true secret of the world, namely, that it is ultimately called in question by the commandments of God. In the deepest sense, then, Hitler was actually setting upon the world to destroy it; his was the truly "alien" approach. Even our entering wholeheartedly and creatively into the affairs of the world is healthy only to the degree that we remain aware of the questionable character of those compromises which the form of this world forces us to make, which are not in accordance with the true will (*voluntas propria*) of God. For this reason it is incorrect to say that the Christian can proceed only at "half-throttle," or that he is so hampered that he cannot compete with the less scrupulous children of the world.

The Christian realizes that he can live only under forgiveness, but he knows too that under forgiveness he really can live and be free in spite of every refraction. This is why he has the distance of objectivity which enables him to discern with quiet realism the problem implicit in all worldly means, and hence to abstain from using them without restriction, i.e., from making a virtue of the world's necessity. This is also why he cannot set up a scale of political morality or try to say precisely and in advance (a priori) which means are legitimate and which are not. He knows that all means, and that all existence in the orders of the fallen world, remain questionable. He knows that here in this world there is no perfect righteousness, but he does not therefore draw the conclusion that everything is under the same condemnation and that everything is thus equally permissible, as though there were no quantitative distinction between reprehensible and less reprehensible, between good and less good possibilities. We have already seen that ethical decision has in fact a great deal to do with the quantitative problem, and that scales of value cannot be eliminated in the name of some abstract ("qualitative") alternative. The weighing of quantitative distinctions is certainly demanded. Bearing in mind the fallen form of the world, however, makes it impossible to assume that whatever action is determined on that basis to be the best possible automatically establishes of itself an ontic righteousness.

Roman Catholicism, however, views the matter differently. Here sin does not permeate all things. An unaffected sphere of authentic "nature" remains intact. This makes it possible to establish a natural law by means of which one may deduce and establish clear correspondences between our conduct and the commandments of God. There can thus be a plain and stated scale of what is legitimate and illegitimate.[17]

This may be seen especially in the involved casuistry of compromise. This casuistry consists of deliberations over the extent to which and the limits within which indisputably legitimate and commanded goals may be sought with questionable means. Here exact distinctions are drawn between means which obviously are morally bad and therefore illegitimate on the one hand, and those which in certain circumstances may lose their illegitimate character on the other.

On the Roman Catholic view, what is involved in the use of such doubtful means is of course a compromise. Emergency situations arise in which the commanded goals cannot be attained directly but are bogged down in a sphere of means which in "normal" circumstances would be open to question and—to put it cautiously—would stand in tension with the divine commandment. But in the emergency this compromise is to some degree sanctioned by the end which it serves; it is thus "justified," even in the theological sense of the term. Because it is thus bound to the situation of the fallen world and encumbered by its available means, it needs no particular forgiveness.

The structure of its doctrine of natural law enables Roman Catholicism to define casuistically this case of "justified" action, and therewith the borderline between unquestionably immoral means on the one hand and sanctioned means on the other. (The very fact that a penetrating casuistry is employed compels us to conclude, of course, that there is no implication in this teaching that the end indiscriminately and always justifies the means. Nevertheless, it remains true that there are certain means which the end does justify, in effect absolving them from the need of forgiveness.)

Victor von Cathrein illustrates the problem in terms of the situation of an innocent man unjustly condemned to imprisonment or death.[18] No one doubts that under these circumstances he is permitted to attempt an escape. To effect such escape, he must needs use certain means which are to be casuistically tested and weighed over against

---

[17] See Chapter 20 above.
[18] See von Cathrein, *op. cit.*, I, 326, n. 2.

one another. It will almost certainly be granted that he may "deceive his guards, break his chains, and rip open the bars." The reasoning here is typical: "If the action itself is legitimate, then the means which are demanded by the very nature of the action and which are integral, or introductory, to it, are also legitimate. . . . The legitimacy of the escape as such would be quite illusory if it did not include also the legitimacy of these means of escape." On the other hand, it is regarded as certainly *not* legitimate for the prisoner to lie, or to kill the guard.

It will be seen here to what subtle distinctions casuistry leads. One may deceive the guard, but not lie to him. Is not every deception a lie, even if it be perpetrated without words? Does it not always use the means of untruth? Do we not have to say, then, that casuistry is engaged in a hopeless enterprise when it works its fingers to the bone in an effort to chisel a "justified" and ontically unimpeachable act out of the hard and intractable material of such a situation? And why stop at the killing of a guard? Might we not see here in the need to kill the jailer a situation analogous to that of the soldier who has to kill the enemy?

It will be perceived that the clear goal which is sought is not to be achieved in this way. All the more plainly, therefore, do we see that, however tortuous the solutions, there is here a real need for casuistry, inasmuch as the doctrine that nature is violated only in part necessarily leads to considerations concerning the extent to which ontically sinless possibilities may be won from concrete reality.

In contrast to this searching about for possibilities whose realization might justify an act, we can only welcome the freedom of Reformation thinking. Although it differentiates among means in respect of their ambivalence, Reformation thinking perceives the vanity and indefensibility of all means. At the same time, however, because it knows the unlimited scope of forgiveness, it is also conscious of the Christian's freedom to engage in such ambiguous action.

It is thus that we are to understand the well-known and dreadfully misunderstood statement in Luther's letter to Melanchthon: "Be a sinner and sin boldly [*pecca fortiter*], but believe even more boldly and rejoice in Christ, who is victor over sin, death and world. We must sin so long as we are what we are; this life is not the dwelling place of righteousness [ontically understood!] but we look, says Peter, for new heavens and a new earth in which righteousness dwells." [19] The essential point here is that the sin committed in compromise is *not* described as the law of this world, a law of tragic character, and so justified. It is rather that this sin takes place in the awareness that Christ is victor over this very sin, and over the world whose structure wrings it out of me, as it were.

This awareness has two consequences. In the first place, the fact of sin, of refraction and ambiguity, is not concealed by means of a pseudo justification supposedly conferred upon us by the envisioned end. We are thus prevented from slipping into that self-righteousness and false security of which we spoke earlier. Compromise never becomes a law— the resultant in a parallelogram of forces—which is vindicated in terms of the propriety and correctness of its origin. It is not a case of compromise being justified by its conformity to natural law, i.e., by the fact that it uses "the means which are demanded by the very nature of the action." [20] The truth is rather that Christ conquers and overcomes the σχῆμα of the world, the structure within which such compromise is necessary. It is not I who justify the compromise; it is Christ who makes it right. I am dealing with an enemy whom I cannot dispatch. For casuistry, on the other hand, what is involved is always a complicated agreement with the enemy, a kind of cease-fire agreement, by means of which I hope to worm my way through. When we know Christ as victor, however, we know also the ferocity of the enemy, with whom no agreement whatever is possible and whom I cannot dispatch. The very fact that a victory by Christ himself is needed shows me this. It shows me how illusory it is to think that I can calmly walk out bearing a flag of truce into a no man's land which is not already secure in enemy hands.

In the second place, awareness of Christ and his justification prevents me from moving about nonchalantly and frivolously on this terrain of the questionable, as if there were no distinctions. I cannot, in the name of the forgiveness which presumably covers and levels all things, simply be indifferent to the various degrees of the bad and the dubious. Are we wherever possible to "continue in sin, that grace may

[19] Luther to Melanchthon, August 1, 1521. *WA, Br* 2, 372; *Luther's Correspondence*, eds. Preserved Smith and Charles M. Jacobs (2 vols.; Philadelphia, 1913-1918), II, 50.
[20] See von Cathrein, *loc. cit.*

abound"? (Rom. 6:1). Because everything is infected by sin, does this mean that we live in a perpetual night in which all cats are gray? Is it not possible for false security to rear its ugly head just as easily in connection with forgiveness as in connection with the legalistic self-righteousness of compromise?

Roman Catholic moral theology likes to raise this objection to the *sola fide* of the Reformation. And it is a charge that has to be taken seriously, for it points us to a real theological secret, namely, that there is no theological principle which cannot be spiritually abused, or, to put it even more sharply, that our spiritual existence can never be sustained much less secured by a theological formulation, however correct. It is not difficult to see how a non-legalistic *sola fide* can lead to unspiritual complacency just as easily as can a legalistic theology. The Law produces false security by the illusion that I can and do satisfy it; but freedom from the Law can conjure up the same *securitas* by way of the false deduction that there are no grades of sin, and that I thus live *in re* in the zone of indifference. Both theological positions may be made the starting point for an abuse.

We know that when the Gospel is not actually seen in its continuing tension with the Law, it does, in fact it must lead to this infamous belief in "cheap grace" (Bonhoeffer), and therewith to complacency and frivolity.[21] The Gospel without the accompanying antithesis of the Law turns forgiveness into a mere state of indifference.

Paul provides the decisive argument against this self-confident view of grace when he says, "Do you not know that if you yield yourselves to any one as obedient slaves, you are slaves of the one whom you obey, either of sin, which leads to death, or of obedience, which leads to righteousness? (Rom. 6:16). Relationship to the crucified and victorious Christ means that "we are buried with him by baptism into death," and that "as Christ was raised from the dead by the glory of the Father, we too should walk in newness of life" (Rom. 6:4; cf. I Pet. 3:21; Col. 2:12). This means that with Christ we are dead to "the elements of the world" ($\sigma\tau o\iota\chi\epsilon\hat{\iota}a$ $\tau o\hat{v}$ $\kappa\acute{o}\sigma\mu ov$, Col. 2:20). We do not belong to them any more. They no longer constitute for us a normative orientation, in the sense in which we could recognize, reverence, and implement the resultant in a parallelogram of forces as a final norm, or in the sense in which the element ($\sigma\tau o\iota\chi\epsilon\hat{\iota}o\nu$) of com-

[21] See Chapter 7 above.

promise, as the primal law of all life, can remain—if we may venture to put it thus—simply an "idol" (I Cor. 7:23).

Christ was hung upon a cross because this aeon cannot endure the absoluteness of God. How then can we belong to, or accept the norms of, an aeon which put to death the Son of God? We can be related to this Son only if we die and rise again with him, and so are subject to his norms.

When we are told that though we still live in the world we are dead to it, and that we live within it only in virtue of the name and dominion of the death of Christ, we are thereby caught up in a spiritual movement. This movement allows us on the one hand to go about our worldly pursuits relaxed and unconcerned. It lets us "sin boldly." It takes from us our long-standing anxiety about getting our hands dirty. On the other hand this movement also reminds us of him from whom we have this freedom of orientation, namely, the Lord who died for this very world and who draws us into his death that we too may be dead to the world.

Thus the power which gives us our freedom for the world is the very same power which also limits this freedom. The word "limit" is, of course, inexact, for it might suggest the kind of freedom which a father grants to his son: "Yes, you may do such and such *but* . . . don't go too far." This is not the freedom we have in view here, for such freedom would imply a limitation of freedom from sources outside the Gospel, specifically from the Law. The "Yes . . . but" would be a complex of opposites which would rob the Gospel of its pre-eminence as being all in all.

We have thus to remember that the Gospel of the cross is at one and the same time freedom and bondage, both rolled into one: "If you yield yourselves to anyone as obedient slaves, you are slaves of the one whom you obey." This means that you can have legitimate freedom in relation to the world only as you are subject to the Lord of the world. Without this subjection there are only two possibilities. The one is the anti-worldly posture of the "religious," who keep aloof from the world out of fear, and who thereby represent indirectly a form of the worldliness which is the exact opposite of freedom. The other is an "enthusiastic" or "fanatical" obsession with the world, i.e., the kind of surrender to the world which makes idols of the world's forms, relating itself absolutely to the relative. This would really be the same kind of bondage, and hence unfreedom, but seen from another angle.

If, however, I have my freedom in relation to the world—and therefore my freedom for the world—only in the name of him who is its Lord and has "overcome" it, then I have this freedom only as one who is bound. Whether I am free or not is not determined by whether I am bound or not. It is determined wholly by who or what it is that binds me. Freedom is a mode of bondage. It is not synonymous with the absence of bondage. The very absence of bondage itself binds me to a law, even though it be only that law of least resistance by which I may live, and by which I am subjected to my impulses and instincts so that, in Kantian terms, I forfeit my autonomy. (The slavery of the prodigal son in the "freedom" of the far country is a classical example.)

Consequently the limitation of freedom is given through and in him who gives me freedom. It arises out of what we have just called a "movement," i.e., a free going forth into the world in the name of him to whom we are bound. Here it becomes perfectly clear that the freedom of *sola fide* is the exact opposite of indifference. We live in the dawn of the new world in which the contours are sharply outlined. We do not live in a dark night in which all cats are gray.

Obviously this does not mean that there can be a casuistically calculated structure of conduct comparable to that which Roman Catholic moral theology sets up in answer to the supposed fanaticism of the Protestant *sola fide*. We are not bound to casuistical calculations. We are caught up in a movement. This fact is brought out even in the very terminology of certain New Testament paraeneses, for example, in the reply of Jesus to a casuistically speculative question of the Pharisees which was designed to pin him down: "Render to Caesar the things that are Caesar's, and to God the things that are God's" (Matt. 22:21). Here Jesus emphatically refuses to map out a legally structured middle course between Caesar and God. On the contrary, he appeals to our awareness of who God is, of who and what—as a result—Caesar also is, and of Caesar's limitations. It is with this awareness of the divine empowering and also the divine relativizing of Caesar that we are to pursue our course. Bound to God, we are free both for Caesar and from Caesar. What this amounts to then is a movement in which we are caught up, a decision which confronts us at every moment. It is by no means a stock solution for the theory and practice of political ethics.

Over against the law of compromise in this world, therefore, we must always keep an appropriate distance, realizing that the fallen world does not bring forth of itself means that are commensurate with God, and that our obedience in worldly matters is always fragmentary, a kind of "alien obedience." In our newly received freedom as the children of God, we may certainly turn actively to a world which is accepted just as it is by the patience of God. Even this relative world may be "ours" because we "are Christ's" (I Cor. 3:23). And yet there always burns within us the petition "Thy kingdom come!" (Matt. 6:10) and: Deliver us from all that is inauthentic, indirect, and provisional.

Evangelical Proclamation: The Problem of the Mass Society

Now that we have noted how compromise invades the religious sphere in Roman Catholicism, a questionable figure like that of the Grand Inquisitor being a good illustration, readiness for self-criticism demands that we should examine the same problem in the context of the Reformation churches as well. We may do this by investigating the particular frictions to which the Christian message is subjected in the age of the mass society. Might it not be that with increasing secularization the question is inescapably forced upon us as to how the Christian message is to be presented outside the ordinary parish situation, on the street-corners, over the back fences, and in the barracks and dormitories?

In the ordinary worship service the message can be presented, as it were, "directly." For here people are gathered who are ready to listen. There is here a congregation in which the Holy Spirit has touched men and acquainted them with the Word. At this level the only compromise involved is of the purely formal kind, which is theologically irrelevant and which may be numbered among the things which are adiaphorous, matters of indifference. For example, here it is only a matter of formally "accommodating" the Word to the structure of a given congregation, be it industrial, rural, or suburban. Some might justly argue—and we are inclined to agree—that with reference to this kind of accommodation, inasmuch as it lacks the essential mark of an adjustment and alteration of content, it is wrong to speak of "compromise" at all. Nevertheless, it is to our purpose to use the term even here because, if we begin with the innocuous formal kind of compromise, we shall later be able to see even more clearly what is involved in the stricter compromise of content.

The moment proclamation becomes explicitly missionary in character—and this is true for Protestants and Catholics alike—the question of adapting content becomes acute. This is particularly true when the message is to be addressed to the largely depersonalized mass man of our day. For mass man is not merely estranged from the spiritual content of the message which is to be addressed to him. He has also little or no relation to words as such, i.e., to words as a means of declaration and communication, to the human word. Will we not first then have to find ways of reaching him apart from words? And if this is actually done, will we not have effected already a considerable compromise with the situation? Hence what we have here is a good case in point which will make vividly concrete our earlier consideration of fundamental principles, putting them necessarily to a practical test.

By the term "mass man," of course, we refer primarily not to the quantity of men involved but to a particular quality of man, specifically to that man who acts not on the basis of free and personal decision but largely on the basis of animal instincts common to all men, on the basis of auditory and visual stimulation of the nervous system by, for example, picture magazines and phonograph records. He acts as an individual, to be sure, but as an individual who is essentially undifferentiated from the rest of the pack, and so determined largely by his herd instincts. Now this situation of the mass man poses a real problem for the church and its proclamation.

When we say that for mass man words have a completely different quality—the "word" has lost its distinctive significance—we are not thinking of such terms as might be used in arriving at a factual understanding between men, between a motorist, for example, and the mechanic who is to repair his car. As thus used, words of course remain unaltered. It is when words are used to convey convictions, however, that they are supremely threatened and subject to profound alteration.

The popular demagogue who sways the masses with his oratory will usually make use of lofty ethical concepts like "willingness to sacrifice," "to get on with the task," or "heroically to shoulder the burden."

He may even appeal to religious convictions. But what he means is something quite different from what he says. He simply uses these normative concepts denoting conviction or attitude for purposes of psychical stimulation, much as a trashy film might introduce a little religious scene, with a bit of organ music and the dim light of a church, in order to produce certain predictable emotional effects but without touching in the slightest the viewer's conscience, the central core of his person, and therefore without depriving him of any portion of the full ninety minutes of stimulation he originally purchased.

Particularly characteristic of this alteration of the substance of words is that decadent phenomenon of language, the catchword or slogan [Schlagwort]. This kind of word is devoid of soul. It has lost all convictional content. It operates like the cues an animal trainer gives his charges when he drills them in their routine. The trainer does not convince his lions. He simply fires his blanks and cracks his whip. He tames the beasts by means of "impressions" incessantly repeated. In the same way, the slogan digs its impressions into the human soul until the desired effect is obtained. For example, the old Nazi slogan, "The Führer is always right," was never intended as a statement of historical philosophy. It did not depend on any convictional content. It was simply designed to implant a suggestion, to make a psychical dent, by its power of penetration and by the rhetorical impact of incessant repetition.

The man who is worked over in this way becomes increasingly unreceptive to the word as a purveyor of convictional content. Even mass man can still retain a kind of scepticism; he can be very intellectual in his own way. Hence when a word of convictional content is spoken, he suspects that ulterior motives and trickery must be involved. For example, he thinks that priests themselves do not really believe in confession, that it is simply a way whereby they can enhance their own power; they are trying to pacify the socially weak by promising them recompense in heaven.

We are thus introduced to the problem which confronts the church in this age of the mass society. For the church, at least in terms of the Reformation understanding, the Word is the focal point of its action and commission. This Word has a decidedly personal character. Indeed, it is itself person—remember that Christ is called "the incarnate Word." The personal character of this Word may be seen clearly from the fact that it is not simply a word that factually informs the hearer, as is, for example, the word of the mechanic explaining to the motorist what is wrong with his car. The Word of the Gospel is an efficacious Word. Christ does not simply lecture on the remission of sins, or on the relation between guilt and expiation, as a philosopher would. He effects remission and judgment: "Your sins are forgiven! . . . Your sins are retained" (Mark 2:5, etc.). The Word which Christ speaks actually brings it to pass for the person to whom it is spoken. It is for this reason that the New Testament speaks of the power and authority of this Word rather than of its logical cogency.

The problem of "the church and the mass society" arises out of the fact that in the Word's field of force two completely different worlds collide. On the one hand there is the Word of the church, the Word of the Christian message, which addresses a man personally, affects the central core of his person, and is efficacious. On the other hand there is the world of empty slogans. The difference between these two spheres is so great that the world addressed by the church does not merely have difficulty in "believing" its Word. That difficulty is nothing new; it has been connected with the essential nature of the Word from the very beginning. The trouble today, however, is that our mass society cannot even know this Word in its character as word, i.e., as a purveyor of convictional content. The world today can no longer grasp the Word as the bearer of a conviction different from its own. For mass man the Word of the Gospel is like Chinese music. He may think that he understands it, regarding it as a relic of a world long since perished. Or he may take it to be nothing but a politically refined ideology, or a trick of the clergy. But either way there is no real understanding.

The question of how the church can make itself intelligible to the world is one of the most urgent questions confronting Christianity. It is not at all the same as the question of how the church can lead men to faith. It is in fact a more elemental, preliminary question. But this preliminary question has about it the urgency of the rudimentary. Of the many different ways in which the church has repeatedly tackled this question we shall mention only two of the more extreme.

On the one hand, the church, or a particular theology, begins with the thesis that this question does not concern it in the least. It has simply to proclaim its Word, trusting that the Word will find its own entrance into the heart, indeed that it will create its own audience. Every attempt to consider methods whereby avenues of access can be opened for the Word is regarded as a supplementary human act of co-operation indicative of a lack of faith. It implies that the authority of the Word is no longer trusted if the Word has to be supplemented by petty tactical measures. Through such an approach the Word is said to be divested of its character as subject and to become the object of human devices. If there is brought against this whole thesis the argument of experience, namely, that the masses are not reached by the Word which is preached thus directly and with no specific goal, that they do not even come within earshot of it, the stock theological excuse is quickly given that this is precisely the "offense" of the Word, the normal reaction whenever the Gospel is faithfully proclaimed.

Behind this thesis there may well be a sound conviction, namely, that we must not intermingle the divine and the human, the eternal Word and the temporal methods of proclamation, that these two spheres are not to be fused in some sort of compromise. By its very concern to keep the Word "pure," however, the church is inadvertently keeping men from being touched by it, with the result that the humanistic secularism the church was trying to prevent is thereby advanced. Moreover, there is here the threat of a magical view of the Word, a view which borders closely on lovelessness insofar as there is hardly a trace of any real compassion and concern for the sheep without a shepherd. Magic has regard only for the power of that which contains the *mana*. For magic, there is no such thing as the problem of "human" contact. But can we really bring men to faith when we ourselves lack the passion of love? Is God really to be found where he is preached in his intrinsic deity? Is he not rather present where the preacher is caught up in the fierce conflict of love, earnestly seeking how he may bring the saving Word to men and make it heard, even though the resultant planing causes all sorts of illegitimate chips to fly?

The second and radically opposed thesis is that, after all, Christ himself left heaven and came down into the distress and ambivalence of the human situation. We ourselves as preachers, it is said, must therefore engage in the same downward movement. Like Christ, we are to seek out men where they are. The secret of the Gospel is condescension. And the essence of condescension is that man is brought back. The bringing back presupposes that man is sought out where he is, the taxgatherer at his miserable trade, the Jew in his desire for miracles, the Greek in his wisdom, the Pharisee in the self-confidence of his ethos. If the man of today has become mass man, if the Word has lost its specific significance for him and he is wholly at the mercy of slogans, then for God's own sake the search for him must begin precisely on this level.

This is the view, for example, of the Salvation Army (and to some degree of Moral Rearmament). The argument is that mass man cannot in the first instance be reached by the convictional appeal and the word proclamation of stately church music. We must rather speak to him initially through the popular melodies with which he is familiar. We must accommodate ourselves in some measure to him and try by devious routes to find him. Since even the very songs of mass man have been reduced to the bare bones of rhythm, with only a skeleton remaining to excite the nerves, the Word of the Christian message must also be brought to the ear of mass man by way of the nerves. It must be reduced to the form of mere rhythm, a form in which it can be taken in—through the ear! Once he has been induced to listen, once he has been roused out of the apathy of indifference, then he can gradually be brought forward from the mere religious impact on his nerves to the central content of Christian conviction.

The weak points in this second thesis are obvious. If such a method is followed, is not the Gospel in danger of becoming merely a matter of psychotechnical calculation? Does not this attempted method of accommodation mean a suspension, at least temporarily, of the essential affirmations of the Gospel, and especially of its form as word? And does this then not amount to a compromise in content? Is it really possible to break free again from the serpentine cunning of such attempted assaults on the soul? Is not the Word in this instance truly robbed of its authority and power, and filled instead with nothing but human, all too human tactics and energies? And has not the love which presumably prompted such attempts in the first place secretly become nothing but a human, all too human love, a love which has forfeited its divine traits because, instead of being the practice of faith, it has brought about the weakening of faith, even usurping the place of faith? Do we not see here the destructive form of compromise, at least when it is dealing with ultimate reality?

The urgent question thus arises whether these two extreme positions represent the only possible attempts to solve the problem of the church and the mass society. If that were the case, the outlook would be very bleak indeed. In fact we would appear to be utterly bankrupt.

One point at least is clear: Jesus Christ wills that the mass be disentangled. He does not view man as merely one member of an ant colony, as merely a particular specimen of his genus or species. He views man rather as a child of God of infinite worth. He views him as one who must die alone, who bears his sins all by himself, and who can only receive forgiveness individually and personally. Each must come to God in a radical solitariness. No one can bear for me, or share with me, my eternal destiny. It is for this reason that on a Christian, and especially a Lutheran view, even the orders of the world, e.g., state, family, or society, do not have the character of a suprapersonal institution in which the individual is done away. On the contrary, the orders are conditioned and sustained by personal love. It is not a case of personhood being overcome but of the person really coming into his own. God does not deal with the mass. He knows only the individual with his eternal destiny. This is why there is "more joy in heaven over one sinner who repents than over ninety-nine righteous persons" (Luke 15:7). Pre-eminence is thereby attached to "the individual"; he is radically singled out from the mass.[22]

To conclude from this that, since its message is designed to overcome the mass society, the church should ignore the mass situation as such, seems to me to be more than doubtful. If the Word which makes man an individual before God and summons him out of the mass is not even perceived, if its claim is not even heard—quite apart from the question of whether men respond to it at all or receive its blessing—then the process of disentanglement cannot even begin. We therefore believe that the church must have "forecourts" to its worship, mission stations, as it were, to which mass man can be "brought back," and in which ways of working will not be despised that are designed to make him hear. These forecourts will be the street-corners and market places, the barracks and backyards. Nor should

[22] Naturally this "individual" is something very different from what so-called individualism understands by the term. What we have in view is probably best expressed by Kierkegaard's category of "the individual."

we be embarrassed at this point to cite the quiet ministry of the Salvation Army as an example of this kind of rescue operation. Jesus Christ has not required that men must have a feeling for the music of the spheres in order to be worthy of heaven. He does not demand that we screw ourselves up, as the mystery religions do with their ecstasies and processes of purification. He sought men where they were. The incarnation of the Son of God, and therefore the Gospel, is—to put it quite simply—nothing else but this gracious seeking and bringing back, this miracle of brotherly solidarity. It is from the incarnation, from the Gospel, that we learn of the duty we owe to mass man, to seek him and bring him back.

Now we ourselves have already raised an objection to this line of thought. We have said there is a risk involved here that the church will cease to operate in a spiritual way that is appropriate to the matter itself, and will instead make use of psychological tricks which betray a lack of faith. The danger is that it will engage in propaganda rather than mission, and that by apparently seeking a compromise between Word and situation it will actually end up compromising the Word. This danger is unmistakable.

It is true, of course, that all Christian work involves an element of danger and is exposed to the risk of intervention and dilution by the devil. There is in propaganda, however, an element which clearly goes beyond this kind of danger and should by all means be avoided. To take an extreme example, if we consider the principles of Goebbels' propaganda we find that it manifests actual contempt for man. It presupposes that man is not a person, but merely a nerve complex to be activated and controlled. There are specific rules for the manipulation of this nerve complex, rules which can be learned, and which ought to be part of the art of governing. The exercise of tyranny is thus a matter of technique, a technique which functions more efficiently when the incalculable factors in man, e.g., his personal decisions of conscience, are set aside, and he is regarded exclusively as a mere complex of nerves. It is then that man becomes subject to the influence of the calculable and calculated mechanics of propaganda. This means that the propagandist must always try to make man increasingly into the mass man, a mere nerve complex. Men of personality and conscience are like sand in the machinery of propaganda. They hamper its smooth operation. This is why propaganda logically involves a deadly hostility toward any ultimate religious commitment.

Once we realize this, it is clear that the effects sought in the "forecourts" to which we have alluded, while they may resemble propaganda, do not possess the specific features of propaganda. For here there is no attempt whatever to cultivate or manipulate a nerve complex; this would be to intensify the mass situation. On the contrary, what is sought is simply an avenue of access into this mass situation which may then be used to open it up from within, or more properly, to let it be opened up by the power and authority of the Word which has now penetrated into it.

We realize that at this point we have touched upon a large number of practical and theoretical problems. Our purpose here is not to delineate these problems in detail, much less to solve them. We are concerned simply to show that the church is in fact confronted here by a real problem.

One further point, however, needs to be strongly emphasized: We must not think that Christianity can be used to disentangle the mass. Christianity can never be "used" at all—though attempts are constantly made in this direction. Some have tried to use it to make people governable, to produce good pliable subjects who will not complain but willingly bear all sorts of social ills for the sake of the recompense they are supposed to receive in heaven. It is on the basis of such thinking that men talk about "the need to maintain religion among the people." But all such attempts are doomed to failure, not simply because the objects of such religious trickery soon discern the true purpose and react accordingly, but also because faith, by its very nature, refuses to be made the means to an end. On the contrary, faith requires a personal decision. Only he who has faced up to such a decision has the right and the authority to demand it of others. If we plug for Christianity merely because it might make good subjects, or because it might fill the urgent need for an ideological foundation for "Christian civilization," or even because it could help to overcome the mass society, this matter of decision is bypassed altogether and we simply look on, calculating and untouched, from the wings.

This is why a religious program for Christianizing culture as a solution to the problem of the mass society is quite out of the question.

For here we are involved in an area where all programs are useless, where in any case they can at best be not the first step but just about the last step of all. The first step is the appeal to individual decision and hearing. Those who in their concern for method would proceed differently are in effect themselves encapsuled in the mass they wish to help. Only those who have become individuals *coram Deo,* who stand in that unique situation of judgment and grace where none can represent them, only they have a liberating word to speak. Before we can issue a liberating call to the age of the mass society, we must first have received and obeyed a call ourselves.

We have discussed this question of "Christian proclamation and the mass society" as an illustration of the wider problem concerning the necessity for compromise at the very heart of the church's life, namely, in its proclamation of the Word. Several points have become clear. There is indeed a certain compromise involved when the word form of the proclamation is played down in favor of making an impact upon the nerves, as in the Salvation Army. This does not mean, of course, that the Word is suspended, even temporarily. For the Word is contained also in the pulsating songs, and is not absent even in the emotional and subjectively oriented addresses. Nevertheless, the emphasis is placed on affecting the nervous system. The message penetrates, as it were, through the spinal marrow rather than the ear. All this represents a concession to the particular situation of those to whom the message is addressed in this age of the mass society, and to that extent it is a compromise.

The particular form of this compromise may perhaps be formulated along the lines of Martin Kähler and Julius Schniewind, who often speak of the church abandoning in its proclamation the principle of *sanctum sancte,* i.e., of preaching sacred things in a sacred manner appropriate to them, of proclaiming the Holy in a "holy" way. This principle presupposes that the "manner" is part and parcel of the "thing" itself and cannot be separated from it. To this extent the compromise would in fact involve a serious defection. The critical question then would be whether the church's proclamation should stoop to the level of trash because that is the level of the mass society it seeks to reach. Would not the church thereby discredit also the content of what it has to declare?

Now if we do in fact regard some sort of compromise as unavoidable—and we ourselves felt forced to admit that there must be such "forecourts"—then we must at the same time reassert the evangelical reservation which exists with respect to all compromises, and not let compromise become, as it were, a law.

First, this reservation is to the effect that we are not to seek here, in the form of a calculable "middle course," a permanent synthesis between the authentic and proper content of the proclamation on the one hand and the limited receptivity of mass man on the other, whereby the prospects of successful results would presumably be enhanced. If the church is true to its message, it cannot allow the scope of that message to be dependent on this type of opportunism. That is to say, it cannot make its provisional mode of address a kerygmatic *norm*.

Second, the church must realize that these forecourts are literally *fore*courts, and that to those whom it invites into them it owes the explicit and full form of that which in the first instance is only implicit—"implicit" in the true sense of the term, that is, "enfolded in" or "wrapped up in" a form which they can assimilate. The church does not of itself determine, according to pragmatic syntheses, what should be the content of its message (as the "German Christians" of the Third Reich clearly tried to do). It simply preserves intact the purity of the Word. And in so doing it becomes aware that the compromise is merely provisional. Indeed, it bends its every effort to lead those who are detached from the mass into the sanctuary itself.

Third, it is paramount for a church which is thus bent on mission that the compromise not be undertaken under the enchantment of calculations about the chances of success, but on the basis of an awareness of the gracious purpose of Christ in the incarnation to bring men back. The compromise thus takes its cue and direction from an action of God himself, an action which we rightly hesitate to describe as a "compromise" (though we shall have to do so later—with tongue in cheek!) because even while coming to men and accommodating his deity to the schema of this world Christ still remains mysteriously aloof. Having this as its basis and direction, compromise is protected against becoming pragmatic and projecting a Christianity conformable to the age of the mass society, a new confession, as it were, which would be determined to dissolve the old confessions. The "German Christians," and in many respects the Religious Socialists as well, show us what compromise is like when it does not have this sort of basis, when instead it takes as the content of its confession certain calculations about possible outcome. Where the condescension of Christ is repeated through an accommodation which takes the form of those forecourts, there remains a lively awareness that what is involved here is really an act of mercy in a fallen world, a world which has not yet recovered from its fall. We are thus prevented from regarding compromise as something absolute, and compelled to recognize instead its provisional character.

Fourth and finally, this purifying and salutary awareness is sustained by the fact that the community of living faith knows how vain are all human efforts and how pitiable all human calculations. It knows the miracle of the Holy Spirit, who alone can open up the Word. The human effort put forth for the sake of accommodating mass man can just as easily lead past the Word as to it. Which of these two possibilities actually takes place is decided solely by the work of the Holy Spirit. Now this cannot mean, of course, that the Holy Spirit condemns us to passivity or absolves us from obligation to cope with particular problems, e.g., this problem of proclamation. For he is indeed, as Paul Gerhardt says, a "spirit of love." It is he who lays upon us the obligation of seeking out our brother. At the same time, however, he makes it clear to us that man is not converted, or opened to the possibility of genuine hearing, through the calculations of mass psychology or an approach to preaching which finds therein its orientation. He thereby draws the poison out of the fangs of compromise. He blunts it. He does not allow compromise to become the mechanical "resultant" of which we spoke.

This should make it clear that the decisive question is not whether compromises are to be made. They arise in fact at every turn. The question is rather how we are to understand them, whether we are able to avoid solidifying them into something ultimate and can comprehend them instead as phenomena of the fallen world, that world which continues to exist under the patience of God. The question is whether we perceive the relativity of compromise and are graciously preserved by this perception from bondage to the law of the world—and all in the name of a day that is coming when we shall be enabled to see him without any accommodations, "as he is" (I John 3:2; I Cor. 13:12; II Cor. 5:7).

27.

Compromise and the Limits of Truthfulness

It is impossible to make lying even a subject of ethical discussion, if the person who tells the lie is acting on the maxim that the purpose of his speaking alone determines the truth or falsity of what he says. Voltaire has something to this effect in his letter of October 21, 1736, to Thiriot: "Lying is a vice only when it effects evil; it is a great virtue when it accomplishes good. So be more virtuous than ever! One must lie like the devil, not timorously or only occasionally, but confidently and constantly. Lie, my friend, lie!" [1] Lying can be the subject of ethical discussion only when we are clear on two points.

In the first place, we must acknowledge that there is genuine conflict between truth and love, between truth and practical necessity. A well-known example used by such diverse writers as F. H. R. Frank,[2] Fichte,[3] and N. H. Søe,[4] involves the question of what one should do when a critically ill woman asks about her dying child, and there is a strong possibility that a full disclosure of the truth will kill the mother. For our present purposes the important thing is not to resolve the conflict but simply to point out that a real conflict exists.

As a matter of fact, the three authors cited give very different answers, in accordance with radically different criteria. Søe thinks that the duty of love takes priority; as a test he commends the consideration that if I withhold the truth from someone, he will come to know this later on when the situation changes, and will then lose all confidence in me. Frank says that the truth should not be withheld because the death of the child is from God, and God can spare the mother any harmful consequences of the disclosure. Fichte says abruptly: "If the

[1] The quotation from Voltaire's *Oeuvres complètes* (Paris, 1818), XXXI, 446, is cited in Victor von Cathrein, *Moralphilosophie* (Leipzig, 1924), II, 90, n. 8.
[2] See F. H. R. Frank, *System der christlichen Sittlichkeit* (Erlangen: Deichert, 1884), I, 421.
[3] This question was put to Fichte by H. Steffens, according to H. Martensen, *Christian Ethics* (*Special Part, First Division: Individual Ethics*), trans. William Affleck (4th ed.; Edinburgh: T. & T. Clark, n.d.), p. 217.
[4] See N. H. Søe, *Christliche Ethik* (Munich: Kaiser, 1949), pp. 277-278.

woman die of the truth, then let her die!" Though the solutions differ so widely, they are all at one in seeing that what is required here is decision rather than casuistic calculation.

A real conflict always raises the question of compromise, namely, which of the two colliding postulates should be given priority over the other. In certain circumstances, therefore, there is in fact a relation between the truth and a specific situation, a relation which I cannot overlook.

The second point on which we must be clear is that there is a relation between "truth and the person" as well as between "truth and the situation." A statement may be objectively correct and still not be "true," if the one who utters it is not qualified to do so. When a schoolboy writes in his essay that Goethe is the greatest of all poets, this may be correct, but as stated by these lips or by this pen it is not "true." That is doubtless what Nietzsche meant when he said that "a toothless mouth no longer has the right to the truth."

In this connection we may also remember Kierkegaard's comparison. When, at the end of a life devoted to the pursuit of knowledge, Faust asserts, "I now see that we can know nothing," this conclusion, formulated at the end of his development, is qualitatively very different from the same statement taken over by a student in his first semester to justify his indolence. What is "correct" about the utterance in both cases is too trivial to deserve being called "truth." Correctness becomes truth only when it is so related to existence as to become the confessional expression of existence, its self-declaration. This is why proverbial truths, when uttered by school children or indolent freshmen, do not stand under the sign of truth. They are characterized either by "precocity" or by trickery, and in either case by an element of mendacity which is the very opposite of truth.

One could cite other examples as well to show that the objective correctness of a statement is not necessarily identical with the truth. Thus a notorious liar can execute a highly refined movement of deception simply by saying for once what is correct. Again, when we are confronted by customs officials who suspect that we are smuggling something over the border, and who begin to close in on their quarry by applying the customary psychological pressures, honesty can often be a most effective device to help us over a moment of alarm and even make possible a rewarding breach of the customs regulations.

Here again we see how truth is bound up with the person. The term "truthfulness" is itself expressive of this connection. Before we discuss the question of lying, and especially of what is called the "white lie" [*Notlüge*], we thus need to consider the prior question of what it means to tell the truth.[5]

Authentic and Inauthentic Accommodation in Telling the Truth

Truth is not linked merely to the agreement between a certain judgment and the objective and verifiable facts, as positivism maintains. For positivism, the criterion of truth consists in the correspondence between what is said and the ontic data. In point of fact, however, we have spoken the truth about something only when we have made, not merely a correct statement concerning the form in which it is actually to be found, but also a relevant statement concerning its meaning or "horizon." For example, if we define man as a "two-legged animal without feathers," this is a correct statement of the data. Nevertheless, this statement does not express the truth about man, for it simply "lumps him with a plucked hen, a kangaroo, and a jerboa."[6] Such a partially correct description really misses the "point"; it fails to express the truly distinctive feature of man. The statement does not use the pertinent "historical" category of singularity but the generalizing category of natural science which divests its subjects of uniqueness and subsumes them under a general law.

The truth about man can be stated only when a decisively relevant statement is made concerning the meaning of his existence, concerning his whence and whither, in short, concerning the "horizon" of his existence, concerning his "nature." This is why a mythical statement, such as that which fixes man's place as somewhere between the Olympian deities and the gods of the underworld, may come much closer to the truth about man than the correct vacuities of positivism. For in mythical diagnoses the nature of man is "truly" grasped—in terms of the Greek understanding of the world to be sure—even though the means used to state this view do not measure up to what

[5] See Dietrich Bonhoeffer, *Ethics*, ed. Eberhard Bethge, trans. Neville Horton Smith (New York: Macmillan, 1955), pp. 326-334.

[6] See Norbert Wiener, *Mensch und Menschmaschine* (Frankfurt-Berlin, 1952), p. 14.

is expected of an objective judgment, namely, agreement of the statement with verifiable data. For objectively, or geologically, Olympus is a protuberance on the earth's surface; it is not, in the same objective sense, the throne of the Hellenic gods. The Greek statement concerning man's position between Olympus and the underworld, however, does get at the very heart of man's existence, and even though as Christians we may define the nature of man very differently, we cannot bring against these diagnoses from myth and tragedy the accusation which must be brought against positivism, namely, that they completely miss the point of the truth in question. Thus it is altogether possible to state the truth with the help of what is objectively incorrect, and to miss the truth with the help of what is objectively correct.

We have also to consider that means which are questionable because they are mythical, means which do not measure up to what is expected of an objective judgment, may be chosen for two different reasons. One possibility is that we may be living in a relatively naïve stage of human history and really believe in the myths (like Homer). The other is that we find it impossible to express objectively the truth of man which transcends external data, and hence willingly resort to the indirect and symbolical statements of the myth or fairy story, even justifying such practice epistemologically.

In the latter case, i.e., where there is a conscious choice in favor of mythical expression even though we live under the sign of the enlightened logos, we have actually effected so far as the form of our statement is concerned a compromise with the images, ideas, and symbols available to man. Indeed the compromise is between the object of our statement, the "beyond" ($\epsilon\pi\epsilon\kappa\epsilon\iota\nu\alpha$) which transcends the objective world, and the means which are oriented to what is actually present in the world. This compromise is effected in order that, by renouncing the correctness of objective statement, we may reach the true goal, namely, the truth as a disclosure of meaning. Sometimes it is possible to translate what is expressed mythically into the language of philosophy. Such a switch in modes of statement may be made with difficulty, for what is intended in such cases is the same truth, even though it is expressed differently.

It may be seen from this that accommodation in the form of a statement does not represent a questionable compromise, because the intention is still to express only the truth, using, however, means which are subjectively at our disposal and also objectively appropriate to the matter itself. Lessing thus expresses a profound insight when he says that it is God himself who effects such an accommodation: God adjusts the truth to the prevailing state of consciousness, e.g., the mythical. No less justly we may say that it is truth which integrates itself into the state of consciousness. In other words, truth—except in the exact sciences—is always historical. It is always bound to a given situation. It is never present in timeless immediacy.

The truth itself is certainly not discredited simply because the means used to express it may not meet the criterion of objective correctness but are instead adjusted to a given state of consciousness. A clear-cut indication of the fact that truth remains unaffected by the means of expression is that it is relatively easy to change the mode of expression without altering what is really said, and that confidence is thus maintained if, for example, a teacher first uses the indirect form of a fairy story and then subsequently a more direct method to communicate to a child the same truth.

To take a very pertinent example, a mother may answer the question where babies come from with the hoary legend of the stork. Then when she is asked why she had to be in bed when little brother came, she may have to seek refuge in the fresh evasion that the stork bit her. But this has nothing whatever to do with accommodation; it is in no sense a white lie but simply an out and out lie, baldfaced, blatant, inexcusable, and tasteless to boot. A lie may be white and excusable only when fellowship with the other person has broken down and he has forfeited his claim to the truth,[7] or when for the sake of the other, out of my love for him, I think I must spare him the hurt that telling the truth might inflict. That lie about the stork, however, is simply a cowardly and egoistic evasion.

Naturally a child cannot yet understand the biological phenomena of conception and birth. There is thus no point in giving him the objective truth about the processes involved. This is why the mother's answer will inevitably involve some kind of accommodation. For the child, truth must take the form of a compromise with that which the child can understand, and bear. The question thus arises what kind of compromise is legitimate in this case. It will certainly not be the

[7] See p. 531.

lazy kind of "cease-fire" compromise represented in the stork fable.

We can make progress in this matter only if we remember that in his question the child is facing for the first time the mystery of man, i.e., the question of his whence and whither. Fundamentally, the question is not meant to be biological! The category of the biological is in fact quite alien to the child. Analogies we might suggest from the mating and birth of animals will throw little light on the subject for him, even though he may not be totally unacquainted with such animal behavior. The child's question is at root a metaphysical one. There may also be discerned in it, perhaps, a rather vague sense of a special relation to his parents, at least to his mother. If the mother answers, then, that little children come from God, that an angel first places them as very tiny beings under the mother's heart for protection, and that they then come into the world when they are big enough, this is an instance of authentic accommodation in answering the question of whence man comes.

As an example of an "explanation" which is not only beautiful and pedagogically correct but also fully in keeping with the truth, I will quote what the mother of Walter Flex said when answering her little son Martin:

You see, my child, people come from God and return to him. Their true home is with him. Before a child comes to earth it is a little angel. . . . When God sees that two people are very much in love, as Daddy and I are, he sends a little angel to them from the heavenly pastures, that it may be their child. They are to love it, and to rejoice over it. . . . But, you know, when a little angel makes the long journey from heaven to earth to become a baby, it finds things very difficult here below. The air is so raw, and everything is so noisy, that it hurts. For a little angel is tinier and more tender than you can ever imagine. It just could not live among us. . . . So God has made other arrangements for it. A little angel which comes to earth . . . must be protected for a long time in a place which is warm and safe, until it is a little stronger and can live on earth. And do you know where there is such a place? Under mother's heart. . . . Every time mother's heart beats, the heart of the little child beats with it. [Then, when the little heart has become strong enough to beat alone, and to be able to stand the raw air of earth,] it can rejoice in all the beautiful things here below; it can love father and mother; it can become pious and good; and so find its way later back to heaven. But many weeks and months pass before it is ready for this, and it is difficult for mother constantly to bear her little child beneath her heart. She must move about quietly so as not to frighten the child. She often becomes tired, because the tiny heart of the child takes all the strength from her heart as it grows. . . . Finally the day comes when the little one can live alone, and then God takes it from under mother's heart and lays it in the cradle. This hurts at first, but later mother has so much joy in her child that she forgets the pain. . . . And you see, when a person dies, he goes back to God. . . . He returns to the place from whence he came as a little child.[8]

Now this is of course a fairy tale. Yet in such fairy-tale language there is contained the truth concerning man, in a form in which it is accessible also to the child. Here is disclosed the fact that man's origin is in God and man returns to God's presence. In his love God has a purpose for our lives. Conception and birth enter in as the means whereby his loving purpose is accomplished. They are thus relativized as mere means which can never express the true reality of human life, means through which this true reality is achieved. Reference is also made here to the nexus of generations, and to the fact that children are bound to their parents. The biological phenomenon of this nexus is again a transparency. Conception and birth are not the "creation" of a new man; they simply represent the reception of a gift entrusted by God. In motherly love there is thus reflected the loving care of him who bestowed the trust. That we are here introduced to the real theme of human life may be seen from the fact that the origin and end of life are seen together, and that they are not regarded as chronological moments, but as the divine origin and end which determine the whole course of life, including the present moment. Little Martin Flex thus comes to learn that the love and care and claim of him to whom he owes his life, is ever about him, and is palpably at hand in his parents. The truth about man's life is contained in this simple mother's tale because it sets forth the horizon of human life, because the biological facts—the truth requires this—are relativized to the status of mere media and veils of this

---

[8] Bernita-Maria Moebis, *Wer Gottes Fahrt gewagt* (Hamburg, 1927), pp. 57 f.

truth, and because the fairy-tale accommodation is thereby made a transparency for the truth. In contrast, the stork fable directs attention away from the real theme, and is thus the very opposite of accommodation.

An essential criterion of the genuineness of such an answer is whether the child has to unlearn it and start afresh when he comes to adult life and experiences at first hand the biological processes of conception and birth. Either his confidence will be shaken in those who previously misled him, or he may be able to fit his new knowledge of the objective facts more or less smoothly into what had already been told him concerning the real theme of his life. When in their earliest "explanations" parents describe the genuine whence and whither of a man's life, using legendary symbols if need be when these illumine the biological data in respect of their true meaning, so that the actual mode of operation can be temporarily set aside, their child will not have to unlearn in later life what they have told him with all the disillusionment and loss of confidence that could entail. The young adult with his knowledge of biology will simply penetrate ever more deeply into the truth which his parents once disclosed in symbolic form. For it is a mark of the truth that one cannot grow out of it; one can only grow into it. What the growing Martin Flex learns of the nature of human life with the help of his more mature reflection can be wholly congruent with the results of a retrospective, existential interpretation of the fairy tale told by his mother. This congruity between the two forms of truth is a kind of proof that what we have, even in the fairy tale, is nothing less than the truth itself, and that the truth maintains its continuity. The accommodating form does not lead to a breakdown in confidence.

Herein may be seen the fact that a compromise with childlike powers of perception does not violate the truth, even though it skips over the truth of certain objective data such as the biological events. By the same test, however, the stork fable is shown to be a lie, a deception, and an evasion on the part of the parents. It is only gradually that an attitude of irony can replace the first shattering of the young person's confidence once his parents' deception has been exposed.

The first and most significant thing to be said, then, with respect to the telling of the truth is this: The truth is not complete simply because there is a congruity between the statement and objective reality, for truth embraces also the telos, the horizon of that objective reality. Truth is always related to the final goal. If we wish to state the truth about men, we cannot be content merely to state the ontic data ("a two-legged animal without feathers"). We must also state the meaning and purpose of human life.

If we wish to state the truth about a historical event, we cannot be content merely to record the objective facts. We must also know the context in which the event takes place, and understand its whence and its whither. Thus a historical work on the Third Reich, even though it be wrong in details, e.g., if it describes Hitler as a family man or Robert Ley as an anti-alcoholic, can still be altogether true if it grasps the real context, the deeper genesis, and the historic significance of this historical apocalypse. In any case it can be "truer" than a work which totals up many irrelevant details with the faithfulness of a news reporter, but which fails to give thematic order to this sum of facts, or relates them to the wrong order. Similarly, historical criticism of the Bible as it was pursued in the nineteenth century undoubtedly saw innumerable details far more correctly than the older and more positive exegetes. And yet we think now that it saw the "truth" of the biblical kerygma far less clearly than those earlier scholars who from the critical standpoint were incomparably more naïve and less learned.

We may sum up what we have said as follows. Truth is not tied to the ontic factualities which are to be enumerated. It is linked rather with the meaning and purpose of these data. In the last analysis, it is linked with their final goal. It is thus linked ultimately with the relation of these data to God. The truth has not been stated until we state the relation of things, forms, and events to their divine "foundation."

Connected with this is the fact that we constantly fail to know the truth, and in principle must fail to know it. We do not know, for example, what "really and truly" happens, when a mother dies leaving a family of small children. We express our ignorance of the truth in such a case by saying that we do not know "why" this should happen, what meaning or purpose may be concealed in it. We can relate to the truth of such an event, not by knowledge, but only by faith, i.e., by believing that there are thoughts higher than ours

(Isa. 55:9), and that the theme of these higher thoughts is love, that love which causes all things to "work together for good to them that love God" (Rom. 8:28). We thus believe in the purpose of a specific event even when we do not know how these "higher thoughts" can find a way to this "good," or what concrete connection there is between the death of the mother and this "good." Believing in this purpose, however, we do participate—by faith—in the truth of the event. For the truth of the event lies in its ultimate goal, in its relation to the final divine purpose.

One might also state the case as follows. Truth is not linked merely to an isolated fact. It is linked rather to the more inclusive system or framework within which, and with reference to which, this fact arises. The sum and substance of the system, however, is God himself, to whom, and with reference to whom, all things are ordered. We thus have the truth—and here we may boldly say, the ultimate truth—only when we are aware of and believe in this subjection and orientation of all things to God. Only then do we get at that system or framework within which all things have their being, and consequently their truth.[9]

The Problem of the White Lie

What happens, then, when a system of truth and a system of falsehood meet in direct confrontation? A wholly new situation arises when in the one camp truth is regarded as a genuine and binding authority while in the other it is desecrated—desecrated not in the sense merely that everything which is said is untrue, that facts are always veiled and twisted and turned into their very opposite, but in the sense that falsehood reigns as a system, truth is scorned in principle as a farce, and whatever benefits the nation is regarded as "true," so that truth is completely subordinated to such things as particular goals and special interests.[10]

---

[9] Cf. Thomas, *Summa Theologica*, part 1, ques. 12, art. 8: Whether those who see the essence of God see all in God?; and 1, 12, 9: Whether what is seen in God by those who see the divine essence, is seen through any similitude? Trans. Fathers of the English Dominican Province (2nd ed.; New York: Benziger, 1920), pp. 137-141.

[10] In 1934 at Nuremberg, for example, the Conference of German Jurists declared that "right is whatever benefits the German nation, and wrong is whatever harms it." See Pierre Boissier, *Völkerrecht und Militärbefehl*, trans. Dirk Forster (Stuttgart, 1953), p. 18.

What happens, then, when these two camps are forced to meet and to have dealings with one another? Can the "camp of truth"—naturally we use this phrase with all due reserve and with tongue in cheek—encounter the "camp of falsehood" in such a way that it is not led astray by it? Will its telling of the truth remain wholly unaffected by the party to whom it is speaking, even when the encounter between them is one of conflict, indeed, when what is involved is a life and death struggle in which the issue is not simply the existence of individuals and families, but whether a whole nation will fall victim to this camp of falsehood and perhaps be ruined for generations to come? It is in this kind of situation that men sometimes feel they are forced to depart from the truth in detail in order that the camp of truth might prevail against its adversary. A case in point would be those involved in the conspiracy which led to the unsuccessful attempt on Hitler's life on July 20, 1944.

White Lies in an Unjust Situation

Now obviously, whatever we may think in principle, or in detail, concerning the legitimacy of the means of deception used in such cases, we cannot speak of them in terms of "lying." To lie is to deny the truth for egoistic reasons. Lying implies that truth has been subordinated to special interest.

What is really involved here is rather a collision of duties. There is first of all a conflict between the truth of individual statements, which one feels compelled to deny, and truth as a system, a rational order, for the sake of which one is actually doing battle. In addition, there is a conflict between the truth of individual statements and one's love for those who are brought into jeopardy by such statements, love for comrades, family, or indeed the nation itself.

Now in such cases objective untruths are indeed uttered on the grounds of a particular interest, but in a way which is very different from that of the liar. The liar lies for his own sake, on the grounds of his own special interest, thus placing interest above truth. Such men as the conspirators of July, 1944, also tell untruths on the grounds of a particular interest, but for them this interest is itself subordinate to the higher postulates of the system, to truth and love. At any rate, there can be no evading the fact that our speaking the truth, and the way in which we do so, cannot be separated from the situation in which the speaking is done. In principle, we made this point already some time ago. In the present context, however, the problem appears in its most acute form.

When we come to speak of the "situation" in which the truth is to be spoken, we may formulate the point at issue rather more precisely as follows: As there are individuals who are not yet ready for the truth, e.g., the child who initially must be conducted along a path which skirts the truth of biological processes, so there are also individuals who have forfeited the claim to truth.

What does it mean to forfeit the claim to truth? When in the totalitarian state I deceive an interrogation officer, I am not just disputing with him concerning the truth of a particular fact. On the contrary, I am resisting an attempt on my life and on all that is dear and sacred to me. The struggle between us does not have to do with the correct subsumption of a particular fact under the norm of a truth which we both accept. We are not arguing about what is to be predicated as truth in the given instance. It is rather a case of our standing on two radically different levels. As one who wills the system of truth, I am fighting for my life against this representative of the system of untruth.

In this encounter the "enemy" has lost all claim to truth. The systematic mistrust with which he conducts the interrogation is an involuntary sign that he has himself surrendered this claim de facto. His appeals to morality are not sincere. He is simply trying to outwit me by cunning or overwhelm me by threats. In so doing he actually admits that he is not engaged in a struggle for the truth but in a power struggle. He may force me to take an oath, but his purpose in so doing is not to touch my conscience or set me before the final court of the numinous. He is simply trying to harness my integrity to his wagon, to paralyze me by destroying my moral self. The oath is nothing but a moral weapon in his power struggle. A positive claim to truth can exist only when the other person subjects himself to the same claim, and we thus speak with one another on the same level. For these reasons, a white lie told in such a situation can no more be called lying in the strict sense than killing in self-defense can be called murder. Whenever an all-out assault is made on me, the attacker forfeits his claim to either life or truth as the case may be.

To sum up, it is obviously impossible on the one hand to deny the truth, i.e., to disregard its authority and to misuse it pragmatically. On the other hand it is also impossible to demand of those to whom truth is sacred that they should "honor" it, if in reality that means that they are to play along in the masquerade of untruth written by their opponents, to the point of condemning themselves to self-destruction under an alien and cynical regime. It is nothing less than this impossible demand which is actually implicit in certain methods of interrogation. The Gestapo and other similar organizations, for example, use truthfulness quite shamelessly as a trick, a mask. By insisting on the truth (in the sense of a correct disclosure of the facts), they seek to bring their victim into a conflict of conscience. By a deceitful appeal to truth—how paradoxical!—they set him against himself and so undermine his moral resistance. By fomenting inner discord on the territory of his own ego, they hope to soften him up, to increase the moral pressures, and thus to effect his downfall, just as the totalitarian state uses the oath—we mentioned this a moment ago—not in order to obligate men to clear and convincing truth but as a kind of moral strait jacket in which to fetter them.

This is also why the ceremonies of administering the oath are generally performed with enhanced solemnity, with a view to increasing their power of suggestion and their spellbinding force. This solemnity is not designed to express the dignity of that truth to which we are obligated. It is rather a psychological attack upon the morality of those thus obligated.

The decisive point is that there is such a thing as a deceitful appeal to truthfulness, an appeal which has as its purpose to plunge me into a conflict of conscience. The initiation of this conflict is to be understood as a purely tactical measure. It is based on the assumption that inner conflicts weaken. Such conflict implies that I am at odds with myself. In this inner struggle, the one side of me demands that in all circumstances I must tell "the truth," while the other side of me tries to give reassurance that the other party, in this case the interrogating officer of the totalitarian state, has forfeited all claim to truth. When the enemy—and this interrogation officer is indeed an enemy! —succeeds in entangling me thus in conflict with myself, he diverts a great deal of my moral energy into this "inner feud," and in so doing turns it away from himself. This is the strategy and the tactic behind the deceitful appeal to truthfulness. It shows us with utter clarity that on this level of interrogation truth is not respected but torpedoed. Truth becomes the mask in a power struggle which is waged with pseudomoral weapons. Under such circumstances, then, we are not merely permitted but actually required not to make our own truthfulness a weapon in this tactical play by conceding to our opponent the claim to truth which he has so obviously forfeited.

In consequence we see that certain needful compromises with the situation itself may be demanded. First, it is possible that an individual or a group has forfeited its claim to truth. This is the case when the sole remaining reason for dealing with the truth at all is to make of it a weapon, when the common ground afforded by a mutual respect for truth no longer exists and all fellowship is accordingly destroyed. Second, there is such a thing as a genuine conflict between truth as a system, i.e., where everything is arranged on the basis of a recognition of truth as authority, and the objective truth concerning certain facts which dare not be disclosed. This is the kind of situation in which a responsible conspirator in a totalitarian state might find himself. Third, there is such a thing as a genuine conflict between the duty of telling the truth and a love which believes it should spare others a crushing truth.

White Lies by Agreement

Even outside the Christian faith—except among those who ideologically champion the law of the jungle—we repeatedly find expressions of the fact that compromises of truth which derive from the exigencies of a situation are in reality a necessity, not a virtue, and that in such compromises there is revealed a secret disorder which is not to be denied as occasion demands but constantly recognized—and characterized as disorder! For obvious reasons, politics, and especially the question of the relation between political ethics and political realism, provides an acute occasion for the discussion of this problem of compromise, as well as a pertinent illustration of it.

Thus Frederick the Great, in a secret testament which was meant only for his successors, can say that the politician, faced by opportunistic opponents who are out to deceive him, has to break his word and to engage in political stratagems himself if he is to survive. According to Frederick, however, men in government should nonetheless "depart as little as possible from the straight and narrow path"; his thought is

that the breach of one's own obligations is excusable—which means that
it is at least questionable—when demanded by the welfare of one's
country.[11] He brings out most clearly and forcibly the disorder involved in these compromises with falsehood—and he himself had made
a great number of them—when in 1742, writing from the standpoint of
the Enlightenment belief in progress, he says in the introduction to his
Denkwürdigkeiten: "One must hope that a more enlightened age will
give to probity the place which is its due. I will not defend the art
of government. I will simply delineate the reasons which in my view
compel princes to pursue a course in which deception and the misuse
of power are accepted practice." [12]

Frederick does not defend all this. He simply points out that this is
in fact the practice, and expresses the hope that with the help of reason humanity will outgrow this miserable interim of compromise with
falsehood. This hope of course is illusory, but it does reveal an awareness of the fact that there is this disorder, that things are not what they
ought to be and there is need for reform. The same cannot be said of
Machiavelli. For, while he may not make a virtue out of the necessity
in which the world finds itself, he certainly does make of it a political
catechism in which there is no place whatever for the concept of that
which ought not to be.

In the sphere of the ideology of "autonomies" [Eigengesetzlichkei-
ten] the thesis that politics is independent of moral values has constantly gained ground. Indeed, Frederick the Great would be sadly
disillusioned if he were to see how the emancipated reason of man has
transcended the conflict under which he suffered, not by creating an
ethical social order from which this conflict has been expunged as a
form of sickness, but by diluting the ethical content to such an extent
that the conflict is rendered quite irrelevant. Strictly speaking, reason
has not risen above this conflict; it has sunk below it.

This later development may be illustrated by means of a modern
statement concerning the diplomatic lie which can be set over against
that of Frederick the Great. "What is diplomacy?" asks Gregory Klimow. His answer runs as follows: "Diplomacy is when two persons look
each other straight in the face and lie to one another, each knowing for

[11] Ludwig Reiners, Friedrich (Munich, 1952), p. 311.
[12] Ibid., pp. 106 f.

certain that the other is lying. If one should on occasion tell the truth, the other does not hesitate to regard it as a subtly refined form of lie, never to be taken at face value. Metternich and Talleyrand would hardly be entrusted with ministerial posts nowadays. The situation today is much more exacting. The progress of human society has introduced a new category of spiritual values. The diplomatic lie has ceased to be an elegant art form of the few; it has become an article of mass production." [13]

Though he himself does not realize it, Klimow is here giving an exact description of the phenomenon which we call the law of "gravitation." This gravitation towards the lie as an article of mass production begins when in the name of the autonomy of the political the word gets around that truth or falsehood are matters of ethical neutrality. Well, once something becomes a "natural law," it is absolutely binding on all. For Frederick the Great there was still something scandalous about having to lie in politics. He explained it on the grounds that reason was not yet sufficiently developed in the epoch in which he lived. Within the schema of the doctrine of progress he was thus aware of the imperfection of the world, or better, of the fact that in its present state the world was not what it ought to be. This is why he regarded the untruthfulness which was incumbent upon princes and statesmen as a kind of vicarious bearing of the burden of the world by a very limited group, a group to which—as in Plato—this dubious "privilege" was restricted.

This social restriction of the political lie is possible only on one assumption, namely, that the political lie is in fact a scandal which must be strictly confined to the sphere of political office and to those who bear such office, and that a strict distinction is to be made between what an office bearer is permitted, or better, compelled to do on the one hand, and what is required in terms of ethics and humanity on the other. Indeed, Frederick the Great distinguishes these two spheres within himself: "I hope that posterity will be able to distinguish the philosopher in me from the prince, and the man of honor from the politician." [14] Recognizing that the political lie is a scandal, i.e., that it ought not to be, and that it cannot be divorced from the ethical, and demanding that it be practiced "as little as possible," he

[13] Gregory Klimow, *Berliner Kreml* (Cologne-Berlin, 1952).
[14] Reiners, *op. cit.*, p. 106.

thereby lays upon it a threefold restriction: first, that it be limited to the political sphere; second, that it be permitted only to those who bear office, who must shoulder the burden, as it were, vicariously; and third, that even within the ego of the office bearer it be confined to the "ministerial" sector and not be allowed to encroach upon the human sector, upon the "philosopher" or "the man of honor."

If this ethical relevance of the political lie is abandoned, then the political sphere is naturalized and the laws which were peculiar to it take on a universal validity. The inevitable result is that the political lie breaks out of its quarantine. It springs the barriers which held it back, and becomes an article of mass production.

As a statesman, Frederick the Great may have been very scrupulous in his use of cunning and falsehood. From Christian tradition he preserved an ultimate awareness of the problematic and provisional character of our present world such that while the wolf was allowed to howl fiercely in the cellar, it was always locked in and never allowed out of that cellar.

Integral to this whole business of quarantine is the fact that another unavoidable aspect of political falsehood, the use of spies and undercover agents, also comes under the verdict of "dirty hands" and is regarded as extremely questionable. There is undoubted hypocrisy involved where this kind of activity is both condemned and used at one and the same time. Yet in this as in all hypocrisy, there may be discerned an awareness of certain norms, and of what is conformable to these norms and what is not, which awareness it is better to have in this contemptible vessel of hypocrisy and self-contradiction than not to have at all.

Here too we see the tendency which immediately sets in once the scandal and ethical dubiety of having spies and agents is no longer perceived or only dimly realized. In the naturalization of politics and the political lie, such as transpires under an ideological tyranny, espionage and intelligence too become "articles of mass production." Here again the dams have given way. The logic of events is once again the same.

As far as the naturalization of politics and the accompanying neutralization of political ethics is concerned, it is Machiavelli rather than Frederick the Great who has decided the fate of the West. The decisive point in Machiavelli's anthropology is that man is a natural force—

nothing more. In Machiavelli's thinking there is thus no place for that which is commonly and justly held to lift man above the nexus of natural causality, namely, the fact that he is able to understand himself and to call himself in question, to set himself a goal in freedom and thereby to repudiate what he presently is, in short, that man is set in the polarity of good and evil and hence also in conflict with himself, in inner cleavage. For Machiavelli, man exists instead in a straightforward and unequivocal movement, the course of which is prescribed for him by his unbridled passions acting as causes. The constitutive elements in this world of passions are "fear" and "cupidity." These are the vital energies of which man "consists." What man does is thus regarded simply as energy in action, and the history he makes thereby is simply a mechanical thing concerning which it is meaningless to put the question of good or evil.

All human conduct, whether private or public, is a blind emanation of man's nature: "As he adjudges the violent acts of a tyrant to be the necessary outflowing of a power-laden nature, so with equal artlessness he adjudges his own adventures in love to be an expression of a guiltless nature. He is not burdened with any conflict of conscience, nor torn by any sense of sin." [15] Since the elements at work in human nature are constant, the mechanical sort of history to which they give rise is something that can be calculated and explained; as in the case of natural processes its rules can be known.

The forces which determine the course of the world may be more precisely differentiated into the subjective power of *virtú* and the transsubjective power of *fortuna*, or fate. How are the two related to one another? The etymological connection between *virtú* and the Latin *virtus* must not lead us to think that what is meant here is ethical virtue. On the contrary, *virtú* is itself also a kind of power, an ability to proceed, a power which can withstand the drag of events and be cultivated and organized by the understanding. The corresponding, transsubjective power of *fortuna*, on the other hand, represents "the irrationality of the factual," [16] the factual which nonetheless gives evidence of certain trends or tendencies.

[15]Rudolf Stadelmann, "Macchiavellis Schicksalsglaube," in *Studium Generale*, II, No. 6 (1949), 303.

[16] Karl Schmid, "Macchiavelli," in *Grosse Geschichtsdenker* (Tübingen, 1949), p. 122.

The relationship between *virtú* and *fortuna* has its own distinctive dialectic, though this has not yet been wholly clarified. One thing at least is clear, that *virtú* is a power which resists the drag which resides in the trends of historical fate. On the other hand, history shows only that "men can help *fortuna*, not that they can resist it. They can weave the threads but not tear them." [17] Hence it is not possible to be simply opposed to the trends of *fortuna*. One can only use the *kairos* of history. Like the "world historical individual" of Hegel, one can satisfy one's passions only in harmony with the tendencies of the times. (In modern terms, one cannot resist the trend towards collectivization in the age of the masses; one can only exploit it. There is consequently no doubt as to which ideology today might appeal to Machiavelli!).

Machiavelli can actually compare fate with a "raging river" which "overflows the plains. . . . All yield to its violence, without being able in any way to withstand it." [18] Even *virtú* itself cannot offer any effective resistance to the natural element of *fortuna*. Nevertheless, it is possible that men, when the weather becomes fair, "shall make provision, both with defences and barriers, . . . so that the force [of the water] may be neither so unrestrained nor so dangerous. So it happens with fortune [*fortuna*], who shows her power where valour [*virtú*] has not prepared to resist her, and thither she turns her forces where she knows that barriers and defences have not been raised to constrain her." [19]

To describe the path here marked out for man between *fortuna* and *virtú* it would be quite appropriate to use the mathematical figure of the resultant in a parallelogram of forces. For what is involved here is a reconciliation of conflicting forces achieved quite apart from any call for decision. Politics, accordingly, is a purely mechanical play of forces. In such an anthropology there can of course be no place for ethical relevance: "The political, conceived of in this way purely as a sphere of ends, does not permit the question of good or evil, but only of effective or ineffective, of right or wrong. . . . Hence the statesman must test his means purely by their effectiveness. If a morally good

[17] Nicolò Machiavelli, *Discourses upon the First Decade of Livy*, II, 2.
[18] Machiavelli, *The Prince*, 25, trans. W. K. Marriott ("Great Books of the Western World," ed. Robert Maynard Hutchins [Chicago: Encyclopedia Brittanica, Inc., 1952]), Vol. 23, p. 35.
[19] *Ibid.*

means seems as though it will be also politically effective, then he should choose it; but if he thinks he needs a morally evil means, then he must use it." [20]

The position here adopted is one which Frederick the Great specifically rejects when he distinguishes the philosopher and man of honor on the one hand from the statesman and politician on the other. Awareness of the provisional character of our still disrupted world, and of the resultant safeguards which prevent political amorality from spilling over into the other spheres of life, forms a highly effective antidote against the naturalistic reduction of man to a mere operating "force." Whoever retains an awareness that in certain fields and situations some crookedness is inevitable is still under a restraint which no longer applies to the man who no longer makes any distinction whatever between crooked and straight, evil and good. "Having to sin" is basically quite a different idea from that of eliminating the concept of sin altogether. Luther's "sin boldly and believe even more boldly" [21] still grasps the horns of the altar, whereas in Machiavelli there is an unending level plain where once the temple stood, and with the temple there has disappeared also the sign, the reminder [Anamnesis] that the world as it is ought not to be, and that we can thus live in it only under forgiveness.

Frederick the Great, of course, did not know a great deal about forgiveness. But even if he did not know the Gospel he did at least know about "the Law." In secularization the Gospel disappears first, and for a period man is left with the Law alone. This is the case with Frederick. From the Law he knows that the political necessity of sinning in politics is not as it should be; something is out of order, and therefore the world too, which contains within itself the impulse towards this partial amorality, stands under the verdict of this disorder. The Gospel with which he consoles himself, and which helps him towards the realistic course of keeping the wound open and of acknowledging it as a wound, is actually only an apparent, a spurious Gospel which he derived from the ideology of progress. It is a Gospel which says man can only hope that one day, when the forces of reason have sufficiently grown, the wound will be closed. This belief itself may be erroneous, but even in the shadow of such an illusion it at least be-

20 Schmid, *op. cit.*, p. 125.
21 See p. 504, n. 19.

comes plain that only in the sphere of judgment and grace is the humanity of man preserved, and kept from sliding all the way down into the amorality of mere natural phenomena—at least for a little while, as long as the echo of the message persists.

Examples of the White Lie

The preceding deliberations have sought to open up in principle the problem of the white lie [*Notlüge*]. If we now deal briefly with a few "cases" it is not because we are entering the sphere of casuistical platitudes but simply for the sake of illustrating the established principles by means of specific examples.

Political Expediency and Diplomatic Cunning

The deception and concealment involved in military cunning—e.g., in the statements of a prisoner of war—and of diplomatic cunning [22] fall very largely under the problems dealt with a moment ago when we touched on the matter of interrogation by an enemy. The untruth involved in both these forms of cunning is mitigated to some extent by the fact that in each case the lie rests on a "tacit agreement"—to use another phrase of Frederick the Great [23]—whereby in politics men usually "do what they must in order to gain an advantage." As a result the situation is one involving mutual mistrust. To speak of "deception" at all in this context may therefore be out of the question, since strictly speaking the term applies only when an existing trust has been maliciously deceived, whereas in the political sphere there is no such prior trust, mistrust having been elevated by "tacit agreement" to the rank of a deliberate policy, indeed a kind of professional virtue.

The law of quarantine would naturally apply to these forms of untruth as well. The knowledge that these things "ought not to be"—and hence the acknowledgment of the norm of truth—means that untruth is restricted, and not merely to the sphere of a particular office but also to certain exceptional situations within this sphere. Winston Churchill saw this very clearly, and in an epigrammatic remark to Roosevelt and Stalin at Teheran he stated it in a way that was both succinct and keenly descriptive of the ethical problem. With reference

[22] On Calvin's defense of diplomatic cunning see Rudolf Schwarz, *Johannes Calvins Lebenswerk in seinen Briefen* (Tübingen, 1909), Letter 707.
[23] See Reiners, *op. cit.*, pp. 106 f.

to the need for vast deceptive maneuvers to cover the invasion of France he noted that "in war-time . . . truth is so precious that she should always be attended by a bodyguard of lies." In his memoirs he notes that Stalin and his comrades "greatly appreciated this remark." [24] He does not say whether Roosevelt and other Western military and political leaders were shocked that he should have spoken in such Machiavellian terms, with no attempt at moral modulation or concealment. But the superficial diplomatic smiles they probably wore would in any case have been indicative of something.

That Frederick the Great is not wide of the mark in his suggestion that certain forms of untruth rest on a tacit agreement to act in such a way as to gain an advantage, may be seen from the common practice with respect to treaties. A treaty between states is in the first instance not an expression of moral commitments but a guarantee of mutual interests. This is a political axiom which any ethics has to recognize. But such interests are subject to historical change, and when the historical situation outgrows the framework of a particular treaty the necessary congruity between treaty and interests is shattered.

Now when this congruity between treaty and interests is broken, both law and moral theology [25] demand that regular notice should be given of the abrogation of the treaty. But it is difficult to harmonize this procedure with the exigencies of practical politics, inasmuch as it might involve the disclosure of certain intentions which for reasons of expediency ought rather to have been kept secret. Traditionally, then, it is an accepted if not a very reputable custom to create artificial situations whereby the other partner is made to appear as the one who has violated the treaty, and thus to give a show of legality to the ending of a treaty which no longer corresponds to reality. This was Hitler's approach, for example, when he wanted to break the treaties with Czechoslovakia and Poland. He used fifth columns to foment within these countries such unrest as would inevitably elicit the kind of police response which he could in turn cite as justification for his own actions.

Theological ethics cannot forbid this age-old practice of putting the

[24] See Winston S. Churchill, *The Second World War: Closing the Ring* (Boston: Houghton Mifflin Co., 1951), p. 383.

[25] See the encyclical of Pius XII entitled *Summi Pontificatus* (October 20, 1939).

blame on the other party; to do so would be to ignore the structure of the political world in which the practice arose. But neither can theological ethics simply sanction such realpolitik. Its task in this regard is rather threefold. It must first bring to light that secret of the world of which this custom is only a symptom. It must also point out the distinction between an understanding of the custom in terms of necessity and guilt and an understanding of it in terms of the morally neutral autonomy of politics, or even of political virtue. Finally, it must discern the theological significance of the fact that, however much statesmen may assert the moral neutrality of their tactical means, they do nonetheless passionately lay claim to and operate in terms of the category of the ethical, if only in this matter of seeking ethically to defame their opponents. We shall discuss this final point in connection with the concept of hypocrisy.

Now the sanction afforded by "tacit agreement" to the cunning and trickery necessary for such realpolitik is not complete, even for so unscrupulous a political realist as Frederick the Great. A painful burden of guilt still has to be borne. This is evident from the deep self-contradiction which such procedures involve, a contradiction which repeatedly causes the charge of hypocrisy to be leveled against them.

First, it is noteworthy that when treaties are broken for reasons of expediency, there is always a concern to give a show of legality and of loyalty to the treaty, so that the norm of the ethical is, as it were, involuntarily proclaimed even in the political sphere. There is as little desire to do away with the ethical altogether as there is inclination to be guided by it.

Second, it is equally noteworthy that, although the constructive politician acts according to the law of expediency, and although he is corroded, or at least spattered, by the acid of Machiavellian unscrupulousness and amorality in political matters, he nevertheless displays an emphatic interest in putting the blame for breaking the treaty—or for war—on the other side. If he were really convinced of the morally neutral autonomy of political action, he would completely avoid the category of "guilt." He avoids it, however, only in relation to his own side. In relation to others, he uses it as a weapon.

It is thus apparent that ethical categories cannot be excluded from the sphere of international power politics. If politics were to be completely beyond good and evil, it would be wholly destructive in its consequences. That this conjecture is correct is shown, for example, by Hitler's attempt to operate with a law of brute interest totally devoid of the corrective of conscience, even that of a bad conscience. By this thoroughgoing amorality of means, he became, as Jacob Burckhardt said of Napoleon, "the personified absence of all guarantees," and thus conjured up against himself the chaos and revolution of fear.

Obviously the ethical phenomenon of trust and reliability cannot be completely done away. A certain amount of verbal credibility is essential if there is to be any kind of continuing intercourse, any alternative to utter chaos. In other words, the uncompromising pursuit of expediency ceases to be expedient at all. For purely pragmatic reasons, a minimum of integrity has always to be preserved. "To deceive [in the name of expedience] is . . . unquestionably a political mistake—if one carries it too far," says Frederick the Great.[26] Blindly to replace the sanctity of treaties by the sanctity of interests is not merely heinous; it is also stupid. We thus see that tacit agreement cannot completely vindicate the diplomatic lie, and that the indelible character of the ethical unexpectedly makes its presence felt even in the nexus of amorality.

Third, we are pointed in the same direction by the fact that the responsible official must always wrestle with the problem of where and how the line is to be drawn between ordinary morality on the one hand and political morality on the other. He must also see to it that the tacit agreement is not allowed to become a warrant for the kind of arbitrariness which is completely inconstant and—even in political terms—quite intolerable. What is here displayed is clearly a fear lest the tacit agreement should awaken a demonic power which cannot be restricted to the allotted sphere of action, namely, the rules of the political game, but will constitute a permanent threat to the person in the central core of his being.

The maxim that "politics destroys character" derives from a consciousness of this demonic threat. The maxim has of course been misused, but its misuse should not prevent us from seeing the element of truth it contains. For it expresses the correct insight that there can be no complete separation between the sphere of a political realism which is supposed to be morally neutral and the personal sphere of ethical obligation. For the Christian this inseparability is one more sign that

[26] See Reiners, *op. cit.*, p. 91.

the world, even the political world, is not just something outside of me or beyond me which after the manner of fate forces certain lines of action upon me. On the contrary, this world is wholly humanized. That is to say, it bears the mark of man, and cannot be set at a distance from him.

If the man in government is a Christian, he is thus aware of the fact that his political activity too is questionable and in need of forgiveness. He is saved from the danger of making a virtue out of a necessity. Moreover, when he does tell the truth—as Bismarck insisted he should, and himself sometimes did—he will not regard this as a virtue or equate it with genuine truthfulness. For the tacit agreement concerning permissible cunning and deception permeates the sphere of politics so completely that when on occasion the truth is spoken it can prove to be misleading in the extreme.

In view of these several factors, we would surely misinterpret political action if we were to brand as "hypocrisy" the attempt to blame others for breaking treaties or starting wars. In one way hypocrisy is of course involved. But this hypocrisy has another aspect: It is an involuntary acknowledgment of the disorder of our world, which as a political world is inclined to apply ethical insights only to the ends which are sought and not to the means which are used to attain them, which thus desires to mark off a sphere of moral freedom or diplomatic license secure from all guilt or punishment, and which is forced to assert with a mixture of reluctance and relief that an unconditional imperative with its unconditional values does in fact permeate the region of means as well. The ethical imperative cannot be restricted to ultimate ends. Its presence in the realm of means is declared at least in the tendency to blame political opponents. While it may only appear as a political weapon—for use against others—it is nonetheless present.

Thus the presence of the ethical norm still attests itself, even if only in the negative mode of hypocrisy. There is still a concern at least to act as if one were impelled by an ethical imperative, and to do so by creating situations which make it seem as if the other party were the offender against this imperative.

It may be that no one active in government can wholly evade the demonic aspects of the political sphere, which is itself an outstanding example of this aeon generally. The Christian politician is protected, not against this *res* form, but against being swallowed up by this world in revolt. He is protected by the fact that he is aware of the antigodly nature of a world which here betrays itself so blatantly, and that he can thus withstand the temptation to make a virtue out of the world's necessity, a justifying norm of action out of what for the world is simply a way out. The believer knows of the high-priestly prayer in which Christ prays, not that his own may be taken out of this world of mortal peril, but that in this world and in this calling they may be "kept from the evil one" (John 17:15; II Thess. 3:3). For the evil one (πονηρός) can achieve dominion only where there is adherence to the law of falsehood. Adherence to falsehood does not have to mean constant lying after the manner of a psychopath. It means primarily an inability to perceive any more the doubtfulness of lying, because in the general and uncontrolled perversion of all values lying has been made a virtue. We are immunized against such adherence, however, if we live under forgiveness, and thus are given the power to keep the final norms in view, i.e., to call a lie a lie without trying to explain, interpret, or surpress it in terms of a pragmatic ideology.

Politeness and Etiquette

We meet this same problem once again, this tacit agreement with respect to the truth, when we come to the matter of conventional politeness.[27] Whenever I try to observe the canons of common etiquette I find myself uttering a constant stream of untruths. By the agreement of common convention I begin a letter courteously with the polite salutation "Dear . . . ," even though I actually regard the recipient as a fool. And I end my letter to a scoundrel with a complimentary close such as "Respectfully yours," even though I have no respect for him whatsoever but only contempt. Indeed I often say "Goodbye," a shortened form of "God be with you," when what I really mean is "Go to the devil."

In a totalitarian system there will be many people who use the politically formulated greeting, e.g., *Heil Hitler*, partly to avoid unpleasantness but partly also to satisfy their desire for honesty by the simple device of raising their hand more or less unenthusiastically as a gesture of ironical contempt or open opposition. This slight discrep-

[27] See Immanuel Kant, *The Fundamental Principles of the Metaphysics of Ethics* (New York: Appleton-Century, 1938).

ancy between the spoken greeting and the accompanying gesture, this little nuance which totalitarian authorities are quick to notice and to stamp out—thereby provoking even more sublime arts of hypocrisy—is extremely important for the ethical side of our problem. For it sets before us the distinction between formulations commonly accepted in ordinary speech and statements which actually say something, whether in word or mime or gesture. The formulae of politeness, among which we may reckon the ceremonial involved in state visits, for example, presuppose the tacit agreement that they are not to be assessed in terms of "truth" or "untruth," that the question of truth is not at all relevant, and that they are simply intended as practical forms of the *modus vivendi.*

Thus as I leave work at the end of the day I may say "Good night" to the watchman and the cleaning woman who are just arriving for duty. I do so without the faintest knowledge of what kind of a night they really have in store. And they nonetheless return my greeting with kindness. They do not demand that I speak the truth honestly and describe explicitly and in detail the kind of night I actually envisage and wish for them.

The forms of politeness make no claim to be true *in concreto.* They are related to truth, but at a different and more basic level. It seems to be typical, and by no means accidental, that they fulfill their pragmatic function, that of regulating social life, by way of formulating our acknowledgment or recognition of others.

In these polite formulae of recognition lies a confession that human society can build only on an attitude which accords respect to others. Such forms thus point to the primal dignity of man. At the same time they also declare the secret of society, which can exist only when this truth about man is respected. If a note of irony creeps in when I address someone respectfully even though I obviously despise him, this irony does not imply that I doubt the original meaning and therefore the truth of the formula used. It simply expresses the fact that the one whom I address does not measure up to the claim on man expressed in the formula. Thus something which *in concreto* appears to be utter hypocrisy may actually contain—in the mode of ironical inversion—a fundamental truth. (We again perceive at this point the ethical significance of irony in the matter of politeness.)

Politeness with its stereotyped formulae is thus misunderstood if we regard it exclusively as a technical device for lubricating the contacts among men. For our dealings with one another are no mere technical affair. On the contrary, they are based on a particular view of man. Hence the tacit agreement which holds that the question of truth in these formulae is not an urgent question, has reference only to the concrete case. It does not refer to the general meaning and content of the formulae.

This is why a critical letter is especially annihilating when it observes the forms of respect rather than resorting to vulgar abuse. For the politeness indicates to the recipient by means of an almost inaudible, formal irony that he does not measure up to the stature suggested by the form of the interchange. Politeness can thus be a deadly weapon. It contains all the possibilities of sublime expression which lift it far above the rank of a mere form or formula and which make of it, not merely a mode of utterance, but the content of an "indirect communication" (Kierkegaard). Extreme politeness exercises a Socratic function, but only for those who stand "above" the form and can thus use it as an instrument. At lower levels it is simply a form of social training much pursued in certain circles of society. As such, it is not merely false but also ridiculous.

Thus politeness can be extremely varied. It ranges from the training exercise of the affected ape to the Socratic thinker's sublime art of expression. It is thus a difficult term to pin down because it stands for such a broad range of heterogeneous attitudes and modes of conduct.

Let us look yet a moment longer at the external form of politeness as the sum total of those formulae which society has agreed upon. As we have already seen, distinction must be made between the formulations commonly accepted as being polite in social intercourse and that which is really indicated whether by word or by gesture, that which goes beyond the formulae and assumes the validity of an independent confession.

To return to the example of the political greeting used in a totalitarian state, the man who regarded himself as an opponent of National Socialism could go along with using the accepted formulation, however much more honorable it might have been for him to have set forth his confession precisely in this world of formulae by refusing even to use the greeting at all. But it then he not only used the greeting but actually did so with enthusiasm and smart precision, reflecting in his eyes the ardor which alone could really give him a moral alibi in the world of fanaticism, he was regarded as a liar and traitor by the subtle standards of that difficult time. And this impression was heightened if along with the raising of his hand and the ardent glance he also gave the shibboleth. (The case involving deception for political reasons, as we discussed it in connection with the problem of offering political resistance—think of the assassination attempt of July 20, 1944—is, of course, exceptional.)

It is exactly the same with politeness. When we go beyond courteous silence, beyond the socially accepted formula, and actually give expression to what we really think, then we cannot claim the simple justification that here the question of truth is not relevant.

The man who thinks that it is simply a socially tolerated flourish of politeness to have his servants tell callers he is not at home, is undoubtedly deceived as to the legitimacy of society's sanction in such matters. For at best what he is seeking to sanction in such a questionable way is actually a human weakness. Weakness is in fact the real issue here. It is not a question of sparing the other person out of considerations of politeness. It is simply a matter of his own cowardly or supine evasion of an unpleasant situation by means of the primitive but highly effective weapon of falsehood. The difficulty which serious-minded people encounter when they find they must bring their servants and children in on such a deception is enough to show that the problem is really one of trust, and this is a sign that what is involved here is real lying.

The line between an ethically neutral formula of politeness and an explicit untruth is frequently not very steady—think, for example, of the matter of "illness" in international diplomacy—so that casuistry is of no more help here than it is elsewhere. What is important is that society's agreement on a certain convention should not be misused as a direct justification of social falsehood. For the Christian the law of the exception and of demonstration remains significant even within the sphere of mere formulae.

Just how little use casuistry is at this point may be seen among other things from the fact that a witty causeur or an eccentric is able, without any particular merit, to handle social untruth far better than a simple and unadaptable person or a child. The Greifswald chemist

X, when he did not wish to be disturbed at his work by callers, used to answer the bell himself and tell the disconcerted group of students gathered at his front door, all dressed in their best dark suits, "No one is at home at the X's." Professor X could do this without any great ethical effort to remain honest, because he was both witty and eccentric.

The same is true of a sexton at whose church theological students frequently did the preaching. He always had three stock answers when they asked with anxious curiosity how they had done. If they had done well he would reply, "The Lord has been gracious"; if moderately well, "The text was difficult"; and if badly, "The hymns were well chosen."

More exact analysis of this example, which is typical of many other situations as well, shows us that the deeper meaning of politeness here finds particularly pregnant expression. For in this case politeness —enhanced by the witty form—is not merely a way of cushioning the disclosure of what may be an unpalatable truth. On the contrary, politeness here displays its original motive, namely, respect for the other person. It is transparent concealment of a truth, and thus becomes "indirect communication" in Kierkegaard's sense. As such it discharges the Socratic function of liberation. It does not say directly, "Your sermon was bad." This would be a discouraging kind of criticism which would "devalue" the person. The politeness expressed functions rather, by means of the puzzling saying, to stimulate the student to find out the truth for himself. It thus turns the criticism into self-criticism. Hence the student is not shamed before the eyes of others, since these other eyes seem to be closed and oblivious of the shame. The student is not made to feel inferior. Instead of being assigned the role of a mere object he is given back the initiative. Respect for the other, and the polite tact by which witness is borne to it, is thus integrated into a complicated interplay of good-natured superiority, ironic or Socratic liberation, and witty formulation. Here telling the truth really becomes an art. As Wilhelm Busch so truly says, "A great mind like that of the simpleton is not so helpless as a small mind."

It is immediately obvious that where there is no such eccentric capacity for expression, or better, for the art of expression, the relation between politeness and truth can be much more complicated. A person who is sincere but not particularly gifted may suffer shipwreck trying to steer a course between the Charybdis of the certainty that "direct communication" can wound severely and the Scylla of obligation to tell the truth. If a child is asked by his teacher whether his father often comes home drunk,[28] and if he proceeds to tell a white lie—comparable to the falsehoods involved in being polite—by replying in the negative in order to protect his father, it is because he has no means of indirect communication or witty evasion at his disposal. He lacks the sophisticated dexterity which could save the truth even in face of such a monstrous pedagogue.

Hence it is impossible to judge all the refractions of truth in the field of politeness in terms of a single criterion, and so be led down the destructive trail of legalistic casuistry. The most that can be said is that, with the possible exception of the most primitive formulae of everyday speech, truth dare not be banished from the sphere of politeness. After all, the social agreement to which appeal is made in justification of the polite lie is an agreement of society. As such it indicates that every proclamation of duties must keep in view the mutuality of the obligation.

If it is true that in the case of those who are intellectually less nimble the demand for politeness may easily violate the duty of telling the truth because they lack the power of indirect communication, then as a matter of genuine politeness, and of the courtesy which goes with it, it is imperative not to embarrass such people by forcing them either to utter such blunt truths as tact would forbid or, precisely out of such tact and politeness, to utter untruth. Many lies of this sort have been provoked by the egoistic impulse reflected in the saying "The world wants to be deceived"; such egoism overlooks the needs and problems of others in its desire to be itself the object of flattery, and thereby to sacrifice the integrity of others. A concert artist who asks a check girl in the cloak room what she thought of the music can hardly be acting from any other motives.

We are thus continually led away from the framework of purely "individual ethics"—which is an inadequate concept anyway!—and directed to the totality of interhuman relationships, indeed, to the whole structure of the world within which I stand. This is the structure of precisely that world which "wants to be deceived," and I am both

[28] This illustration is used by Bonhoeffer (*op. cit.*, p. 330).

the agent and the victim of that deception. It is always the total system which is involved in the truth and falsehood of the individual, because it is the system within which the individual stands and meets his vis-à-vis. No one can completely dissociate himself from the honesty or the deceit of his social milieu, his age, his world. But the relationship between them and the individual is, as we have seen, a reciprocal relationship. As I cannot abstract myself from dependence upon the practices and malpractices of the world around me and of the person who stands over against me, so I cannot regard these practices and malpractices as a transsubjective destiny which overtakes me from without. On the contrary, I am implicated in the state of my world as one who bears his own measure of responsibility for it.

For this reason I cannot claim to be just an unwilling victim. I can never invoke the social agreement as a full and sufficient justification. If I take the commandments of God as my standard, then I am acknowledging a standard which is not tailored to the possibilities of this world—and hence is neither formed nor deformed by what society decides—but which in the name of the divine transcendence calls in question both me and "my" world, nailing me to the impossibility of any self-justification. Only when I keep this standard in view do I remain faithful to the real meaning of politeness, namely, respect for the other person. Only thus am I aware of the limits of any particular form of politeness. Setting the duty to truth above all forms, I am able to see soberly and realistically the continual "revolt of means," namely, the revolt of the form of politeness, which may easily be a cynical denial of the other, against the real purpose of politeness, which is that he may be respected.

In this sphere too what is involved is an issue greater than that of purely technical instruction in human intercourse and in the forms which facilitate and enhance it, for example, in social etiquette. Here we are confronted by nothing less than the sign of the fallen world in which there yet shines forth the commandment and the promise of love for one's neighbor.

The Truthfulness of a Physician

We shall proceed in this section as we have done before, analyzing closely a typical situation which can reliably serve to illustrate a larger complex of questions. To illustrate the conflict between the duty of sparing others and the duty of telling the truth, we shall examine here the problem of the physician at the bed of his patient. We shall begin with certain basic insights which either were expressed in our earlier deliberations or may be deduced from them.

We have stated that truth is not adequately defined simply in terms of the agreement between a statement and objective reality. On the contrary, it became clear to us that, if we are to define a statement as true, account must also be taken of the sphere within which that statement is made. In fulfilling my duty to tell the truth I cannot abstract myself from the situation in which I speak.

It is also clear, however, that truth depends on the person who does the speaking. Naturally this is not true of the statements of exact science.[29] But it does apply to truths which belong to the human and personal sphere. I can communicate only that which I have and am. I can state a truth only in the act of confessing, i.e., only as I am myself related to the truth in question.

He who himself lives in untruth covertly changes even the objective truths which he happens to speak into untruths. What is formally true, true in form, can be falsified by what it is that actually dwells in this form. Pornographic literature, for example, may contain objectively correct statements concerning erotic and sexual processes.[30] Yet in pornography the parcels of truth are integrated into a perverse system of values and thus become unequivocal lies. Pornography falsifies the place of sex in humanity. It depersonalizes, biologizes, and psychologizes it. Out of the "true" stones of individual processes it makes a perverted mosaic. It uses "truths" but tells lies. It does so because the pornographer, although he uses correct and in some sense "true" details, is not personally related to the truth of the whole. In his own person he has missed its real meaning and purpose. So it is that the untruth of the person makes the statement to be untrue.

An analogous process may be seen in the temptation of Jesus in the wilderness. The tempter appeals to the words of Holy Scripture.

[29] See Kierkegaard's "Reflections on Christianity and Natural Science," in his *Eine Literarische Anzeige,* trans. Emanuel Hirsch (Düsseldorf: Diederichs, 1959), pp. 125-129.

[30] This is true even of the various Kinsey reports, which stand on the three-corner frontier of zoology, pornography, and anthropology.

He thus appeals to "truths," to things which are correct in detail. But on the lips of the satanic liar these become lies because they are spoken in the context of a false relationship. To use an arithmetical figure, these individual truths are set within a bracket before which there stands a minus sign. The negative thus robs them of their positive value as truths. The devil who so uses them is falsehood in person. Hence that which is originally correct becomes false; the truth becomes a lie.

The general statement of Luther that it is "the person who makes the works" [31] may here be reformulated to the effect that it is the person who makes the statement true or false. Only he who is subject to truth tells the truth. The liar always lies, even when he uses truth. Either way, man is always making confession of the ultimate relationship in which he stands. Confession is the only way in which "human" truth can be stated.

These preliminary but very basic observations, which in part take up and develop points already made, will serve then to introduce us to the specific problem of this section: What should the physician do when his patient is incurably ill or at the point of death? Does he owe his patient the truth or should he practice deception? [32]

It is obvious that this situation creates a genuine conflict. In the weakened condition of those who are seriously ill, information concerning the gravity of the sickness may come as a shock and really prove fatal. If we assume that there is still a slight hope of improvement, this prospect may be shattered by the imparting of such information, whereas the encouragement afforded the patient by a positive, though untrue, diagnosis of his condition may in fact spark his will to recover, and thus snatch him back from the critical point.

To recognize this possibility is to come directly up against a fundamental fact in the whole problem of truth. In certain circumstances truth is not just the sum and substance of a correct determination and delineation of the facts. On the contrary, the statement of a very questionable truth can itself conceivably bring into being facts which were not really present at the moment they were proclaimed in the

[31] WA 39^I, 283; cf. pp. 60 ff. above.
[32] The same question could of course be put to the pastor, but for reasons of method which will emerge later the example of the physician is more suited to our present purposes, since it forces us to consider how the situation of illness works out in terms of a variety of philosophical positions.

name of truth, facts which could not at that time have comprised the content of a "true" statement. This is what happens, for example, in "fortunetelling." Here the significance of the pronouncement is not objectively to state something, but magically to bring it to pass. In other words, the prognosis in fortunetelling, while it does not state an existing truth concerning the future (since to do so would be to presuppose the existence of the future itself), it often does make the stated facts true. It causes them, as it were, to become true. This may happen, for example, when the one who is under the influence of such a prognosis actually reckons with the facts contained in it, whether by looking forward hopefully to something which is welcome, or by allowing some fearful prospect to gain magical power over him, like a tree which has such power over someone learning to ride a bicycle that he is magnetically attracted by it.

Only when we grasp this point do we appreciate the full severity of the problem by which we are confronted. If there is some chance that the optimistic, though dubious, prognosis of the doctor may become true, ought he not to risk it? Ought he not to tread the narrow and slippery ridge which separates truth from falsehood?

How then is the doctor to decide between the destructive lie and the constructive truth? The deception of patients has, in fact, made its way into the stronghold of medical practice to some extent under the rubric of "suggestion therapy." We need only think of the saline injections given in lieu of a drug which has to be withheld, like morphine. Psychosomatic interaction may make it imperative that certain psychical preconditions be created in order to make somatic effects possible. These preconditions are impossible unless wool is pulled, as it were, over the eyes of the psyche, i.e., unless it is deceived. This is undoubtedly a working rule of the doctor's art which the patient, without any loss of confidence, will regard as justified once he is in condition again to make appropriate judgments. Indeed the deception is practiced precisely in order to bring the patient to this condition. In the situation of the patient it certainly does not amount to a lie, because it does not affect the person in the central core of his being, as it would if he were misled about his proximity to death, or if during a long siege of cancer he were deceived into neglecting important internal or external matters which he would have wanted to care for had he been given a more realistic appraisal of his situation.[33]

Then too in the latter case concealment of the true situation is inevitably followed by a chain reaction of further deceptions, so that finally we have, not a single structure of lies, but a whole colony of such structures. A pact of secrecy and deception must be sworn by a whole group of nurses, attendants, relatives, and friends. To avoid the terrible collapse of these finely spun illusions, and the attendant shock and radical collapse of confidence which must go with it, the deception is sometimes sustained even in the very last stages of the illness by means of narcotics, a chemically induced kind of personal irresponsibility.

These examples show that the concept of "a working rule of the doctor's art" cannot be a blanket authorization for all the various deceptions which, rightly or wrongly, are advanced as being necessary for medical reasons. The line between what is possible and what is impermissible is very difficult to determine. Indeed this cannot be done a priori. There are, however, certain indications that this line does not lie merely at the point where the deception relates to the issue of life and death. This is shown, for example, by the therapeutically as well as morally dubious mock operations which are sometimes performed on neurotics under the category of "suggestion therapy." [34]

In our further discussion of this medical conflict between telling the truth and sparing the patient we shall now apply to the situation of the physician what we have laid down concerning the dependence of the truth that we speak on the truth that we are as persons. We may begin with the negative proposition that the untruth in which a person, in this case the doctor, may find himself can show itself at the sickbed of one who is on the point of death, or very likely to die, in the following forms.

[33] This example is suggested by Richard Siebeck, *Medizin in Bewegung* (Stuttgart, 1949).

[34] The case reported by Jores, for all the humor of its outcome, nonetheless illustrates the point. A very wealthy and pampered woman, severly neurotic, constantly complained about headaches caused by the sofa in her head. Various remedies were tried but without success. Finally a mock operation was performed. When the patient awakened from the anesthetic, back in her room with her head generously wrapped in bandages, she noticed on her night stand a tiny doll-house size sofa. Looking at it but a moment she said merely, "My sofa was green." Arthur Jores, "Arzt und Lüge," *Universitas*, X (1949), 1198 f.

The physician may be motivated by fear in respect of the dying person or his relatives. The anguish, the psychologically understandable reluctance, which keeps him from "shattering" the other person with a grave pronouncement, leads to out and out lying if in fact it triumphs over the truth. For in this case the motive is not to spare the other; it is to spare oneself. Hence the doctor does not merely speak untruth; he speaks out of untruth, i.e., out of a situation which is itself untrue. In appearance he seems to be impelled by the motive of a love which desires to spare; in fact, however, he is impelled by the very opposite.

It is again the untruth of the person which is involved when the doctor fears that he may have to share these last moments with a patient who is incurably ill and knows it. In other words, the doctor is afraid that he will have to disclose the secret of his own failure. This particular fear is twofold.

In the first place, there is the natural and understandable defensive reaction against the fact that in every visit and consultation and treatment he must publicly demonstrate the limits of his ability, and perhaps even his human failure. Any concession to this defensive reaction is obviously a concession to untruth. What is involved here is not only an untruth of speech but also an untruth of being. The reaction is not merely psychological but also philosophical. It is based on the fact that the doctor does not clearly understand the fundamental relation between medical treatment and the nature which is treated, between medical ability and its limits, between the courage to act and the humility to surrender. It is based too on ignorance of the fact that sickness and death are not just a fate which befalls a man but a task to be taken up—by doctor and patient together. When the doctor is not clear about these basic questions he stands in a false and illusory relationship to the fundamental realities of existence. It is no wonder, then, that what he says on the basis of this untruth should also be untrue.

In the second place, and closely related to what has just been said, there is the fear that in face of death, and of the mystery of life therein declared, he can only betray his complete "inner" helplessness, and so be forced to give a false hope instead of saying something which conveys real help, enlightenment, and "knowledge." Not to know the truth about death always means lying in face of death.

If the physician has not himself faced the mystery of death, he must evade the dying. None of us can give more than he himself has.

Having stated the matter negatively, we may now put it in positive terms and say that the doctor is in the truth, first, if he himself has a relation to death, or, more accurately, if he has the true relation to death, i.e., if he knows the Lord of life and death. The man who is clear about the basic realities of life and death, who has faced up to these realities, will also be able to find the word of truth. There are no routine statements which can fill the gap when this existential condition is not met. Truth is constitutively linked with the confession of that which we have come to know in our own existence.

Second, the doctor is in the truth if he sees the "true man" in the other person. This "true man" is characterized by the fact that he is not merely the activating center of certain vital functions, a physiological "it," a bundle of woes which must be spared as many pains and disturbances as possible. On the contrary, his life between birth and death is one of decision and responsibility; in face of the borderline situation of death he is thus summoned to win his life, or to lose it, to hold on or to fail, in short, to face the basic realities of his existence. Only when the physician respects this situation does he stand in a relation of truth to the other person. If he does not, the relation is one of falsehood, whether he actually tells lies or merely follows the evasive tactic of silence.

Third, the doctor is in the truth if he respects the truth of illness and of the fact that man must die, i.e., if he does not regard sickness merely as a physiological and somatic event but realizes that man, the whole man, is sick. He must note well the terminology of the sufferer, which more often than not says quite specifically "I am sick," rather than just that "something is wrong with" this or that particular part of the anatomy; and he must know the reason for such phraseology. The patient's use of the first person means that man is the subject of his illness. Sickness is not a calamity which befalls him, reducing him to the level of an object. We need not call upon psychosomatic medicine—which is of course very emphatic, determined, and perhaps even extreme on this score—to find medical men who appreciate this point and who interpret the phenomenon of illness along these lines even from the medical standpoint.

This understanding means that being sick is a matter of the whole person. Consequently the truth of sickness can be grasped only when it is regarded not as a hostile force from without, but as a component part of man himself with which man must grapple and come to terms as he does with himself.

For the doctor this means that he must be ready to help the patient to accept illness, and in certain instances its incurability, and also death, and to understand these things as a task to be tackled. The truth of sickness and death is not perceived if we simply allow the instinct of self-preservation to dominate the field, and hence take a purely negative attitude toward illness. There is of course an un-natural element in it which must be combatted. Yet illness also has about it a providential aspect which is to be accepted and brought into harmony with the person.[35]

The struggle which is demanded of the patient in face of sickness and death can never be an externally oriented insistence that this cannot and must not happen to me. It must be rather an inwardly oriented struggle with oneself, a struggle for an answer to the question who and what this person is who is thus stricken, how to deal with the new demands that such a blow brings, how through endurance and acceptance it can be made meaningful and hence differentiated from an accidental calamity. The question is not merely why did this have to happen to me, but for what purpose.

Fourth, the doctor is thus in the truth to the degree that he knows the truth, i.e., to the degree that he knows the meaning and purpose of the blow and lives it out in his own life. Only according to the measure of his relation to this truth can he give true information on the actual situation of the patient in respect of the blow that has befallen him. Where the doctor is not related to this truth, or where the relationship is a negative one, he will be forced to lie, and at the same time to resort to the false excuse that he does so out of a desire to spare the patient, that he is acting out of a love which is here in conflict with the duty of truth. In face of death there can be no openness and honesty unless at least one of the two parties, doctor or patient, knows the meaning of that which is about to happen. In such a case, however, experience repeatedly confirms the fact that the truth makes free (John 8:32), and that fear of the unknown gives way to peace. Truth and peace indeed belong together.

[35] For the dialectic involved here see pp. 257 and 263 above.

From what we have said, it is obvious that the age-old problem of truth at the sickbed is completely misunderstood if it is regarded simply as a question of psychological tactics: to speak or not to speak. The question is not one of method at all but goes right to the heart of the matter involved. The point at issue is that of "existence," primarily the existence of the physician, but secondarily also that of the patient. It is a matter of relationship—especially that of the doctor —to the truth. It is a matter, too, of communication between the medical man and the sick man in face of this truth.

Hence this is a problem which cannot be solved for medical students in the usual way, simply by giving them some additional working rules. It is not a question of their learning just one more therapeutic technique, that of administering carefully measured doses of non-injurious information to the incurably ill and the dying. The problem is incomparably deeper! It involves a question directed to the doctor as a living person, and the demand that he be clear himself on the basic issues of life. The question of the physician, "What am I to say to my patient?" recoils upon himself in the form of the more basic question, "Who am I that I should have authority to speak here at all?"

The task hereby posed is one which the medical schools cannot solve alone. They must at least perceive the problem, however, in order that they may seek the resources of other faculties in the university which can take over this task of training whole men, or at least take a supplementary part in it.

Now that we have stated the basic points—but not before!—we can turn to the pedagogical question of how in the borderline situation of incurable illness or death those who know the "truth" can impart it to the patient. If we tackle this question too soon, we fail to see the existential problem of the physician, which is the main point at issue, and treat the matter instead as one simply of tactics or technique. Something which touches the very roots of our understanding of man is thus reduced to a matter of simple working rules.

Even here, however, we again do only partial justice to the situation of incurable illness or death those who know the "truth" can certain method or technique, or if we speak merely of the required tact and necessary sympathy. For here too the problems are not just psychological but existential.

We come up against this problem when we ask why a certain method is needed at all. There are obviously some truths which can be communicated by means of a method which is wholly factual and not at all psychological. The question as to the truth of the Pythagorean theorem requires proof, and hence also a method of "proving" it, whereas the truth about death clearly demands a method of "communication" which embraces the human element.

Obviously this must mean that this particular truth is not one I can just blurt out at any time. The physician may come right out with it, of course, when he is dealing with a colleague who is called in as consultant; to him he can discreetly communicate the realization that there is nothing more to be done. In relation to the patient, however, such directness of communication is not possible. Whereas the colleague who takes part in a medical consultation can receive such news with factual objectivity, or with emotion but of a very different kind, for the patient himself it could entail a fatal shock.

The fact that the same truth can have such very different effects on different individuals is clear indication of the fact that in respect of this kind of truth—as distinct from that of the Pythagorean theorem! —men can have very different relationships to it. We may formulate the distinction as follows. Relationship to the Pythagorean theorem is timeless and hence unhistorical; that which relates to this truth is the abstract understanding which is independent of situations and of a particular existence. Relationship to the truth of death and of whatever else befalls me, however, is historical. This truth becomes free only in concrete encounters with the blows that strike, in the conflict and experienced problems of life. It is a truth which in the course of things either comes to light or is increasingly obscured. Its corollary, then, is either a developing knowledge or a hardening ignorance.

The development of this knowledge is expressed in the concept of "enlightenment" or "wisdom," that which lies at the end of the process. Wisdom is not something that can be imparted at any moment, like the Pythagorean theorem. It has its own time, its own hour and moment. To the degree that it is the object of wisdom, therefore, the truth cannot be taken over from others, like the truths of mathematics, or even like certain communicated facts, e.g., of history. Truth can only be appropriated. It can be attained to only in the process of growth. It becomes free only as I am confronted with it, only as these confrontations actually take place in my life. A child can, of course, repeat the sayings of Plato or of Paul concerning death. He can even bring to them a measure of intellectual apprehension. But he cannot "appropriate" them because he has not yet been confronted by death or by our "being unto death." He must simply wait upon this knowledge until it is released within him.

But this means that we have departed from the purely psychological level of how to tell the patient and have pressed on to the heart of the matter itself, namely, the truth which is to be communicated. The truth at issue here is the truth of death, not the fact that a certain concrete event is about to take place, but the truth about death itself: the truth that we must die, we must depart this life, and that our life has a goal (Ps. 39:4-6).

This is a truth which I cannot communicate to the one concerned in a direct way, or at any time whatever, as I might communicate truth to a colleague who is not personally involved. The inability to do so obviously means that in practice I must to some extent withhold and conceal the truth, giving it only slowly and in small doses. Am I lying, then, when I do this?

We can make headway with this question only if we distinguish sharply between the psychological and the existential standpoints involved in the situation. The psychological answer to the question why a direct and unmitigated communication is impossible is that the suddenness of the communication (since it could presumably take place "at any time") does not befit the psychical state of the patient and could produce quite a shock. The existential answer is that such a communication would not be in keeping with the nature of this truth. By its very nature, the truth about life and death is a developing and ripening truth; it is not timeless and hence accessible at any time. The communication of this truth must consequently include that process in the course of which the truth can become free and the physician can contribute to its becoming free. Our concern here is with a truth to which we must be led, a truth which is not a presupposition but a goal.

For this reason it is not a lie, nor is it sabotage of the truth, if the doctor is at first very slow and hesitant about communicating, if he steers clear of the subject of death in order to approach the patient from a wholly different angle and prepare him for the real theme of his situation. Whether the physician lies or not does not depend on whether he momentarily avoids the situation or directly acknowledges it. It depends rather on whether he regards his first momentary denial as the initial or preliminary stage of that process which ultimately leads to the truth, a process into and through which he is determined to lead his patient *or* whether the denial is final, made with the intention of "sparing" the patient to the very end any certainty as to his condition, so that he can finally be conjured over the last threshold in a state of unconsciousness, with the help of narcotics as a chemical means of concealment.

In the latter case the method can only be the expression of a particular matter, the actual relationship to death. In terms of the specific relationship in question, it is held that beyond death there is nothing but unconscious nothingness. Hence it is deemed appropriate on this side of the threshold too to induce unconsciousness. A man's life has no specific significance since it bears no relation to anything beyond itself which could force him to make certain basic decisions. As a result, the last phase of a man's life is filled up entirely with the wounded and hence anguished impulse of self-preservation. Since man is helpless in face of death to forestall that lethal attack upon his instinct to survive, the only remaining possibility is to put an end to those final torments which it causes. Logically, then, the acute problem is one not of existence but of chemistry. The question of death has been reduced to zero.

If, on the other hand, the doctor is resolved to lead the patient into that process of disclosure, into that borderline theme of existence, then his action is not out of keeping with the truth even if in the first instance he conceals rather than reveals. The doctor does honor to the truth by making himself a companion of the patient in this entire process, and by therewith deciding to communicate with him in face of the ultimate reality.

Perhaps it might be objected that the physician has no right to embark upon this long-drawn-out process of preparation, that his duty is instead to speak out directly if the patient has already attained to mature knowledge concerning death, and on the basis of this enlightenment wants a straightforward answer. Yet even here there is no obvious justification for direct communication, nor does it constitute an obligation for the physician. This thesis may sound surprising and even strange. It is thus necessary that we state our reasons for it.

The truth concerning human existence [*Dasein*] is of such a nature that our relationship to it will vary according to whether we are indirectly or directly affected by it. To turn again to the example of the colleague already mentioned, the doctor who treats the dying patient is only indirectly affected because, although death is the theme of his life too, this theme does not have for him the same degree of relevance as it has for the patient who is directly affected. To this degree the knowledge which the patient has appropriated concerning the truth of human existence [*Dasein*] does not merely go through a process of liberation "in general," whereby "one"—in the third person—suffers, and dies, and is faced by decisions. On the contrary, truth attains to freedom only in the immediacy whereby it affects this particular man, so that he has to speak in the first person.

It is in this sense that Heidegger speaks of "Dasein's public way of interpreting," in which "it is said that 'one dies,' because everyone else and oneself can talk himself into saying that 'in no case is it I myself,' for this 'one' is *the 'nobody.'* 'Dying' is levelled off to an occurrence which reaches Dasein, to be sure, but belongs to nobody in particular." [36]

It would be wrong and overfacile to give a mere psychological explanation of this remarkable phenomenon that we never relate to ourselves the representations of death which are objectively set before us. Such an explanation might run as follows. Man has an astonishing ability to react against, and to suppress, the impression of menacing powers of reality; his instinct of self-preservation controls a mechanism of suggestion therapy whereby he is set at a distance from whatever might logically seem to involve him too. Here as elsewhere, however, the purely psychological explanation leaves much to be desired. For the psychical process of which it speaks is itself conditioned by the prior realities of existence. What are these realities as regards the case in point?

It is of the nature of man's personal being that man cannot see him-

---

[36] Martin Heidegger, *Being and Time*, trans. John Macquarrie and Edward Robinson (New York: Harper, 1962), p. 297. See also my own *Tod und Leben* (3rd ed.; Tübingen, 1947), pp. 70 ff.

self to be determined merely by laws, nor can he subsume himself under these laws. Indeed, he ought not to do so. For to be a person is to have the task of freedom and the capacity for it. This freedom, however, sets the person in a relationship to the reality which encompasses it which is a relation of decision, not merely effect. Within the sphere of its possibilities the person has to "act," i.e., to be a subject. And within the sphere of that which comes upon him as ineluctable destiny, e.g., death, the person has to "relate," which means again to be a subject.

Personal freedom thus has the ability to set itself at a distance, and it is for this reason that man is able—also psychologically able!—to refrain from subsuming himself in the first person under the natural law of death in the same way as he subsumes a dead body which is before him, a "thing" of which he objectively takes cognizance. Being there objectively and being subjectively involved are, as it were, two different dimensions of being, each of which has its own way of seeing and putting things, and they cannot be harmonized with one another.

Naturally it would be a mistake to use this freedom to foster the illusion that we need not be subsumed under the law of death. Freedom confers distance all right, but along very different lines. It actually has reference to how we relate to death, and only thus does it refer to our subjectivity. Nevertheless, the illusion that we are able to remain aloof does rest on the fact of personal freedom, albeit a misapplied freedom. The psychological fact that man can resort to this shelter of personal non-involvement would be inexplicable as a psychological fact if we were not to take into account the existential possibility of aloofness which is afforded by freedom.

We are thus confronted by an important and basic fact of human existence which we have now observed in many different connections. Man's special prerogative of being endowed with personal freedom is wholly ambivalent. It is oriented toward genuine responsibility before the divine Thou, toward man's relationship with God; but on the dark side it also possesses the other possibility of using a gift of grace to evade the claim of that which is sent and ordained by God.

We have said that the truth of death really becomes free only when I can speak in the first person: "I die." And we have tried to explain this statement in terms of its negative counterpart. Only when the death at issue is my death does it cease to be a general and hence only half-true phenomenon and become a personal event in which I experience the mystery of my existence. Thus death is something that must grow upon me. In the words of Rainer Maria Rilke, it must grow "in me."

It was in this sense that Karl Holl, when he lay on his death bed and the doctor wanted to help him over the last hour with narcotics, stated in reply, "I will not be robbed of *my death!*" It goes without saying that what moved him in this desire to remain conscious was not the curiosity of a scientist wanting to observe the extinction of life "from within," but the mystery of existence, to which one's own death belongs and which is experienced in this personal and intimate way. Because love, the meaning of life, and human extremity all reach their end and climax in death, it is here that the real substance of life first becomes fully apparent. A person really comes to know himself at death. This is why death is not merely physical or natural but quite "personal." Its truth is revealed fully only when "*I* die."

It is important to bear in mind this fact that the truth about death is something which "grows" in life. It is not something that we always have at our disposal and can communicate at any time. It is also important to realize that in this process of growth there is always a gap between knowledge of "death in general" and knowledge of "my own death" in particular.

Now the relation to death—or lack of relation—in which the patient stands is something which cannot be known in advance. Consequently the character of the disclosures which the doctor must make is linked with the situation, with the person addressed, with the doctor's own knowledge of death, and in all of these with that particular form of truth which is involved in man's "being unto death." Certainly the fact that this truth is one which grows and develops must be expressed also in the character of the communication made to the person who is incurably ill or on the point of death. This communication must itself be a gradual disclosure, a preparatory leading of the patient into a discovery of his own.

On these grounds compromise—the merely temporary compromise! —between the postulate of "sparing" the patient, i.e., the postulate of a gradual process, on the one side, and factual diagnostic truth, on the other, is perfectly legitimate. This must be said even though the suspension of factual truth remains a genuine burden for the conscience which is constantly at war with itself and always tempted to yield to illusion. Even in this case the rift in the structure of the world is too obvious to miss, namely, the divergence and conflict between the different forms of truth. If the peace with God which prevailed at the beginning were still extant, there would be no conflict as there is now between the privilege of going home and the necessity of departing. There would be no painful severing of ties, because there would really be only one tie. And we would not be wandering between two worlds, existing in a state of cleavage.[37] But then no "preparatory leading" would be needed, and consequently no compromise such as that involved in the suspension of factual truth. Thus in the depths of every human situation we discover evidence of the fact that our world is a fallen world still awaiting its redemption.

In conclusion, we may repeat the decisive point of all we have been saying. No man can bring another to what he himself does not have. No man can liberate in another a truth which he himself does not share. The final issue is not the psychological technique used in such disclosures nor the problem of formal factual truth, i.e., the agreement between what the doctor says and the actual state of things at a given moment. What is ultimately at stake is the appeal to the fact that the physician is himself a human being who must come to truth and to maturity. This is his task.

There is much talk about the patient being a person, a whole man who is to be treated therapeutically in his personal being, in his humanity, and whose illness is not to be regarded as a somatic or biological accident, something which has merely come upon him. All this is true. But it is equally true—and unfortunately this has not been seen with the same clarity—that the physician too is a person, a whole man, and that his skillful deeds of diagnosis and therapy are not merely a matter of technique, but are expressive of his humanity, which is herewith challenged to its very foundations.

[37] See pp. 281-284 above.

28.

God's "Compromises" with the World

As we discussed the various forms of compromise we came again and again to see that compromise as such is a tribute which must be paid to the fallen world. The structure of this world in which we are enclosed—not in the sense of κατὰ αἰῶνα but of ἐν αἰῶνι—is something we cannot get away from, even though we now belong not to the world, but to the Lord who has overcome the world (John 16:33; 17: 15, 16; I John 5:4; Rev. 12:11). Because it is a tribute to the fallen world, however, compromise is something which does not correspond to the "true" will of God, to his plan in creating the world. To this degree, we cannot justify compromise merely on the ground that it is "necessary." For the world which makes it necessary is "my" world; it is the objectification—if you will, the institutionalization—of my rebellion against the plan of creation.[1]

The need to make compromises, then, the world being what it is, does not have the significance of—to use Luther's terminology—a "compulsion" which is laid upon me from without, so that I am overpowered and have no option. It has rather the significance of a "necessity," which includes the fact that what I am necessarily will out.[2] The confession that this world "demands" compromises implies the fact that this world, and my own action within it, stands in need of forgiveness.

There was thus good reason to speak of Christian restraint in respect of compromise, and to use the image of "gauze in the wound" to express the fact that we must always preserve a sense of the ultimate ambiguity of compromise.

On the other hand, compromise is imposed upon us as a task. For without it we fall victim to a fanatical radicalism which ignores and destroys the form or "schema" of this world, and which thereby unleashes chaos rather than serving the divinely willed perpetuation of

[1] Cf. pp. 434 ff. above.
[2] See p. 297 above.

the world (Gen. 8:22). Yet how can something be both the subject of ethical restraint and the content of an ethical task?

To understand this, we have to be clear on two points. First, the making of a compromise is in a certain limited sense a reflection and re-enactment of what God himself has done in patiently bearing with us. In condescension he "accommodates" himself to the nature of the world and sustains it in the sphere of its remaining possibilities until the last day. Second, this analogy is broken down by the divine prerogatives which show God to be the "wholly other," the incomparable one, and which make it clear that if we would describe the accommodations of his grace as "compromise" we can do so only with tongue in cheek.

Nevertheless, we have to think through the relatively valid analogy because only thus can we discover the link in the indicated conflict between "restraint" and "task." We shall therefore study the divine accommodation from two angles, first in respect of the Law, and then in respect of the Gospel.

God's Law as "Compromise"

As concerns the Law, the most important insights have been dealt with already in Part Two. As regards the factual material, it will thus be sufficient merely to recapitulate the main points, and then to evaluate them in relation to the question whether this divine accommodation can be subsumed under the rubric "compromise."

Man in his primal state is forbidden to eat of the tree of the knowledge of good and evil (Gen. 2:16, 17; 3:3). He lives, as it were, in an eternal liturgy. His face is unequivocally towards God. He does not know the cleavage into which he will inevitably be betrayed should evil become acute on the horizon of his life in the form of genuine temptation. It is only when this actually happens that he is set in a situation where he must decide. The necessity of decision is itself a sign that cleavage has arisen, and that the unequivocal unity is broken. It is a sign that the serpent is seriously challenging God for mastery over man, and that in this conflict man is rent asunder. To be sure, it is God's will that in this battle which has broken out man should decide in favor of his Lord. What is equally certain, though, is the fact that the situation in which this decision is demanded is a perversion of

God's real plan for creation.[3] All human ethics stands finally under the sign of this ambiguity. It is an emergency measure with respect to the fallen world.

According to Genesis 1 and 2 the knowledge of good and evil is a mark of the fall. All the more significant, then, is the truth that the Law of God does not merely judge man in respect of the cleavage, but that it also tells man—in the name of God!—what is good and what is evil, and what the Lord requires of him (Mic. 6:8; Deut. 11:26; 30: 15).[4] The Law thus establishes itself on a soil composed of the facts created by the fall of man. The Law itself tells us what we had previously been forbidden to know.

This is a circumstance which has repeatedly given rise to objections against Luther's doctrine of the two kingdoms. For the "kingdom on the left hand"—at least to uncritical observers—seems to be a kind of institutional expression for the concessions which God makes to a world which has "gotten out of hand," indeed for God's virtual capitulation to this world. For this reason it is most important, if we are not to fall victim to terrible misunderstandings, that we should know precisely what is meant by the phrase "on soil composed of the facts." What follows here then is implicitly a contribution to the proper understanding of Luther's doctrine of the two kingdoms.

The "patience" of God does not negate the substratum of this encounter between Law and world; that would be to condemn the world to destruction. On the contrary, the patience of God goes down into these very depths. This is its accommodation. The world would be broken and consumed by the "true" will of God represented in the plan of creation. It would wither before the immediacy of the divine majesty. Hence God relativizes his own demand in order that man may live, in order that man may be granted the *kairos* of God's ongoing salvation history.

This accommodation of grace, however, breaks down the analogy of human compromise by virtue of the fact that it is undertaken in freedom, in the freedom of grace. Man cannot live without compromise but God could, for—contrary to what the mystics say—God could live without man. Man is forced into compromise by the structure of the world about him. Still he must concede that this world which thus

[3] See pp. 281 ff. above.
[4] This is the so-called *usus elenchticus* of the Law.

compels him is his world, and that accordingly he must regard the situation of compromise in terms of guilt rather than of fate. God, on the other hand, is not forced into accommodation, nor does he share in the guilt of the world which precipitated the cleavage. Neither fate nor guilt occasions the divine accommodation. When God stoops down to meet man on his own level, this is the very opposite of the "necessity" which forces accommodations and compromises on man. It is the miracle of the divine act of grace which is not necessary and which cannot be postulated. It is the miracle of God's gracious refusal to let man fall. It is at this point that the limit of the analogy with human compromise is brought out with the utmost clarity.

The accommodation of God in the sphere of the Law is also seen in the fact that one must speak of a "refraction" of the divine *nomos* by the medium of this aeon. We have already drawn attention to the change which takes place between the command [*Gebot*] of creation (Gen. 1:28) and the law [*Gesetz*] of the Noachic covenant (Gen. 9:1 ff.).[5] In creation, when man is told to "have dominion," he is given first place in the hierarchy of undisturbed and God-serving creation. His rule is a reflection of the divine rule. But when in the Noachic covenant man is told to exercise dominion, this rule is linked with fear and threats, and with violent self-assertion. Once wrong has entered into the world, rule can no longer be an expression of hierarchical rank. It is bound up instead with self-assertion and hence the law of might. The revolt of creation must be held in check. A sign of this ordering force which has to be set over against the force of disintegration is the command of God in Genesis 9:6: "Whoever sheds the blood of man, by man shall his blood be shed; for God made man in his own image." Here the accommodation of God is plain.

The force which has broken loose in the world is not wholly negated. On the contrary, it is restrained—and by means of this very same force harnessed in obedience. God deals with the world with a kind of higher homeopathy. He uses the principle of "like for like." In patience, all that is strange about himself and his righteousness he hides behind that which is not strange but common. Only thus can the stricken world bear his presence. There are, of course, many indications in the Noachic covenant that what is expressed in this accommo-

[5] This is another use and form of the Law, namely, the *usus politicus.* Cf. pp. 147 ff. above.

dated form is not the true divine will. It is only now, for example, that man becomes carnivorous (cf. Gen. 1:29). It is only now that the groaning of the creature begins. But the enjoyment of meat is granted to him only to the degree "that he does not touch the blood, which on the ancient view was the special seat of life [the life which belongs to God]." [6] This restriction is a reminder of the true will of God which declares that the world belongs to God and not to itself. In any case, it is only by the gracious act of God that the world does not fall victim to the forces of destruction, but is preserved with the help of these very same forces working in the opposite direction. In capital punishment, for example, we may say that the will of God which is at work is not his original will but a will which is altered for the sake of man, which is relevant to man and has regard to man's possibilities, and which is thus "patient."

This alteration is seen most clearly when Jesus sets aside the divine permission of Mosaic divorce (Matt. 19:7; Mark 10:4). He points out that "from the beginning" of creation it was not so. Divorce does not correspond to the true will of God or the true plan of creation, but is simply a concession of the divine patience to the "hardness" of men's hearts which sometimes makes divorce indispensable (Matt. 19:8; Mark 10:6). For if the command of creation, according to which man and woman are definitively related, were maintained intact, this would lead in a world of weakness and hardness of heart to intolerable conditions. In this world and on this level the radical command of creation could only issue in terror and chaos. The world is so alienated from the divine plan that it would be shattered and destroyed by it, did not God cause the grace of his accommodating intervention to triumph over the legitimacy of his original claim.

We thus misunderstand this divine accommodation completely if we regard it as a license for "this aeon" which gives legitimacy to the resultants in the parallelogram of forces, the forces which lift men up and those which pull men down. What is involved here is not a sanctioning of what is, but a continuing reminder that what is ought not to be. The grace of the divine patience must be seen against the background of judgment, of the divine No, that negation which has not yet been implemented but is "still" held in abeyance. (This is the No of

[6] Gerhard von Rad, *Das erste Buch Mose* ("Das Alte Testament Deutsch," No. 2 [Göttingen, 1949]), pp. 109 f.

the father to the path his prodigal son has taken into the far country. But even as he is saying No, the father awaits the return of the lost son and is open—his arms are open!—to receive him.) If we do not see the divine accommodation in terms of this dialectic of judgment and grace, of Law and Gospel, we obviously do not see it correctly.

It is important to note again, however, the precise point at which the analogy to all human accommodations and compromises breaks down. That the divine accommodation lacks the element of necessity characteristic of all human compromises may be seen from the fact that the accommodation of the divine requirements never is, or purports to be, a kind of timeless law. It is limited rather to the interim emergency period of this aeon. As the Law which has been altered out of consideration for the hardness of men's hearts did not exist "from the beginning" of the world, so it will not last into eternity. It will disappear when the kingdoms of this world have been supplanted by the second coming, and when the petition is answered: "Thy kingdom come!" (Matt. 6:10; 20:30; I Cor. 15:24).

The expectant Christian knows that this time of grace which is "still" in effect is in fact limited. He thus sees that the divine accommodation is restricted by that from which the world comes, and also by that to which it goes. He knows that the "will"of God which is hidden under "alien" accommodations is not his true will.[7] He knows that God has not made a compromise peace with the world whereby it is allowed to continue, but that the Noachic covenant is only a kind of "cease-fire" or "breathing space" during which, figuratively speaking, God deals with his opponent and confronts him with the decisions which will lead either to real peace or to the breaking off of negotiations. The divine accommodation in the Law thus expresses quite literally the ambivalence of this time of decision, this *kairos* in which the possibility is still open which is soon to be closed.

Theologically, then, it is all important that we see this accommodation localized in a specific epoch of salvation history and that we do not make it into a timeless law or a sweeping validation of the *modus vivendi* of this world. Only when we see this can we appreciate the distinction between the divine accommodation and every human situation of compromise. Only thus do we see with final clarity what we were earlier concerned to show, namely, that regard for this divine

[7] See p. 148 above.

form of accommodation radically changes man's attitude toward compromise. Compromises on the human and historical level can no longer be regarded as justified in terms of a necessity inherent in the laws of life. In the law of compromise there is now seen instead the permitted interim of this aeon. There is seen the fact that the world of creation no longer exists, and that the coming kingdom has not yet arrived. In other words, compromise is seen in terms of its ambivalence and its need of forgiveness.

There thus arises a wholly new form of dialectical analogy which to some extent comprehends the distinction between the analogy of being [*analogia entis*] and the analogy of faith [*analogia fidei*]. For now we can no longer see the divine accommodation of grace in analogy to human compromise, as if God too took the all too human line of least resistance. Quite the contrary, we must now understand human compromise in terms of what we know of the divine accommodation. We learn to understand it as a compromise which is conditioned and limited, which has no intrinsic justification, which lives only by the divine patience and is therefore highly dubious *in re* but possible *in spe*.

The Gospel as "Compromise"

The gracious accommodation of God is encountered also, indeed particularly, in the form of the Gospel.[8] For the heart of the Gospel is the condescension of God in Jesus Christ. The Law bore witness already to this movement of divine mercy, for behind it stood the motive of patience. Patience too is grace. But it is grace which lives concealed in expectation, and which thus creates its own preconditions. It is grace waiting for the moment when it can come into action and be powerful and direct. The Gospel, on the other hand, is the immediacy of grace itself. In it the divine heart is visibly revealed. Only as we look back from the Gospel is it apparent that this same heart was also beating under the concealment of the Law.

The accommodation present in the Law achieves immediacy in the Gospel through the fact that Jesus Christ comes in the form of man, which means in the form of the sinner (Phil. 2:7). "For our sake he made him to be sin who knew no sin, so that in him we might be-

---

[8] See Dietrich Bonhoeffer, *Ethics*, ed. Eberhard Bethge, trans. Neville Horton Smith (New York: Macmillan, 1955), pp. 79 ff.; Heinrich Vogel, *Christologie* (Munich, 1949), pp. 143 f.

come the righteousness of God" (II Cor. 5:21). He comes down to the level of sinners. God does not remain aloof from the lost son. Jesus Christ comes to the very place where the human heart is exposed to the three basic temptations, and he exposes himself to temptation (Matt. 4:1 ff.; Heb. 2:17; 4:15). He enters the zone between God and Satan. He partakes of suffering and death. In all this he is the Son of God incognito (Kierkegaard), "born in the likeness of man" (Phil. 2:7), and "in the likeness of sinful flesh" (Rom. 8:3). This is his accommodation.

Yet in this accommodation he is mysteriously preserved from solidarity with human compromise. For one thing, the motive here is again the freedom of love. The Gospel does not stand under the law of necessity which compels some coming to terms. It is sustained rather by the miracle of free grace. "Compromise" is wrung out of us by the state of the world. But God, if we may put it thus, wrings his "accommodation" out of himself. His love is stronger than his justice. Grace overcomes the law of retribution. Accommodation is the victory won in God's conflict with himself. That which is "necessary"—if we wish to use such terminology—is the law of distance. This law expresses the ineluctability with which the holiness of God reacts to the sacrilege of man's revolt. What is "necessary" is the wrath of God. This necessity is the very opposite, however, of that which applies in the case of human guilt, whereby, beginning in freedom, there is set off at once an uncontrollable chain reaction such that guilt becomes a suprapersonal fate which for its part sweeps the guilty along with it. The necessity of the divine reaction, on the contrary, is wrapped up in the majesty of God himself, who stands over against this guilt as a hostile power.

For this reason, what we have called the divine "accommodation" is not the result of a compromise with man. It is not a kind of agreement on a *modus vivendi* without which everything would break down. God is not dependent upon man; things will go along quite well without him. On the contrary, the divine accommodation is a miracle "in" God, a victory in God's conflict with himself. It is the triumph of the heart over holiness, as Luther describes it in forceful imagery. Luther's antithesis of Law and Gospel preserves this very thought, that the Gospel is a free miracle which stands out in bold relief against the dark background of the Law and cannot be postulated. The only thing that can be postulated is the judgment or negation which is so clearly reflected in the mirror of the Law that kills.[9]

In the second place, the accommodation of Jesus Christ is different —and hence bears no real analogy to human compromise—because he is without sin (Heb. 4:15). In II Corinthians 5:21 we are told not that God made him to be a "sinner," but that God made him to be "sin." "The consoling mystery in the mystery is that he was made sin, but not a sinner. This is the paradox of this substitution: Precisely by not becoming a sinner, he took the place of sinners. If he had become a sinner, he could never have taken the place of sinners." [10]

Thus the Pauline formulation safeguards the fact that Christ always transcends the accommodation. He is always manifested as the one who actively gives himself to it. He is the subject and not just the predicate of the accommodating act, and hence is not absorbed or exhausted by it. Jesus Christ is not just the one who is humbled, but also the one who does the humbling. Theologically this means that the Logos is more than the humanity which is assumed.[11]

Already in the Christmas story there is an element which is more than accommodation, and its transcendent character comes out even more clearly in the cross and resurrection. The death of Jesus is more than a mere accommodation to human death. There is lacking here the essential human connection between sin and death (Rom. 6:23).[12] Death does not swallow him up. On the contrary, the innocent one gives himself up—vicariously—to this correlation on behalf of those who are engulfed by sin and death. In dying he takes upon himself the guilt of others. He allows himself to be smitten by the judgment of God—for others.

The resurrection brings this transcendent element fully to light, not only because here the accommodation of lowliness ceases, but supremely because here there is an abrupt change in the orientation of the accommodation. For now Christ accommodates his own to him. He causes them to participate in his conquest of death and his resurrection life. To become his disciple is to be taken from the realm ruled

9 See pp. 117 ff. above.

10 Vogel, loc. cit.

11 We are not referring here merely to the so-called extra calvinisticum, for there was always discussion of this point in Lutheranism too, e.g., in the discussions of Brenz and Chemnitz. Cf. Gottfried Thomasius, Christi Person und Werk (Erlangen: Bläsing, 1857), II, 399 ff.

12 See my Tod und Leben, pp. 139 ff.

by the powers of death and to stand beside the prince of life who has already passed through death into life (John 5:24; 17:3; Rom. 8:38; I John 3:14). It is thus clear, first, that the divine accommodation to men does bear a certain analogy to compromise; second, that this analogy is limited and breaks down; and third, that the position of man in relation to compromise is thereby altered, so that compromise ceases to be a justified law of the world.

We cannot make this third assertion, however, without adding at once that a false justification may find expression not merely in the matter of compromise but also in its opposite, namely, in radicalness. Both attitudes and lines of actions have their own problems.

Radicalism, as a form of fanaticism,[13] is an anticipated eschatology which sets aside the given state of the world and the given interim of salvation history. It rests on an attitude of self-justification because it refuses to see the close connection between man and the laws of this world, which is man's world. It claims it can free itself from such connection by radical acts which depart from all previous patterns.

The spirit of compromise is the opposite of this radicalism. But as such it is to be distinguished from the attitude which we have described as Christian realism in face of the emergency order of compromise, a realism which recognizes the ambivalence of the world of compromise, its need for forgiveness, and its interim character. The spirit of compromise seeks to achieve self-justification by arguing that it is a natural law of the world to follow the line of least resistance, to pursue the art of the possible, and to follow the course indicated by the resultant in a parallelogram of forces. By so doing, one submits to the law of nature and achieves integration into the economy of this world's forces. This view might also be expressed as follows: As in the forest we are justified in howling with the wolves, so in the world we are justified in living according to the world, in making the law of compromise a norm. Whereas radicalism is an anticipated eschatology, the spirit of compromise thus seeks an illegitimate prolongation of the world. It does not recognize that the world has its limit, its end, nor does it know him who calls the world in question.

Both radicalness and compromise are necessary, each in its own way. They are evil, however, if they are conceived as programs which derive either from man's contempt for the world or his attachment to it, and

[13] See pp. 348 ff. above.

which in both cases lead to self-justification. But they are good if those who act, whether in compromise or with radicalness (not an "-ism"), realize that their action is of no avail before the last judgment, if they live in expectation of the coming kingdom and see that what they do in this aeon is set under the forgiveness and also the patience of God, who does not abandon this world but allows it to continue and to prosper in the sphere of its possibilities and its ambivalences until the last day. In such case both alternatives are "good." God causes them to be good, for those who live in justification are right before God.

THE BORDERLINE SITUATION
OF EXTREME CONFLICT

29.

Meaning and Analysis
of the Borderline Situation

As we pointed out earlier,[1] theological ethics usually makes the mistake of taking the "normal case" as its standard for measuring reality. The result is the illusion that by providing certain Christian directives we have actually solved the problems. In ethics, however, the situation is similar to that in medicine. The problems do not arise with the ordinary cases, but with the borderline cases, those involving transitions or complications. It is the abnormal rather than the normal case which brings us up against the real problems. Hence the real test, even in respect of foundational principles, is whether an ethics has been proved in the crucible of the borderline situation and emerged with even deeper insights.

Thus a doctrine of the orders will receive its inner movement and depth from the very fact that it deals with concrete disorders. Similarly, the doctrine of justification and sanctification will have to prove itself by submitting to the acid test of whether and how far it can be practiced in an underground movement of resistance against a perverted government, a movement which cannot get on without lying, deception, and falsehood. If our doctrine of the divine commandments applies only in a Christian culture or a democratic society, but is helpless when the times are "out of joint," what we have is not a doctrine of God's commandments at all but simply a religious and ideological superstructure extrapolated out of the "normal" situation.

The Theological Relevance of the Borderline Situation

On the drawing board all ethical problems may be represented with relative ease. This may be seen from the available handbooks. The ease of representation is due to the fact that reality is first of all illegit-

[1] See Chapter 23.

imately reduced to that which is normal. To put it even less charitably, reality is robbed of the element of caricature; it is divested of its perversion, of its disorderliness. This means that ethics is deprived of its usefulness, inasmuch as the general tendency is for men to turn to ethics for guidance in critical and unusual situations, when available schemes provide no answer to the question "What shall we do?" (Acts 2:37). But the impairing of its usefulness is not the most serious consequence. Beyond that, ethics is also divested of those questions and problems which are theologically decisive.

When the "world" is seen with the normal spectacles of the nature aesthete it is all too easy to contrive a natural theology. But how does this business of natural theology, and its underlying principle of the analogy of being, stand up in the face of earthquakes, air raids, and a few other riddles of history?

It is easy, almost dangerously easy, to speak of the "orders of creation"—and of obedience to "authority" as one such order—when what we have in view is the "ideal" case involving a Christian society and a ruler who is at the same time the foremost member of the church. But what happens in an age such as that of Hitler and Stalin? Do we not have to recognize that the firm stance taken in actual practice at the time of these critical situations could have basic implications for theory as well, and that theology and ethics could have been thrown at this point into a crucible of massive dimensions?

This is the more to be expected because it is in the borderline situation of injustice, where sin is institutionalized as it were, that we learn in no uncertain terms that "this aeon" is not shot through with stable "orders of creation" which have a kind of indelible character and can serve as our standard of measurement. The possibility of total corruption is an indication that there is no such thing as those ontic relics of creation, and that the orders are simply a macrocosmic reflection or objectification of the human heart.[2] And just as a firm line cannot be drawn between creation and sin in the individual heart, so it is impossible to draw any such line in the macrocosmic dimension.

This means that institutionalized evil, the borderline case in which injustice has taken on structural form, is not just a marginal possibility. It is not just an exception, a deviation from the true nature

[2] See pp. 434 ff. above.

of this aeon. On the contrary—and this is a far cry from all ideology of the orders—it is simply an extreme example of a constitutional perversion of the aeon. So it is not a proclivity for dramatizing the exceptional which leads us to begin with an analysis of the borderline situation. In adopting such an approach we are simply being objective, true to the facts, the fact of the "world" as it is.

It is perhaps because our ethical and theological theories are here forced to undergo an unprecedented death and renewal that treatments of ethics have constantly avoided this issue like a hot potato. There seems to be an unwritten law that one does not speak about certain problems, e.g., the theological problems of an underground movement. Such problems are too painful and shocking. They must be relegated to the private sphere of the quiet chamber, where the most personal decisions of conscience are made. But this chamber is supposed to be a place of prayer, not just a closet for junk or for depositing whatever the theological professor wants to discard from his classroom, whatever he does not wish to be seen—or cannot find a place for—in the sacred precincts of academic theology.

Accepting, then, the thesis of Tillich that the border or boundary [Grenze] is the most fruitful place for gaining knowledge, we first turn to the question of what is meant by the "borderline situation" [Grenzsituation].

The Meaning of the Term "Borderline"

The concept itself derives from Karl Jaspers, and, although we do not mean to take over all that he has said, we would like to adapt for our theological purposes certain significant elements in his thought.[3]

For Jaspers borderline situations are those in which I cannot live without conflict and suffering, in which I take upon myself unavoidable guilt, in which I must die. What happens when I accept and endure such situations is that I enter into confrontation with myself. I cause being to "surge up" in me. Apart from such situations I do not exist; I simply vegetate. "We become ourselves as we enter into the borderline situations with open eyes . . . to experience borderline situations and to exist are one and the same thing."[4] Only as I stand

[3] See Karl Jaspers, *Philosophie* (Berlin, 1932), II, 201 ff., esp. 220 ff.
[4] *Ibid.*, p. 204.

up to such situations do I cease to live merely in the moment, staggering blindly from one moment to the next along my line of time. Only thus do I grasp myself as a totality, since it is only from the perspective of my boundary or border—only from guilt, suffering, and death, the "outermost" limit of my existence—that I can see myself as a whole. Whether I consciously lay hold of my existence or fail to do so is determined wholly in view of this boundary and solely from the perspective it affords. Only as I stand at this outer limit does my existence become a question for me, and therewith the object of inquiry. It ceases to be the non-objective wrapper which envelops me so tightly that I simply vegetate inside without any awareness of it.[5] Now it seems to me important that we should apply to the mode of being of the whole aeon these thoughts about the boundary or border which Jaspers has expressed in respect only of individual existence. This we could do, of course, without even mentioning the concepts of Jaspers. For the thesis that the boundary is the place of knowledge, and hence—in modern terms—the only place at which I can consciously lay hold of my existence and experience it as the object of decision, is no new discovery. As regards knowledge of the world, for example, it is quite plainly found in the Bible. For what the world is, and what man is within the world, is apparent—in the strict sense—only in the light of the boundary or limit, i.e., of the fact that the world comes from the hand of the creator and ends on the last day. In other words, it is revealed eschatologically. No statement can be made concerning the world, its guilt, freedom, orders, or history, except with reference to this beginning and this end.

In the light of this it would seem obvious that the "boundary" is not to be understood simply in terms of a temporal horizon embracing the beginning of the world and also its end. It involves also a qualitative horizon, as it were, first in the sense of what God has invested in the world by way of creation, by making man in his own image and by making the phenomena of creation to serve as transparencies for "his eternal power and deity," and second in the sense of what the world has become after the fall, in the extreme perversion of the orders and in times of crisis when sin is not merely present in individual acts but has become crystallized, as it were, into the very form of the world itself.

5 *Ibid.*, p. 209.

Here too, then, we are confronted by two limits or boundaries of the world, in the light of which it may clearly be seen in its totality, i.e., in its ultimate possibilities and hence also in its very nature. On the one hand, there is that which was posited in the world as the possibility of creation, that wherein it should attain to its authentic being, truly actualizing God's original plan. On the other hand, there is that which it bears within itself as the extreme possibility of decline and fall, of chaos, and of misappropriation of that wherewith it was endowed at creation.

Somewhere between these two limits stands the world as it is. This is why we think it makes sense to apply to the world, to the aeon in its totality, that which Jaspers describes as the "borderline situation" of individual existence. In other words, there is value in seeing the world in the light of this extreme limit of its possibility, whereby it is confronted with the theme of the destiny which it has missed but which is still set before it, and consequently with the task of understanding itself and of adopting an attitude towards itself.

The borderline situation is thus an instructive example in terms of which to study the fact of the fallen world and to put the problem of ethics in its sharpest form. Here it will become clear whether we did right in rejecting the normativeness of the original order and its natural law on the grounds that we could not find any "formula of transformation" whereby to apply the laws of man's primal state before the fall to the completely different world after the fall.[6] Here it will also become evident whether we were right to follow the New Testament and the Reformation in understanding the fall in such radical terms, as wholly characterizing the order of this world, an order which thus stands in need of forgiveness, and hence to interpret the normative directives given to the world as merely negative imperatives.[7]

This ethical situation is expressed particularly well in the underground movements of the Second World War, especially those in Holland and France, in which Christians took a leading part. The constant problem of underground ethics is that of "behavior in illegality," the question of how to act obediently where injustice reigns supreme. It is the question of how one can expect and claim for himself the

[6] See pp. 401 ff. above.
[7] See pp. 440 ff. above.

forgiveness of God in a world of lies in which it is apparently no longer clear at all what truth means and how it is to be told, a world of the "black market" in which illegal methods of doing business are routinely followed, etc.

The possibility, indeed the necessity of breaking the law of human order—and not just the law of the land but in some cases even the moral law—in order to obey the Law of God, such "legitimate illegality" presupposes a state of injustice in which the fallen nature of the world assumes a particularly extreme and paradigmatic form. "Here law and justice may be so far apart that the former still insists upon being dutifully observed even when it is in fact nothing but codified injustice." [8]

The most distinguishing feature of this state of injustice and illegality is to be found in the inescapable conflict of values for which there is no clear solution, so that whichever way I go I am guilty. It is immediately apparent that in *this* situation Roman Catholic canon law is of no help either, with its principles that "necessity knows no law" and that "necessity makes lawful that which is not by law." [9]

The Analysis of Borderline Situations

The analysis of borderline situations is best accomplished with the help of the most pertinent possible examples. As a criterion in the selection of these examples we may use the question of how committed Christians—men who were not only aware of the theological responsibility of their decision but also reflected on it—faced up to classical instances of the borderline situation, even though they may have perished in so doing.

We choose this method because of the danger that a purely theoretical and detached approach may involve an oversimplification which can be wholly misleading. To discuss the borderline situation purely in terms of principle, to sketch it, as it were, on the drawing

---

[8] Dietrich Oehler, "Die Achtung vor dem Leben und die Notstandshandlung," in *Juristische Rundschau* (1951), XVI, 1.

[9] *Decretum Magistri Gratiani, Pars Tertia (de consecratione), distinctio* I, canon XI: "*necessitas legem non habet*"; *Decretales Gregorii Papae IX, Liber* V, *titulus xli (de regulis juris), capitus* IV: "*Quod non est licitum lege, necessitas facit licitum.*" *Corpus Iuris Canonici*, ed. Aemilius Friedberg (Graz: Akademische Verlagsanstalt, 1955), I, 1297; II, 927.

board, is to reduce the actual frictions to which it is subject. Here, if anywhere, we must draw on both sources of knowledge: the perspective afforded by the "existential" threat involved in the borderline situation (which is more than merely a threat to physical existence) and the more aloof and disinterested "theoretical" point of view. The two modes of consideration supplement one another, since each alone has its own particular failings.

The perspective of existential threat is insufficient of itself, for there is a danger that elements of personal weakness will impair the objectivity of decision. The decision may be prejudiced, for example, by anxiety, by a misunderstanding of the situation, or by tactical considerations. When it comes to the question of how Christians ought to decide in a given instance, however, these defects in the concrete situation should not be allowed to deceive us into thinking that the solutions attempted by so-called "prominent" Christians can simply be taken over and made normative, for this would be to misuse them in the manner of legal casuistry. The so-called "prominent" Christians were far too clearly aware of their own fallibility—and of the need that they be forgiven for the decisions they did make—to allow us to treat them casuistically as models to be imitated.

On the other hand, aloof theoretical consideration is also insufficient of itself, because it cannot take into account specific factors which have also to be decided on the spur of the moment. Thus it is very easy (and can lead us almost to the verge of triviality) to subscribe in the abstract to the saying of Theodor Storm: "One man asks what will be the result, another asks simply is it right; and the difference distinguishes the free man from the slave." This saying has a radically different ring when "the result" involves the lives of many innocent people, when time or new possibilities of negotiation may sometimes be won by compromise, or when out of regard for the prestige of others all doors are not slammed shut but the possibility of retreat is left open in order that the other person may not lose face.

This is why, after 1945, many Germans felt it was far too facile and hence quite illegitimate for neutrals, and for others who had not had to live under a dictatorship and face the dangers involved, to pass sweeping judgments on the "collective guilt" of Germany. Even the factual correctness of this kind of judgment was necessarily felt to be pharisaical by those whose objectivity had been exposed to un-heard-of frictions, not merely in terms of such non-objective emotional elements as anxiety and the instinct of self-preservation, but also in terms of conflicts within the region of actual facts, the confusing complications in the way that problems were posed, and the inability to find any obvious criteria for the decisions which had to be made. Even where they themselves were conscious of guilt, they were not disposed to allow this guilt to be foisted upon them by voices from "off stage."

True appreciation of the situation in which decision must be made is thus possible only when we approach it from both standpoints. The existential perspective and the theoretical point of view are both essential for a truly adequate analysis.

Concrete Forms of Conflict

The borderline situation is characterized above all by the fact that in it one is confronted by an opponent who is known to be bent wholly on the exercise of power, and who is obviously on the side of evil. The best examples of such a situation are thus to be found in countries which have been occupied by the representatives of an ideological tyranny. For while the countries in which the tyranny originates un-dergo a gradual development under the despotism so that they tend to become accustomed to dictatorship, to lose the clarity of judgment afforded by distance, and to be deceived as to its final goal by in-numerable tactical tricks, the countries occupied at a later date at least have the chance of arriving at what is essentially a more dis-passionate diagnosis.

To fight such an opponent is an obvious duty. But if this is so, the duty is one which can be fulfilled only as we adopt the methods of the opponent. One must necessarily—and this means willingly going against one's own will—share in the depravity of these methods, i.e., get one's hands dirty. The church as such may be obliged to suffer wrong rather than do wrong; it can hardly be said to have the right of putting up political resistance. But by the same token the individual Christian who is called to political responsibility is in-eluctably forced into a situation in which he must act within a frame-work of injustice. As Gebhard Müller, minister-president of Baden-Württemberg, has put it, "He cannot sit back with folded hands and watch the forces of the abyss devastate and destroy his land, the people committed to his care, and the values which he holds sacred."

In what follows we shall present the analysis of Alexander Miller, a Scot who was ordained a Presbyterian minister in New Zealand, went to London in 1938, served among the dock workers in the East End of London, and finally returned to New Zealand after extensive journeys through Canada and the United States. In 1945 Miller wrote a Christian evaluation of Karl Marx.[10] In later publications he tackled the questions of political ethics and of the social responsibility of the Christian. He was not an active fighter in the resistance, but his accounts are significant as contemporary reporting and commentary.

In an incomparably realistic analysis, Miller examines the ethical problem of the Christian resistance movements in countries under National Socialist occupation. Although his basic principles are anything but clear, and he is often confused theologically and politically, his factual analysis is nonetheless devastatingly penetrating. He argues that as far as the resistance movement is concerned

> . . . not to resist Nazism was to acquiesce in it. There was no living alternative at all. Yet to resist Nazism was to be plunged into the same chaos. For to resist one must stay alive: and one could stay alive only by forgery and deceit. Ration books and passports must be forged or stolen. Propaganda and organization must be carried on clandestinely and by trickery. . . . Even within the Christian groups the traitor or the potential traitor must be liquidated without hesitation if not without compunction, since not only might the lives of the group members themselves depend upon it, but the good cause itself. . . . There is the old story of Francis Drake and his mutinous lieutenant. Drake commanded his execution for the sake of discipline and the survival of the crew, but partook of Holy Communion with him in the greatest fraternity before giving the order for his death.

But drive this to its logical limit and where does it take us? Presumably if a man may be liquidated as a danger to the good cause, the same man may be tortured to make him yield informa-

[10] See Alexander Miller, *The Christian Significance of Karl Marx* (New York: Macmillan, 1947).

tion vital to the good cause. If he resists torture himself, would it not be more effective to torture his children before his eyes? Without any doubt many honest Nazis used this kind of argument and resorted to torture with real "veracity" of motive. Torquemada's inquisitors were not dishonest men, nor necessarily vicious beyond the rest of us. Within a certain frame of reference—or lacking a certain frame of reference—this is irrefutable logic, and it is not only Nazis who begin to feel the force of it.

If forgery, liquidation, and torture may be used, then "Is everything permitted?" Miller's answer is necessarily twofold: "Yes; everything is permitted—and everything is forbidden." [11]

It may thus be seen that the essential marks of the borderline situation are as follows. First, the struggle against the blatant representative of injustice is not a personal struggle against a personal enemy but a struggle to preserve orders, values, and the lives of men from external destruction and internal perversion; to make this struggle is thus an inescapable duty. Second, the struggle against the blatant representative of injustice can be carried on only if to a certain degree, which cannot be calculated in advance, one is prepared to use his methods (and thus to incur a measure of guilt), methods which have a logic of their own and a tendency progressively to limit one's freedom of action. Third, the whole sphere of methods or means is thus shown to stand in need of forgiveness. Fourth, the ability to act in this sphere without hatred implies that while one's external actions are indeed bound by the logic of methods, he himself as a Christian never looks upon the opponent as a mere agent of these methods but—in the manner of Francis Drake—as a child of God who has been bound by the chains of evil and thus gone astray; there thus persists a human sphere which is not swallowed up in the autonomous machinery of the conflict.

Having thus surveyed the fundamental structure of the borderline situation, we may now try to gain some impression of its most important variations. There is in fact a certain typology of the borderline situation which arises out of the different values at issue in the conflict.

11 Alexander Miller, "Is Everything Permitted?" *The Student World*, XXXVIII, No. 4 (1945), 288-289.

The Conflict between Life and Truth

The experiences encountered in rendering illegal assistance to the Jews during the National Socialist persecutions afford classic and terrifying examples of the conflict between life and truth. Such assistance sometimes took the form of striking names from the proscribed list, or of adding them to groups which were not earmarked for liquidation. A high price was exacted by the Gestapo for these steps. Men adopted such dubious means with a sense of anguish about having thus to plunge into direct association with evil without any knowledge of how it might all turn out. But it seemed more important to extend a brotherly hand to the persecuted than to avoid at all costs any personal contamination by association with criminals.[12]

Important from the standpoint of the character of the borderline case in ethics are the means used and the law of logical progression in respect of these means. For example, it was first arranged with an eye clinic to receive a proscribed girl on the staff without her having to report to the police or the rations office. But to get there she had to have a pass which would enable her to secure a railroad ticket. Arrangements were then made for another woman to lend her own pass for this purpose. But this was not enough. The pass could be used only if it were counterfeited. Its owner insisted, too, that the name and number be altered slightly. So a counterfeiter had to be sought. In addition, arrangements had to be made with the underworld to purchase ration cards, which in some cases had themselves been stolen. In addition, it was often found necessary in dealings with others to connect the assumed name with a detailed life history, which in many cases had to be very far from the truth in order to appear natural and credible. Often those who helped also had to act out a part. In their association with illegality, however, they found their true existence, while the externally inviolate part of their life became increasingly unreal. So it was that they came to be involved in a complete inversion of truth and falsehood, as these were normally understood. The borderline situation can thus bring about a complete reversal of all values, a reversal which is itself nonetheless genuine.[13]

[12] See Helene Jacobs, "Illegalität aus Verantwortung. Dr. Franz Kaufmann zum Gedächtnis," *Unterwegs,* III (1947), 10 ff.

[13] Cf. also Gertrud Staewen, "Bilder aus der illegalen Judenhilfe," *Unterwegs,* III (1947), 20 ff.

The Conflict between Life and Life

There are many forms of the conflict between life and life. The best-known is perhaps in relation to the medical problem of a dangerous pregnancy, namely, whether to save the life of the mother or the life of the child.

In view of the techniques of interrogation used by the modern totalitarian state, which involve both torture and the administration of drugs, the same problem arises in the form of the question whether one who has "fallen among thieves" and is in danger of being reduced by chemical means to a depersonalized talking machine is justified in taking his own life. Is it right to sacrifice one's own life in order to protect the lives of others whom he might involuntarily betray by talking? The fact that Carl Goerdeler, for example, when he was in a deranged state and personally no longer responsible, supplied hundreds of names of fellow conspirators and thus condemned them to certain death, shows that what is involved here is a real conflict. We cannot lightly pass judgment on the readiness to make such sacrifice in terms of its being "suicide." "If a prisoner takes his life for fear that under torture he might betray his country, his family or his friend, or if the enemy threaten reprisals unless a certain statesman is surrendered to them and it is only by his own free death that this statesman can spare his country grievous harm, then the self-killing is so strongly subject to the nature of sacrifice that it will be impossible to condemn the deed." [14]

This form of the borderline case, involving the conflict between life and life, is most clearly displayed in an example which is used by Eugen Kogon: "When the SS demanded that the political prisoners in the concentration camps should quickly determine which of their fellow prisoners were 'unfit'—that they might be summarily executed —and threatened that any delay in making the selection would result in an even harsher system of sanctions, there had to be *a readiness to shoulder guilt* [italics by Kogon]. The only choice was between active participation and an abdication of responsibility which, as experience showed, could only bring down even greater evil. The more tender the conscience, the harder was the decision." [15] According to Kogon,

---

[14] Dietrich Bonhoeffer, *Ethics*, ed. Eberhard Bethge, trans. Neville Horton Smith (New York: Macmillan, 1955), p. 126.

[15] Eugen Kogon, *Der SS-Staat* (Berlin, 1947), pp. 343 ff.

Christians were usually unable to make it: "Since it had to be made, and made quickly, it was better that the less sensitive spirits should make it, lest all should become martyrs and none be left to bear witness. Who would dare to condemn such comrades?"

We can correctly assess the dilemma involved in such a conflict situation only if we compare it with the externally similar case of a doctor condemned at Nuremberg.[16] A doctor in a mental institution listed a portion of his patients for extermination in order to make sure that the SS doctor who would otherwise replace him might not condemn all the patients to death. Here too what is involved is the readiness to do wrong in order to prevent an incomparably greater wrong.[17] Yet the case is quite different in principle from that described by Kogon.

In the first place, the command to exterminate the mentally ill was an administrative one, which the doctor concerned could have avoided by resigning his position. Strict obedience to medical ethics, which demands the saving of life rather than its destruction, would certainly have had the force of a demonstration which might have encouraged other doctors to make similar resolves, even though in this particular case the result may have been the ruthless bloodletting of the SS doctor. Readiness to do wrong in order to "prevent something worse" is a very dubious principle, because it implies that the end justifies the means, and because it reposes no confidence in one's own confession and its creative power really to prevent that "something worse" or to bring it under condemnation. Oehler is quite right when he says, "If one believes that the killing of people affords the only way out of a tragic situation, then one must be ready to accept the legal responsibility for it. After all, expiation implies supreme recognition of man as a person; for only man may make expiation, beasts cannot. To take from man the legal responsibility for such action would be to replace guilt by fate."[18] For this would be, not

---

[16] See Oehler, *op. cit.*, pp. 3 f.

[17] Many of the church's social missions institutions were confronted by a similar problem (e.g., Bethel, Bielefeld). In order to protect imbeciles, mental patients, and epileptics, they falsified lists, forged death certificates, used every kind of delaying tactic . . . but in the end had to send a hundred, two hundred, a thousand, two thousand to die in order—as it turned out—to rescue ten thousand others from a similar fate. See Kurt Pergande, *Der Einsame von Bethel* (Stuttgart, 1953), p. 157.

[18] Oehler, *loc. cit.*

to avert guilt—as "something worse"—but to falsify the whole affair, turn it into a metaethical natural process. The matter of punishing the guilt incurred under such pressures is, of course, a wholly different question.

This case of the doctor can thus be subsumed under fairly clear legal norms. But, as regards the legal and moral criteria, the situation involving killings in the concentration camp is far different. Here it is quite impossible for those compelled to assist in the killing to "resign." Nor is it possible to save the condemned by refusing to co-operate. In view of the total and unrestricted powers of the camp commandant there is in fact the certain prospect that the others will be killed too, and that the command to "survive," which goes far beyond the mere instinct of self-preservation, will thus be violated.

Here too, of course, personal confession and the readiness to sacrifice oneself and all the rest rather than to share in doing wrong is a genuine ethical possibility, in some cases even a binding command. We cannot overlook this aspect of the gruesome situation. For we owe even to the worst of criminals the preaching of judgment, trusting that the Spirit of God still holds open the possibility of repentance. Yet what preaching of judgment could be more massive and forceful than that done through the sacrifice of one's own life and the lives of others, through the witness of refusing to participate in the killing of people?

This particular line of action, however, cannot be required of everybody. For the contrary decision also has its ethical justification. Furthermore, to take just an example, it is up to each individual to decide whether beyond a certain point the criminal who has rejected every appeal to repent should yet be given one more opportunity, or whether instead he should now be made solely responsible for what must take place. However this individual may decide that question, there can be no doubt that every such decision stands under an ineluctable burden of guilt which apart from forgiveness would be quite unbearable.

The Conflict between Our Obligation to People in General and a Particular Oath of Loyalty

It may be that I can preserve a life, the life of my nation or neighbor, only by trying to overthrow the total order which threatens this life and which is intrinsically corrupt. It may be that I have to become a revolutionary, with all that this entails. It may be that, among other things, I have to break my oath of allegiance.

Military and civil oaths of allegiance pose a particularly complicated ethical problem. I can of course argue that such an oath always involves a mutual obligation. If the state to which I have given it becomes a criminal state it thereby dissolves the oath *de facto,* thereby releasing me from obligation, so that in the strict sense it is quite impossible to speak of my "breaking" an oath at all. This would be a case analogous to that in which all claim to truth was forfeited.[19]

Yet it is not often that the problem arises in a form so simple and comparatively clear. As a rule the oath is given to a state which is already manifestly unjust, so that the state's injustice cannot exempt me from keeping my oath but actually makes my guilt very real if I do break it. The real ethical decision thus takes place at the point where with open eyes I take an oath which I cannot keep with a good conscience.[20] Here there is a clear ethical situation—if this business of keeping my "eyes open" is really that clear and simple. In practice, even this is always a very complicated matter. For a state which is manifestly evil is never totally evil. It discharges a certain function of order which I cannot wholly ignore and which carries with it a claim to my support. This support is also to some degree unavoidable inasmuch as the totalitarian state is usually hermetically sealed against escape. Indeed within it there is not even a desert to which I can resort and where I can hide. Unless I am prepared to condemn myself and my family to destruction, and to leave only the evil men and supine collaborators alive, I must work out some kind of *modus vivendi.* Concretely, this means that I cannot rule out altogether the venture of taking an oath.

We are not to overlook, of course, the two conditions on which such a compromise—for that is what it is—seems to be possible. First, there is the explicit proviso, "to the extent that my Christianity will allow." (This condition could be fulfilled vicariously through the church's making a corresponding declaration, as was done in con-

[19] See pp. 529 ff.

[20] An example is the swearing in of the German *Volkssturm* near the end of the Second World War, when the corruption of the system which exacted the oath was no longer a potential future development but an accomplished fact.

nection with the oath taken by Karl Barth in 1934.) Second, there is the readiness to apply this restriction in concrete and flagrant cases of injustice.

The conflict in respect to oaths may thus be described as a conflict between the life entrusted to me (that of my nation or my neighbor) and my oath of loyalty. Am I not set here between the Scylla and Charybdis of two responsibilities? On the one hand, I am guilty if I break the oath for, although I may be personally justified or even under obligation to do so, I thereby help to destroy the public validity of the oath, its power to establish social order. On the other hand, I am plunged into guilt in a different way if, bound by my oath, I look on passively while a group of criminals which has seized power leads the life entrusted to me to physical and moral destruction.

Christians in the German underground movement were faced by this most difficult conflict. On neither side does there seem to be any easy solution, at least *in re*. Deliverance which is purchased by weakening the validity of oaths, and which will have a disruptive and unsettling effect, especially among those who do not responsibly undergo the rigors of the conflict, is in any case a very doubtful deliverance. But loyalty to a dubious oath, which allows me to be a passive spectator while life goes down to destruction, is also a very doubtful loyalty.

At the Nuremberg trials power politics and propaganda may have caused the thesis to be propounded that morally the only possible course would have been to resist by breaking the oath. Such a judgment, however, should not blind us to the fact that what is involved here is a genuine conflict, one for which no stock solution is possible, and that for this reason the solutions actually proposed and implemented through the personal decision of various individuals could be and were quite different.

The pastoral situation involved in such cases of conflict will always confront the counselor with the fact—and I speak from experience— that he does not have authority to prescribe a specific solution, even though he himself may personally decide in favor of a particular course. He would be a hireling rather than a shepherd if he were to use such authority as he has with the inquirer to make binding upon him this purely personal decision, a decision which cannot be transmitted to anybody else because it is not based on objective criteria. What for one is a response in obedience, a decision ventured in the assurance of forgiveness, might well be for the other a reckless adventure, an act of irresponsibility, or even the line of least resistance.

The typology of the borderline situation, which we have been describing with the help of typical examples, is still incomplete. Many other forms of the conflict might be added, e.g., the question which became only too common as foreign troops began to occupy Germany during World War II, whether a father should kill his wife and daughters rather than allow them to be ravished. But what has been said will suffice to give us some impression of what is actually involved in the situation of thoroughgoing injustice. To the detailed observations already made, and the comments supplied in part by persons who were themselves actually involved, we shall now attempt to add a theological exposition of these phenomena.

A Theological Interpretation of the Borderline Situation

The first point concerning the borderline situation, as has been repeatedly suggested, is that it does not leave open any way of escape. For example, when I am under hostile interrogation, and the fate of a great cause and the destiny of living men lie in my hands, I have to make a choice. Either I "mislead" the opponent or I surrender my cause and my friends. The concept of "misleading" can here include, as we have seen, a wide variety of ethical qualities.

The demonic power, with its usual perspicacity, has also recognized the impossibility of any direct escape. One clear indication of this is the fact that ideological states nowadays artificially create a conflict situation as a tactical means of crippling their opponents. Their diabolical analysis thus leads them to the same conclusions as those of the theological observer. The following example may be adduced.

In staging elections to the Reichstag the technique of National Socialism—and it seems to be that of all totalitarian states—was to couple together on the ballot two different questions which could not be answered separately. The one might be a matter of foreign policy, e.g., the desirability of union with Austria, while the other sought approval for the regime as such, and especially for its leading figures. An artificial tension was thus created for the voter. If he wished to endorse the foreign policy, he had also to approve the system which proposed it and was trying to implement it. If he wished to vote against the regime, he had also to "sabotage" an individual political policy which might have seemed desirable or even necessary. If he adopted the obvious course suggested by this obscure—or all too perspicuous—maneuver and abstained from voting altogether, he would simply be leaving the decision to others. Such a course at any rate did nothing to improve matters, for the non-voter had no opportunity publicly to state his reasons for abstaining. We must also remember that in a totalitarian state there was usually a high degree of personal risk entailed in such abstention, and that the official propaganda publicly turned the motive for not voting into its very opposite, namely, rejection of the particular policy put forward for decision.

Even though all these factors might not finally deter the Christian from taking a position of symbolical aloofness from this whole plane of corruption—and in many cases even this external possibility did not exist—the fact must still be faced that this solution of the conflict is not in any sense what may be called a direct and unequivocal way of escape.

In all this inauthentic conjoining of electoral questions what is involved is a psychotechnical refinement of the exercise of power. Obscuring the issue weakens the capacity for action. Aloofness from the whole plane of decision works in the same direction. In this way the responsible intellectuals who stand out above the common herd, the elite whose opposition might be particularly dangerous, are backed into a corner and subjected to the initiative of the regime.

To the degree that there is no clear inward determination of action, the chances are increased that all direction will be from without, i.e., that guidance and control will be exercised by the apparatus of the dictatorship. Such a result is the more to be expected because of the primal inclination of ethical man to seek moral grounds for his action, and therefore to seek justification for the inactivity to which the conflict situation has brought him. The regime is willing to provide this justification for him in the form of arguments which in effect serve only to weaken his resistance all the more, namely, that he can with a good conscience vote for world peace, or for the outlawing of the atom bomb, or for reuniting Germany—or for the Austrian *anschluss*.

These artificially constructed conflict situations from which there is no escape deserve careful consideration if we are to recognize the refined tactics of such political maneuvers and gauge their power to weaken and debilitate. They surely reflect a high degree of demonic insight into the nature of man, indeed, what one might almost call a "demonically perverted theological anthropology."

The second point is that the guilt which we take upon us in the borderline situation constantly increases, that it necessarily gives rise to fresh evil. It is a truism that one untruth sets off a chain reaction of further untruths; a web of lies is never complete but always requires continued spinning.

To return to our previous example, if the voter in the Soviet Zone should desire the unity of Germany, then a vote in favor of such union is also necessarily a vote for the next step in that direction, i.e., to approve the particular power which has seized the initiative towards this unity as it conceives it, namely, Communism. The conflict situation always issues in a suprapersonal nexus of guilt which not only fixes the course of my own continued sinning but also forces others—perhaps children and relatives—to pursue the same course. Inevitably therefore the time must come when an ideological dictatorship can establish itself, when the personal rot and social decay are such as only a universal guilt could have produced, and these in turn with the passing of time themselves call forth a new sharing of guilt which is greater at every stage than what preceded, intensified by a process of demonic dialectic, a devilish circle.

The process involved here points essentially to a theological circumstance which will repeatedly occupy our attention, namely, that wherever there is a borderline situation we must speak not only of a fate which comes upon us but also of a guilt in which we are implicated. This guilt consists in the fact that our origin and point of departure is always to be found in a complex of wrong decisions, those made by ourselves and by others.

The fact that conflict situations and conflicts of values exist at all in this world is not due to the "structure" of the world itself. More accurately, it is not due to the character imparted to the world by creation, as though "from the beginning" (Matt. 19:8) this included tragic features. It is due rather to the complex of wrong decisions which lie behind us, which have their ultimate root in that primal decision recorded in the story of the fall.

It is not that "the world" is so perverse that it constantly thrusts us into borderline situations. It is rather that man has perverted his world, and no man is exempt from sharing responsibility for this human character of his world, this demonic humanizing of the world. No man is merely the object of a conflict situation. We are all part of the human subject that has helped to bring it about.

Every sin passes over somehow into the suprapersonal sphere and assumes an "aeonic" form. It becomes part of the structure of the world. It becomes, for example, the institutional lie, which atmospherically envelops and pollutes the individual attitude toward truth. As "the best of us cannot live in peace unless it pleases his wicked neighbor," so I am implicated in falsehood and disloyalty if the form of the world—which I have helped to create—secures power over me. Thus freedom and necessity, volition and ineluctability, are all wrapped up in a single package labeled, not "tragedy," but "original sin."

It is clear too that in this aeonic situation one can live only under forgiveness. I stand under forgiveness not merely in respect of what is traditionally called "actual sin," but also in respect of the whole structural nexus of my situation. I bring to the reconciling sign of the cross not merely my own "individual sins"—indeed such an abstraction is impossible—but "the guilt of the world." For I am a participant in the guilt of the "one man" Adam (Rom. 5:12), and I cannot be "in Christ" until I have first seen myself to be "in Adam" (I Cor. 15:22). It is only because God has loved the world (John 3:16) that he has also loved me, and it is only because he loves me that he loves the world also. For the two are inextricably intertwined. The separation of them takes place only as I place myself under forgiveness. For to stand under forgiveness means that the aeonic nexus of guilt has lost its claim to dominion over me (Rom. 6:14). That is, I am now characterized and determined, not by being "in Adam," but by being "in Christ." We have in fact died with Christ; we are crucified, and dead to the onslaught of sin (Rom. 6:5-12). God no longer sees us as representatives of this Adamic world. He sees us instead as brothers of his Son, oriented to the new life of the resurrection. And because in the eyes of God we are other men, we really are. For the ontological character of our being is determined by its reference, i.e., by the *coram Deo,* not the *in me ipso.*

Hence the secret of forgiveness consists in the fact that we are taken out of the nexus of guilt, and that as a consequence we are determined, not by the past of the world, but by the future of Jesus Christ. *In re* we are still encapsuled in the nexus of guilt, and to that extent we are "still" sinners. But *in spe* we are oriented to a new future. What we are in the eyes of God we really are. And for God we are no longer "in Adam."

The third point is that, so far as our guilt *in re* is concerned, in the borderline situation we are in danger of succumbing to the law of the inclined plane. By "inclined plane" we mean that our descending to the level of illegality, our readiness to incur guilt, or the hard and unwelcome realization that we must incur guilt, entails unforseeable consequences, not merely in the sense that guilt has disruptive sociological and institutional effects but also in the sense that it is itself subject to the law of progression. To illustrate the downward movement which it initiates, we may recall the example of torture adduced earlier by Alexander Miller.[21] In this connection Miller distinguishes two logical steps.

The first is the resolve to use torture. This resolve has its origin in the conflict between two considerations. On the one hand, we may treat our opponent with humanity, and in so doing perhaps miss getting the vital information which might have been secured under the pressure of torture. On the other hand, we may decide to use torture in order to fulfill our human duty to our neighbors and the obligations of service to our cause.

The second step consists in the readiness to torture the family of the prisoner before his eyes, if he is able to resist the dreadful torture himself. Once one descends to this level in the first place, there is a logical progression which finally makes us willing to use the most extreme and dreadful forms of psychical pressure. The first decision in this direction makes it unnecessary that all subsequent steps should be the subject of new and reiterated decision. Instead they are simply logical consequences of the first decision, having the character of automatic advances. Here too the freedom of personal decision is to be found at only one point, namely, in the initial act of decision.

We can also go beyond Miller to distinguish a third step in this downward movement, namely, torture for torture's sake, torture as an end in itself. When humanity collapses, so do other protective walls of

[21] See pp. 586-587 above.

division, and there open up desolate background areas in which sinister powers roam unchecked, e.g., the sadistic potentialities of the human psyche. These spirits are not invoked. They are contrary to the original purpose. Yet once they have been unleashed, they cannot be exorcised. They seize the initiative.

The logical progression thus leads us to a fourth and final stage which is particularly acute in respect of the question of the generations. When torture is first used, it is still conceivably possible that those who resort to it may yet be conscious of the dreadful conflict and of the guilt involved in it; at least in their own thinking torture can still have the character of a last resort. There can still conceivably be a sense of good and evil, and hence of an ultimate norm. But in the younger generation which grows up within this unjust order the situation is radically different. For this generation no longer knows the initial conflict. It simply receives what for it is already established norm, namely, the state of lawlessness, the moral flexibility, the inconstancy of all ethical values, the ambivalence and haziness of the world structure. The world order is not an "order" at all. It is a pulp of values through which each must work his own way. It is a scrap heap of rusted norms, a dump of discarded values, a resting place for that which obviously is no longer valid and perhaps never was valid. The only valid law is the law of self-preservation. Thus everything is indeed permitted—and without any safeguarding sense at all of guilt and of forgiveness.

The youth problem in the totalitarian state is the most direct witness to this law of progressive ethical decline. To protect itself from the prevailing terror, the older generation practices petty hypocrisies, dissimulations, and repressions of individual conscience, along with related compromises on a broader front. But it is at least immunized to a certain degree against some of the destructive consequences of all this by the fact that at one time it learned certain norms, that it still knows these norms, and that it has hidden them away deep down inside where it can secretly preserve them.

The problems involved here are so obvious that they do not need to be particularly stressed. In any case, that is not our present concern. The point which we are making is that the life of the younger generation in this state of injustice is radically different. It has never even known the institution of law. It has never known lying to be publicly outlawed. It has never known a manner of life in which an open statement of one's own mind is demanded, and hypocritical silence is repudiated as dishonorable, if not downright cowardly. It has never been taught by public example to lead the life of a real human being. It has never been shown how such authentic personal existence is the indispensable precondition for all human society. It is no longer oriented to changeless values. The world has begun to slip and each one must see to it that he himself is not upended by such a slippery surface. Each one must try to force or maneuver his own way through. This is the vexing problem—and it is by no means confined just to Christian parents in totalitarian states.

During the Second World War, I once conversed with a high school student who had to write an essay on the subject, "Why We Shall Surely Win." Since he was an intelligent boy, who knew the facts of the military situation and whose parents had kept him an alert critic of the regime, I was particularly interested to see how he would handle this delicate task. My astonishment was all the greater, then, when I found that the essay contained eloquent and enthusiastic passages about the gifted German *Führer*, that it underscored all the arguments pointing to a German victory, and that it systematically rehearsed all the familiar phraseology of the war propaganda. When I asked him how he could be so hypocritical before his teacher, he answered quite unperturbed, "So what! The teacher regards it all as nonsense too; he is convinced that we shall be defeated." What a ferment of disruption has been started in this young breast, a ferment that is only beginning to work! What is still only a mutual agreement —accompanied on both sides by knowing smiles—to dissemble concerning a truth of which both are aware, is bound to develop increasingly into a cynical playing with truth and ultimately into an utter contempt for truth, i.e., a complete denial of its existence.

One of the saddest aspects of the "black market" just after the war was to be seen where it involved young people. Older citizens resorted to the black market because of pressing need or to assist a neighbor. They acted with real pangs of conscience. The youth hanging around the railroad stations, however, stood wholly beyond the conflict situation. For them necessity had become a virtue, and illegality, a law of life. The unnatural—the black market, smuggling, prostitution—had become "second nature" to them.

It is thus of fundamental importance that the ethical decision never be regarded as an "isolated" decision affecting only the one who makes it. There is no self-contained ethical subject. My decision is surrounded by a kind of zone of radiation which affects—perhaps adversely—the decisions of others, and which thus entails for me a co-responsibility for their decisions. We have here a fresh indication of the fact that there is no such thing as individual ethics. For there is no individuality which can occupy the realm of decision alone without conveying its movements to other partners in the total system, like a fluid in communicating vessels. In the Bible this concept of the communicating character of ethical decisions is expressed in two main forms.

We find it first in connection with the doctrine of original sin (Rom. 5:12 ff.; I Cor. 15:21 f., 45, 55), which teaches that the first decision taken against God creates an evil field of force and causes all subsequent decisions to participate in the preceding original decision. "Adam" is thus responsible, not merely for his own time, but for the time of all others which follow, because all time is caught up in the law of the chain reaction unleashed by him.

We then find the same thought expressed by Paul in connection with the question of food offered to idols (I Cor. 8:10-13; Rom. 14:15, 21). The decision of the mature Christian may be that he "personally" can eat such food without hesitation. Since for him the idol does not even exist, his eating could not make of him an idol worshiper. Nevertheless, even though his responsibility "for himself" is not affected by the eating, he is bound by his responsibility for others. For the inner, personal reflection which allows him to eat is not accessible to, and may not be attainable by, other persons. These may want to imitate him in eating such food, but they lack the requisite freedom to do so. Eating with a belief in the reality of the "idol," they yield themselves into a demonic bondage; they do not have God as their only Lord. Thus one man's decision affects that of others. It is never made in isolation.

When the effect of our action on the decisions of others is detrimental, Paul calls it an "offense" (Rom. 14:13; 16:17; II Cor. 6:3). The very possibility of offense thus points to the communicating character of all ethical decisions. The borderline situation is a fruitful theological illustration, not least of all because of the way it draws attention to the law of progressive ethical deterioration and the "inclined plane," i.e., because it brings out with great illustrative power the communicating significance of ethical decision.

In all this we again see, though this time from rather a different angle, that Christian ethics cannot have the task of prescribing or even recommending casuistical solutions to borderline situations, not even in the sense of offering insights by means of which we may discern in any given instance "the lesser evil." It is of the very nature of the borderline situation that it resists being pigeonholed in any larger or prior framework, and that it refuses to be subsumed under natural law.[22] Here the venture of deciding, under the judgment and the grace of God, is demanded of us. Here we are also given the power to make such a decision.

The task of theological ethics at this point is to teach us how to understand and to endure—not "solve"—the borderline situation, and that from the twofold angle of Law and Gospel, of judgment and grace. We may now sum up our theological analysis of this matter by presenting the decisive insights in the form of theses.

When we undergo and endure the borderline situation this means in the first place, from the angle of Law, that this situation can never be merely a matter of fate, in the sense of being rooted in a suprapersonal cosmic structure which is tragic from the beginning. On the contrary, the borderline situation is provoked through "Adam." It has its origin in a communal guilt which already stands behind our situation and which we for our part ever increase and intensify. We are thus both the object and the subject of this communal guilt: it lays us under the constraint of necessity,[23] yet at the same time we are summoned to act in freedom, responsibly. We are both channels and perpetrators of cosmic guilt. Our own action needs forgiveness, but so too does the situation within which it occurs.

When we thus relate the borderline situation to our own guilt, and thereby rob it of its character as fateful tragedy, we actually understand it as judgment. For it is an essential aspect of divine judgment that what is initially a relatively small evil is condemned by the very fact that it gathers momentum like an avalanche, proceeding under its own power and in its own way. For the voter even to consider

22 See pp. 409 ff. above.

23 In this respect we should, of course, pay careful heed to Luther's distinction between "compulsion" [*coactio*] and "necessity" [*necessitas*]; see pp. 296 f. above.

the restricted alternatives put before him by the totalitarian regime, as we discussed them a moment ago, seems at first to be a relatively small evil. But the next moment he is the prisoner of the alternatives in which he has acquiesced.

What this means, fundamentally, is not only that injustice calls for judgment, but also that judgment intensifies injustice. Paul expresses this when he speaks of God "giving them up" [παρέδωκεν] in Romans 1:24, 26, 28. The sin of the builders of the tower of Babel (Gen. 11:1 ff.) is so judged that it has effects not merely on the vertical level, between man and God, but also on the horizontal level, between man and man. For when men no longer take God as their norm they come under the dictatorship of their own instincts. They thus become alienated from one another, unpredictable, objects of fear. There thus breaks into human society that centrifugal tendency which leads to disruption and scattering (Gen. 11:7 f.). In this context this is both a judgment and an intensification of the guilt itself.

The same point emerges, for example, in the problem of taxation. Those who for reasons of self-protection—not in the primitive sense of personal survival but in the sense of protecting their business—seek illegal tax advantages gain a step on their competitors momentarily, but as a result they force the others to act in the same way. What was initially a relatively small injustice is thereby condemned by the fact that it necessarily brings about an unforeseeable intensification and magnification. The economic order as a whole, which was what seemed to make the first injustice necessary, is as a result itself made even more unjust. Concretely, it necessarily brings about a further tightening of the taxation screw in order to make up the sums illegally lost. Repeatedly we are the "victims" of our initial fault, but at the same time initiators of new faults.

In the second place our dealing with the borderline situation is something that cannot be restricted to the sphere of personal decision. It always puts us also on that downward path which will help to determine not only our own future action but also the action of those around us, not least of all that of the next generation. In some cases it may even lead to basic ethical "mutations." Thus we do not merely act on the *basis* of the suprapersonal nexus of cosmic guilt from which we come. We also for our own part influence and effect it, and thus help to determine the milieu in which others must live and act. "Real" decisions are made only by those who are aware of these factors of action and interaction, and who thus see beyond merely the pressing issue of the moment.

To undergo and endure the borderline situation also raises questions from the standpoint of the Gospel. The ambivalence which attaches to the Word of God in its form as Law and Gospel [24] attaches also to the conflict situation when seen from both these angles. It gives to the conflict situation an ambiguity which cannot be resolved theoretically. This is the same ambiguity as is inherent also in judgment, which as *krisis* implies a crossroads from which the one way leads to life and the other to death. In this sense the borderline situation bears the sign, not only of the fallen world, but also of the grace which bespeaks promise within this world. It does this in two respects.

First, when I stand in the borderline situation as in God's presence, i.e., when I am aware of its guilt and need of forgiveness, then it cannot harm me or destroy me. By virtue of the power of forgiveness I can admit its guilt, and in so doing I remain "whole." Under forgiveness I am free from guilt. I can thus stand aloof from the situation and look at it objectively. To be dead to sin in the Pauline sense (Rom. 6:7, 10, 11) is to have no more interest in denying sin or in trying to falsify its true nature dialectically by viewing it as a transitional stage or as tragic fate.

To put it epigrammatically, one might speak of an ethical correlate of the concept of "disinterested pleasure" in Kantian aesthetics.[25] This would be the Christian's "disinterested displeasure" in guilt. But we could use such a phrase only in the sense that the Christian has no axe to grind in the recognition and estimation of his guilt. He gives up all attempts to argue that it is not really guilt, or to persuade himself, for the sake of his own self-respect, that it is only fate. For forgiveness alters the basic character of self-respect.

Self-respect is no longer possible on the basis of self-justification, i.e., on the basis of an existence *in re* in which our moral actions outweigh the immoral. When we base our self-respect on such calculations we become the victims, as it were, of a drive for moral self-preservation which is supremely concerned that the balance of our actions should be positive, and which is thus forced to resort to every possible manip-

[24] See Chapter 7 above.
[25] Kant, *Kritik der Urteilskraft* (Berlin, 1869), p. 214.

ulation and reinterpretation. Such moral calculation leads inevitably to pharisaism (Rom. 11:20).

The self-respect or "honor" of the Christian who lives under justification rests on the fact that God is the one who forgives us, that he has received us into fellowship, and that guilt can no longer tear us out of this fellowship. The fact that this fellowship is imparted to us "through faith alone" liberates us from the tendency to produce a moral alibi of "works," according to which good works are merits and bad works are the ineluctable product of a coercive situation which is itself derived from the tragic structure of the world. Forgiveness allows us to be realistic. This is why it keeps us "whole."

I am reminded of a former student, a pastor's son, one who had proved himself in the church conflict and who was later killed. Once when home on leave from the Eastern front he greeted his father with the words, "I can no longer look a Christian in the eye. Before I change clothes and go out you must give me the Lord's Supper." What was his father and pastor to do? Should he read him a chapter of moral theology and tell him that he ought to have acted thus and so rather than as he did in that terrible borderline situation? Possibly he thought of doing just that, but if he did, he kept it to himself. For in so doing he could only have said things of which his son was already keenly aware in his conflicts and distress of soul. There was no point in telling the boy what he already knew—what he *could* know only because he already knew of the divine forgiveness, so that there was no need to bring home to him his guilt. What was required was simply to declare to him what he could not give to himself, namely, the royal pardon of God and fellowship at his table. The father must have reasoned: We can speak later about what he has done, after in forgiveness he has won release and been enabled to breathe freely again. For more important than insight into the correctness of action—important as this is—is a continuing awareness of the corruption and guilt of action, and hence the sustaining of a hunger and thirst for righteousness.

It is only insofar, and so long, as I exist under forgiveness and take my orientation from it, that I can know the royal freedom of the children of God which is repeatedly promised me, the gratitude of the prodigal son who falls down before his father, and the divine table at which I am welcome even though I cannot break out of my impasse. Since this table and fellowship are the true meaning of life, everything depends on my not missing this contact. In the final analysis this may well be the meaning of the statement that whatsoever is not of faith is sin (Rom. 14:23), and vice versa. For the implication is that whatever guilt I may have incurred must fade into the background, and can no longer be regarded as sin and separation, if I keep before me this goal of maintaining faith and of being at peace with God through Jesus Christ. Then all things do truly "work together for good to them that love God" (Rom. 8:28).

This can only mean that things take on a "ministering" significance, no matter what they may be *in re*. If I love God, even guilt serves me —Augustine even dares to speak of guilt as blessed (*o felix culpa*)— for it causes hunger and thirst for righteousness to burn within me, thereby keeping me in contact with the promise of "being satisfied" which is pronounced in the Beatitudes (Matt. 5:6). If I love myself, however, guilt merely serves self-righteousness; it stirs me to argue and dispute away my sin dialectically (Rom. 6:15).

Guilt is a force on which I have to declare war, and in face of which I must be ready to make the clearest possible decision. For once it has occurred—as it does a thousand times for all of us—it is an open question how I will cause it to serve me, i.e., whether I will so fear, love, and trust in God above all things as to claim the promise that guilt too will work for my good.

When I perceive beyond the borderline situation the divine offer of forgiveness, I remain oriented to spiritual poverty. Indeed, this borderline situation is a demonstration of such poverty (Matt. 5:3). In brutal fashion it robs me of any chance for justification by works. It is thus a "crude" model of the world which simply does not allow of such works. Its very crudity is in fact symbolic, for the borderline case is an example or type of "this world" in its most concentrated form. This is why it has such a close affinity to spiritual poverty. It sets before us the tribulation of the world (John 16:33), and the fact that "this aeon" is incapable of giving us peace (John 14:27 f.). Here there can be no possible misunderstanding—provided that as the recipient of forgiveness I remain "objective" to the situation, aware that I am not obliged to create my own moral self-respect and thereby deceive myself.

A second promise which the Gospel brings to the borderline situation is that I am secured against the law of progressive ethical decline on the strength of my being forgiven. We have already spoken of this law. It implies that personal decision is involved only at the first step and that from then on I am in danger of being delivered up to the autonomy of its consequences. If, for example, in the matter of helping persecuted Jews I alter the name of someone who is threatened, I must then enter also into alliance with the counterfeiters of the underworld, and beyond that continually invent new stories.

As we have already seen, and as we cannot emphasize strongly enough, this downward path which looms before me does not consist primarily in the consequences that might follow, but in my evaluation of these consequences. The threat to myself—and it is even more of a threat to the next generation which does not know the initial conflict—is that I shall simply say, "Such is the world; it likes to be deceived; that is its way." If this is my view then I am not simply caught up in a trail of consequences; I am taking the plunge for myself. "If I yield myself to anyone as an obedient slave, I am the slave of the one whom I obey" (cf. Rom. 6:14). If I obey the law of the world, according to which, for example, might always triumphs over right, then I am the slave of the world, implementing its law of progressive decline. My obedience may be of two sorts: Either I consciously and intentionally go along, attempting to act as the world does and therefore in a supposedly realistic fashion, or I comply unconsciously and in ignorance, because I am too bereft of standards to recognize the progressive decline for what it is, as in the case of Hitler. On the other hand, if I am the slave of the Lord from whom I receive forgiveness, then I cannot be a blinded slave of the downward path at all, because I do possess a standard by which I can always see it for what it is.

All this is not merely a matter of knowledge and the clarity of perception. It is also a matter of action. The borderline situation may offer no avenues of escape. But if that ultimate standard is present, I am protected against the leveling out of alternatives, I shall not despise the nuances, and I shall never reach the state of indifference which allows me to say that in the blackness of this world's night, in the darkness of the borderline situation, all cats are gray. The Christian will always draw back from this downward plunge that looms before him. In opposition to all worldly wisdom, he will again and again be inconsistent. He will demonstrate to himself and to others that he is subject to no law κατὰ σάρκα. To return to the example of hostile interrogation, he will carefully consider whether the impending threat is to himself alone or also to others, and if it is only to himself he will not "automatically" deceive and lie on the ground that the enemy's claim to truth has been forfeited. The necessity for this course may be great, and he may indeed have the right to lie if the claim to truth has in fact been forfeited. Still he may make his demonstration "once more," not in terms of this right or according to the law of the inevitable consequences but by bearing witness to the authority of the truth, witnessing to the enemy through the very fact that he brings himself into jeopardy by this witness.

30.

The Commandments of God and the
Borderline Situation

Let us now return to the main point with which we began our discussion of the borderline situation, namely, to a consideration of the
borderline situation as the crucial test of ethics.[1] For the thesis that
there are situations in which man must shoulder guilt cannot be our
final word, unless we are willing to submit to a purely tragic understanding of the world.

The Ambiguity of the Commandments in This World
To begin with we therefore pose this question: Do we not have
clear commandments concerning things that we may not do under any
circumstances in the conflict situation? Am I not told as unambi-
guously, unqualifiedly, and unconditionally as possible that I may not
steal, kill, or bear false witness? Does not this mean that direct and
simple obedience is here demanded of me? Am I not instructed to
leave all responsibility for the results of such obedience in the hands
of God alone?

We have seen already that this absoluteness can be shaken. We
have only to recall the saying of Jesus with regard to Mosaic divorce.[2]
From the standpoint of our present problem the significance of his
saying is that the command of creation, i.e., that the two should be one
flesh (Gen. 2:23-24; Matt. 19:5-6; Eph. 5:28-31), can be robbed of its
absoluteness. Under the circumstances of this aeon in which "hardness
of heart" prevails, and concretely in the case of adultery or certain degrees of adultery, the union of man and woman which was designed

---

[1] In his preface to the third edition of *Die letzten Dinge* (Gütersloh, 1926),
Paul Althaus writes, "He who writes on eschatology must betray almost all his
theological secrets: his understanding of history, his Scripture principle, his
Christology, etc." We have seen that he who thinks through the borderline situation is similarly forced to betray almost all his dogmatic and ethical secrets: his
doctrine of justification, his concepts of the world, of history, and of the Law,
and his views on the nature of sin and on natural law.

[2] See p. 148 above.

to be a blessing and a promise can become such a curse and burden that a continuation of the marriage is no longer to be expected. Similarly, when Paul in I Corinthians 7:12-16 "gives the Christian partner the right to dissolve a marriage if hostility to one's Christian conviction leads the other partner to pull out," [3] we are again confronted by the fact that marriage, while it is a christological likeness (Eph. 5:25 ff.) having the force of a command, may in its concrete form come up against the limit of its parabolic character, and therewith alter the nature of the command.

The borderline situation is an extreme form of this crisis of the commandments, this unavoidable crisis. Because of its extreme character, it is, as it were, the example par excellence of the whole situation of conflict. Yet it would be theologically mistaken to argue that, because absoluteness often seems to be lost and we are compelled, often in the midst of complete confusion and uncertainty, to search for clarity, it is therefore up to us to decide for ourselves by our own contingent resolve which of the conflicting commands we regard as binding, e.g., whether we should obey the command which requires the preservation of life or the command that we speak truth and be in truth. Arbitrariness in this respect could lead to real excesses of spontaneity, of perverted desires, and of no less perverted anxieties, so that finally an unrestrained and truly contingent eclecticism holds the field.

No matter how we proceed, however, or fail to proceed, there can be no doubt that here decisions do have to be made, that the situation is in the true sense problematical, that it is not unambiguous, and that it consequently involves real distress of conscience. The ambiguity of the situation is manifested in the fact, not merely that decisions are required, but also that—the pertinent commands themselves being in conflict—the criteria which must underlie such decisions are also equivocal.

To deny the conflict situation is to deny decision, that decision which, being threatened by the lack of criteria, is threatened at its very foundations. But to deny decision is to deny the most profound of all the plights of conscience which confronts us, and to regard it as based on mere imaginations, on unredeemed chimeras and complexes. To think thus is to contest a fundamental phenomenon of our spiritual life, the life which Paul has in view when in speaking of the broken reality

[3] Adolf Schlatter, *Die christliche Ethik* (Stuttgart, 1929), p. 400.

of this aeon he says, "We do not know what we should pray for as we ought." Even in prayer we do not know which way to turn. This is why we must be hesitant with our own petitions. This is why we must retreat behind the Spirit, behind God himself, who leads us out of the dilemma (Rom. 8:26).

The dilemma, the ambiguity is not due to any lack of clarity in the divine commandments themselves. It is due to the mists of this aeon, in which a clear beam of light becomes a diffused cloud of light. The sun itself is not darkened. It is simply concealed by the clouds in the atmosphere of this aeon.

The Noachic world conceals the true order of creation [*Schöpfungsordnung*]. If we overlook this concealment, if we count upon an unequivocal world which does not exist, or no longer exists, we inevitably become fanatics. Figures like Thomas Münzer and Leo Tolstoi are hewn out of such wood.

We are thus warned not to overlook the aeonic roots of the conflict situation. We are reminded to take seriously the problem of ambiguity in the commandments in our world, and to recognize the essential element in an "ontology of the fallen world." Thus prepared, we may now take up the question whether some clarity of action may not still be possible, whether there are not after all some criteria on which to base our decisions.

The Problem of Means and Ends

Methodologically, we may begin with the presupposition that the ambiguity which we have described, which is to be taken seriously, is not so great as that of the "muddy waters" in which men proverbially "fish." That is to say, it is not such as to commit us to irresponsibility or indifference. For in this conflict the competing norms are arranged in a certain order, even from a purely temporal standpoint.

By way of illustration we may adduce the whole matter of "denazification" in Germany after 1945. It provided a rather unwelcome occasion for exploring all kinds of basic questions in respect of law, ethics, and politics. As a result it came to possess an exemplary significance which went far beyond the sphere of its immediate application. It is of value to us too as we seek here to express what we mean by the temporal sequence and substantive order of the commandments.

Thus the university administration might find that to keep the wheels of education turning at all in the critical postwar period it would have to go along with the commanded denazification procedures which required "permanent dismissal" of all professors who were suspect in the slightest, or perhaps even formally charged (though not yet tried and convicted). It would have to fill such vacancies (and thereby sanction what might have been an illegitimate dismissal). But this raises a searching problem of conscience: Can "blessing" rest on the fulfillment of one command, namely, that of service (providing a sorely needed education under the most difficult of circumstances for the vast postwar influx of eager students, whose leadership in the years ahead would be indispensable), if such fulfillment is attained at the expense of violating another command, namely, that of standing up for what is right (protesting the questionable denazification procedures by refusing to fill the teaching vacancies they had created)? In the more precise formulation of ethics, can the means to an end be "blessed" when the means is itself the breach of a command?

For ethics, the decisive point is brought out in this formulation. Once the two demands, namely, the rendering of service and the standing up for what is right, are made the subject of ethical and juridical reflection, they are by that very act set objectively in the relation of means and ends (not intentionally of course but by virtue of what is actually involved in the nature of the case). Then it becomes obvious which command must take precedence so far as I am concerned, namely, the command to stand up for what is right even though it impedes my opportunity for service.

This implies quite simply that even in the name of an end which seems to be commanded I cannot disregard the means; I cannot scorn the commands which restrain or the safeguards which impede me, as though the situation of conflict could that easily be resolved. I may of course have to face up to the decision whether to "sin boldly," and "believe even more boldly in the forgiveness of Christ." [4] I may have to face the possibility that what is involved here is a borderline situation which does not allow of any easy solution. But I can reach such a decision only by going through the conflict and enduring it, not by evading it in the name of some kind of perfectionism (which robs me of justification in the judgment) or of indifferentism (such as we find in fatalism and pantheism).

[4] See p. 504, n. 19.

Within the confines of our particular problem there are two such attempted evasions. The first stresses the commands in respect of means, the second stresses the commanded ends. The evasion in the name of the commands having to do with means holds that I must never under any circumstances steal, etc., even if I am in peril of death. The evasion in the name of commanded ends holds that in the name of the service intended I may forget about doing right, e.g., I may maintain the full complement of urgently needed faculty by replacing those teachers who have been wrongfully dismissed (thereby doing them an additional injustice).

Both attempts at evasion derive from decidedly non-Christian ideologies. The first issues from a leglistic moralism which stops with the letter of the law, refuses to face up to problems or make decisions, and seeks its "security" in servile obedience to the letter. The second is attempted in the name of a non-legalistic pragmatism which is based on the illusion that the end justifies the means. But the legalism and the pragmatic violation (not transcendence!) of legalism are in antithesis to the freedom from the Law which prevails "where the Spirit of the Lord is" (II Cor. 3:17). For the Spirit of the Lord makes me free, the subject of my own decisions and not the unfree functionary or impersonal organ for implementing some "ism," be it nomism or perfectionism. "Where the Spirit of the Lord is," I can accept the risk of decision, even of a doubtful decision because the question of my abiding in fellowship with God is no longer a matter of my own achievement and rectitude but of divine grace. Hence I am free to carry responsibility. I can be myself, stand on my own two feet. I no longer need either nomistic safeguards or pragmatic evasions.

To be sure, I know and can never forget in whose name I have this freedom. My freedom does not have its source in libertinism—that would be another "-ism." On the contrary, it is given to me. I am "bought with a price" (I Cor. 6:20; 7:23). It is thus a freedom which binds rather than releases. I have it only as a "slave of obedience, which leads to righteousness" (Rom. 6:16). All that we do—that *we* do, i.e., in freedom, in personal responsibility, and not as mere functionaries of the Law—is now to be done for the Lord and not for pragmatic human ends (Col. 3:23; cf. I Cor. 10:31; Col. 3:17). Hence the freedom of our decision, and the risk we run in making it in the conflict situation, is shown to be in every respect the freedom of those who are "bound."

In the conflict situation, then, we seek to arrive at a legitimate decision, to take an authentic stand with respect to its humanly insoluble problems. And as we do so we are not simply conjuring up that night in which all cats are gray, in which it is ethically a matter of "indifference" or little consequence what we do, and license is given for every caprice. On the contrary, we are seeking for the possibility of obedience, not the obedience of the legal functionary which roots in servile fear but the "obedience in freedom," the obedience of the redeemed, an obedience under the grace which both binds and liberates, but which is always faithful to us.

It is just because we are not left to our own caprice that so much depends on our first establishing the point—we repeat our thesis— that even in the name of an end which seems to be commanded I cannot disregard the means, or scorn the commands which restrain or the safeguards which impede me. This proposition may be supported by two considerations from the Bible.

First, it is inherent in the New Testament concept of time that I am always referred to my neighbor, to what is close at hand, to what is directly before me, to what I can do at the moment. Neither the priest, the Levite, nor the Samaritan was asked about the (doubtless significant) purpose of the journey he would have to interrupt if he were to risk his life by helping the man who had fallen among thieves (Luke 10:30-37). Each was simply referred to that which was right there at hand, in this case quite literally the man who lay there before him. Love is a capacity for improvisation directed to the moment. Even to ask concerning the result should we obey this duty in respect of our neighbor is to incur guilt. The priest and Levite do, in fact, incur such guilt. This is why we are commanded not to be anxious about tomorrow (Matt. 6:25 ff.; Phil. 4:6; I Pet. 5:7; I Tim. 6:6; Heb. 13:5). We must be content with the bread rations for only one day, today (Matt. 6:11). Even the Word of God which illumines my path is only a "lamp to my feet" (Ps. 119:105) which lights up the next step; it is not a searchlight enabling me to see great stretches of the way that lies ahead.

This means that we are not to "calculate," or, more precisely, that we are not to be driven by the promptings of anxiety and the dictates of fear to calculate, what will happen, what will be the result of an action we must undertake. (We are obviously not thinking here of weather forecasts or of graphs projecting, for example, population increases or decreases.) For the future does not develop in the fashion of an entelechy out of an immanent law which controls the world. It lies rather in the hands of God, and ends at his throne. This is why we are forbidden to ask what the result will be when I keep this or that commandment, e.g., the command to do what is right even though the long-range possibilities of service seem thereby to be reduced if not shattered. We simply have to do what is right in the situation which is before us.

At this point it has also to be shown, of course, that there is no pat solution. We cannot calmly commit ourselves to a certain course knowing that it will automatically bring us out at the right point. This kind of consolation, namely, false security [*securitas*], is always a sign of the Law.

In fact, the problem is made all the more acute by the fact that I am usually not asked at all concerning the ultimate consequences. On the contrary, obedience to a commandment, e.g., the command to tell the truth, may be identical with simultaneously delivering hundreds or even thousands of others up to certain death, as in the examples taken from the ideological tyranny of the modern totalitarian state. In such a case what is involved is no longer a matter of calculating the future and distinguishing between the short-run and the long-run effects. For the whole point of such an interrogation, and consequently of my answers, might well be to make possible and to justify this committal of hundreds and even thousands of others.

It seems to us that in such extreme cases untruth cannot be regarded as a "commanded" way of escape, for if we regard it as commanded, what we have is again an evasion of the conflict situation, and this is always wrong. It is rather that in such cases a man is prepared to accept the guilt of untruth or of forging papers (when he illegally assists persecuted Jews, for example). He is willing to take this guilt upon himself not in the name of the tragic, but in the name of forgiveness.

Second, the Bible will not allow us to overlook the "command along the way," the command in respect of means, because, as we might put it, violation of this command alters and destroys the end it was intended to attain. When I set aside the obstacles which block my path, I may seem to reach my goal more quickly, but in fact the goal itself has suddenly been changed. It has been disfigured and defaced. It has become something that is not my goal at all but its very opposite. By way of illustration we may take two examples, one from the New Testament and the other from our own time.

The story of the temptation of Jesus offers an exact illustration of what we have in view (Matt. 4:1-11; Luke 4:1-13). When the devil takes the Lord to a very high mountain, shows him the vast expanse of the world, and offers him dominion over it, Jesus seems to be given the chance to attain his goal more quickly. By using the means of earthly power granted him by the devil, it would seem he could bring about the rapid Christianization of the earth and the actualization of the kingdom of God.[5] But to reach this quicker and easier result he would have had to break another commandment, namely, that the Son of Man must suffer (Matt. 17:22; 20:18; 26:2, 45), that he must accept lowliness and poverty (Matt. 8:20), that he must remain in the unassuming form of service without laying any claim to the demonstration of his power (Luke 4:3, 9; Matt. 26:53; Matt. 16:23).

If, however, Christ had accepted the offer, and evaded the intervening commandment in order to reach his goal the more quickly and easily, that goal itself would have been altered in substance. The "kingdom of God" attained along these lines would have been a mere caricature of what the phrase was really supposed to mean. For the kingdom of God would have emerged from the concealment which makes it an object of faith. It would have become an object of sight, a demonstrable reality. Then it would no longer be sin to reject it. It would be palpable folly. (For what is more foolish than to refuse to recognize an established world empire replete with all the emblems of its power, or even to try to deny its existence?)

The kingdom of God would no longer be the object of decision. For why should I decide that something exists and is in operation when I can see it in every detail, when at every step of the way it is a matter not of judgment but of perception? Opposition to something so plain is absurd. The kingdom of God would no longer be the majestic sphere of the Gospel into which I could be integrated by the gift of the Spirit. It would instead be a kingdom of sight and of Law,

---

[5] See my *Zwischen Gott und Satan* (2nd ed.; Tübingen, 1946).

which would more or less automatically incorporate me into it by the sheer fact of the way things are. What was "previously" called sin would now be merely an offense against the state, something that could also be established objectively, whereas sin is something that must be believed inasmuch as he from whom sin separates is himself an object of faith. The "theology of the cross" would have become a "theology of glory" (Luther).

We have said enough to make it plain that if Jesus had evaded the command having to do with means—to the effect that his glory be hid under suffering—the realized goal of the kingdom of God would have been changed in substance. It would have been turned into its very opposite. The goal is essentially dependent on the means used to reach it.

The second illustration of this fact, the example from our own times, is again taken from the instructive sphere of the ideological power state. Around 1934 there arose within the Evangelical church in Nazi Germany certain lines of thought which are astonishingly similar to those in Soviet Russia and its satellites, and which run more or less as follows.

It looks like totalitarianism is going to be around for a long time to come. This seems to be indicated, not merely by the concrete measures the totalitarian powers have taken to consolidate their rule, but also by certain observable features of this age of the mass society, in which there is almost an inherent tendency towards dictatorship. We have thus to plan for the long haul. If we do not try to work out a *modus vivendi* with dictatorship, in which there can be some kind of fusion and synthesis in the form of a national Christianity or a communist Christianity, we shall sooner or later be listed among the enemies of the state and be condemned to destruction. We may lose our public influence, of course, particularly upon young people, but at least we shall be allowed to grow old naturally and to die by the normal processes of old age (Alfred Rosenberg put it just that way). On the other hand we do have a binding command to preach the Gospel. In terms of our present context the preaching of the Gospel is a commanded end. But we cannot keep this command if we reject the indicated fusion and *modus vivendi*. We can attempt to keep it only if we enter into this synthesis "in order to prevent something worse." We must act strategically, circumventing the hampering commands in respect of means.

Compromises, it is held, must consequently be made which will involve initially the breach of some commands. We might mention only two of these commands which seem to have played a particularly important role in such considerations: first, that the church ought to take seriously its duty to instruct young people and to build them up in the fellowship of faith, using to that end such means as are in keeping with the kerygma; and second, that the church ought to reserve for itself alone the right to train, test, and supervise the preachers of the Gospel without any outside interference. These commands in respect of means, which take precedence of the commanded end of preaching the Gospel, must now presumably be relaxed a bit if the commanded end is to remain in force and be attainable.

Concretely, this relaxation of the commands having to do with means will have the following implications. The church's own youth work and youth organization will have to be dissolved or transformed beyond recognition because, first, they would otherwise sooner or later be forbidden, second, their voluntary dissolution would be a proof of loyalty, and of agreement with the postulate that the training of youth is entrusted to the state, and third, this demonstration would feed the hope that what was lost in intensity and depth could be made up in scope as the youth of the entire nation came to be permeated by the leaven of Christian influence. These were and are the arguments which usually accompany Christian capitulations and the hoisting of the white flag.

Much the same arguments are used in respect of the second command, namely, that preachers of the Gospel should be exclusively under the control of the church. This means concretely that once the authorities of the church, its administration and its theological faculties, are brought under the dominion of the ideological tyranny, for example, are controlled by the "German Christians," the church has then to proceed to establish its own seminaries and educational requirements. The consequences may easily be foreseen, indeed they followed at once (as they always will) in Nazi Germany. The unofficial theological schools were not recognized by the state but were actually forbidden. The graduates of such institutions were not allowed to proceed to the vocational goal for which they had prepared.

While the sheep were left without shepherds, young preachers were dismissed and condemned to inactivity. Under Hitler, pastors who had received such irregular training were classified not as ordained men but as theological students; they were restricted in their pastoral activity, and promptly drafted into the armed forces when the war broke out.

In face of this situation, the question was repeatedly asked whether we can allow such a situation to go on, the sheep being without shepherds. Is it not necessary to break the commands in respect of means—that preachers of the Gospel should be under the legitimate control of the church—in order to be able at all costs to attain the commanded goal, namely, that the Gospel should be preached?

Here in respect of the preachers, as in respect of the youth, it is apparent that the goal itself is partially determined, or at least affected and modified, by the means whereby one hopes to attain it. The goal cannot be attained by a path which is strewn with the remnants of broken commands. In the case of the youth, the leaven effect did not materialize because the young people were no longer in the Word's field of force but simply entrusted to the "Christian spirit" of particular persons. In the case of the preachers, the message which was supposed to be preserved along with or by the intact ecclesiastical apparatus was watered down to nothing—think of how the Old Testament was handled in instruction—under the aegis of non-ecclesiastical directives and "German Christians" consistories.

The conflict in which the church finds itself when confronted by decisions such as these is more acute than this sketchy presentation can describe. Nor should we condemn too summarily the solutions which were less radical or involved a greater readiness for compromise. The host of factors involved is much too great to allow of discussion, even by illustration, in the present context.

Undoubtedly these church conflicts, even under totalitarianism, are less "difficult" than those of the underground situation which we described earlier. For in them the church has only to face the decision whether it will obey the commands which are in force today with respect to means, or whether it will anxiously attempt to predict the fateful results which may follow tomorrow from such obedience, results which may seem to portend a day of martyrdom, and even worse, the complete silencing of its message. The solution to this

619

church conflict is at least clearer than in the case of the underground situation.

The comparative ("clearer") is adapted to remind us that, when ethics is dealing with concrete decisions, it always moves in the sphere of the quantitative, of what is more or less right. It would be a false abstraction, an illegitimate ignoring of the conflict situation, to think that we are always faced by clear-cut alternatives, or, in view of our total sinfulness, to cultivate ethical indifference by supposing that all acts come under the same condemnation. With reference to this ethical problem of the quantitative, Dietrich Bonhoeffer says quite rightly (and he uses a series of comparatives to say it): "What is worse than doing evil is being evil. It is worse for a liar to tell the truth than for a lover of truth to lie. It is worse when a misanthropist practises brotherly love than when a philanthropist gives way to hatred. Better than truth in the mouth of the liar is the lie. Better than the act of brotherly love on the part of the misanthrope is hatred. One sin, then, is not like another. They do not all have the same weight. There are heavier sins and lighter sins. A falling away is of infinitely greater weight than a falling down. The most shining virtues of him who has fallen away are black as night in comparison with the darkest lapses of the steadfast." [6]

Ethical Decision in Concrete Conflict Situations

Having considered the crucial difference between the commanded ends and the commands in respect of means, we may now re-examine the typical examples of the borderline situation which we drew from the experience of the underground movements. We did not choose these examples because of their dramatic quality or for the sake of their anecdotal interest. We chose them because they represented a certain type of conflict, one that had been selected on the basis of theological criteria. As we come back to them now, after surveying the whole range of fundamental problems, we hope to be able to reach some real conclusions.

Life in the Concentration Camp

To prevent something worse, such as the application of an even

---

[6] Dietrich Bonhoeffer, *Ethics*, ed. Eberhard Bethge, trans. Neville Horton Smith (New York: Macmillan, 1955), pp. 3-4.

harsher system of sanctions for the entire prison, and so to spare the lives of others, the intervening command "You shall not kill" is set aside. Christians, however, are too soft at this point, according to Eugen Kogon.[7]

Now it goes without saying that we cannot pass any simple judgment on this matter in terms of theological theory. Nor is it necessary to offer reasons why we cannot. Our theoretical judgment would be of no help anyway to someone who should find himself in a similar situation. As we have repeatedly insisted, theological ethics cannot casuistically prejudge all possible decisions. Its task is rather to clarify their content, and thereby make them true decisions. It has to show what is at stake in the decision, what is the ultimate issue. The task of ethics is not to make decisions, to relieve its devoted followers from assuming that responsibility for themselves. A decision that we take over from somebody else is an enfeebled decision. It bears the marks of legalism. For the person who is summoned to decide is himself no longer the acting subject. He is not allowed to be free, to make his own decision (Philemon 14).

Ethics does not solve problems; it intensifies them by showing what is the point of any particular decision. Its task is that of clarification. It does not seek to spell out precisely what must be done. It simply lays bare the full implications of that "what," so that the underlying issue of the particular decision is heard as a claim. Ethics in effect complicates matters by analyzing the contents of the decision. But there is a counterbalancing simplification for the one who then decides: When it comes to the point of actually taking concrete action his decision is simplified by a gift of the Holy Spirit.

We cannot judge what Kogon says because so many factors enter into a decision, and because a decision seldom has in practice the relative simplicity of an illustrative example. Thus, in the example of the concentration camp, as always in everyday life, even where no borderline case is involved, the decision is complicated by the fact that it is not the Christians alone who have to decide. Christians live in company with a relatively preponderant number of non-Christians who also bear responsibility—perhaps the main responsibility— for the decision. In such a decision there is consequently not only a substantive problem involved but a social problem as well: May we,

[7] See p. 589, n. 15 above.

must we, sabotage the agonizing decision which others have reached to proceed with the liquidation of the unfit? And even if one Christian believes that he is in fact called to bear a symbolic witness of protest, can we make of this a general law in the sense that in such instances theological ethics can countenance no other course?

This obvious interplay of heterogeneous and complicating factors, while it is particularly clear in the borderline case of the concentration camp, is present everywhere. It may be seen with classical pregnancy in the ethical problems of the Second World War. Many Germans ultimately recognized that this war was unleashed by a madman, and theoretically they ought to have concluded that they would have no more part in it. But for each individual, the general, the company commander, the non-comissioned officer, this terrible diagnosis was accompanied by the question whether he could now abandon his comrades, whether he had any right just now in the midst of this deteriorating and guilt-ridden situation to forsake his nation rather than acknowledging his solidarity with it.

Is there not needed here some understanding, not only of the courageous symbolic protest of the conscientious objector, but also of the conscientious determination to "sin boldly"?[8] Do we not have to respect the fact that under the shadow of forgiveness different decisions are possible and different liberties and loyalties may exist? What church administration would have the freedom to proclaim in a legal sense—quite apart from tactical considerations—that this way, and this alone, is commanded? This is of course a rhetorical question. It carries with it its own obvious answer, even though individual pastors may be obliged to say to individual parishioners that this, and this alone, is indeed the commanded way *for you*, and that it will be sin for you to refuse *your* allotted task (whether it be that of solidarity or of conscientious objection).

In respect of all these complicating factors which enter into the situation, the following thesis may be advanced. The essence of the required decision does not consist merely in deciding in favor of practical obedience to what is recognized to be right and consequently "obligatory." The essence of the decision is to be found rather at a prior stage than that of practical action. For it consists already in a recognition of the factors at work and an investigation of the cri-

[8] See p. 504, n. 19 above.

teria which must be determinative of such a decision. Hence the essence of ethical decision lies, not merely in the obedience of action which implements it but in the obedience of cognition, and therewith in the struggle for determining the ethical point at issue.

But let us look a little more closely at the theological side of the problem. As we have seen, the problem consists in the fact that it is seldom if ever posed without theological ambiguity. In other words, there is nearly always a whole group of alternatives, not a single either/or.

This very fact makes it difficult to decide the matter "from the outside," from the standpoint of Law. It rules out the possibility of there being any a priori rule of casuistry which could apply to the given situation. The concrete situation can seldom if ever be subsumed under an existent rule, because there are always, or almost always, a number of rules under which it would have to be subsumed.

This is what ultimately gives to the situation of conflict its "background" quality. This is also why we have to struggle for decisions; we do not just find them ready at hand, in the prefabricated form of existent rules. Finally, this is why a spiritual decision can and should be authoritatively influenced only in connection with individual pastoral counseling directed to the particular situation, i.e., only to the degree that someone is present who accepts solidarity with the individual concerned in the actual situation, in the given complex of alternatives. The pastor who regularly accepts solidarity with his parishioner's situation may be said to be "on the inside," whereas rules and laws are obviously "on the outside." There is thus a radical difference between them.

"Outside"—and this is also where theological ethics stands—the situation is complicated by consideration of the many alternatives. All the factors at work have here to be surveyed and evaluated in order to get at the point of decision. By the guidance of the Spirit, however, the one who has to act is led from multiplicity to simplicity. He who is brought to court in the name of Jesus need no longer be anxious about "how he is to speak or what he is to say" (Matt. 10:19). When the time actually comes, in the ᴵ of trial, he need no longer consider the multiplicity of factors involved. He has the promise that at the decisive moment everything will be simplified by being reduced to the single alternative of Christ or Antichrist. Simplification is not always bad. There is such a thing as the simplification by grace. But this gift is usually enjoyed, not when we theorize from without, but when we are thrust into the midst of the crisis.

This means that we cannot pass theological judgment from without on the decision made in the concentration camp, involving as it does a substantive question, plus the problem of solidarity, and the further consideration that those concerned included both Christians and non-Christians. All that we may have been forced to say on this issue in our present context has been no more than a mere question. To be true to the facts of the case the question must indeed be put in the form of a pastoral inquiry which would have to run as follows.

Might not things have been the very reverse of what they seemed to be in Eugen Kogon's account of the SS concentration camp? He says the Christians did not have the necessary strength and courage to make such decisions concerning the death of their comrades, and hence to shoulder guilt. We are compelled at least to pose the question whether it did not in fact demonstrate quite the contrary, precisely an unparalleled strength on the part of those Christians who in a situation requiring witness left to others the decision on whether or not to kill—perhaps even to do the murder themselves—and who took that stance even at the risk that they would themselves be the next victims, and that the result of their stance might be a real fiasco rather than some kind of success. In this kind of attitude might there not be an incomparably powerful testimony to the sanctity of human life? And though the question of results is not the decisive one, though it can actually obscure and blunt the basic point, the further question is perhaps not illegitimate whether such a witness through suffering might not arouse attention and even fear, whereas the desire to "prevent something worse" would possibly—perhaps even certainly—have encouraged the spirit of inhumanity. For this inhumanity might easily have found in the collaboration of Christian prisoners the assurance of a certain legitimation. Has not Gandhi given us an impressive example of the fact that the symbolical renunciation of force is not merely a "Platonic" act of confession but a factor of realpolitik which can have historical influence?

The same problem occurs in a less acute form in the church conflict of the 1930's, and there are again very close analogies in the struggle of the church against Communism. When pastors were imprisoned by the hundreds, there were always those who typically argued, "In order to prevent something worse, in order to prevent a complete silencing of the Gospel, we must make concessions." But to yield thus was to legalize the wrong that had been committed, to make earlier resistance unconvincing, and to stimulate the persecutors to even more drastic measures.

The slogan "to prevent something worse" is alway ethically destructive because it subjugates our action to a non-Christian pragmatism. This pragmatism is the more sublime—and the less easily recognizable —the more this "something worse" consists not in personal danger but in danger to the sacred cause, e.g., through the silencing of the message. To touch once again on the question of results, we may say that Martin Niemöller is still probably right in his statement, "The more that get locked up, the sooner they will all come out again." In the long run bravery and steadfastness are the most prudent form of action, because they demonstrate the fact that our cause is truly convincing—that we really believe in him to whom "all authority has been given" (Matt. 28:18) and hence do not credit the possibility of our being surrendered *in toto* to that which is "worse"—and thereby make of this cause a factor which has to be taken into account.

According to the more recently published testimonies of those in power, the reason why the Evangelical church fared so badly under the Third Reich was not just because genuine offense was taken at its message, as many in romanticizing retrospect liked to think. To an equal and perhaps even greater degree, the church was persecuted because there was so much contempt for its pliant and unconvincing character. What is to be deduced from this ought to be addressed to the conscience with all due reserve and appropriate humility, i.e., as a true question in face of the situation in the concentration camp to which we have referred.

Stealing as a Matter of Life or Death

The point at issue here is whether the commandment not to steal may be broken in cases of extreme urgency, e.g., where it is a matter of life or death. We must first insist that so far as other men are concerned [*coram hominibus*], the meaning of my action is to be found in its character as testimonial or witness (Matt. 5:16; 7:20), even where the action does not specifically involve words (I Pet. 3:1). It is from this angle that we must judge each instance of stealing in extreme emergency, or any kind of illicit helping of oneself such as is almost unavoidable in times of institutionalized disorder. Two attitudes are possible at this point, and the question is whether each may not be a demonstration of faith. The one is that I would sooner go hungry, or bankrupt and leave my employees to their fate, than steal. The other is that I would sooner use illegal means to help me over such an emergency, even though in so doing I become a part of and contribute to the evil situation.

In the latter case, to be sure, we have to add the solemn proviso that this possibility of action too must take on the form of a demonstration of faith. But this is possible only on one condition, namely, that with my contribution to and participation in the evil situation I combine a public confession which will furnish the equivocal action with a word form which is not equivocal.

It will be seen at once that such confessions are not easy. Indeed, they may be very dangerous. But there is no promise that Christians can set forth their confession of faith without danger. The only real question is this: What is to be the critical point in my confession? Is it to lie in my silent endurance and readiness to perish, or is it to lie in an illegal resistance which openly confesses itself to be such, and which thereby also explicitly establishes the evil of the situation as a whole? It may be pointed out that a demonstration of Christian faith can be made in either way, that either may involve an aloofness which simply will not go along with "the world."

The question as to the concrete form of such a demonstration cannot be answered a priori, and is in any case a question of secondary importance. The real point is that whatever its form, this confession must be made. If it retains firmly its character as a demonstration *coram hominibus*, then in the dimension *coram Deo* it will guarantee that the wound is kept open, and that the downward movement toward the normalization and legalization of guilt will never even begin.

Helping the Persecuted Jews

Here too we shall have to restrict ourselves to just a few questions. The situation was so terrible that we are reluctant to use the scalpel of analysis at all, much less to take up the accuser's stone, for we are speaking here of men who practiced deceit out of *agape* and at the risk of their lives.

In this connection we are reminded of the attitude of Jesus to the woman whose sins were many, but were forgiven because she loved much, and who conversely could love much—and express that love in highly unconventional ways—because she was forgiven much (Luke 7:47). It is true that these "many sins" are not made good by the counterbalancing act of love. They remain *peccata in re*. But concerning them the promise is given that they will be remitted. We are also reminded of Paul's wish "that I myself were accursed and cut off from Christ for the sake of my brethren" (Rom. 9:3). Might not this readiness to be anathema for the sake of the brethren be connected with the readiness to shoulder guilt out of *agape*, in the constant realization that this is really to stand under an anathema, really to be dependent altogether upon a miracle which cannot be taken for granted, the miracle of forgiveness and royal pardon?

Apart from this, the question has to be asked how the unjust situation came into being in the first place. As we have seen, this question of origin is always the theologically decisive question in matters of guilt, since it has in view the personal-suprapersonal character of guilt, and thus assures from the very outset that sin will not be moralistically reduced to a mere resolve of the will. If this origin is not to be understood in the amoral or morally neutral sense of tragic fate—a view which we cannot accept—then it can be understood only as deriving from certain guilt-ridden decisions which constantly and inevitably give rise to fresh evil once we yield ourselves to, or are caught up in, their mechanism. Although we cannot at this point attempt a historical investigation of the genesis of the unjust situation, we can at least touch on the aspect which is of decisive concern to the church.

This particular aspect is brought out in the question whether there is not reflected in the emergency assistance given to persecuted Jews —by individuals who were acting under great stress of conscience because of the illegality of what they had to do—a failure of the church to make the kind of confession which was incumbent upon it. How would matters have gone, might they not have gone differently, if the church had been less preoccupied with itself—a charge which has been justly leveled even against the confessing church—and in an act of public confession had specifically branded the persecution as evil and wrong? [9] How would matters have gone if, when the injustice could no longer be arrested, the church had publicly identified itself with Jewish Christians at the Lord's Table, using this way of preparing the victims for whatever lay before them. Would this not have been a witness which would at least have made it impossible for the persecuting state to argue that it was only dealing with "subhumans" and that it could therefore legitimately move against them with a free hand? Would not the state have found itself confronted instead by martyrs? And would that not have served as a reminder of what killing really means? Who can say whether the state might not have hesitated before this final barrier? This much at least we can say, that such action by the church would have had the quality of witness, and been covered by divine promises.

Why does the secularist strategy of the totalitarian state always recoil before a true confessional stance in the same way as the devil flees from holy water? The official reason is that the situation "Here I stand; I cannot do otherwise," [10] is to be evaded on tactical grounds because it represents the element of extreme rigidity. Hitler himself repeatedly insisted that people of opposing views must be overcome "by installments," i.e., in small stages, none of which will seem significant enough to merit going all out for. In this way it would be possible to get around the firm confessional stance by means of a gradual erosion. But behind this tactical consideration there is concealed a much deeper concern which can be understood only in theological terms. This is the fear of being confronted by genuine martyrs, and hence by that claim—made convincing by their suffering—which challenges the exclusivism of totalitarianism and provokes the kind of uncertainty that is feared most by the fanatical representatives of ideological tyranny, both in respect of themselves and of their people.

Even if all this had been externally useless, the further question arises whether the church should not have publicly acknowledged the injustice of the situation and its own readiness to give illegal help. While this would indeed have been very difficult, and while it is

[9] We are thinking, for example, of the impact which the church could have had if the penitential sermons which followed the anti-Semitic outbreaks of November 8, 1938, had been much more specific than—with very few exceptions—they actually were.
[10] Cf. LW 32, 113.

much easier and perhaps gratuitous even to suggest it after the event, the question still remains: Would not the illegality thereby have taken on a very different aspect? Would not that which had the dreadful appearance of the underworld, at times even of unrestrained melodrama, thereby have been transformed into the distorted countenance of this aeon crying out for redemption? We can do no more than ask these questions. But ask them we must, and that for two reasons.

For one thing, they bring out the fact that the guilt men shouldered in illegally helping the Jews cannot be isolated. It does not belong only to those who, acting out of love, dared to do something rash, questionable, and illegal. The church of Jesus Christ is implicated in this circle of guilt. For it is asked whether it leaves those members who are afflicted in conscience to stand alone with their guilt, or whether it comes to their assistance by making what we have described above as the indicated act of "confession." If a questionable oath of loyalty to the tyrant must be given, if parents are forced to entrust their children to an atheistic youth movement, if the integrity even of responsible citizens is mercilessly assailed in time of extreme economic pressure, those concerned must not be abandoned to their own individual guilt-laden decisions or to their failures in respect of the impossible demands of conscience. It is then that the community of Christian believers is summoned to exercise a ministry of vicarious participation and witness. It is summoned, for example, to the legitimation of what is illegal. Surely we have made it sufficiently clear that this legitimation does not mean absolution. It is simply a matter of bearing witness before God and man to the compulsion involved in the situation. Only thus can the conscience of the individual be so guided and restrained that all his "wild" and spontaneous acts of illegal assistance do not plunge him quite unnoticed onto the downward path of evil to which we have referred.

It is wrong to let one's neighbor grapple all alone with a decision which is quite beyond him, involving as it does institutionalized injustice, for such a decision if he makes it will inevitably implicate him in guilt. To a degree which someone other than I must assess, the offense involved in such a case—and there is great offense even in illegally helping the Jews—falls upon those who far too often allowed these helpful acts of love to become wild and uncontrolled improvisations simply because they themselves would not face up to their own responsibilities on this front line of decision.

Theological ethics is here faced by one of those not infrequent cases where it cannot say what is "permitted" and what is not. Most ethical systems are only too ready with their censures and casuistries, but in cases like this where men are forced by *agape* to be anathema themselves, theological ethics can only say what is required of the believing community—which is the sphere within which the personal sense of urgency and the guilt to which it gives rise have their origin—namely, that the church by its act of confession must hasten to the assistance of those who are thus threatened and assailed in conscience. The moral insensitivity of the pious to what some Christians have dared to do in this illegal situation is in any case totally inappropriate. A confession of personal guilt would at least be truer to the facts. Furthermore, the vicarious assistance the church provides does not have reference only to the threatened members of its own community anyway, but avails also for "the children of this world," indeed the nation as a whole, whose conscience the church may rightly consider itself to be.

The second reason which forces us to ask these questions is also paraenetic in character. Since our generation and those that come after us will always be brought by various routes into situations in which there is oppressive institutional injustice, e.g., some new dictatorship, the question inevitably arises how far we ourselves have helped to bring about these situations. When we are told to resist such injustice in its first beginnings [*Principiis obsta!*], then we must know that these beginnings [*principia*], these first tendencies, are already present now, and that they have to do with my present decisions. It may be that at this very moment there is some political misuse of law, some tendency towards impersonal collectivization in legislation, some movement towards the curtailing of freedom or toward state omnipotence, some trend towards the biologization of sex, which is unobtrusively working in the direction of what will one day finally confront us as a full-blown tyranny, as institutionalized materialism or some other kind of demonism.

The ethical problem is imperfectly posed if we merely ask what is to be done in the borderline situation. The more urgent and very much related question is how to prevent the borderline situation from arising in the first place. Where are the seeds hidden today which will inevitably produce the underground movements and illegal activities? The lying and deception practiced in helping the persecuted Jews, for example, which while it stems from *agape* is nonetheless an ethically dubious kind of illegality, is simply an active and painful representation of that which others have brought about by compliance, by their *failure* to "resist in the first beginnings," which makes them indirectly and implicitly responsible. Which of the two forms of wrongdoing will be able to stand before him who looks upon the heart rather than the person (Mal. 1:9; Acts 10:34; Rom. 2:11; Eph. 6:9; I Pet. 1:17) or the institution? Neither form! But which of the two can refer to the love which has the promise of manifold forgiveness? (Luke 7:47). "Let him who is without sin among you be the first to throw a stone at her" (John 8:7).

Murder, Prostitution, and the Concept of "the Exception"

One can hardly speak of the problems connected with the borderline situation in our generation without thinking of Dostoevski, and especially his great novels *Crime and Punishment*, *The Brothers Karamazov*, *The Idiot*, and *The Possessed*, in which all that we have been saying finds classical condensation in artistic form.

We shall consider only the first of these novels, from which we may derive—particularly through an analysis of Sonia—a new concept which we have not hitherto discussed, namely, that of the "exception." There is thus good reason to add this analysis to our list of examples.

In *Crime and Punishment* two offenses are committed, or two commandments broken, namely, "You shall not kill" and "You shall not commit adultery." In both cases the wrong act stands under the shadow of a "higher" end which seems to demand it. We are thus confronted by the basic structure of the conflict situation.

We shall turn first to the murder committed by Raskolnikov, in respect of which there may be distinguished a logical downward progression in the matter of motivation. Raskolnikov kills a rich old woman pawnbroker, Alyona Ivanovna, who is living on usury. Her sister, Lizaveta, arrives at the decisive moment, and he must kill her too. He resolves on this murder initially because of the financial straits to which he has been reduced as a result of his charitable impulses to give all his money away—he was so distressed by the misery of the poor. The murdered woman, on the other hand, is a bloodsucker who thrives on misery. The deeper, underlying motive, however, has its source in the consideration that situations can arise in which a man, for the sake of his great objective, must "step over corpses." [11] In other words, for the sake of the "higher" end he must set aside the intervening command about the sanctity of human life and property. And is not the destruction of "a vile noxious insect" a great end? [12] This concept of the great end which justifies the means is the first phase in the downward movement of his motivation:

> I simply hinted that an "extraordinary" man has the right . . .
> that is not an official right, but an inner right to decide in his
> own conscience to overstep . . . certain obstacles, and only in
> case it is essential for the practical fulfillment of his idea
> (sometimes, perhaps, of benefit to the whole of humanity).[13]

Raskolnikov would believe that the murder is justified in view of the conflict situation. He is thus convinced that it has been divested of its ontic character as wrongdoing. He arrives at this idea of self-justification quite apart from all thought of forgiveness. He vindicates the act exclusively in terms of the acting subject who commits it and of the peculiarity of the situation. There is at work here a kind of automatic process of exculpation which seems to be a constitutional mark of the conflict situation. In other words, Raskolnikov no longer sees the guilt involved. He no longer sees it in its factual ineradicability. He is closed to what we have described as the *sine qua non* of all ethical action (in the sense of theological ethics).

We may now observe just how this vindication in terms of the autonomy of the conflict situation works itself out. Raskolnikov argues—and this is the second phase in the downward movement as regards motivation—that one cannot concede to all men these higher ends which justify the means. What would be the result if all men claimed to be acting in accordance with the demands of the conflict situation? What chaos there would be if these higher legitimations crowded out the laws of God, if moral sovereignty triumphed over divine authority! No, the end which justifies the means is not a law which confers general license upon "all." It applies—and this brings us to the key word

---

[11] Fyodor Dostoevski, *Crime and Punishment,* trans. Constance Garnett (New York: Macmillan, 1948), p. 231.

[12] *Ibid.,* p. 457.

[13] *Ibid.,* p. 230.

of this second phase—only to the "exceptional man." It is not by accident, then, that Raskolnikov refers to such exceptional figures as Kepler and Newton, Napoleon and Lycurgus.

> I maintain that if the discoveries of Kepler and Newton could not have been made known except by sacrificing the lives of one, a dozen, a hundred, or more men, Newton would have had the right, would indeed have been in duty bound . . . to *eliminate* the dozen or the hundred men for the sake of making his discoveries known to the whole of humanity . . . legislators and leaders of men, such as Lycurgus, Solon, Mahomet, Napoleon, and so on, were all without exception criminals, from the very fact that, making a new law, they transgressed the ancient one . . . and they did not stop short at bloodshed either, if that bloodshed . . . were of use to their cause . . . I maintain that all great men . . . capable of giving some new word, must from their very nature be criminals—more or less, of course. Otherwise it's hard for them to get out of the common rut, and to remain in the common rut is what they can't submit to, from their very nature again, and to my mind they ought not, indeed, to submit to it.[14]

Starting from a vindication of guilt within the situation of conflict, the train of thought thus leads to a vindication of exceptional men who may claim this freedom from guilt, this right personally to assess and end the conflict. Hence the third phase in the downward movement is that in order to attain to such freedom of action a man must categorize himself as "exceptional" and claim for himself the prerogatives attendent thereto.

> What if some man or youth imagines that he is a Lycurgus or Mahomet—a future one, of course—and suppose he begins to remove all obstacles. . . . He has some great enterprise before him and needs money for it . . . and tries to get it. . . .[15]

When I claim the predicate of the exceptional man, I am free for this law of the privileged. The genius, and finally the complete man who is no longer determined by the mere mechanism of reproduction, lives according to his own laws. Ordinary legal and moral obligations

[14] *Ibid.,* pp. 230-231.
[15] *Ibid.,* p. 234.

—this is the underlying thought—are for the average man. The result is that the murder and robbery which the exceptional man must commit for the greater glory of his ends are not to be regarded in terms of guilt, since the law with reference to which they would have had to be construed as guilt has been shown not to apply to these higher men. In the strict sense of Nietzsche, these acts are "beyond good and evil." But what is to hinder me from claiming for myself this license of the exceptional man, from making the identification characteristic of this third phase?

If we look at the matter very carefully we realize that serious ethical decision—and Raskolnikov is a classical example—is characterized by an underlying dialectic. For it is made by the individual alone. It is decision in the strict sense only insofar as in making it I am not imitating somebody else, and so letting him decide for me. It is also characterized by the fact that I do not sink into the slough of anonymity, doing whatever "men" do in such cases, but that I confront alone the law which addresses itself to me here and now at this very moment.

This loneliness of ethical decision is what makes me an "exception" in that moment when the decision is made. For ultimately the commandment claims me as an "individual." I cannot appeal to others in respect of the decision which is required. I myself must be the "subject."

In this concept of the "exception," which is integral to every serious ethics, there is also contained, however, the possibility of ambivalence. For, being an exception, I may evade the claim and so rob ethics of its seriousness. The loneliness of decision carries with it the tendency to understand my "self" as the exceptional phenomenon, so that I apply the concept of the "exception" not merely to the momentary subject of the particular decision but to my constitution generally, to my type, and claim for it too the law of the exception.

When I investigate as I must, however, the laws of my entelechy, when I ask what is befitting for me, I covertly become my own legislator. I exempt myself from all superior laws, and grab for myself that license to which we have referred.

The culminating point of ethics, therefore, the point at which the "exception" arises, is at the same time—dialectically—the crossroads where ethics may enter upon the downward path which threatens it with total destruction. Raskolnikov traversed this critical juncture.

This is the ultimate secret of the fact that in the very seriousness of decision he was led to include himself in the category of the "exception," whose prerogatives he had earlier asserted in respect of the great figures of world history.

There is yet a final step. After guilt has been ruled out in the conflict situation and license granted to the exceptional man, what follows in the last phase is the idealizing of the exception. This implies supremely that we are to strive after the ideal of the exception, and thus make a virtue out of the necessity of incurring "guilt." This then is the end of the way which begins with a failure to recognize guilt, indeed an inability to recognize it because we live beyond the mystery of forgiveness. Guilt is the mode of life of the exceptional man. And the exceptional man is the ideal which we are to seek to attain, the ideal which forces us to prove to ourselves that we are indeed exceptional, not stunted and subordinate nonentities living in garrets and eking out a miserable existence by teaching, but that we have the nobility of the bold.

> I could have earned enough for clothes, boots and food, no doubt. Lessons had turned up at half a rouble. Razumihin works! But I turned sulky and wouldn't. (Yes, sulkiness, that's the right word for it!) I sat in my room like a spider. You've been in my den, you've seen it. . . . And do you know, Sonia, that low ceilings and tiny rooms cramp the soul and the mind? Ah how I hated that garret! . . . Then for the first time an idea took shape in my mind which no one had ever thought of before me, no one [recall what we said about "loneliness" and the attendant movement toward the "exception"]! I saw clear as daylight how strange it is that not a single person living in this mad world has had the daring to go straight for it all and send it flying to the devil! I . . . I wanted *to have the daring* . . . and I killed her. I only wanted to have the daring, Sonia! That was the whole cause of it! [16]

We must now turn to the second main character in the book, Sonia. Sonia too does wrong for the sake of a "higher end." She sets herself above the commandment against adultery. But the parallel between her case and that of Raskolnikov breaks down in a way that is decisive

[16] *Ibid.*, pp. 367-368.

and highly characteristic. First, let us consider the facts. Her father is a notorious drunkard. Her mother and family live in indescribable and almost bestial squalor. To protect them against the worst, Sonia becomes a prostitute, earning far more than "a respectable poor girl can earn by honest work" as her father is quick to point out.[17]

Sonia herself, however, and those around her—even her corrupt father—adopt towards the foul sink in which she lives an attitude very different from that of Raskolnikov toward his situation. They do not justify and idealize it. Her father, realizing that his is the greater fault, since it was he who plunged her into this life, says that his daughter will be forgiven because she has sinned only for her parents' sake.[18] But this means that the sin is acknowledged. It is called by its name. It is not camouflaged or covered up. It stands under the hope of forgiveness, the acknowledged miracle of God. How Sonia's father seeks to understand this miracle of forgiveness—in antithesis to the causality of guilt and retribution—is expressed in the fact that he views his own eternal destiny very differently from that of his daughter: God will address him as a "swine, made in the Image of the Beast," who knows that for him there can be no acceptance.[19]

Raskolnikov too seems to be aware of some comparable distinction. He cannot think that this ethically grounded prostitution, this self-surrender out of childlike *agape*, can have the same good issue as his own fault. He cannot see any beneficent dialectic to make the evil good, and thus to break through the causality of good and evil. He sees in Sonia that which he does not see in his own ethically vindicated wrongdoing, namely, that guilt has its own law of logical progression.

There are three ways before her, the canal, the mad house, or . . . at last to sink into depravity which obscures the mind and turns the heart to stone.[20]

In any case, Raskolnikov is sure that the way of Sonia can lead only to disaster. And one of the forms of this disaster, which he represents under the figure of a "sink of iniquity," consists in the fact that the secret guilt of prostitution, even though it be to some extent ethically motivated and thereby justifiable, plunges one headlong into an open,

[17] *Ibid.*, p. 15.
[18] *Ibid.*, p. 20.
[19] *Ibid.*
[20] *Ibid.*, p. 286.

blatant, and excessive guilt, which involves the death of the heart and the severest self-judgment.

But strangely enough it is precisely the flagrant prostitute, Sonia, who is not the victim of that law of progressive deterioration which Raskolnikov discerns (discerns in others, though not in himself). On the contrary, Sonia is like a saint who moves mysteriously untouched through the darkness, and of whom one is inclined to say that she is a virgin prostitute. This woman of the streets reads to the murderer Raskolnikov, who is driven half insane by qualms of conscience, the story of the resurrection of Lazarus,[21] and she explains to him what it is to have to die, to perish, and be judged, but also what it is to know him who can make all things new. At another point she whispered rapidly, forcibly, while glancing at him with suddenly flashing eyes and squeezing his hand, "What is man without God? What should I be without God?"[22] The implication is that without God she would be hopelessly plunged into that uncontrolled descent into the depths of depravity. She would be one who could be described *only* as a harlot. She would be one who in a supremely sublime and dialectical manner would be entangled in that downward movement by the very fact that she regarded her prostitution as ethically ennobled, that she ventured to understand and to glorify herself as a heroine of daily life, as a harlot out of *agape,* and as a martyr to her environment. Without God, she would be delivered up to such horrible self-confidence, and for that very reason to the downward plunge into the abyss.

She obviously does not see how she can raise herself up out of the profession into which she has sunk at the command of her parents, half insensibly, and under the dreadful compulsion of unparalleled domestic need. There is, in fact, no automatic self-purification of guilt. On the contrary, the only thing which is automatic and sure is the causality of guilt and retribution, of harlotry and destruction, of prostitution and the death of the heart. Quite apart from any formal instruction or lectures on ethics, she knows what her profession is. Hence she does not try to make life possible by camouflaging what it really is, by pointing to the inherent virtues which blossom forth from it like marsh flowers. On the contrary, she lives exclusively on the basis of an "alien righteousness." Seeing her profession realistically for what it is, she

21 *Ibid.,* pp. 289-291.
22 *Ibid.,* p. 287.

lives in the name of a miracle, the miracle of forgiveness. What should I be, she asks, without the God who raised up Lazarus, who in a wondrous miracle broke through the causality of life and death?

In exactly the same way, Raskolnikov learns at the very end to say that everything must begin afresh in his life now that everything has collapsed,[23] i.e., now that there has been no regeneration through the dialectic, the teleology, which he followed in his action and which underlay the whole pattern of his life. The only law that exists is that of guilt and retribution, and this causality is broken only from without.

That this is how matters really stand with Sonia, and that she liberates Raskolnikov by thus drawing him into her own inner life, is indicated above all in the pastoral way Sonia reacts when he first confesses the murder. He makes the great confession to her and her alone, whereupon she jumps up, seizes him by the shoulders, and with eyes suddenly shining that had been full of tears cries:

> Stand up! Go at once, this very minute, stand at the cross-roads, bow down, first kiss the earth which you have defiled and then bow down to all the world and say to all men aloud, "I am a murderer!" Then God will send you life again.[24]

At this very point in her life in the abyss, she thus preaches the resurrection. She does not preach the indelible character of humanity, according to which some humane core is presumably preserved intact through all the vicissitudes of inner and outer death. She preaches instead the ineluctability of death, which is implicit in the fact that a person is forced to confess, "I am a murderer—nothing but a murderer." Only in face of this death can resurrection come into play as the miracle which opposes death. A dialectical connection between the two does not exist. There is no such thing as confession for the purpose of effecting resurrection. In the language of dogmatics there is no teleological relationship between Law and Gospel.

Thus Sonia rises to the point of a mysterious, substitutionary bearing of guilt. To be sure, it is not the prostitution itself which becomes vicarious, in the sense that through her profession she bears vicariously the social guilt of society, that she bears symbolically, as it were, in her violated body the wildness of the male sex or the guilt of her toper

[23] *Ibid.*, pp. 481-482.
[24] *Ibid.*, p. 370.

of a father. There is no question of any symbolical transfigurations of this kind. The point is rather that during her sisterly pastoral conversations with Raskolnikov she bears the guilt of her human brothers and sisters as one who stands in ultimate humility before God. In the discipleship of Jesus Christ—though in a way which is very different from his!—she bears it as her own guilt, accepting solidarity with murderers and usurers.

We must insist again, however, that she does this not as a harlot who might argue in purely legal fashion that materially and morally prostitution is similar to murder or usury. To argue thus is to achieve only the solidarity of the Law. What Sonia has in view, however, is the solidarity of the Gospel. For she aligns herself with the murderer, Raskolnikov, and with the murdered usurer, in the name of the cross of Jesus Christ: They all live only by his death. In a despairing conversation with the murderer, who is now at the end of his tether, she asks him, "Have you a cross on you?" And since he did not understand her question, she continues:

> No, of course not. Here take this one, of cypress wood. I have another, a copper one that belonged to Lizaveta. I changed with Lizaveta: she gave me her cross and I gave her my little ikon. I will wear Lizaveta's now and give you this. Take it . . . it's mine! It's mine, you know. We will go to suffer together, and together we will bear our cross! [25]

Thus all four—the harlot, the murderer, the usurer, and the feeble-minded sister—are linked together by the same cross. We cannot insist often enough that they are not linked by their sin. Sin isolates. Otherwise a prison would provide the highest form of communion this world can offer. No, they are united by the grace which is promised to sinners in the cross of Jesus Christ. It is in the name of this promise that the pardoned harlot refuses to leave the murderer, both in his trial and in his exile. And if we have had to speak earlier of being carried away by the downward movement of guilt, then under the cross and beside the empty tomb we have also to speak of another kind of carrying away [raptus], that of miracle and grace. Whereas grace is first of all that which lays hold of the harlot and the murderer in the midst of the abyss, whereas Sonia is enabled by this grace to endure the abyss,

[25] Ibid., p. 372.

grace now possesses a new carrying power all its own inasmuch as it does not simply make possible endurance in the abyss but actually leads one out of it. As Sonia and Raskolnikov stand under the cross and beside the empty tomb, they are gradually lifted to a wholly new plane of action. They become different even *in re*.[26] In this sense the novel can say concerning the exile which they endured together:

> . . . that is the beginning of a new story—the story of the gradual renewal of a man, the story of his gradual regeneration, of his passing from one world into another, of his initiation into a new unknown life.[27]

We cannot conclude our treatment of this novel without one final observation. While the figure of Sonia is indeed real and true to life, still she moves in a remarkable and mysterious zone of withdrawal. For Sonia is a saint. We cannot misunderstand her more seriously than by interpreting her in terms of the Law instead of the Gospel, e.g., by trying to draw the theological conclusion that in the shadow of forgiveness one might ethically follow her example and be a harlot, as though one were perfectly free to choose one's own form of obedience, or as though one might circumvent the commands in respect of means in order to achieve the commanded end, in this case the support of one's family.

Theologically, we must remember precisely at this point that *coram Deo* each of us is unique. The law of imitation and copying, even as far as Sonia is concerned, is absolutely and radically banned from the kingdom of God. For imitation is itself legalistic. This is why Luther rejects the "imitation of Christ," at least as an expression of our fundamental relation to Christ. For imitation, Christ is an "example," and as such the content of a Law which commands. The truth is, however, that Christ is for us "the exemplar" (the original prototype or representation) of the righteousness of God which is granted to us, not commanded but given.[28] (It is surely not without significance that we constantly feel compelled to comment on Sonia and Raskolnikov in terms of citations from Luther.)

[26] See above pp. 126 ff.

[27] Dostoevski, *op. cit.*, p. 482.

[28] See above p. 186, n. 9. In his 1519 Commentary on Galatians Luther writes: "Imitation does not make sons, but sonship makes imitators" (WA 2, 518).

Sonia is protected in her abyss by the miracle of grace, and she grows to the stature of a saint. But only by a miracle! Miracle cannot be taken for granted, however. It has the character of singularity. In its own way, and in notable contrast to the prideful use of the term as we found it in Raskolnikov, miracle is thus an "exception." This is why it is something we cannot count on. To return to our earlier metaphor of the parallelogram of forces, we cannot make guilt and forgiveness the vectors in a parallelogram, from which we could deduce—in the sense of a resultant—what is the right way to live. To do that would be to make the Gospel into Law, miracle into causality, and forgiveness into a means of rendering innocuous the unconditional commandment.

Dostoevski constantly expresses the singularity of Sonia "between the lines," as it were, after the manner of a poet. He impresses upon us that it is not Sonia herself, in her ontic being, so to speak, that is an exception. This is what Raskolnikov ventured to think of himself, and this was his mistake. On the contrary, Sonia is an "exception" because she encounters grace, because she is visited by grace. Only by grace is she "singled out," and to that extent—but only to that extent—made a true ex-ception. She had this character of the exceptional, not *in re*, but as the object of grace, as the bearer of an "alien righteousness." Only on this level of singling out, of true "exception," can there be such a bold, daring, and original phenomenon as a charismatic harlot.

This is why there can be none like her. Her life is not one which can be taken over, as if it were in the ordinary human sense a "possible" existence. From a human standpoint such an existence is absolutely "impossible." To put it in terms of the New Testament: "With men this is impossible, but with God all things are possible" (Matt. 19:26; Luke 1:37)—even the aura of sanctity about the face of the harlot, even the charisma of grace in the midst of the underworld.

We are thus summoned to think of the New Testament category of the exception, in which the limit and the transcending of the Law find their most profound expression. There are those who have "made themselves eunuchs for the sake of the kingdom of heaven." This too is a unique action. It is not a general rule, something to be imitated. "He who is able to receive this, let him receive it" (Matt. 19:12). Our receiving and understanding always relate to what is calculable, and hence also possible. But what happens here is "impossible" to men because it involves a particular call of God which is addressed to the individual and singles him out. This is also why it cannot be understood. And for this reason it can never be made into a law, and "imitated."

Now this does not imply by any means that there can be no theological access to this exception, or to the exception of the "pardoned harlot." For as surely as the charisma, this singular thing, is a miracle, it is not just any gift of grace, for example, in the loose sense in which we might speak of a "gifted artist." It is the gracious gift of God, the grace which is specifically defined and essentially characterized by the cross and resurrection of Christ. This is brought out in a monologue of Raskolnikov, in which he wonders about how Sonia the prostitute can go on living; we would say: how she can go on living as a Christian in prostitution.

> What held her up—surely not depravity? All that infamy had obviously touched her only mechanically, not one drop of real depravity had penetrated to her heart.[29]

Translated into the language of theology what this means is that she is not upheld by a blessing inherent in sin, which itself lies in the far country, nor by a productive dialectic immanent in negation. On the contrary, sin affects her only outwardly. In truth it has no authority over her. The poison has been removed from the fangs of the serpent.

Raskolnikov is expressing here what one might almost call a "docetic" quality in Sonia. She is not absorbed into the filth. Instead, the muddied waters flow off her as off a duck's back. This is how the Christian appears—and this is the decisive point—"from the outside," and Raskolnikov, who does not know the ultimate secret of Sonia, sees her typically "from the outside." He seems to stand before a docetic image. Yet this is quite untrue! For the waters do not flow off her because of anything in herself, because of the structure of her character, because of some property or disposition [habitus]. They flow off her because she is raised again, because she is one whom "nothing can separate from the love of God," because the sin to which she is bound in re does not characterize her, because she is snatched—has been snatched—from sin as one who stands under the cross.

What takes place here can again be expressed most succinctly in a well-known formulation of Luther concerning immortality. If we outlast death—or, as one might say here, guilt and prostitution—this is not

because of an enduring substance in our own souls such as was postulated in the Platonic concept of immortality. On the contrary, it is exclusively because God has entered into a history with us, because he
does not break off this history, and because, as partners in this indestructible history, we are to this extent immortal, and therewith immune both to death and to the sin which would destroy our relationship to God. To put it even more precisely, it is not we ourselves who
are immune; it is the history of God with us which is immune. The
continuity and immunity of my own person is wholly a matter of God's
faithfulness. I go on living in the name of an "alien righteousness." [30]

Yet Raskolnikov's observation "from the outside" was correct. The
secular man always sees something which is true about the pardoned
sinner, namely, that his soul remains inviolate in the underworld. But
in this form such an observation is only a statistical conclusion which
is falsely explained and just as falsely "evaluated" in terms of a certain
quality of soul. The inviolability of soul actually has only one basis,
that it is hidden in Christ, and that the Lord of this soul will not allow
other powers to have dominion over it. It is not some mysterious thing
about the soul itself. It is the mystery of the soul's preservation, which
comes to it from without. What is involved here is not the *proprium*
of my inner constitution, but the *alienum* of divine justification. The
relevance of this inviolability therefore is not psychological but theological.

The Limits of Authentic Conflict in an Unjust Situation

Although from the standpoint of justification "all things are possible,"
in human affairs there are from the human standpoint certain limits
which cannot be transgressed. These limits obtain even in the extreme
forms of the borderline situation. While this is not to admit casuistry,
it is to admit a kind of "casuistical minimum." There are no casuistical
norms but there are casuistically demonstrable limits.

That we have to reckon with such limits is grounded in the fact that
at certain points in life there are confrontations with transcendence in
which it is impossible for us to regard that which is encountered as on
the same level with various other possibilities of decision. Now it is of
the very essence of conflict that two possibilities which are relatively
co-ordinated and of relatively equal validity stand over against one

[30] See the Luther quotation documented at p. 232, n. 7 above.

another. If, however, one of these possibilities is transcendence itself, it loses its character as one "possibility" among others. It then becomes an "impossibility" to think that the choice before us is still wide open and that we have the possibility of decision. In such a case the situation is precisely not one of conflict. It is these limits which are imposed upon the very existence of the conflict situation—and the consequent need for and possibility of decision—to which we make reference when we employ the phrase "casuistical minimum" as a technical term. We propose to illustrate this direct confrontation with transcendence by means of two examples.

First, there is no such thing as an authentic case of conflict in which I am set before the possibility of denying Christ or blaspheming God. That is to say, it is quite out of the question by means of such denial to attempt to redeem my life or my freedom to the end that I may perchance have opportunities in the future for proclaiming him whom I have thus denied. Such a possibility does not even exist. I may argue that such denial is altogether necessary, on the ground that if I do not thus spare my own life, or that of others, the Gospel will necessarily be silenced and left without a witness. But this is not an authentic conflict at all. When it comes to the point of actually confessing, i.e., when a confessional stance is called for by virtue of what we have called a direct confrontation with transcendence, all such pragmatic considerations fall to the ground, and that for two reasons.

In the first place, I totally misunderstand my role in the kingdom of God if I think that the continuance of the kingdom depends on my continuing to live and to speak. "God is able from these stones to raise up children" (Matt. 3:9). And the silence which results through my being silenced can in fact be a judgment which God himself inflicts.[31] In the pragmatic consideration which leads to denial there lurks the blasphemous misconception that the kingdom of God is grounded not in the sovereign will of God but in the activity of active man.

In the second place, such denial involves me in guilt with respect to the persecutor whom I have allowed to force me into it. I am guilty of hardening him in his conviction that the Lord to whom I am subject is an "imaginary lord" whom I can give up "if need be." The offense thereby given is not an authentic offense. It is not occasioned by

[31] Cf. the παρέδωκεν of Romans 1:18 ff. ("God gave them up"); also Amos 8:11 ff.; Ps. 104:29.

the truth of God, but rests on the offensiveness of my human weakness and unbelief. The "casuistical minimum" which obtains in the matter of such denial is stated in Luther's hymn:

> Let goods and kindred go,
> This mortal life also;
> The body they may kill:
> God's truth abideth still,
> His kingdom is forever.[32]

The second example of direct confrontation with transcendence which dispels the conflict situation is to be found wherever that which is at stake or in question is the personhood of my neighbor. For the person of man as the image of God, as one who is "bought with a price" [I Cor. 6:20; 7:23], as the bearer of an alien dignity, is the direct representation of transcendence. Of man's personhood too we may say, figuratively, "He who touches it touches the apple of God's eye (Zech. 2:8; Deut. 32:10; Ps. 17:8).

Among the borderline situations in which this problem arises we may list those extreme cases to which we have already referred, in which everything (the fate and success of our movement, perhaps the lives of our companions, wives, children, and even of our very nation) depends upon the obtaining of certain information. In such cases the question inevitably arises whether we may obtain this information by means either of torture or of procedures of interrogation involving the use of certain truth drugs. In what we are about to say concerning "torture" we shall really have in mind both these methods, inasmuch as there are certain basic similarities between them in respect of what they do to the personhood of man.

Torture "touches" the personhood of the other inasmuch as its purpose—seen from a theological standpoint—is to force and break the conscience which is commanding him to keep silence. It is a kind of negative counterpart of temptation by desire. Yet there is a radical difference. In its initial stages, temptation by desire must always proceed by way of the man's own decisions. By contrast, temptation through pain—at least as far as torture is concerned—negates its own character as temptation. The refined scientific methods customarily

32 Martin Luther, "A Mighty Fortress is Our God," trans. Frederick H. Hedge, in the Lutheran *Service Book and Hymnal*, No. 150.

used by modern police systems under the guidance of their physiological and psychiatric consultants drive the victim to a point where he is no longer the subject of a decision to speak or not to speak. The supradimensional pain simply eliminates the ego altogether. Torture does not force a decision out of him by promises, threats, and bribes, by arousing fears and hopes, by an appeal to the instinct of self-preservation. Torture simply bypasses the sphere of decision altogether by making it impossible. This is true at any rate of the modern and scientifically "sophisticated" forms of torture which stem from a perversion of psychiatry.

Among them is the truth drug, which chemically destroys the person in the central core of his being. It is inhuman not because it is so sinister but because it de-humanizes. To this extent both torture and the truth drug, in all their various contemporary forms, go much beyond temptation, even beyond coercion. For temptation, at least in its first stages, is a frontal attack on a person. So too is coercion. But torture—and this is true also of the lie detector and narcoanalysis—"bypasses" humanity. It takes away the personal right to decision. To this extent it "touches" the personhood of man. But since man's personhood is a representation of transcendence, such attack upon it does not involve the constraint of an authentic conflict situation.

There is also the further point that where the personhood of man is not respected he is reduced to the level of a mere means to an end. The "infinite worth of the human soul" (von Harnack) is replaced by a worth in terms of what the market will bring, and there is a consequent readiness to violate and even liquidate the person for the sake of certain specified ends. In face of this defacement of the human image, anxiety reigns supreme. There is anxiety as to the unfathomable "possibilities" which lurk in man, his uncontrolled will to power, and his unpredictability when a desired end drives him to excesses.

In face of this possibility that a man is "capable of anything," the Christian is summoned to adopt a confessional stance. For the Christian owes to the world the public confession that he is one who is committed, "bound," and hence not "capable of anything." If we make ourselves fundamentally unpredictable, i.e., if as Christians we think that torture is at least conceivable—perhaps under the exigencies of an extreme situation—we thereby reduce man to the worth of a convertible means, divest him of the *imago Dei*, and so deny the first commandment. This denial can never be a possible alternative. It can never be the way out of a conflict. The case here is identical with that discussed a moment ago concerning denial of Christ. This is no accident, for in both cases what is involved is that direct confrontation with transcendence.

If Christians, then, are found championing as a fundamental possibility, or perhaps themselves using, torture or forensic narcoanalysis, the faith of those who see this is necessarily jeopardized and exposed to extreme offense, because for them the communion of saints is thereby given the appearance of a devilish caricature. Eternal as well as temporal issues are thus at stake here. But the eternal destiny of those about us can never be weighed on the same scales as the temporal fate of, for example, companions, wives, and children, as if both were on the same level. In such a situation, then, there can be no genuine conflict.

Our deliberations have made it clear, however, that reasons have to be given why this is so, and that these reasons are not immediately apparent, at least so long as we cannot see how to rid existence of genuine and deeply inscrutable conflicts. Finally, it is a sign of the spiritual situation that theological ethics is compelled to think through the problem of torture. This has probably not been done for centuries, if ever. Indeed torture may well continue to be a subject of ethical discussion for some centuries to come.

Extreme though the problem is, it shares with the problems of authentic conflict and with the borderline situation in general a symbolical participation in the ambivalence of the fallen world, a world which is freighted with unfathomable forces, and which is saved from self-destruction only because the divine promises hold fast (Gen. 8:22; 9:14-15) and the divine patience does not cease (Rom. 3:25; I Pet. 3:20; Exod. 34:6; Nah. 1:3). Thus the borderline situation is, to be sure, on the border or boundary [Grenze] of the world. But this does not mean it is an atypical or unique case. It can mean only that the mystery of the world is disclosed precisely here. For "the boundary is the most fruitful place for gaining knowledge."

31.

The Guidance of the Holy Spirit in
the Given Situation

Throughout this volume we have repeatedly had occasion to point out that decisions cannot be made a priori, that future decisions cannot be arrived at on the theological drawing board. The three main reasons which till now have been given in the most varied connections may now be summarized.

Situational Ethics

First, to decide ethical questions in advance is possible only within the framework of an illusory natural law which pictures the world as an orderly construct, permeated by eternal norms laid down at creation, the structure of which may be demonstrated. If the world is viewed in this way, then, with the help of "subsumptions" and "conclusions" all sorts of "circumstances" may be brought into relationship to final norms and arranged hierarchically. The world then becomes a hierarchical cosmos, and in principle every case which arises and every decision which may be required can be prejudged morally by reason of the hierarchy of values.[1] This type of cosmic order presupposes a specific interrelationship between the original state and the fallen world, between creation and sin, such as is asserted in Roman Catholic theology.[2] This relationship is determined by the fact that sin violates creation only in a very peripheral way. In Reformation thought, on the contrary, the world is seen to be so totally permeated by both creation and sin that we are prevented from supposing the fallen aeon to be still determined by the "orders of creation," and hence from espousing the illusion of a hierarchical cosmos of natural law. It is only logical, then, that on the Reformation view there can be no advance decisions, and that ethical casuistry is consequently done away.

The second reason why Reformation theology cannot ethically pre-

[1] See Chapter 20 above.
[2] See Chapter 11 above.

judge future cases consists in the fact that such advance decisions are possible only within the framework of a "legalistic" rather than an "evangelical" view of things. The view would be "legalistic" to the extent that it does not let man be the acting subject in the making of his own decision. He instead becomes merely the object, agent, or executor of a decision which has been made already by others, e.g., those in places of authority. Hence the decision is not his own original act. It is like a suit that he puts on as ready-made, after the tailoring and fitting already have been done (by somebody else). It is a characteristic feature of legalism that it does not let man be a subject. Instead it impels him "from without," and thus makes him the object of this impulsion from outside himself.

The third reason why decisions cannot be made in advance, or advance decisions simply taken over, is closely connected with the second. When the Law moves man from without, it does not move him totally. Even at best—i.e., when he acknowledges the Law and does not obey it merely out of constraint or servile fear—it divides him into two parts, namely, the part which accepts the Law, and the part which resists it and must be overcome.[3] The only obedience which is whole and undivided, and in which the ego has the significance of an acting subject wholly committed to its action, is love. For love does not compel itself to do something in the sense that there would yet remain that non-participating element which is the reluctant object of such compulsion. On the contrary, love is marked by spontaneity and *promptitudo;* it arises only where there is a "complete transformation" of heart.[4]

The reality or power which makes me the subject of love, and thus makes possible full commitment in my "new obedience," is the Holy Spirit.[5] Accordingly, the *pneuma*—as distinct from the Law—does not alienate me from myself by causing something other than me to speak and act in and through me. Instead the *pneuma* enables me to be "myself" in my new existence as a Christian. He grants me a new capacity to be a subject, and hence "I myself." This is what Paul means when he says that, where the Spirit of the Lord rules and the servitude of

[3] See pp. 63-65 above, and especially note 23.
[4] See pp. 63 and 65, n. 26 above.
[5] See pp. 84 ff. above.

THEOLOGICAL ETHICS – FOUNDATIONS

the Law is ended, there is freedom (II Cor. 3:17; cf. Gal. 2:4; 5:18, 23; Rom. 6:14; 7:1 ff.; II Cor. 3:17-18).

If, then, there can be no legalistic ethics with preformed decisions, decision must be made within the framework of each existing situation. The ethics of Law is replaced by a kind of "situational ethics" [*Situationsethik*].

What this term means is best illustrated once again by the borderline situation of extreme conflict, which demonstrates particularly well the absurdity, not merely of arriving by way of theory at advance decisions, but also of attempting in practice to subsume the "circumstances" under hierarchically ordered norms. It thus makes good sense that we should conclude our discussion of the borderline situation with this investigation.

Concretely the question is how I can face up to and endure the situation of conflict—and hence any situation whatever—without the regimentation of legalistic casuistry and in the freedom which is given to me by the Spirit. Our mode of procedure will be to take examples from the Bible in which the three main aspects of the problem are brought acutely to our attention: the simplification of action in the "moment," our prayer that we will do what is right, and God's promise for insoluble ethical problems.

Improvisation in the Moment

We have already pointed out that, in a kind of antithesis to the theoretical reflection on ethics, the moment of action confers a certain "simplification." There is reference to this in the commissioning of the Twelve in Matthew 10:19-20: "When they deliver you up, do not be anxious how you are to speak or what you are to say; for what you are to say will be given to you in that hour; for it is not you who speak, but the Spirit of your Father speaking through you."

Anticipatory reflection on the situation of being "delivered up" plunges us into "anxiety." This anxiety relates to the question whether we will objectively see through to the heart of the situation, whether we will correctly recognize that a confessional stance is here demanded despite the many factors which would seem to conceal or minimize its importance, in short, whether we will be able to distinguish between the spirits (I Cor. 12:10; 14:7). However, the anxiety relates also to the question whether we will actually say and

650

do the right thing, make the right confession, i.e., whether we will be personally equal to the situation.

According to Matthew 10:19-20 this complication—which goes with such anticipatory and analytical reflection—will through the help of the Spirit be resolved at the decisive moment by a clear simplification, i.e., by the manifestation of the single unequivocal choice with which we are confronted. The simplification resides in the fact that the antithesis between the kingdom of God and the power of the Antichrist is plainly disclosed.

This implies two things: first, that I am here set before the ultimate —and hence unequivocal—choice; and second, that God himself is present to assume responsibility for his own cause, and therewith also for the answer which in obedience and trust I openly give. Whether the situation will be met and endured does not depend on the support I give to the kingdom of God. The kingdom pleads its own cause. Mine is not the starring role. I am just a supernumerary in this conflict. The kingdom of God "uses" me. The Spirit himself is present to meet the assault. What we have here is a kind of repetition of the trial of Jesus, in which he "made the good confession," and in which I now stand alongside him (I Tim. 6:13). "Anxious concern as to which word might affect the human judge is not the business of the disciples. God gives the word. His Holy Spirit speaks in the disciples' stead and through their lips." [6]

Our exposition of the saying in Matthew 10 would be inadequate if we were to regard the interposition of the Spirit merely as a kind of supernatural "inspiration" [Einflüsterung]. What is meant by this rather hyperbolical expression is legitimate only if this subjective "inspiration" is seen against the background of the objective strategic situation. This means that here in the microcosmic moment of my individual situation the two cosmic powers are ranged against one another on a common front, and I am mobilized as, so to speak, the last soldier on the side of the kingdom of God, which is itself present and active however on its own account. I am allowed to take my place in the front lines of the kingdom of God. This is why the Spirit speaks in my stead.

I know that I am granted this simplification even in the situation of

---

[6] Julius Schniewind, *Das Evangelium nach Matthäus* (Göttingen, 1937), p 126; cf. also Phil. 1:19; II Tim. 4:16 f.

conflict, which is the most complicated form of this cosmic encounter. I know it, however, only because it is promised to me, and hence only to the degree that I believe in this promise. This is why for the believer all concern is taken away, at any rate all concern in the sense of "anxiety." His concern for clarity, which is part and parcel of ethical reflection and hence of any analysis of the situation, of course continues. The anxiety forbidden in Matthew 6:25 ff. does not refer to planning and providing for the coming day. It refers merely to anxiously dwelling upon events to come, under the prideful illusion that this coming day is in my hands rather than God's hands. Ethics itself, however, belongs to the legitimate sphere of concern.

As we have said, this legitimate concern is destroyed not only by anxiety but also by the false notion that such concern can take the form of figuring out our decisions in advance. The promise that at the decisive moment the Spirit will be our advocate and bear witness in our stead gives us freedom for this moment and confers upon us the power of improvisation. This freedom is not, of course, a blanket license to do whatever we want. It is not libertinism. Similarly, improvisation does not imply a mere vegetating indifference in respect of the moment. The best extemporaneous speakers are those who have prepared best and who, out of the clarity which they have reached in meditation, give themselves to the moment and leave their notes behind. Hence Christian improvisation does not stand in antithesis to careful reflection. It is rather that freedom for the urgent claim of the moment which has been won by way of such reflection. It has nothing to do with a laissez faire approach or attitude.

Only when the thought of improvisation is thus safeguarded theologically does it enable us to see what this Christian freedom for the moment is, namely, a freedom which the Spirit, and he alone, can give. This freedom is done away wherever the assistance of the Spirit is lacking. For it is then crowded out by that which is incalculable and unpredictable, e.g., the blind, spontaneous reactions of our instincts, or by that which is quite predictable, e.g., a deliberately calculated program which is in effect a law.

Illustration of this may be found in the story of the Good Samaritan (Luke 10:30-37) which is to be understood wholly and exclusively in terms of the guiding principle of improvisation. For none of the three men passing that way was the battered victim listed on the day's agenda of appointments. He represented in every instance a demand for improvisation. We are not told what prevented this improvisation in the case of the priest and the Levite. There is freedom to imagine two possible explanations. On the one hand, they may have had important duties to perform. The scribe was possibly delivering a lecture that very evening in Jericho on the subject of "brotherly love." They were thus involved in a conflict between the claim of the moment and the claim of prior commitment. By weighing these conflicting claims—by calculating—they decided to set aside the moment, and to pass by both the person and the task closest to hand. Thus improvisation was hampered by calculation. On the other hand, they may have been impelled along the lines of laissez faire. They may have been unprepared to decide because they had devoted no careful reflection to the question of the neighbor at all. Hence anxiety and the instinct of self-preservation carried the day. In either case the Spirit, who is bent upon action, was given no room to operate.

True improvisation means the assurance and belief that after I have done all the careful reflecting I can do God will take over as his own that cause in which I am seeking to serve him. He will "himself" intervene, and at the decisive moment enable me to see clearly the borderline of demarcation between his course and that of his adversary. I know that in such believing improvisation I cannot be separated from him (Rom. 8:35), even though I constantly fall into actual sin in the borderline situation, particularly there! For the fact that the issue no longer turns decisively on my calculations is grounded in the fact that I am no longer under the Law but under the freedom of the Gospel, i.e., that my salvation is no longer linked with conditional clauses, e.g., the condition that only if you act in such and such a way in the given situation can you remain in fellowship with God, and that if you act otherwise you will cut yourself off from him.

There is a link between the Law and anxiety: The Law tells us to be careful that we do not surrender the conditions of our salvation by wrong decisions or by secret disobedience. And there is a similar link between faith and freedom: "Love, and do what you want" (Augustine); "Sin boldly, but believe even more boldly and rejoice in Christ" (Luther).

We can hardly conclude this section on Christian improvisation, however, without mentioning a very serious objection that might be made. The doctrine of improvisation is grounded, to be sure, in a specific doctrine of the Spirit. This pneumatology in turn teaches us that the Spirit will be our advocate, that God will himself plead and protect his own cause, and that we may know ourselves to be on the side of his assailed and actively engaged kingdom (which is the source of our certainty as to the simplicity of the final choice). Now the critical question arises at this point, whether we have not hereby overshot our target. If we wish to maintain the concept of the borderline situation (as we obviously do), and to take it for granted that in such a situation there is no direct way of escape, and if we are to insist upon recognizing the existence of some *de facto* "crooked" element in every proposed solution, then can we at the same time really have this unequivocal assurance that we ourselves stand in the front line of the embattled kingdom of God? To put it another way, do we not encumber the kingdom itself, and the leading of the Spirit, with this *de facto* "crooked" element? Have we not said ourselves that we can undergo and endure the borderline situation and its inescapable conflicts only under forgiveness? But if we have to claim forgiveness for our actions, then we are confessing with all possible clarity that this action is itself in conflict with the will of God, and that consequently it is wrong and not in good order. But if it is wrong, then we surely cannot chalk up our questionable solutions and evasions to God's account as if they were his doing. After all, Yes cannot be No, or No Yes! We are thus plunged into the innermost core of the theological difficulties posed by the ambivalence of the borderline situation. As a result we have no recourse here but to stop and think through these difficulties.

Of course the question is one on which we have had to touch already in other contexts. The need to distinguish in Luther's sense between the "proper work" and the "alien work" of God [*opera Dei propria et aliena*] points to a similar problem. For when God in his condescension comes down to the sphere of facts posited by the fallen world, he accomplishes a kind of "accommodation of patience" and thus gives himself up to a conditional self-alienation. We recall our earlier discussion of the institutions of the state, law, divorce, and of the whole complex of questions involved in the Noachic covenant.

In respect of these institutions too we are confronted by the same dilemma. We act politically, exercise judicial justice, perhaps impose the death sentence, we kill as soldiers, and in all these things we vindicate our conduct in terms of God's "emergency order" [*Notverordnung*], his "alien will," after having first carefully weighed our duties *coram Deo* within the schema of these orders lest we should simply be carried along by the autonomies of these spheres of life. We thus venture to think that our action has its source in the will of God, and that there is a certain legitimacy in the motto on the German soldier's uniform, "God is with us"—which there is, though the phrase has often been abused and stands in need of close definition. Abuse arises when it is regarded as an unrefracted legitimacy, as if the divine will were unreservedly "on my side," as if it were incarnate, so to speak, in my action. In this case there would ensue a mythical sacralization of my deeds which would no longer have anything to do with obedience. The god who in that sense is "with us" would become the god of war or the god of iron. He would bear a closer resemblance to Mars than to the Father of Jesus Christ. Indeed, the authority which is legitimate within the framework of the emergency order may actually have been supplanted by a deified government, the totalitarian state.

Along with the certainty that I am on God's side when I act within those orders there must always be kept in view the relativization of God's will which is indicated by the distinction between his "proper will" and his "alien will." What is involved is that will of God which has voluntarily restricted itself in face of the fallen world. This is why a Christian in the armed forces, while he may with a "good conscience" champion his cause so far as this is possible, risking his life for his country and for his wife and children, can never simply "approve" of war. He knows that even a war which—given things as they are—is "just," must always stand in need of forgiveness. Thus he will act and fight in the name of God, knowing as he does so that the very plane on which his action necessarily proceeds is itself questionable, because it is that of the world after the fall.

While Luther's doctrine of the two kingdoms allows for some

ambiguity in this respect,[7] his doctrine of justification clearly implies
an understanding of the world such as we have presented, an understanding which views the world after the fall as an ambivalent
quantity. On the one hand, it is comprised of the orders of the
Noachic covenant as orders of the divine patience. On the other hand,
it is at the same time an objectification of man's powers of revolt in
the situation after the fall. But if this is so, then the *sola gratia* and
sola fide have reference to all our action, including what we do within
the orders. Only on this understanding does Luther's doctrine of
justification achieve its comprehensive significance. It is true that
for Luther the heart of theology, his doctrine of justification, did not
yet send the blood coursing, as it were, into the extremities, into his
cosmology and his doctrines of the orders and of society. Here one
can still find fingers and toes that are cold and numb. But it would
be untrue to the facts—Catholic and Protestant interpreters of Luther
have long argued about this—to appeal to these clammy and un-
quickened members of Luther's theological corpus in support of the
claim that his doctrine of the orders is either the expression of a
"double morality" (Troeltsch, Wünsch) or a Scholastic doctrine of
"natural law." [8]

But even with this understanding of the world we have not yet
plumbed the depths of the matter. For now two further problems
emerge. The first is how I can know myself to be "at one" with God
if the "unity" involved has reference only to his "improper will." The
second is how the unity of my ego is affected if my motivation is
divided between this questionable "union" on the one hand and the
confession expressed in my prayer for "forgiveness" on the other. For,
after all, in such a prayer I am confessing that I am not at one with
the will of God; I am asking God to overlook that fact.

Both questions find a common solution in the realization that the
so-called alien will of God [*voluntas Dei aliena*] is very inadequately
described if it is regarded merely as his "improper" will, as it were,
his "depraved" will. Everything depends on the question why God's
will thus emptied itself in a kind of self-alienation. We have already
seen that this need not be viewed as some kind of a divine "com-

[7] See Chapter 18 above.
[8] See Franz Xavier Arnold, *Zur Frage des Naturrechts bei Luther* (Munich, 1937).

promise" with the world, as though God were no match for the rebellious world and were trying with the help of concessions to bring about a kind of armistice.

The motive behind the alien will of God is rather that of his condescending love, which has in view the continuation—the *kairos*—of the rebel. For this reason the alien will cannot be regarded dynamically as compromise. It must rather be understood soteriologically as divine condescension. Thus it can be explained only in terms of the primal fact of this condescension, namely, the fact of Christ. It is the Father of Jesus Christ who already in the Noachic covenant sets up a preliminary and prophetic sign of the condescension which is to find its fulfillment in Jesus Christ.

Consequently, when we have to do with God's alien will, we are not dealing with an impropriety stemming from weakness. On the contrary, we are dealing with an expression of God's true being, with his patience and his grace. The mode of appearance itself cannot demonstrate this, any more than the appearance of the crucified Christ can of itself convince us that what is involved in the cross is the triumph of grace rather than a disgraceful bankruptcy. To be at one with the will of God thus means to be at one with his love. More precisely at this point it means to be at one with the purpose and goal which move God both to condescend to the need of the world and then to accept the service we render within the framework of this altogether questionable arrangement.

But this throws light on the second question, namely, that of the threat to the unity of my ego if it must vacillate between a supposed union with the divine will, on the one hand, and the recognition of separation from him and of the need for forgiveness, on the other. From what has been said, we know that the relation between these two acts or attitudes is not one of simple contradiction. For to be at one with God's alien will cannot meaningfully imply union with the factualities of the fallen world to which God comes. It can only imply union with the motive which causes God to come to this world and to its factualities. And to be at one with this motive of love can for its part only imply faith in, and adoration of, that which Luther calls "God's turning to me of his fatherly heart," that heart which sets the rainbow of reconciliation above the "emergency orders" of the Noachic world (Gen. 9:13 ff.), and which opens up the way to forgiveness through the sacrifice of the Son.

We are not to interpret God in terms of the course and structure of the world. To do so would be to make him into a god of war and vengeance who sanctions our own acts of war and vengeance. Quite the contrary, we are to interpret the course and structure of the world in terms of personal union with God's fatherly heart. Then the apparent cleavage in our personal being is resolved. Then that which is unity in God's heart is unity also in us. As it is one and the same motive which impels God to keep this questionable world in being, indeed lays upon us the task of acting in his name precisely within the framework of this questionable world, and at the same time moves him to make us see that such action itself stands in need of forgiveness, so there is in us a corresponding unity between work we do in a fallen world by divine commission and our prayer for God's forgiveness of this very work.

This unity, of course, is never static and complete. Even pychologically the Christian soldier who does his duty in a "just war" does not have this unity as a peaceful possession. On the contrary, he is thrust into a movement. His affirmation of the deeds of the moment, i.e., his cheerful and confident doing of his duty, is continually forced to expose itself to a deeper question. It may be that the war itself is necessary, under the circumstances even imperative, and hence a "just war." But the circumstances which thus necessitate it are themselves unquestionably not just. For the underlying premise behind all of them is man's fall into sin. The Christian in uniform may seek the counsel of the Most High in all his acts and then go unperturbed to whatever awaits him, and what he does on the counsel of the Most High he may actually do in the name of the Most High. But he will do it also in the name of God's pardoning grace. His will not be simply a "good" conscience, but a "comforted" conscience.

Thus there is no straddling these two positions. The two lines do not intersect. What is involved is rather a movement, a flight from God to God. It is a movement in elliptical orbit around two foci, the unity of which I cannot see but only believe.

It is a movement like that between the Law and the Gospel, where again I cannot objectively see and fix the unity. Theologically I cannot make the unity of God an objective matter in which the author of the Law and the author of the Gospel are seen to be identical; I can only believe in this unity.[9]

The unity in God, and therewith my own unity of person, is not objectively accessible to me. The affirmation of duty whereby I "act in God's Name" and the calling in question of such action through the "prayer that God will forgive me for it" cannot be reconciled. They always remain as the two foci around which faith constantly moves. The fact that there is no neutral ground between them is what keeps the Christian from becoming a fanatic and from losing himself in some this-worldly task. The "enthusiasm" of the Christian soldier for his task will always be tempered by this deeper question concerning the task itself. His enthusiasm will thus always be kept under control. He will always act with a certain detachment. His realism is thus protected against the blinding power of fanaticism. Indeed this is what makes it realism in the first place.

What we have been saying may now be applied to the conflict situation. Even though the direction in which I allow myself to be led by the Spirit is in no sense a "direct way of escape," even though it is not right and just *in re*—and what is in this fallen world? —God still makes it, despite its *de facto* imperfection and inadequacy, *his* affair. This is why I can tread this dubious path, which is disclosed only after much searching of conscience and with prayer for the help of the Spirit, in the confidence that I venture upon it in the name of God. In other words, I venture upon it in the assurance that God confesses *it* even as I confess *him* by treading it. To confess him is to confess the God of condescension who has taken upon himself my temptations (Matt. 4:1 ff.; Heb. 4:15), who in my place has borne my sins in his body on the tree (I Pet. 2:24), who has made my hardness of heart the occasion for his accepting my service within the framework of the rebellious world and making it "just and right" before him. Indeed, what else is the meaning of my justification? My certainty that acts done under the guidance of the Spirit are, despite their "crooked" form, done in God's name as *his* affair, and that at the same time they nonetheless stand in need of forgiveness— a forgiveness which has actually been given—this certainty is thus grounded upon a deeper unity in God himself.

To the *simul justus et peccator* that relates to the doer there thus

[9] See above pp. 117 ff.

corresponds a *simul justum et peccatum* in respect of the thing done. As I act in his name I am right before him, and he makes the deed to be his own. My affair thus becomes the affair of the Lord who inclines to me in all my doubtful ambivalence. That he inclines to me means that he comes to me in love, which is his "proper work." But it means too that in this same movement of the divine heart he embraces with forgiveness the very ambivalence which caused him to come, and maintains an unmerited faithfulness toward the sinner.

Praying to Do What Is Right

In prayer we ask that the Spirit should come to help us and give us clarity. The Spirit is thus the object of our prayer. Anxious and calculating concern is dispelled by the confidence and the plea that the morrow will come to me from the hands of God and that I will hence be given only what is appropriate for "today" (Matt. 6:11, 34; Phil. 4:6).

The Spirit, however, is not only the object of prayer; he is also its acting subject. In this dimension too he stands in my place and acts in my stead. "Likewise the Spirit helps us in our weakness; for we do not know what we should pray for as we ought, but the Spirit himself intercedes for us (or represents us) with sighs too deep for words. And he who searches the hearts of men knows what is the mind of the Spirit, because the Spirit intercedes for the saints according to the will of God" (Rom. 8:26-27).

Here again, ethically speaking, we are taken out of the sphere of calculability and casuistry. For ultimately, as those who pray and therewith as those whose deeds follow their prayers, we are not the acting subjects. On the contrary, the Spirit of God himself intervenes and represents us.[10]

Why does the Spirit intervene? Because "we do not know what we should pray for as we ought." Origen finds here two difficulties in prayer: we do not know what to ask for, and we do not know how to bring our requests to him who is eternal.[11] But since the problem of the "as we ought" can hardly refer to the matter of liturgical form

---

[10] That what is intended here is not such a pure and simple negation as would threaten personhood should have become clear already in what we said above on pp. 648-650.

[11] See Ernst Gaugler, *Der Brief an die Römer* (Zürich, 1945), I, 316.

—after all, there is no prescribed form; according to the instruction of Jesus we are simply to approach God "as beloved children approach their dear father" [12]—it must be understood wholly in terms of the object of our prayer, i.e., in terms of the fact that we do not know what to pray for, and consequently we do not know how to act either.

Why do we not know what to pray for? Because prayer requires a reading of the situation. I pray for daily bread, i.e., for food and clothing, house and home, because I regard these as necessities of life, and hence as "masks" beneath which God exercises his providence and preservation. Thus I may pray for what I need, or, more precisely, for what I think I need, as my Father in heaven knows and is concerned about this need (Matt. 6:8). It is in this sense that I pray for healing, for the safe return of a loved one from the war, for fresh air and clement weather.

The phrase "what I think I need" is meant to express the same ultimate proviso as is expressed in the Lord's Prayer: "Thy will be done" (Matt. 6:10), and in the prayer of Jesus in Gethsemane: "Not my will, but thine, be done" (Luke 22:42). My request might rest on a false reading of the situation, and therefore on false hopes and fears. God's blessing might come in a very different way, and lead me into very different and unexpected paths.

When it is said that God's thoughts are higher than my thoughts (Isa. 55:8-9), this does not denote a purely quantitative superiority, as if God's thoughts were continuous with my own postulates, expectations, and interpretations and simply carried them forward. What is meant is rather that God's thoughts are wholly other than mine; there is a qualitative distinction between his sphere and mine. I know only the theme of these thoughts, namely, that they are "thoughts of peace, and not of evil" (Jer. 29:11). But the relation between this theme and the concrete events of life is hidden in the heart of God. Hence I cannot know or read God's thoughts. I have to trust them. As a result I stand in constant readiness to have my request corrected by God's fulfillment, to understand God's response, whatever it may be, as a definite fulfillment of my petition, even though it may take a quite contrary form. For the petition had reference to my need; and the fulfillment is a meeting of this need.

[12] Martin Luther, The Small Catechism; BC 346 (cf. Matt. 6:7 f.; Rom. 8:15; Gal. 4:6).

If I take seriously the saying of Jesus that my Father in heaven knows what I need (Matt. 6:8), this gives me the assurance that in principle there can be no prayer that is not answered. For in every case God fulfills the intention of my request, by supplying the people, things, and situations that I need. It is simply that his fulfillments surpass, and consequently relativize, those wherein I think to see these intentions realized, or better, wherein I actually see them realized.

In the strict sense, therefore, the prayer "Thy will be done" does not imply a revocation of my previous petitions. Theologically, it simply makes the distinction between the intention of the prayer and the specific requests used to implement that intention. The fourth petition of the Lord's Prayer consequently leaves it to God to judge whether or how far I have rightly interpreted, in my actual petitions, my real intention that he should meet my need. God's higher fulfillments do not deny me the necessary thing for which I pray; they simply revise my understanding of the need. The apparent discrepancy between petition and fulfillment is a noetic rather than an ontic problem. Ontically, the need which underlies my petition is actually met. Noetically, I subject my understanding of this need, and consequently my self-understanding generally, to God's "higher thoughts."

We can express this decisive insight as follows. In principle my prayer never goes unanswered, so far as its true point or goal is concerned, namely, the supply of that to which my need, or the need of those for whom I pray, refers. There is only a revision of the "ways" which I propose for deliverance from the distress.[13] I have to trust in the thousands of ways which God has at his disposal. The prayer "Thy will be done" does not express a "love of fate" which accepts with resignation whatever befalls. On the contrary, it is the expression of a "love for God" which fulfills an act of confidence.

We thus maintain that the incongruity between the content of my petition and the divine answer results from the fact that God knows

---

[13] In commenting on Romans 8:26-27 Luther can even say, "It is not a bad but a very good sign if the opposite of what we pray for appears to happen. Just as it is not a good sign if our prayers eventuate in the fulfillment of all we ask for" (*LCC* 15, 240). If everything were to go "the way I want it," I would end up in that kind of false security [*securitas*] which is really an instrument of the divine judgment. The greatest assault of temptation [*Anfechtung*] is to have no doubts or perplexities [*Anfechtung*] at all. And God may well be most angry precisely when he appears not to be angry at all (cf. WA 3, 420).

my real need better than I do myself. I have thus to allow my stupid, all too human, and hence dilettante interpretation of my situation to be corrected by his higher, more percipient, and helpful understanding. At the same time, I may be confident that God will do what is ontically right to meet my need.

The external and internal incongruity between our petition and its fulfillment has also—it should now be clear—an ethical aspect which is extremely relevant for the conflict situation. For the uncertainty in our interpretation of the situation—which is the root of our ignorance in prayer—finds its exact counterpart in our conduct as well. Again and again we do not know how to act "as we ought." We know this least of all in the borderline situation of extreme conflict. Here too there applies the promise that the Spirit will come to us as we pray for a "solution" to a confusing situation, that he will give us wisdom and accordingly lead us into the appropriate modes of action. And even if the institutionalized wrong within which we have to act is so great as to rob us of simplicity and leave us in doubt as to what really is right, within the magnetic field of the Spirit who acts in our stead no ambivalence can "separate us from the love of God" (Rom. 8:35).

God's Promise for Insoluble Ethical Problems

The Bible brings before us two almost insoluble conflict situations which may be considered in the order of their difficulty. In their respective ways, both situations bring out the ambivalence of the conflict, the promise with respect to this ambivalence, and the overcoming of it in prayer.

The first of these situations is that of Peter when in a vision preparatory to his encounter with Cornelius he is summoned to eat unclean animals (Acts 10:1-20). The conflict in which Peter finds himself consists in the fact that God seems here to contradict himself (Acts 10:14; cf. Ezek. 4:14). Peter feels bound to the requirement that he should honor the divine majesty by abstaining from what is "unclean." And now the author of this very requirement commands him to do the very opposite.

Now some might say that the element of conflict in this situation is limited by the fact that God removes his former command and causes a new aeon to dawn (Acts 10:28, 34-35, 43). In that case there is of course no absolute self-contradiction on the part of God. But such a limitation is certainly not perceived by Peter in his actual situation. He simply stands before the riddle of a contradiction in God, and is "perplexed" and confused (Acts 10:17). But he obeys the direct and concrete command of God with which he is immediately confronted (vv. 28-29) and leaves it to God to resolve the contradiction or to work it out in terms of successive redemptive epochs. And as he does so, he opens himself up to that saving decree of God which comes to him precisely in this concealment, this acid test of his faith.

A far more difficult problem is posed by a second conflict situation in the Bible, namely, the offering up of Isaac.[14] Here the self-contradiction in which God sets himself by the command to offer up Isaac is objectively insoluble. The divine promise that through Abraham all the nations of the earth will be blessed is linked with his son, Isaac (Gen. 12:3; 22:18). If Abraham must offer up the bearer of the promise, then—quite apart from the terrible nature of the act from a human standpoint—he is tempted to assume *either* that God will not honor his promise and is thus a liar, *or* that he, Abraham, must not jeopardize the promise by sacrificing its bearer but must instead disobey the command. Thus either God is discredited, or Abraham is discredited as a believer because this faith of his can be preserved only through disobedience.

This hopeless dilemma is grounded in the fact that the command of God conflicts with the promise of God. The Holy seems to be playing a cynical game, and the man Abraham to have gotten into a situation of authentic tragedy. For tragedy is characterized by the fact that the conflict between the ultimate powers of being is fought out on the battlefield of man, that man is dragged into the battle quite without his will, and that in a state of innocent guilt he is responsible for the rift which cleaves all being. Man is only a passive representation of these ultimate powers of being which are locked in conflict. He is in fact simply a victim of being. The origin of the tragic is in the transsubjective sphere, not in an internal dispute of conscience such as might be involved in, for example, a conflict over norms.

[14] See Luther's commentary on Genesis 22 in *LW* 4, 91 ff.; cf. Soren Kierkegaard, *Fear and Trembling*, trans. Walter Lowrie (Princeton: Princeton University Press, 1941), pp. 9 ff. and 83 ff.; and Werner Elert, *Morphologie des Luthertums* (Munich: Beck, 1932), II, 24 ff.

Thus the conflict of Abraham seems to be tragically insoluble. As the conflict is seen from without—and the tragic is always grounded in the aspect "from without"—he will be guilty no matter which way he decides. If he offers up his son, he will no longer be taking seriously the promise which is bound up with the existence of this son and with his having "descendants"; he will cease to be a believer. But if he refuses to offer him up because of this faith, he will be disobeying the command. The interrelationship between indicative and imperative, between Gospel and Law, which is the constitutive element in existence *coram Deo,* is shattered. The very possibility of such existence is therewith destroyed. The horizon of the world is tragically or even nihilistically closed.

The biblical account says nothing about Abraham's reflecting on this conflict. It tells of no attempt on the part of Abraham to balance these conflicting norms, to validate the precedence of one over the other, and to make this precedence the criterion of his decision. The journey to the place of sacrifice is marked by a brooding and painful silence. The hint of conversations between father and son during the journey only makes us the more painfully aware that the real theme was too hopeless to be a subject of reflection or conversation.

Abraham endures the conflict by obeying after the manner of improvisation. Ethical reflection cannot be the basis of his decision. For the point at issue is not a cleavage between that which is godly and that which is ungodly, but a cleavage within God himself. Hence Abraham resolves upon the obedience of improvisation. He obeys the command which is closest to hand, with which he is immediately confronted. He does this—as Luther understands it—because he leaves it to God to resolve what to him is inevitably a contradiction. But this is precisely the quintessence of faith, namely, to trust to God that behind what seems to be a hopeless situation there is hidden God's own way, that behind what seems to be meaningless there is God's meaning, that behind what is insoluble to our thoughts stand the "higher thoughts" of God.

Abraham thus ventures to believe that God will not prove a liar in respect of his promise, that he will instead uphold that promise even through that which seems in the eyes of Abraham hopelessly to jeopardize it. Abraham casts himself into the arms of God, not seeing or sensing how these arms will enfold and deliver him, but confident only that they will. He does not believe in a tragic cleavage in the ultimate power of being. He believes in the unity of God, even though he does not see it. Indeed, it is because he does not see it that he really believes.

The end of the story shows how God, having tested his servant, then delivers him. That it was only a test, that God's face was hidden under the form of its opposite, would never have been clear to Abraham if he had not ventured to be obedient in faith, and if he had not thereby provoked the miracle in which God emerged from the apparent self-contradiction and manifested himself as the One in whom there is no cleavage.

Faith has thus also a noetic function. But this function does not precede faith. It is not as though we had first to see the resolution of the contradiction and so be disposed to believe, i.e., to trust in the unity of the divine person. On the contrary, the noetic element follows faith. When we trust the God who seems to contradict himself, then, like Abraham, we come to know his unity.[15] There thus stands over the conflict situation the promise that the very God who is hidden in the conflict will manifest himself as the One who he is, and who has a purpose in view in submitting us to such tests.

To be sure, the special situation of apparent conflict in God, through which Abraham had to suffer and believe, cannot except with the strictest reservations be regarded as being on the same level with what we have called the borderline situation. For the borderline situation, being a case of extreme conflict, is characterized by the fact that, in the form of institutionalized injustice, it implies an extreme concentration of the forces of revolt which are at work in this aeon. To this degree it involves a conflict with God, not a conflict within God.

Nevertheless, the borderline situation does coincide with that of Abraham in two respects. First, it seems to confront us with the choice between two forms of sinning. And second, God wills to make himself known precisely at this border or limit of life. Here he shows us who he is, the gracious God who blesses empty and helpless hands

[15] On this matter of sequence cf. the prayer of Anselm (*Proslogium,* ii): "Lord, do thou, who dost give understanding to faith, give me, so far as thou knowest it to be profitable, to understand that thou art as we believe; and that thou art that which we believe." *St. Anselm: Proslogium,* trans. Sidney Norton Deane (La Salle, Illinois: Open Court, 1954), p. 7.

when they are lifted up to him, who is waiting right here on the borderline with divine possibilities which begin only where all human possibilities are at an end, and who in the dilemma of our moment drives us to a prayerful expectation of his hours: "Thy kingdom come." In this sense it is right and fitting that we should conclude this most difficult portion of our ethics with this consideration of Abraham.

INDEXES

Authors

Albertus, Magnus, 30
Althaus, Paul, 70 n. 32, 141 n. 16, 216 n. 30, 265 n. 13, 609 n. 1
Ambrose of Milan, 240
Anselm, 302 n. 3, 666 n. 15
Aquinas, Thomas, 30, 75 n. 2, 161 n. 8, 204, 204 n. 12, 236 n. 11, 246 n. 25, 340 n. 6, 342 n. 11, 344 n. 13, 345, 346, 347 n. 20, 408 n. 2, 409 n. 3, 411, 529 n. 9
Arnold Franz Xavier, 444 n. 6, 656 n. 8
Augustine, 29, 59 n. 10, 303, 338 n. 3, 339 n. 5, 340 n. 8, 341 n. 10, 346, 456, 653

Barth, Karl, 28, 67, 67 n. 30, 95 n. 3, 99, 100, 100 n. 8, 103 n. 14, 106, 107, 107 n. 25, 108, 109, 110, 111, 112, 113, 113 n. 46, 114, 114 n. 48, 115, 116, 167 n. 17, 265 n. 14, 275 nn. 18 and 19, 321 n. 1, 322, 323 n. 3, 326 nn. 12 and 13, 368, 368 n. 18, 369 n. 20, 430, 433
Barth, Peter, 325 n. 11
Beales, A. C. F., 445 n. 8
Beyer, Herman, 359 n. 1
Blanke, Fritz, 80 n. 9
Beissier, Pierre, 529 n. 10
Bonhoeffer, Dietrich, xv, 421 n. 2, 477 n. 19, 522 n. 5, 550 n. 28, 573 n. 8, 589 n. 14, 620 n. 6
Brunner, Emil, 29, 156 n. 5, 167 n. 18, 321 n. 1, 325 n. 11, 326 nn. 12 and 13, 422 n. 3
Bultmann, Rudolf, xvi, 70 n. 32, 102, 435
Burckhardt, Jacob, 543
Busch, Wilhelm, 549

Calvin, John, 55 n. 4, 94 n. 1, 103, 113 nn. 43 and 44, 120, 121, 122, 122 n. 70, 134, 135
Churchill, Winston S., 541 n. 24
Cochrane, Arthur C., 145 n. 19
Cullmann, Oscar, 100 n. 9, 335 n. 2

Dehn, Günther, 433 n. 6
De Menthon, M. Francois, 386

De Quervain, Alfred, 477 n. 20
Denzinger, Henry, 196 n. 1, 209 n. 19, 222 n. 1, 235 nn. 8 and 9, 239 n. 14, 392 n. 6
Deutelmoser, Arno, 12, 370, 370 n. 23
Dibelius, Otto, 359 n. 1
Diekamp, 199 n. 3, 235 n. 10
Diem, Harald, 8 n. 5, 359 n. 1, 371 n. 28
Diem, Hermann, 95 n. 3, 359 n. 1
Dostoevski, Fyodor, 88 n. 14, 494 n. 2, 495, 496, 631, 632, 632 nn. 11 and 12, 633, 635, 636, 637, 638, 639, 640, 642

Ebeling, Gerhard, 134 n. 7
Eichrodt, Walther, 105
Elert, Werner, 55 n. 4, 30 n. 7, 134 n. 7, 664 n. 14

Fichte, 520
Franck, Sebastian, 351 n. 21
Frank, F. H. R., 520 n. 2
Frank, Franz, 134 n. 7
Frederick the Great, 535, 543

Gaugler, Ernst, 660 n. 11
Gerhard, Johann, 215, 215 n. 28
Gloede, Günther, 325 n. 11
Gloege, Gerhard, 108 n. 27, 115 n. 52
Goethe, xxi, 188
Gogarten, Friedrich, 28 n. 1

Heidegger, Martin, 466 n. 1, 469, 563
Hellbardt, Hans, 100, 104, 104 n. 18, 105
Hermann, Rudolf, 228 n. 6
Herrmann, Wilhelm, 6
Hesiod, 396
Hessen, Johannes, 390 n. 4
Hirsch, Emanuel, 106 n. 22, 325 n. 10, 444 n. 6
Hofmann, Hans, 479 n. 22
Holl, Karl, 10 n. 6, 64 n. 25, 77 n. 7, 118 n. 59, 141 n. 16, 220 n. 34, 338 n. 4, 346 n. 18, 352, 352 n. 22, 374 n. 33
Horton, Walter M., 422 n. 3

671

Scripture References

Names and Subjects

Ability, 328
Absolute, 402, 609, 610
Absolutizing, liberal, 348-354
Accidental, 388, 389
Accidents, 217
Accommodation (of Christian message),
498, 499, 571, 572, 573, 574, 575,
576
Accusation, divine, 300, 301
Act
ethical, 51, 52, 54, 290, 467, 468,
485
initiatory, 40
Action, 17, 18, 21, 22, 23, 24, 33, 34,
39, 40, 46, 47, 57, 60, 62, 65, 77,
86, 87, 88, 91, 92, 131, 140, 172,
176, 220, 259, 291, 294, 302, 315,
343, 346-348, 355, 376, 431, 432,
451, 455, 456, 460, 622, 625, 626,
640, 644, 656
aggressive, 360
Christian, 5, 13, 17, 18, 21, 22, 26,
43, 46, 53
moral, 24, 44, 92, 412, 604
political, 7
Acts, level of, 19, 21, 27
Adam, x, 98, 99, 160, 161, 281, 287,
288, 296, 434, 597, 598, 601, 602
Address (divine), 164, 165
Adiaphoron, 91
Adultery, 31, 44, 60, 88, 609, 631-643
Advance, 128, 130
Advent, 102, 103
Aeons, xviii, xx, 39-44, 46, 103, 109,
112, 358, 372, 373, 380, 381, 474-
481, 571, 572, 573, 577, 579, 580,
581
Agape, 627, 630, 631, 636, 637
Age (new), x
Age (old), x, 4
Agere (of law), 255
Agnostic, xv
Alien work, 654 (see also Opus
alienum)
Alienation, 262, 488
Alienum, 171, 177, 184, 190, 191, 193,
195, 196, 209, 226, 228, 233, 234,
237, 241, 643
Althaus, Paul, 216, 609

Ambiguity, 604, 609-611, 623
Ambivalence, 252, 564, 634, 660, 663
Ambrose, 240
Amoralism, 254, 543
Analogia entis, 181, 197, 208, 251, 253,
276, 327, 369, 390, 447, 463, 494,
497, 573
Analogia fidei, 573
Anamnesis, 139, 146
Anfechtung, 75, 96, 180, 315, 662 n. 13
Angebot, 94
Anima, 203
Anknüpfungspunkt, 252, 321, 498 (see
also Point of contact)
Anlage, 202, 286
Ansbach Ratschlag, 365, 366
Anselm, 302
Anthropology, xxi, 166, 238, 327, 337,
421, 449, 450, 536
biblical, 82, 83, 312, 435
Roman Catholic, 342
theological, 596
Antichrist, 62, 651
Antinomians, 130, 133
Antinomy, 83, 85, 93
Anxiety, 26, 40, 188, 248, 285, 309,
312, 315, 316, 386, 469, 584, 610,
614, 646, 650, 652, 653
Apocalypse, 4
Apocatastasis, 97, 100, 116
Apologetics, xxiv, 38
Apology to Augsburg Confession, 62 n.
17, 65, 66, 67, 74, 75, 77, 212, 238
Aquinas, Thomas, 161, 204, 240, 340,
342, 344, 347, 387, 400, 408, 409,
414, 529 n. 9
ἀρετή, 19
Aristotle, 9, 247, 388, 400, 420
Art, xxi, xxiii, 7, 480
Ascension, 44, 336
Asceticism, 472
Assurance, 76, 82, 96
Atheist, xv, 312
Atonement, 113
Attitude, 87, 92, 162, 352, 377, 432,
469
Attribute
divine, 179, 237, 238, 280

human, 159, 177, 179, 180, 181, 192, 235
Aufgabe, 52, 55, 69, 94, 458
Augsburg Confession, 53, 54, 63 n. 19, 69, 71 n. 33, 72
Augustine, xii, 29, 338, 339, 340, 346
Aussenschaften, 156, 157, 178, 180
Authority, 35, 43, 63, 88, 96, 136, 137, 143, 144, 269, 272, 277, 375, 457, 579
  moral, 13, 23
  political, 361
Autonomy, 6, 7, 8, 9-12, 13, 14, 18, 25, 28, 29, 32-36, 38, 58, 74, 78, 82, 146, 339, 361, 367-371, 375, 376, 431, 472, 534
  moral, 13, 32-33, 34
  political, 12, 376, 535, 542
  sovereign, 24

Baptism, 39, 40, 42, 71, 112, 184, 241, 320
Barth, Karl, 37, 67, 98-117, 120, 275, 321-324, 325, 326, 364, 367, 368, 369, 430, 433
Barth, Peter, 325
Barthians, 100, 101, 364
Beatitudes, 52
Becoming, 133
Being, xvii, 44, 62, 66, 79, 291, 292, 387, 597, 658
  new, 46
  order of, 388, 398, 400, 403, 410, 428, 429, 443, 447
Being-as-it-is, 299
Being-in-relation, 470
Being-in-the-world, 6, 434, 469, 470, 477, 479
"Being unto death," 565
Believer, xiv, xvii, xviii, 38, 40, 55, 57, 71, 73, 92 n. 15, 112, 115, 130, 136, 182, 184, 198, 227, 265, 378, 479, 665
Bengel, Johannes, 336
Bible
  historical criticism, 528
  See also Scripture, Holy
Birth, 200, 389, 391, 526
Bismarck, Otto von, 500
Blasphemy, 62, 306, 644
Bliss, eternal, 324, 325
Body
  of Christ, 184, 191, 229, 245

physical, 90, 160, 161, 197, 199, 203, 205, 207, 212, 435, 471
Bonaventura, 347
Bondage, xii, 31, 32, 59, 87, 89, 90, 103, 131, 138, 151, 174, 176, 284, 316, 465, 506, 507, 519
Bonhoeffer, Dietrich, xv, 620
"Borderline cases," x, 460
Borderline concept, 445, 446
Borderline situations, 459, 461, 463, 559, 578-585, 587, 588, 589, 594-608, 609, 609 n. 1, 610, 620, 630, 643, 645, 647, 650, 653, 654, 666
Boundary (*Grenze*), 580, 581, 647
Brunner, Emil, 29, 252, 321-324, 325, 384
Bultmann, Rudolf, xvi, 102, 117, 435
Business, xix, 130
Busse, 128

Calling, xvii, 10, 31, 131, 377, 478, 500, 641
Calvin, John, 113, 120-125, 134-135, 279, 325, 326
Calvinism, 120
Canon law, 583
Capability, 321, 328, 485
Capital punishment, 148, 412, 571
Capitalism, 41, 448
Cases, individual, 409, 410, 411, 415, 428
Casuistry, xx, 20, 31, 44, 72, 92, 347, 354, 410, 444, 455-459, 494-508, 548, 550, 623, 643, 650, 660
  nomistic, 350, 584
Catechetical instruction, 28, 498
Categorical imperative, 6, 18, 23, 33, 34, 51, 388, 484
Causality, 54, 58, 63, 67, 68, 69, 82, 89, 636, 637, 638
Cause
  formal, 400, 401
  secondary, 366
Certainty, 406, 407, 411
Chaos, 276, 416, 438
Character indelebilis, 159, 167
Christ-event, 40, 43, 53, 109
Christian, x, xi, xiii, xiv, xv, xvii, xxiii, 39-44, 47, 51, 55, 71, 79, 90, 114, 126, 133, 136, 141, 142, 145, 146, 189, 192, 194, 198, 270, 273 n. 16, 307, 354, 362, 363, 375, 431, 432, 479, 485, 487, 501, 543, 584, 590, 595, 601, 621, 624, 646

INDEX

Christianity, 4, 23, 35, 40, 41, 42, 70, 126, 128, 184, 196, 352, 363, 461, 470, 474, 511, 516, 518
Reformation, xxvi
"Christianity of Action," 17
Christmas, 39, 103, 112, 328
Christology, 45, 108, 110, 111, 112, 113, 180, 184, 185, 275, 326, 369
Christomonism, 115
Church, ix, xiv, 8, 9, 42, 130, 210, 229, 230, 265, 270, 273, 367, 388,, 431, 433, 449, 450, 495, 510, 511, 512, 514, 515, 517, 518, 618, 619, 625, 627, 628, 629, 630
confessing, 627
outside the, 269, 270, 271, 272, 273, 276, 277, 278
teaching office of, 407
Churchill, Winston, 540
Circumstances, concept of, 410-419, 424
Citizen, x, 420
Civil code, 141
Claim, God's, xviii, xix, 8, 52, 152, 153, 154, 299, 564, 621
Claims, xx, 10
Clericalism, 145
Coexistence, 371-373, 381 (see also Two kingdoms)
Cogito, ergo sum, 408, 472
Cognition, 199
Collectivism, 448, 538
Command, xii, 384, 485, 612
Commandment, radical, 359-364
Commandments, xviii, 4, 9, 10, 25, 32, 34, 87, 184, 196, 256, 258, 259, 260, 262-272, 283, 314, 338, 340, 346, 347, 357, 367, 370, 376, 377, 383, 390, 392, 431, 440-447, 451, 463, 486, 499, 500, 502, 551, 578, 609-611, 615, 634
Commands, God's, 6, 27, 34, 35, 41, 54, 64, 70, 84, 85, 94, 128, 140, 146, 147, 149, 150, 151, 155, 156, 158, 196, 259, 281, 284, 326, 332, 339, 340, 487, 570, 615, 618, 663, 664, 665
Common law, 385
Communication, indirect, 547, 549, 550
Communism, 423, 424, 596, 624
Community
Christian, 144, 232, 375, 379, 436, 450, 473, 519, 629, 630
civil, 430

Compromise, 380, 381, 460, 485, 486, 487, 489-493, 494, 495, 500, 502, 504, 505-506, 508-509, 513, 515, 517, 518, 519, 521, 523, 524, 533, 534, 565, 567, 568-577, 584, 592, 618, 619, 656-657
Compulsion, 34, 54, 297, 602 n. 23
Concentration camp, 620-625
Conception (human, biological), 526, 527
Concern, ethical, 258, 652
Conclusion, concept of, 406-409, 411
Concupiscence, 219, 235, 401
Condescension, divine, 657, 659
Conduct, 27, 37, 39, 44, 79, 83, 362, 459, 482, 499, 500, 537
Confession, 591, 638, 644
public, 626, 627, 629, 630, 646, 650
Confessions, Lutheran, 60-69, 72-73, 77, 92 n. 15, 97, 211
Confidence, false, 10
Conflict, xviii, xix, xx, 460, 521, 532, 533, 580, 585-594, 595, 643-647, 654, 664, 665
Conflict situation, 596, 597, 604, 609, 610, 611, 612, 614, 620-647, 650, 652, 659, 663, 664, 666
Conformity, 192, 471
Conscience, 12, 25, 29, 35, 76, 78, 96, 122, 160, 190, 196, 197, 210, 294, 298, 299, 300, 313-320, 321, 322, 323, 324, 328, 329, 330, 332, 333, 370, 378, 379, 383, 387, 393, 401, 465, 500, 515, 531, 532, 537, 543, 592, 600, 610, 612, 629, 645
accusation of, 301-302
autonomous, 13, 18, 333
Christian, 305-313, 333
natural, 320, 323, 325, 328, 329, 332
unredeemed, 301-305, 308, 309, 310, 311
witness of, 299, 301, 317
Consciousness, Christian, xiv
Constituents, elementary, 197
Contingency, 391, 420
Continuity, 40, 42, 43, 44, 45, 72, 320
of ego, 230, 231, 232
Co-operation, 58, 70, 86, 96, 164, 185, 209, 247, 316, 326, 497, 498
Cor accusator, 96, 125, 128, 139, 300, 315, 319, 328
Cor defensor, 300, 319, 328

INDEX

263, 271, 310, 313, 320, 415, 431,
437, 488, 500, 501, 502, 503, 504,
514, 539, 544, 545, 567, 576, 577,
582, 583, 587, 591, 594, 597, 602,
604, 605, 606, 607, 622, 631, 632,
635, 636, 638, 640, 641, 654, 655,
657, 658, 659, 660
Form, 201, 219, 242, 324, 400, 475
Fornication, 87, 88, 89, 90, 91, 92
Fortuna, 537, 538
Fortunetelling, 554
Frank, F. H. R., 520
Frederick the Great, 533, 534, 535, 536,
539, 541, 542
Free love, **41**
Freedom, 12, 35, 160, 176, 199, 202,
203, 217, 277, 281-290, 393, 402,
404, 427, 430, 465, 537, 563, 564,
569, 574, 581, 598, 602, 633
  Christian, x, xii, xiv, 28 n. 1, 34, 88,
  89, 91, 135, 136, 137, 139, 151,
  174, 455-459, 461, 463, 479, 480,
  503, 506, 507, 605, 613, 652, 653,
  ethical, 465
  original, 281
Freedom of the will, 29, 203, 465-466
Fruits, 20, 53, 54, 55, 56-69, 83, 181,
266
Futility, 312
Future, 172, 173, 180, 184, 469

Gabe, 52, 55, 69, 94, 458
Gandhi, Mahatma, 624
Gebot, 94, 147, 570
Generation, biological, 389
Genius, 633
Genocide, 386
Gentiles, 277
Gerhard, Johann, 215, 216-217
"German Christians," 325, 365, 518,
618, 619
Germany, 9, 88, 365
Gesetz, 147, 570
Gift, 52, 53, 55, 69, 70, 83, 85, 94, 152,
153, 158, 159, 457, 458
  superadded, 215, 217, 234
Gifts, praeternatural, 204, 205, 401
Gloede, Günther, 325
Glory (of God), 173, 176, 177, 180
Goal, 155, 324, 528, 529, 616, 617, 619
  ethical, 18, 21, 27, 53, 54, 61, 70,
  172, 483, 484, 485, 486, 487, 497
  pneumatic, 275
God, xvii, xviii, 10, 14, 20, 25, 26, 27,

31, 32, 34, 35, 38, 40, 43, 47, 53,
61, 65, 66, 67, 68, 70, 73, 74, 75,
76, 77, 78, 80, 81, 82, 84, 85, 86,
89, 91, 92, 95, 96, 97, 98, 99, 104,
107, 108, 109, 110, 111, 112, 113,
114, 115, 117, 119, 121, 123, 124,
126, 127, 129, 131, 133, 139, 142,
143, 144, 145, 146, 148, 152, 156,
157, 161, 163, 165, 166, 175, 177,
178, 188, 203, 206, 207, 209, 210,
211-221, 223, 224, 231, 234, 237,
238, 245, 246, 247, 263, 279, 280,
300, 303, 309, 310, 313, 314, 327,
353, 368, 370, 441, 442, 512, 526,
529, 568, 574, 603, 658
God-world identity, 280
Goethe, 398, 400
Good, 6, 7, 9, 27, 142, 212, 279, 282,
283, 341, 353, 391, 408, 430, 447,
484, 538, 539, 599
  absolute, 283, 402
  original, 283
Good works, 53-55, 56, 58, 60, 62, 62
  n. 17, 65, 67, 68, 73, 74, 82, 83,
  84, 302, 305, 326, 605
Goodness, 9, 97, 106, 156, 319
Gospel, xii, xxv, xxvi, 13, 26, 27, 30, 31,
36, 52, 65, 67, 70, 76, 89, 94, 95,
96, 99, 103, 110, 117-125, 126,
127, 153, 179, 186, 228, 248, 252,
253, 254, 311, 315, 320, 328, 344,
369, 374, 436, 437, 474, 506, 512,
513, 539, 572, 573-577, 602, 604,
606, 639, 640, 653
Government, spiritual, 371
Governments (*Regimenten*), 371, 372,
376
Grace, xiv, 12, 29, 30, 56, 60, 65, 75,
76, 77, 80, 82, 86, 95, 96, 97, 99,
100, 110, 112, 117, 118, 119, 122,
125, 126, 129, 139, 146, 148, 154,
164, 177, 179, 180, 182, 193, 195,
196, 202, 207, 208, 209, 210, 214,
218, 219, 221, 224, 225 n. 3, 232,
233, 234-249, 252, 253, 262, 264,
266, 270, 271, 278, 279, 303, 315,
317, 337, 342, 345, 374, 382, 391,
392, 497, 498, 505, 517, 570, 571,
572, 573, 574, 602, 604, 624, 639,
641, 642, 657
  actual, 239
  antecedent, 239
  co-operating, 239
  efficacy of, 324

INDEX

104, 105, 106, 108, 109, 120, 121, 122, 124, 276, 300
Omnipotence, 199, 370
Ontologism, 239
Ontology, 155, 156, 159, 160-170, 171,, 188, 192, 194, 195, 197, 198, 199, 200, 204, 206, 208, 209, 210, 211, 212, 213, 215, 216, 217, 218, 219, 220, 222, 226, 228, 232, 236, 238, 239, 242, 243, 245, 247, 249, 251, 266, 329, 337, 342, 347, 348
Opponent, 585, 598
Opus alienum, 366
Opus operatum, 239, 241
Order
normative ethical, 7
original, 402, 403, 405, 416, 582
temporal, 361
Orders (*Ordnungen*), xii, xiii, xiv, xv, xxi, 5, 9, 10, 11, 12, 13, 19, 30, 36, 37, 38, 40, 43, 47, 145, 357,, 359-364, 366, 370, 372, 376-378, 379, 380, 381, 383, 385, 387, 389, 403, 417, 419, 430, 433, 434-440, 450, 472, 475, 476, 499, 500, 501, 514, 578, 579, 580, 581, 582, 587, 656
emergency (*Notordnungen*), 440, 447, 655, 657
Orders of creation, 265, 274, 275, 276, 439, 450, 480, 579, 648
Origen, 340, 660
Original sin, 151, 207 n. 16, 212, 214, 290-297, 445, 597, 601
Original state, 150, 198, 204, 238, 279, 283, 356, 381, 394, 403, 404, 405, 415, 416, 419, 445, 648
Orthodoxy, Protestant, 19, 143, 165, 179, 213, 214, 215, 265
Osiander, Andreas, 78, 79, 80, 81, 82
Ought. xix

Paedagogus, 255, 256
Pantheism, 279
Paradise, 283
Paradox, 55
Paraenesis, 52, 70, 86, 266, 507
Parousia, 43, 106, 277, 334, 335, 379
Participation (in likeness), 172-184, 185-194, 195
Particula exclusiva, 244, 247, 317
Pascal, Blaise, 167
Passivity, 352, 371
Past
national, 469

personal, 469
Patience of God, 273, 278, 378, 431, 439, 500, 508, 519, 569, 571, 573, 577, 654, 657
Paul, xii, 52, 57, 58, 69, 70, 83, 84, 85, 87, 88, 89, 90, 91, 114, 136, 137, 149, 158, 173, 176, 177, 208, 262, 284, 299, 601
Peace
with God, 188, 189, 190, 191, 193, 248, 309, 310, 313, 315, 317, 319, 328, 558, 566, 606
political, 483, 484, 572
Peasants' Revolt, 30
Pecca fortiter, 41, 503
Peccator in re, 80, 94, 98, 128, 172, 231, 315, 436
Peccatum in re, 196
Peccatum originale, 290 (*see also* Original sin)
Pelagianism, 29, 186
Pelagius, 185
Penitence, 129
Pentecost, 39
Perception, 462
Perficere, 324
Perfection, 202, 340, 341, 345, 391
moral, 9
Perfectionism, 32, 310, 612
Permission, divine, 571
Person, 60, 61, 62, 63, 83, 140, 151, 164, 176, 190, 222-233, 239, 264, 267, 286, 288, 289, 292, 295, 297, 330, 346, 353, 355, 362, 377, 446, 465, 466, 467, 468, 510, 511, 514, 515, 521, 522, 549, 551, 552, 553, 555, 556, 558, 564, 565, 566, 646
secular, 363
"Person-in-relation," 236
Personal, realm of, 286, 287
Personal, the, 294
Personalism, 199, 417
Personalistic mode of thought, 197, 199, 208, 216, 217, 226, 227, 237, 238, 239, 243, 245, 249, 251, 329, 337
Personality, 160, 424
Personhood, 465-467, 473, 514, 645, 646
Peter, 39. 40
Pharisaism, 86, 253, 605
Pharisees, 20, 302, 333
Phenomenology, 461, 463, 470
Philosophy, xiii, xxiv, xxvi, 268, 269, 462, 481

691